Landscapes and Landforms of Terrestrial and Marine Areas

Landscapes and Landforms of Terrestrial and Marine Areas

Editors

Mauro Soldati
Federica Foglini
Mariacristina Prampolini
Alessandra Savini

MDPI • Basel • Beijing • Wuhan • Barcelona • Belgrade • Manchester • Tokyo • Cluj • Tianjin

Editors
Mauro Soldati
University of Modena and
Reggio Emilia
Italy

Federica Foglini
National Research Council,
Institute of Marine Sciences
Italy

Mariacristina Prampolini
National Research Council,
Institute of Marine Sciences
Italy

Alessandra Savini
University of Milano-Bicocca
Italy

Editorial Office
MDPI
St. Alban-Anlage 66
4052 Basel, Switzerland

This is a reprint of articles from the Special Issue published online in the open access journal *Water* (ISSN 2073-4441) (available at: http://www.mdpi.com).

For citation purposes, cite each article independently as indicated on the article page online and as indicated below:

LastName, A.A.; LastName, B.B.; LastName, C.C. Article Title. *Journal Name* **Year**, *Volume Number*, Page Range.

ISBN 978-3-0365-1653-0 (Hbk)
ISBN 978-3-0365-1654-7 (PDF)

Contents

About the Editors

Mauro Soldati is Full Professor of Physical Geography and Geomorphology at the Department of Chemical and Geological Sciences, University of Modena and Reggio Emilia (Italy), where he teaches courses of 'Geomorphology', 'Georisks and Civil Protection' and 'Ecology and Global Changes'. His research is focused on landslide hazards in mountain and coastal environments with special emphasis on the effects of climate change. He is also active in the field of geoheritage and geoconservation. His research has mainly been carried out in the Italian Dolomites, Northern Apennines and Maltese Islands, and secondarily in Greece, Spain and Svalbard. He is currently a member of several editorial boards of international journals, including Water. He is author or co-author of about 200 papers. Currently, he is President of the International Association of Geomorphologists (IAG) and Council Member of 'Deep-time Digital Earth' (DDE)—IUGS Big Science Program.

Federica Foglini is a senior marine data scientist at the Institute of Marine science (CNR-ISMAR) in Bologna with a primary interest in seafloor mapping and geomorphology, marine cartography, habitat mapping technologies, multi beam swath bathymetry acquisition and processing, marine spatial data management, Geodatabase design and implementation. She has participated in more than 20 oceanographic cruises as supervisor of geophysical data acquisition and processing and as chief scientist. She is co-author of several ISI papers and technical reports on implementation and design of Marine Geodatabase and GIS mapping and multi beam bathymetry processing. She teaches GIS for Habitat mapping at Bologna University. She is involved in the GEBCO (General Bathymetric Chart of the Oceans) as member of the TSCOM (Technical Sub-Committee on Ocean Mapping) and Chair of the Metadata working group. She has been member of the Ocean BIG Data Working group since 2018 for the European Marine board.

Mariacristina Prampolini is a post-doc researcher at the National Research Council - Institute of Marine Sciences (CNR-ISMAR) of Bologna (Italy). She is a geomorphologist and works on marine habitat mapping topics in the framework of European and National projects. Her work consists of acquisition and analysis of multibeam data, geomorphological and sediment characterization of seafloor for habitat mapping purposes, and geomorphological carthography. She is member of the Italian Association on Physical Geography and Geomorphology (AIGeo) and one of the coordinators of the section for Italian Young Geomorphologists.

Alessandra Savini is Associate Professor at the University of Milano-Bicocca, where she teaches introductory courses in Submarine Geomorphology. She is a marine geoscientist with expertise in seafloor imaging and mapping using geophysical data (bathymetry, backscattering and seismic) and RGB still images and videos. She investigates the interaction between submarine geomorphic processes and benthic organisms, over long- and short-time scales. Emphasis is given to marine bioconstructions (from cold-water to tropical corals) and cold seep systems, to decipher the response of submarine landscapes and landforms to Pleistocene and Holocene climate changes. She is a member of the IAG Submarine Geomorphology Working Group that contributes to the establishment of submarine geomorphology as a scientific discipline. She is a member of the Scientific Committee of the MaRHE center (www.marhe.unimib.it), coordinating the annual training course on mapping technologies in coral-reef environments.

Preface to "Landscapes and Landforms of Terrestrial and Marine Areas"

In the last decade, the interest in integrated studies of terrestrial and marine landscapes and landforms has been expanding, thanks to technological improvements and also to the development of innovative methods for the collection and analysis of geospatial data in the so-called "white ribbon." Onshore geomorphological data have been increasingly integrated with submerged datasets for paleo-environmental reconstructions and land management purposes. Modern advances in geo-acoustic and optical full-coverage mapping of both seafloor and near-shore areas allow geoscientists to acquire a large amount of remote data that can be combined with onshore data and direct ground-truthing information, with promising outputs. Hence, the production of detailed and accurate maps imaging both the seabed and the coastal morphology can foster integrated studies of emerged and submerged landscapes and landforms.

Investigating emerged and submerged landscapes as a continuum began to be applied mostly in geomorphological research concerning volcanic islands or in marine geohazard assessment (e.g., large coastal landslides). Recently, studies on coastal geohazards induced by global and climate changes lato sensu took advantage of coupling terrestrial and marine spatial datasets to model and quantify impacts in terms of vulnerability and risk scenarios. Additionally, both reconstruction of Late Quaternary geomorphological evolution and geoarchaeological research in coastal areas have been highly favored by the combination of onshore and offshore datasets. Merging data from coastal and nearshore areas has also supported the recognition of geodiversity and geosites, and the study of those ecological aspects that disclose the system connectivity across the coastline. This makes coordinated plans for the management of terrestrial and marine areas more effective and capable of addressing complex problems, preserving and managing coastal environmental resources and ecosystems.

The aim of this Special Issue of Water on "Landscapes and Landforms of Terrestrial and Marine Areas" was to collect contributions showing how the integration of emerged and submerged datasets is unquestionably beneficial in scientific research and environmental management. The volume includes a review paper showing the reasons why the integration of terrestrial and marine datasets would be desirable in geoenvironmental research and outlining the fields of geosciences that have mostly benefitted from land-and-sea dataset integration. The core of the Special Issue consists of 13 articles dealing with landscapes and landforms of the Mediterranean region. However, a case study from coastal areas of the Indian Ocean is included, too. The articles focus on Late Quaternary landscape evolution, geohazards, geomorphological mapping, geoarchaeology, geoheritage, and geodiversity and marine benthic habitat mapping. Vulnerability of coastal areas was also examined, taking into account the consequences of extreme events enhanced by climate change. The majority of the research carried out took advantage of surveying, modeling, mapping, and analyzing different types of datasets.

The editors would like to thank the authors of the articles for having enthusiastically contributed to this volume, and for their kind cooperation during the whole editorial process. In particular, the involvement of members of the Italian Association of Physical Geography and Geomorphology (AIGeo) is acknowledged. We are grateful to May Zheng, section managing editor, for having so professionally and kindly supported us throughout each step of the Special Issue planning and management.

The publication of this Special Issue is part of the activities of the 'Submarine Geomorphology' Working Group of the International Association of Geomorphologists (IAG).

Mauro Soldati , Federica Foglini , Mariacristina Prampolini, Alessandra Savini
Editors

Editorial

Landscapes and Landforms of Terrestrial and Marine Areas: A Way Forward

Mauro Soldati [1], Mariacristina Prampolini [2,*], Federica Foglini [2] and Alessandra Savini [3]

[1] Department of Chemical and Geological Sciences, University of Modena and Reggio Emilia, Via Campi 103, 41125 Modena, Italy; mauro.soldati@unimore.it

[2] National Research Council, Institute of Marine Sciences, Via Gobetti 101, 40129 Bologna, Italy; federica.foglini@bo.ismar.cnr.it

[3] Department of Earth and Environmental Sciences, University of Milano Bicocca, Piazza della Scienza 1 e 4, 20126 Milano, Italy; alessandra.savini@unimib.it

* Correspondence: mariacristina.prampolini@bo.ismar.cnr.it; Tel.: +39-0516398936

check for
updates

Citation: Soldati, M.; Prampolini, M.; Foglini, F.; Savini, A. Landscapes and Landforms of Terrestrial and Marine Areas: A Way Forward. *Water* **2021**, *13*, 1201. https://doi.org/10.3390/w13091201

Received: 8 April 2021
Accepted: 21 April 2021
Published: 26 April 2021

Publisher's Note: MDPI stays neutral with regard to jurisdictional claims in published maps and institutional affiliations.

1. Introduction

In the last decade, the interest to jointly analyze landscapes and landforms of emerged and submerged areas has been expanding [1]. A major contribution in this direction has been provided by technological improvements and the development of innovative methods for collecting and analyzing geospatial data in the so-called "white ribbon." A key advance has undoubtedly been the innovation in robotics, remote sensing, and computer vision technologies that have fostered the ability to easily construct high-resolution and even photorealistic terrain models as a base surface for 3D mapping in the underwater environment [2].

Different fields of geosciences have recently benefited from integrating land-and-sea spatial datasets (cf. [1]), mainly as a result of the following:

- An improved understanding of coastal and marine landforms and processes generating them;
- A more efficient assessment of risks and impacts in coastal and nearshore areas, in the context of the ongoing climate change and the development of coastal areas;
- The promotion of more sustainable management practices for coastal and marine environments and related resources.

The aim of this Special Issue is to collect contributions that demonstrate how the integration of emerged and submerged datasets is unquestionably beneficial, in a wider perspective, for geoenvironmental research.

2. Investigating Terrestrial and Marine Landscapes

Investigating terrestrial and marine landscapes as a *continuum* began to be applied mostly in geomorphological research concerning ocean volcanic islands—in order to investigate the effects of volcanic eruptions on volcanic apparatus (i.e., landslides and collapses; e.g., [3])—or in marine geohazard assessment referring to large coastal landslides (e.g., Nice landslide; [4]).

Recently, studies on coastal geohazards induced by global and climate changes *lato sensu* took advantage of coupling terrestrial and marine spatial datasets (e.g., elevation data) in order to model and quantify impacts due to, for example, coastal subsidence, sea-level rise, and extreme weather events (i.e., storms), in terms of vulnerability and risk scenarios [5–10].

Curiosity and attention on this approach spread also toward other scopes. Both the reconstruction of Late Quaternary geomorphological evolution and the geoarchaeological research in coastal area have highly been favored by the combination of onshore and offshore datasets [11,12]. Indeed, the continental shelves have undergone several sea-level

1

oscillations, and recorded a sea-level fall at ca. 130 m below the present msl during the Last Glacial Maximum (LGM). Thus, large portions of the present continental shelves were exposed to subaerial processes that shaped landforms lying on the seafloor nowadays. These areas possibly hosted prehistoric and historic human settlements—as witnessed by archaeological remains, pollens, bones, or other remains in the sediments, caves, and so forth—which can be discovered and investigated in support of paleo-environmental reconstructions.

Furthermore, merging data from coastal and nearshore areas helped the identification of geodiversity and geosites and the study of marine and landscape ecology considering their connectivity across the coastline. This makes coastal environmental management practices more effective and capable of addressing complex problems, especially in the present context of climate change and sea-level rise. A coordinated plan for the management of terrestrial and marine areas, such as the Protocol on Integrated Coastal Zone Management, is indeed fundamental to preserve and manage coastal environmental resources and ecosystems [13,14].

All these themes can be faced in different ways, and the combined knowledge of terrestrial and submerged landscapes can shed light on cases of geomorphological equifinality between land and sea features.

Finally, the ever-growing scientific and industrial interest in integrating land-and-sea datasets is strongly promoting technological advances and new approaches to picture/depict/represent land- and seascapes in a single view (cf. [15]). Prampolini et al. [1] draw the state of the art of the methods and remote sensing technologies that underpin the current ability to produce seamless digital elevation models (DEMs) and/or digital terrain models (DTMs) for coastal areas, with a comparison of their convenience, limitations, and costs, including some new promising techniques (e.g., structure from motion (SfM) and satellite-derived bathymetry (SDB)).

3. The Special Issue

The papers of this volume deal with two main themes, which fully reflect the initial aim and purposes of the Special Issue:

1. Terrestrial and marine landforms in nearshore areas, coupling terrestrial and marine datasets for geomorphological reconstruction, hazard assessment, benthic habitat mapping, environmental conservation, coastal management, and marine spatial planning.
2. Landscapes and landforms of recently submerged areas (e.g., during the post-LGM marine transgression) and comparison with present terrestrial areas.

The articles focus on Late Quaternary landscape evolution, geohazards, geomorphological mapping, geoarchaeology, geoheritage, and geodiversity and marine benthic habitat mapping (cf. Table 1). Furthermore, a review paper [1] is included, showing the reasons why the integration of terrestrial and marine datasets would be desirable in geoenvironmental research. Prampolini et al. [1] outline the fields of geosciences that have mostly benefitted from land-and-sea dataset integration, showing the topic breakdown of the large number of scientific papers analyzed. The most explored topic is geohazards (34.5% of the papers), followed by geoarchaeology (16.2%), coastal planning and management (15.5%), Late Quaternary changes of coastal landscapes (9.9%), geomorphological mapping (8.45%), marine and landscape ecology (8.45%), and geoheritage and geodiversity (7.0%).

Table 1. Papers published in the Special Issue grouped according to their theme and content.

Paper	Topic/Theme	Geographic Context	Content
Mattei et al. [17]	Geoarchaeology	Campi Flegrei (Italy)	- Detailed reconstruction of the coastal morphoevolutive trend and rsl [1] variations of the last 2 kyr and insight on the submerged remains of a Roman harbor
Perotti et al. [18]	Geoheritage and geodiversity	Sesia Val Grande UNESCO Global Geopark (Italy)	- Conceptual model and procedure for evaluating geodiversity connected to water resources
Coratza et al. [19]		Portofino Natural Park and MPA [2] (Italy)	- Identification and assessment of land-and-sea geosites for tourist improvement and fruition
Biolchi et al. [20]	Geohazards	Premantura Promontory (Croatia)	- Monitoring movement of coastal boulders and simulation of storm wave height
Rizzo et al. [21]		Gozo (Malta)	- Coastal vulnerability assessment by means of a vulnerability index that integrates physical exposure and social vulnerability
Meilianda et al. [16]		Sumatra Island (Indonesia)	- Investigating morphological resilience of barrier islands to secondary effects of seismic activity of the Sumatra–Andaman subduction zone
Guida and Valente [22]	Geomorphological mapping	Cilento Promontory (Campania, Italy)	- Geomorphological study and mapping of land-and-sea landforms
Hasan et al. [23]		Novigrad and Karin Sea (Croatia)	- Acoustic map and geomorphological analysis of river canyons in the NE Adriatic Sea
De Gioiosa et al. [24]	Late Quaternary changes and coastal landscapes	Torre Guaceto MPA (Italy)	- Reconstruction of Late Pleistocene landscapes and sea-level oscillations
Novack et al. [25]		Northern Adriatic Sea (Italy–Slovenia)	- Reconstruction of Quaternary sedimentary succession and related processes
Savini et al. [26]		Cilento Promontory (Campania, Italy)	- Identification of sea-level oscillations in marine terraces on land and at sea to reconstruct geomorphological evolution of the area during the LGM
Deiana et al. [27]		SW Sardinian continental shelf (Italy)	- Reconstruction of Late Quaternary geomorphological evolution of continental shelf areas
Marchese et al. [28]	Benthic habitat mapping	Southern Adriatic Sea (Apulia, Italy)	- Mapping the distribution of coralligenous buildups in shallow coastal waters

[1] rsl: relative sea level; [2] MPA: marine protected area.

Most of the papers included in this volume deal with landscapes and landforms of the Mediterranean region. However, a case study from coastal areas of the Indian Ocean is also included [16]. Special Issue contributions took into consideration both the present sea level, its past oscillations, and possible future variations in terms of sea-level rise.

Vulnerability of coastal areas was also faced under different points of view, taking into account climate change extreme events, such as storms causing boulder displacement. The majority of the research carried out took advantage of surveying, modeling, mapping, and analyzing different types of data (e.g., acoustic data and seabed samples for analyzing sediments and deposits) (Figure 1).

Figure 1. Word cloud of the Special Issue "Landscapes and Landforms of Terrestrial and Marine Areas" providing a visual representation of the most recurrent words in the full texts of the 14 articles published in the volume. Keywords and focal issues for further investigations on the land-and-sea interface appear to be sea level, submerged features, data, mapping, modeling, waves, changes, systems, and so forth.

3.1. Topic 1: Terrestrial and Marine Landforms in Nearshore Areas

The papers concerning the analysis of terrestrial and marine landforms in coastal and shallow water specifically deal with geomorphological mapping and landscape evolution [22,23], geoheritage and geodiversity [18,19], geohazards [16,20,21], and marine and landscape ecology [28].

Guida and Valente [22] studied the geomorphological aspects of onshore and offshore areas of the Cilento coast (Campania, Italy), correlating land and sea landforms in order to outline their evolution. They produced a digital geomorphological map that supported a physical-based modeling of geomorphological evolution.

Hasan et al. [23] mapped and analyzed the morphological features of the river canyons linking the Karin, Novigrad, and Velebit channels (NE Adriatic Sea). The latter are characterized by tufa barriers that had a role in the Holocene flooding dynamics of the canyons and karst basins. The authors outlined the evolution of the karst estuarine seafloor and reported about the flooding of semi-isolated basins due to sea-level rise.

Perotti et al. [18] proposed a GIS qualitative–quantitative methodology (defined as hydro-geodiversity assessment) to assess and evaluate the geodiversity of both superficial and underground water resources in the Sesia Val Grande UNESCO Global Geopark (Italy). They produced a hydro-geodiversity map of the study area, reporting a weighted evaluation of zones distinguished by a remarkable relationship between surface and underground hydrodynamics.

Coratza et al. [19] identified and evaluated geosites located both onshore and offshore the Portofino Natural Park and MPA (Liguria, Italy). Their analysis was meant to recognize the most suitable sites for tourist improvement and fruition, outlining connections between the land and sea systems. Scientific, ecological, cultural, and aesthetic values were considered, as well as site accessibility, services, and economic potential of the geosites identified.

Biolchi et al. [20] monitored the recent displacement of boulders located on the southernmost coast of the Premantura Promontory (Croatia), possibly due to the storm Vaia that occurred in October 2018. The authors carried out aerial and underwater photogrammetric surveys to identify boulders' movement onshore and submarine landforms associated with boulder detachment from nearshore seafloor. Furthermore, they modeled a wave height that could have been responsible for boulders' displacement.

Rizzo et al. [21] assessed the coastal vulnerability of the NE coast of the Island of Gozo (Malta), which is characterized by a high economic value due to tourism activity and the presence of natural protected areas (i.e., Natura 2000 sites). The authors propose an overall vulnerability index combining physical exposure and social aspects, and a cost-effective method that can be reliably applied in areas affected by climate- and marine-related processes (namely, coastal erosion, landslides, and sea-level rise).

Meilianda et al. [16] performed a spatial analysis on multisource datasets to investigate barrier islands' morphological changes along Sumatra Island coasts (Indonesia), as a result of the secondary effects of the seismic activity related to the Sumatra–Andaman subduction and the Great Sumatran Fault system. In particular, the authors documented that the 2004 megatsunami irreversibly changed barrier islands on the investigated stretches of the coast.

Marchese et al. [28] analyzed biogenic landforms (i.e., coralligenous builds-up) and mapped their distribution in coastal waters along the Apulian continental shelf (Italy) through a semiautomated GIS-based methodology relying on geomorphometric techniques. They calculated the area and volume of these mapped bioconstructions, which represent a hotspot of biodiversity for the Mediterranean Sea and a key carbonate producer in temperate water (cf. [29]).

3.2. Topic 2: Landscapes and Landforms of Recently Submerged Areas

The outputs of research on Late Quaternary landscape changes are presented, with different approaches, in a number of papers of this Special Issue. Mattei et al. [17] supported their research by means of archaeological markers; De Gioiosa et al. [24], Deiana et al. [27], and Savini et al. [26] analyzed terrestrial and marine landforms to reconstruct past sea-level oscillations and paleolandscapes; and Novack et al. [25] took advantage of the acoustic characterization of stratigraphic records.

Mattei et al. [17] performed a multidisciplinary investigation of the submerged Roman harbor at Nisida Island (Campi Flegrei, Italy) and set up a multiscale dataset with the goal of reconstructing the coastal and marine geomorphological evolution of the area. They were also able to outline relative sea-level changes in the last 2000 years through a new type of archaeological sea-level marker.

De Gioiosa et al. [24] analyzed LiDAR data, spectral images, and data from onshore surveys and scuba dives, with the aim of depicting land and sea morphologies and reconstructing Late Quaternary environmental changes of the Torre Guaceto Marine Protected Area (Apulia, Italy). The authors inferred climatic phases and related morphogenetic processes, identifying evidence of past sea-level oscillations.

Novack et al. [25] acoustically characterized the Quaternary sedimentary sequence of an alluvial plain of the NE Adriatic Sea that was submerged after the LGM. The paper shows that the sediment grain size is the main factor influencing sound velocity in shallow areas, the overload effect being negligible.

Deiana et al. [27] analyzed a large amount of multisensor data (seismic data, MBES and SSS data, dives, and UAV data) from the continental shelf off San Pietro Island (Sardinia, Italy) to reconstruct the geomorphological evolution of the area during the LGM. The authors outlined the complexity of a variety of coastal paleolandscapes, which also hosted Mesolithic population.

Savini et al. [26] spatially analyzed data on marine terraces both on land and at sea. Geomorphometric analysis on submerged terraces helped to figure out the depth range distribution of submarine terraces and associate their origin and evolution to the geological and structural setting of the Cilento Promontory (Campania, Italy). The work focused on highlighting the importance of submerged terraces formed on an outcropping bedrock (wave-cut or abrasion platforms) to decipher the signal of sea-level changes along the promontory and offshore, and to reconstruct the tectono-geomorphological evolution of the area.

4. Future Perspectives

Significant contributions have been collected in this volume, showing a variety of approaches and techniques used to integrate terrestrial and marine spatial datasets. The listed outcomes not only show how innovative and advanced geomorphological mapping techniques can support a more informed sustainable management of coastal environments, but also pave the way for other studies that rely on an improved efficiency in providing 3D landscape visualization from remote multisource and multiscale data.

The newly available geomorphological mapping tools and techniques are strongly impacting research on remote landscapes and those environments that are challenging to visualize, model, and analyze by means of traditional techniques. High spatial, temporal, and spectral resolution data from planet, moon, asteroid, and comet surfaces in our solar system collected by means of satellites, landers, and rovers are significant examples. Data visualization and processing in planetary geomorphology are indeed similar to those employed for terrestrial and marine data, and the procedure for recognizing landforms and processes that model planet surfaces can be compared with those acting on the Earth. An improved 3D visualization is also the basis for a more efficient modeling of those physical and abiotic processes that feature environmental changes and consequently can critically improve the associated simulation of future scenarios.

Hence, a strong collaboration among scientific communities working in the field of subaerial, submarine and planetary geomorphology is strongly needed to improve our understanding of geomorphic processes and the links between process and form on earth, seafloor, and even planetary surfaces.

Author Contributions: All authors participated equally in the preparation and writing of this editorial. All authors have read and agreed to the published version of the manuscript.

Funding: The study was carried out in the frame of the project "Coastal risk assessment and mapping" funded by the EUR-OPA Major Hazards Agreement of the Council of Europe (2020–2021), Grant Number GA/2021/08 n_689165 (Resp. Unimore Unit: Mauro Soldati).

Institutional Review Board Statement: Not applicable.

Informed Consent Statement: Not applicable.

Data Availability Statement: Data sharing not applicable.

Acknowledgments: This volume is part of the activities of the Working Group on "Submarine Geomorphology" of the International Association of Geomorphologists (IAG) chaired by Aaron Micallef (Malta) and Sebastian Krastel (Germany). We would like to thank the authors of the articles included in the Special Issue for having enthusiastically accepted to submit their manuscripts to this volume and for their kind cooperation during the whole editorial process. We are grateful to May Zheng, section managing editor, for having so professionally and kindly supported us throughout each step of the Special Issue planning and management.

Conflicts of Interest: The authors declare no conflict of interest.

References

1. Prampolini, M.; Savini, A.; Foglini, F.; Soldati, M. Seven good reasons for integrating terrestrial and marine spatial datasets in changing environments. *Water* **2020**, *12*, 2221. [CrossRef]
2. Hartley, R.I.; Zisserman, A. *Multiple Views Geometry*, 2nd ed.; Cambridge University Press: Cambridge, UK, 2003; p. 655.
3. Moore, J.G.; Normark, W.R.; Holcomb, R.T. Giant Hawaiian landslides. *Annu. Rev. Earth Planet. Sci.* **1994**, *22*, 119–144. [CrossRef]
4. Mulder, T.; Savoye, B.; Syvitki, J.P. Numerical modelling of a mid-sized gravity flow: The 1979 Nice turbidity current (dynamics, processes, sediment budget and seafloor impact). *Sedimentology* **1997**, *44*, 305–326. [CrossRef]
5. Field, C.B.; Barros, V.; Stocker, T.F.; Qin, D.; Dokken, D.J.; Ebi, K.L.; Mastrandrea, M.D.; Mach, K.J.; Plattner, G.-K.; Allen, S.K.; et al. *Managing the Risks of Extreme Events and Disasters to Advance Climate Change Adaptation*; A Special Report of Working Groups I and II of the Intergovernmental Panel on Climate Change (IPCC); Cambridge University Press: Cambridge, UK, 2012.
6. Hoegh–Guldberg, O.; Jacob, D.; Taylor, M.; Bindi, M.; Brown, S.; Camilloni, I.; Diedhiou, A.; Djalante, R.; Ebi, K.L.; Engelbrecht, F.; et al. 2018: Impacts of 1.5 °C Global Warming on Natural and Human Systems. In *Global Warming of 1.5 °C*; Masson-Delmotte, V., Zhai, P., Pörtner, H.-O., Roberts, D., Skea, J., Shukla, P.R., Pirani, A., Moufouma–Okia, W., Péan, C., Pidcock, R., et al., Eds.; An IPCC Special Report on the Impacts of Global Warming of 1.5 °C above Pre–Industrial Levels and Related Global Greenhouse Gas Emission Pathways, in the Context of Strengthening the Global Response to the Threat of Climate Change; Sustainable Development, and Efforts to Eradicate Poverty; Intergovernmental Panel on Climate Change: Geneva, Switzerland, 2018.

7. Intergovernmental Panel on Climate Change. Summary for Policymakers. In *Climate Change and Land: An IPCC Special Report on Climate Change, Desertification, Land Degradation, Sustainable Land Management, Food Security, and Greenhouse Gas Fluxes in Terrestrial Ecosystems*; Shukla, P.R., Skea, J., Buendia, E.C., Masson-Delmotte, V., Pörtner, H.-O., Roberts, D.C., Zhai, P., Slade, R., Connors, S., Diemen, R., et al., Eds.; Cambridge University Press: Cambridge, UK, 2019.
8. UNISDR (United Nations International Strategy for Disaster Reduction). *Sendai Framework for Disaster Risk Reduction 2015–2030*; The United Nations Office for Disaster Risk Reduction: Geneva, Switzerland, 2015; Available online: https://www.preventionweb.net/files/43291_sendaiframeworkfordrren.pdf (accessed on 7 April 2021).
9. European Commission. *An. EU Strategy on Adaptation to Climate Change*; The European Commission: Brussels, Belgium, 2013; Available online: https://eur-lex.europa.eu/LexUriServ/LexUriServ.do?uri=COM:2013:0216:FIN:EN:PDF (accessed on 7 April 2021).
10. Directive 2007/60/EC of the European Parliament and of the Council of 23 October 2007 on the Assessment and Management of Flood Risks. Available online: https://eur-lex.europa.eu/LexUriServ/LexUriServ.do?uri=OJ:L:2007:288:0027:0034:EN:PDF (accessed on 7 April 2021).
11. Bailey, G.; Sakellariou, D.; Members of the SPLASHCOS Network. Submerged prehistoric archaeology and landscapes of the continental shelf. *Antiq. Proj. Gallery* **2012**, *86*, 10.
12. Harff, J.; Bailey, G.; Luth, F. *Geology and Archaeology: Submerged Landscapes of the Continental Shelf*; Geological Society Special Publications 411; Geological Society of London: London, UK, 2016; p. 294. ISBN 978-1-86239-691-3.
13. Schlacke, S.; Maier, N.; Markus, T. Legal implementation of integrated ocean policies: The EU's marine strategy framework directive. *Int. J. Mar. Coast. Law* **2011**, *26*, 59–90. [CrossRef]
14. UNEP/MAP/PAP: Protocol on Integrated Coastal Zone Management in the Mediaterranean. Split, Priority Actions Programme. 2008. Available online: http://iczmplatform.org/ (accessed on 7 April 2021).
15. Prampolini, M.; Foglini, F.; Biolchi, S.; Devoto, S.; Angelini, S.; Soldati, M. Geomorphological mapping of terrestrial and marine areas, northern Malta and Comino (central Mediterranean Sea). *J. Maps* **2017**, *13*, 457–469. [CrossRef]
16. Meilianda, E.; Lavigne, F.; Pradhan, B.; Wassmer, P.; Darusman, D.; Dohmen-Janssen, M. Barrier Islands Resilience to Extreme Events: Do Earthquake and Tsunami Play a Role? *Water* **2021**, *13*, 178. [CrossRef]
17. Mattei, G.; Troisi, S.; Aucelli, P.; Pappone, G.; Peluso, F.; Stefanile, M. Sensing the Submerged Landscape of Nisida Roman Harbour in the Gulf of Naples from Integrated Measurements on a USV. *Water* **2018**, *10*, 1686. [CrossRef]
18. Perotti, L.; Carraro, G.; Giardino, M.; De Luca, D.A.; Lasagna, M. Geodiversity evaluation and water resources in the Sesia Val Grande UNESCO Geopark (Italy). *Water* **2019**, *11*, 2102. [CrossRef]
19. Coratza, P.; Vandelli, V.; Fiorentini, L.; Paliaga, G.; Faccini, F. Bridging Terrestrial and Marine Geoheritage: Assessing Geosites in Portofino Natural Park (Italy). *Water* **2019**, *11*, 2112. [CrossRef]
20. Biolchi, S.; Denamiel, C.; Devoto, S.; Korbar, T.; Macovaz, V.; Scicchitano, G.; Vilibic, I.; Furlani, S. Impact of the October 2018 Storm Vaia on Coastal Boulders in the Northern Adriatic Sea. *Water* **2019**, *11*, 2229. [CrossRef]
21. Rizzo, A.; Vandelli, V.; Buhagiar, G.; Micallef, A.S.; Soldati, M. Coastal vulnerability assessment along the north-eastern sector of Gozo Island (Malta, Mediterranean Sea). *Water* **2020**, *12*, 1405. [CrossRef]
22. Guida, D.; Valente, A. Terrestrial and marine landforms along the Cilento coastland (southern Italy): A framework for landslide hazard assessment and environmental conservation. *Water* **2019**, *11*, 2618. [CrossRef]
23. Hasan, O.; Miko, S.; Brunović, D.; Papatheodorou, G.; Christodolou, D.; Ilijanić, N.; Geraga, M. Geomorphology of canyon outlets in Zrmanja River estuary and its effect on the Holocene flooding of semi-enclosed basins (Novigrad and Karin Sea, eastern Adriatic). *Water* **2020**, *12*, 2807. [CrossRef]
24. De Giosa, F.; Scardino, G.; Vacchi, M.; Piscitelli, A.; Milella, M.; Ciccolella, A.; Mastronuzzi, G. Geomorphological Signature of Late Pleistocene Sea Level Oscillations in Torre Guaceto Marine Protected Area (Adriatic Sea, SE Italy). *Water* **2019**, *11*, 2409. [CrossRef]
25. Novak, A.; Šmuc, A.; Pogljen, S.; Celarc, B.; Vrabec, M. Sound velocity in a thin shallowly submerged terrestrial-marine Quaternary succession (northern Adriatic Sea). *Water* **2020**, *12*, 560. [CrossRef]
26. Savini, A.; Bracchi, V.A.; Cammarosano, A.; Pennetta, M.; Russo, F. Terraced Landforms Onshore and Offshore the Cilento Promontory (South-Eastern Tyrrhenian Margin) and their Significance as Quaternary Records of Sea Level Changes. *Water* **2021**, *13*, 566. [CrossRef]
27. Deiana, G.; Lecca, L.; Melis, R.T.; Soldati, M.; Demurtas, V.; Orrù, P.E. Submarine Geomorphology of the Southwestern Sardinian Continental Shelf (Mediterranean Sea): Insights into the Last Glacial Maximum Sea-Level Changes and Related Environments. *Water* **2021**, *13*, 155. [CrossRef]
28. Marchese, F.; Bracchi, V.A.; Lisi, G.; Basso, D.; Corselli, C.; Savini, A. Assessing fine-scale distribution and volume of Mediterranean algal reefs through terrain analysis of multibeam bathymetric data. A case study in the southern Adriatic continental shelf. *Water* **2020**, *12*, 157. [CrossRef]
29. Basso, D.; Granier, B. Calcareous algae in changing environments. *Geodiversitas* **2012**, *34*, 5–11. [CrossRef]

Review

Seven Good Reasons for Integrating Terrestrial and Marine Spatial Datasets in Changing Environments

Mariacristina Prampolini [1], Alessandra Savini [2,*], Federica Foglini [1] and Mauro Soldati [3]

[1]   Institute of Marine Sciences, National Research Council, 40129 Bologna, Italy;
     mariacristina.prampolini@bo.ismar.cnr.it (M.P.); federica.foglini@bo.ismar.cnr.it (F.F.)
[2]   Department of Earth and Environmental Sciences, University of Milano Bicocca, 20126 Milano, Italy
[3]   Department of Chemical and Geological Sciences, University of Modena and Reggio Emilia,
     41125 Modena, Italy; mauro.soldati@unimore.it
*   Correspondence: alessandra.savini@unimib.it; Tel.: +39-02-64482079

Received: 10 June 2020; Accepted: 3 August 2020; Published: 6 August 2020

Abstract: A comprehensive understanding of environmental changes taking place in coastal regions relies on accurate integration of both terrestrial and submerged geo-environmental datasets. However, this practice is hardly implemented because of the high (or even prohibitive) survey costs required for submerged areas and the frequent low accessibility of shallow areas. In addition, geoscientists are used to working on land or at sea independently, making the integration even more challenging. Undoubtedly new methods and techniques of offshore investigation adopted over the last 50 years and the latest advances in computer vision have played a crucial role in allowing a seamless combination of terrestrial and marine data. Although efforts towards an innovative integration of geo-environmental data from above to underwater are still in their infancy, we have identified seven topics for which this integration could be of tremendous benefit for environmental research: (1) geomorphological mapping; (2) Late-Quaternary changes of coastal landscapes; (3) geoarchaeology; (4) geoheritage and geodiversity; (5) geohazards; (6) marine and landscape ecology; and (7) coastal planning and management. Our review indicates that the realization of seamless DTMs appears to be the basic condition to operate a comprehensive integration of marine and terrestrial data sets, so far exhaustively achieved in very few case studies. Technology and interdisciplinarity will be therefore critical for the development of a holistic approach to understand our changing environments and design appropriate management measures accordingly.

Keywords: terrestrial geomorphology; submarine geomorphology; white ribbon; paleo-geography; coastal management

---

1. Introduction

Ongoing climate changes are producing remarkable impacts worldwide [1]. From the melting of polar ice sheets, and the consequent sea-level rise, to the increasing occurrence of extreme weather events [2], climate changes are notably modifying and/or threatening Earth's environment and ecosystems. Considerable effects have been observed especially in coastal regions [3], which host more than 10% of global population [4]. Coastal erosion, flood risk, increased landslide occurrence and wetland loss are expected to intensify in the coming decades [2], posing serious threats for inhabited areas and environmental assets. A major issue is the uncertainty about the extent and timing of climate-driven impacts, which has often reflected in a "non-immediate" adoption of effective prevention measures and in a non-sustainable coastal management, producing negative socio-economic consequences [2,5].

Coastal zones are the interface between the terrestrial and marine environments and together with nearshore ones are affected by multiple physical processes that originate in both the terrestrial (i.e., fluvial) and marine (i.e., waves and tides) environments. These processes drive geomorphic change, which determines increasingly hazardous conditions (e.g., unstable cliffs, lowlands susceptible to floods) in coastal areas in relation to climate changes and economic development [3,4]. A comprehensive understanding of environmental changes taking place on coastal regions, whose ecosystems are highly productive and very susceptible to changing environmental conditions, relies on an accurate integration of both terrestrial and submerged geo-environmental datasets. This practice is often lacking in coastal management and that still need to be addressed in many regions of the world, where climate change, rising sea levels, tectonic and marine geohazard of different nature are pressing harder year by year and resources need to be more carefully managed (e.g., [6]).

The absence of a seamless spatial data framework prevents in particular a standard procedure for locating and referencing spatial data across the land-marine interface. Different accessibility and investigation costs are the main causes for the huge disparity in size and quality among terrestrial and marine spatial datasets used in environmental research, especially in geomorphological investigation. Landscapes and landforms of terrestrial and marine areas have been traditionally investigated separately, and scientific communities used to working on land—coastal zone included—or at sea independently. On land, geomorphological mapping has been extensively used as a primary method to visualize and analyze Earth surface features ever since early geomorphological research. Taking advantage of the cartographical potentials of Geographic Information Systems (GIS) and the increasing availability of remote sensing tools, data, and products—especially high-resolution aerial and satellite imagery and derived datasets such as Digital Elevation Models (DEMs) and Digital Terrain Models (DTM)—geomorphological maps have been largely produced for many emerged sectors of Earth's surface. Besides being crucial for georisk assessment and land management, detailed geomorphological maps provided reference data for a variety of applied sectors of environmental research, such as landscape ecology, forestry or soil science and spatial planning [7,8]. In the terrestrial domain, the need to share and integrate spatial data for more efficient resource information management has been recognized for over a decade. On the contrary, although the latest developments in submarine acoustic remote sensing [9] have offered increasingly detailed DEMs for the submerged domain of the Earth's surface, only less than 18% of the world's seafloor has been surveyed with a resolution of 30 arc-second, while less than 9% of world's seafloor has been mapped with high-resolution multi-beam sonar [10–15]. Undoubtedly, new technologies are revealing more of the seafloor than ever before. The recent development of marine robotics, along with the critical improvement in optical underwater imaging systems (among which the use of hyperspectral cameras in the submarine realm, [16]) and computer vision made it possible to obtain, for the first time, 3D optical reconstruction and a real perception of underwater environments. This allowed for the first time to make more accessible the scientific understanding of submarine processes and environments, not only to scientists but even to the community, with a renewed appreciation of their heterogeneity. Nevertheless, a comprehensive characterization and categorization of submarine landforms and processes is still in its infancy, notwithstanding the important applications for the whole spectrum of submarine geomorphological research [9].

The attention recently paid to the integration of terrestrial and marine spatial datasets for different purposes is notable, although the generation of seamless DEMs and DTMs, based on a reliable integration of onshore, nearshore, and offshore data, is often a challenge for the most part of the coastal regions, especially due to technical issues often associated with the integration of multisource data (i.e., differences in resolution, precision, and accuracy among the datasets available). Addressing this process relies on the development of innovative approaches in using methods and techniques traditionally developed for the investigation of terrestrial environments, for exploring and imaging submarine landforms. A variety of challenging and promising techniques capable of providing a homogeneous and continuum representation of the Earth's surface from the land down to the deep seafloor has been recently tested. They basically rely on the collection of high-quality data for the integration of multiscale elevation datasets coming from different sources (i.e., bathymetric Light

Detecting And Ranging (LiDAR), photogrammetry, and echo-sounders—Figure 1). The application of photogrammetry based on the Structure from Motion technique (using a variety of platforms such as scuba diving, uncrewed surface vehicles—USV—and/or uncrewed aerial vehicles—UAV) has gained in particular new attention as a valuable tool for obtaining bathymetric measurements in very shallow environment, where acoustic devices cannot be safely employed, demanding high costs as in the case of LiDAR surveys [17]. The intertidal and nearshore zones are indeed of critical importance to be surveyed to bridge the gap between the land and the sea and to obtain a reliable integration of marine and terrestrial spatial dataset (Figure 1).

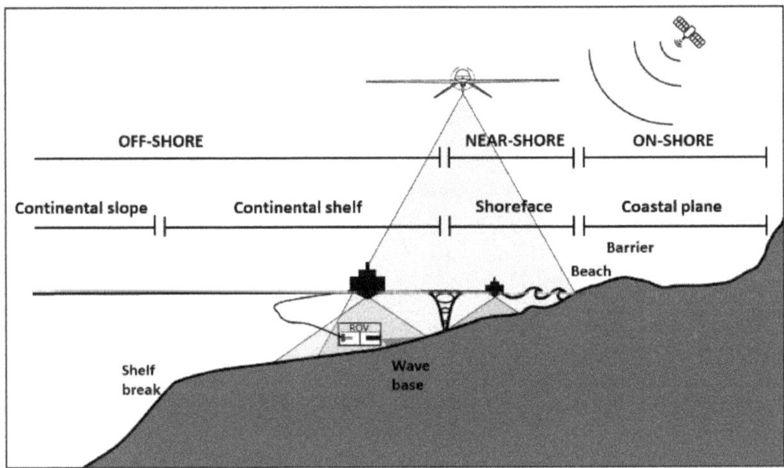

Figure 1. Different technological solutions for collecting high-resolution terrestrial and marine data exploiting different platforms: satellite imagery, airborne Light Detecting And Ranging (LiDAR) and bathymetric LiDAR, Uncrewed Aerial Vehicles (UAV)-mounted instruments, and keel-mounted and ROV (Remotely Operated Vehicle)-mounted echo-sounders to obtain high-resolution imagery and derived datasets such as DEMs and DTMs. Note the comparison of different coverages between traditional bathymetric survey vessels (equipped with acoustic systems, such as MBES-MultiBeam EchoSounders) and airborne optical sensors (i.e., bathymetric LiDAR), both in deep and shallow water. The advantages in time and costs of using the latter is obvious, although LiDAR systems provide a lower spatial sampling rate that generates much lower accuracy and precision than that offered by acoustic systems, especially when approaching extinction depths (see Section 3.2 and Table 1 for further details on *pros* and *cons*).

This paper aims at showing the reasons why the integration of terrestrial and marine dataset is beneficial for (i) an improved understanding of coastal processes and landforms, (ii) a more effective assessment of coastal risks and impacts, (iii) a sustainable management of coastal environments and associated resources. Progress in the fields of marine acoustic and underwater optical imaging that are contributing to obtain a seamless 3D reconstruction from the land to the sea, are also discussed.

2. Why Integrating Terrestrial and Marine Datasets?

The generation of a seamless digital terrain model that spans the terrestrial and marine environments is a key process for the mapping, modelling, and forecasting of the impact climate-driven changes to coastal geomorphic processes and environmental responses that may occur. This process has in particular the potential of making achievable a more integrated and holistic approach to the management of the coastal zone and the establishment of a timely disaster response and adoption of mitigation measures, especially for those climatically sensitive coastal areas where the changing climate is already severely pressing coastal population and their economy.

Hereafter are seven main distinct reasons why a wider integration of land and sea datasets is of great benefit for the associated implications in the context of global environmental changes. They refer to geomorphological research, and to all those applied sectors of environmental research (e.g., marine spatial planning, archaeology and landscape ecology) that focus on spatial planning and sustainability (Table 1 and Figure 2).

Table 1. Outline of the most relevant scientific literature showing the outputs of integrated terrestrial and marine research with reference to the 'seven good reasons' mentioned above.

	Field of Applications	**Relevant Examples**
1	Geomorphological mapping	Miccadei et al. [18–20]; Leon et al. [21]; Gasparo Morticelli et al. [22]; Mastronuzzi et al. [23]; Prampolini et al. [6,24]; Brandolini et al. [25]; Furlani et al. [26]; Campobasso et al. [27]; Genchi et al. [28]
2	Late-Quaternary changes of coastal landscapes	Bridgland et al. [29]; Pujol et al. [30]; Rovere et al. [31]; Westley et al. [32]; Micallef et al. [33]; Kennedy et al. [34]; Greenwood et al. [35]; Aucelli et al. [36]; Foglini et al. [37]; Benjamin et al. [38]; Furlani et al. [39]; Furlani and Martin [40]; De Gioiosa et al. [41]; Lo Presti et al. [42]
3	Geoarchaeology	Antonioli et al. [43]; Bailey and Flemming [44]; Harff and Lüth [45,46]; Fisher et al. [47]; Benjamin et al. [48]; Westley et al. [32]; Bailey et al. [49]; Furlani et al. [50]; Anzidei et al. [51]; Evans et al. [52]; Westley et al. [53]; Bailey et al. [54]; Aucelli et al. [36,55]; Harff et al. [56]; Cawthra et al. [57]; Benjamin et al. [38]; Benjamin et al. [58]; Furlani and Martin [40]; Mattei et al. [59]; Sturt et al. [60]; Veth et al. [61]
4	Geoheritage and geodiversity	Orrù and Ulzega [62]; Orrù et al. [63,64]; Brooks et al. [65]; Rovere et al. [66]; Mansini Maia and Alencar Castro [67]; Veloo [68]; Gordon et al. [69,70]; Coratza et al. [71]
5	Geohazards	Moore et al. [72]; Mulder et al. [73]; Assier-Rzadkiewicz et al. [74]; Caplan-Auerbach et al. [75]; Gee et al. [76]; Masson et al. [77]; McMurtry et al. [78]; Zhang et al. [79]; Dan et al. [80]; Scheffers and Scheffers [81]; Chiocci et al. [82]; Baldi et al. [83]; Parrott et al. [84]; Casalbore [85]; Violante [86]; De Blasio and Mazzanti [87]; De Gange et al. [88]; Sultan et al. [89]; Casalbore et al. [90,91]; Chiocci and Ridente [92]; Lambeck et al. [93]; Mazzanti and Bozzano [94]; De Jongh and van Opstal [95]; Della Seta et al. [96]; Knight and Harrison [97]; Masselink and Russel [98]; Mastronuzzi et al. [99]; Minelli et al. [100]; Carracedo [101]; Cazenave et al. [102]; Mataspaud et al. [103]; Mottershead et al. [104]; Biolchi et al. [105]; Yonggang et al. [106]; Antonioli et al. [107]; Aucelli et al., [108]; Zaggia et al. [109]; Casalbore et al. [110]; Di Paola et al. [111]; Moore et al. [112]; Obrocki et al. [113]; Pennetta et al. [114]; Urlaub et al. [115]; Biolchi et al. [116]; Buosi et al. [117]; Mucerino et al. [118]; Toker et al. [119]; Rizzo et al. [120]
6	Marine and landscape ecology	Hogrefe et al. [121]; McKean et al. [122,123]; Tallis et al. [124]; Wright and Heyman [12]; Vierling et al. [125]; Brown et al. [126]; Leon et al. [21]; Marchese et al. [127]; Prampolini et al. [6,128]; Harris and Baker [129]
7	Coastal planning and management	Cicin-Sain and Belfiore [130]; Sarda et al. [131]; Schultz-Zehden et al. [132]; Cogan et al. [133]; Ehler et al. [134]; Watts et al. [135]; Meiner [136]; Schlacke et al. [137]; Smith et al. [138]; Qiu and Jones [139]; Kerr et al. [140]; Ramieri et al. [141,142]; Barbanti et al. [143]; Domínguez-Tejo et al. [144]; UNEP/MAP [145]; UNEP-MAP PAP/RAC [146]; Decision IG. 22/1 [147]; Decision IG. 22/2 [148]; Sustainable Development Goals [149,150]; UNEP/MAP PAP [151]

In detail, integrated research merging terrestrial and submarine data can substantially contribute to:

1. Development in the field of geomorphological mapping and coastal morphodynamics thanks to innovative mapping techniques and products for coastal and nearshore environments;
2. Deeper understanding of Late-Quaternary changes of coastal landscapes and environments, with positive implications for prediction of future risk scenarios;
3. New approaches for geoarchaeological research development in coastal and nearshore environments;

4. More comprehensive recognition and assessment of geoheritage and geodiversity;
5. Wider assessment of coastal geohazards and vulnerability in the frame of disaster risk reduction, thanks to the development of models of different processes (e.g., coastal hydrodynamic modelling);
6. Critical support to other disciplines involved in generating key-data for the sustainable management of marine resources, such as marine and landscape ecology;
7. Establishment of more sustainable development objectives in planning and management of coastal areas, also in the framework of Integrated Coastal Zone Management.

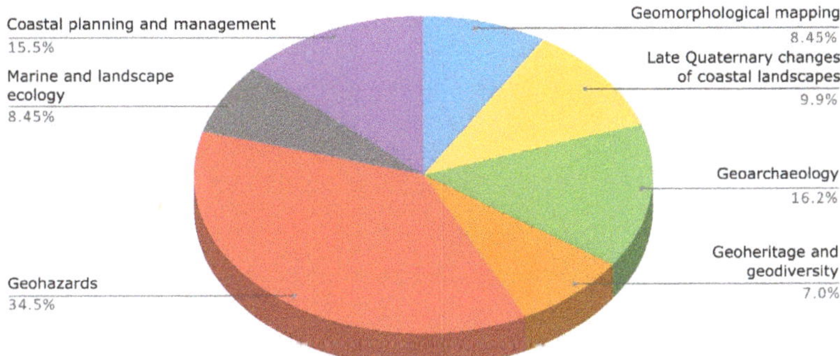

Figure 2. Thematic distribution of the main scientific literature referring to the integration of terrestrial and marine datasets according to the fields of application shown in Table 1.

2.1. Geomorphological Mapping

Analyzing and representing terrestrial and marine data in an integrated perspective allows us to provide a comprehensive picture of a coastal landscape, taking into account the variety of processes that contribute to landform development through time. The most appropriate tool for the representation of emerged and submerged landscapes and landforms in coastal areas (or, in general, in transition environments) is an integrated geomorphological map that relies on the availability of elevation data from the onshore to the offshore settings. Scientific literature started to report consistent examples toward this effort only during the last decade (e.g., [18–20]).

A traditional (terrestrial) geomorphological map generally represents landform genesis, distribution, morphometry, and state of activity, offering a dynamic vision of the investigated coastal landscape (e.g., [23,152,153]). According to the mapping approach used, information on lithology, geological structure, and age of surficial deposits can be included. The value of such maps is unquestionable for both scientific and applied research, aiming at environmental conservation and management, and hazard and risks assessment. The same applies to marine geomorphological maps [9,154].

The main differences between terrestrial and marine geomorphological maps and the hurdle in producing geomorphological maps integrating the representation of land and seafloor features in coastal areas and shallow waters concern:

- Representation scale: terrestrial geomorphological maps are more easily drawn at fine scale (e.g., 1:5000) than submarine geomorphological maps, because of the higher ease of data collection on land than underwater;
- Standardized terminology and classification schemes: terrestrial landforms are codified at an international level, and their definition and representation are generally shared, either worldwide or country-wide, while for submarine landforms—apart from the main physiographic classification [155]—different terms and classification schemes are used (e.g., see how bedforms are differently categorized in: Rubin and McCulloch [156]; Ashley [157]; Wynn and Stow [158]; Stow et al. [159]);

- Standardized symbology: standard symbols are codified for terrestrial geomorphological mapping, though they may vary from country to country, while they are lacking for submarine geomorphological mapping and for integrated terrestrial and submarine geomorphological mapping;
- Coverage of lithological and chronological information: in the terrestrial environment, the coverage of lithological and chronological data is generally much higher than for the seafloor, thanks to the wider availability of maps, DTMs, scientific literature, and datasets;
- Acquiring elevation data in marine regions poses significant challenges and some limitations which make the process more complex, from a technological point of view, than in the emerged system. Customized surveys and dedicated technological solutions are indeed required to apply specific corrections that can address all measurements errors created in particular by hydrodynamics (especially tides and wave motion that must be always severely taken into account in hydrographic survey carried out by mean of echo-sounders) and the physical variability of the water column (which has a strong impact on the sound velocity/refraction of beams, creating at places challenging conditions, such as in the case of fresh water influx at the mouth of a river). Cloud coverage, turbidity, water surface glint and breaking waves can also create challenging environmental and operational condition that may prevail during optical remote sensing surveys in shallow water.

Nevertheless, the scientific community has recently paid increasing attention to the above-mentioned issues and different solutions were introduced to overcome them. First of all, technological improvements in acoustic remote sensing has now made it possible to acquire high-resolution elevation data in shallow environments, ranging from 0.5 m [160] to 0.05 m of resolution [28,161], and also with precision and accuracy close to those achievable for terrestrial landscape investigation. In addition, even the deep environment (deeper than 200 m, [162]) can be now surveyed at high resolution thanks to the progress made by underwater robotics. Some compromises are still needed for shallow-water data acquisition, but recent progress has definitely produced unprecedented results (see Section 3.2 and Table 2 for additional details). On the contrary, the standardization of terminology, classification schemes, and symbology is definitely still an unresolved issue, even though many efforts have been made globally to offer practical and applicable solutions in different contexts. In December 2015, a partnership originated from seabed mapping programs of Norway, Ireland and the UK (MAREANO, INFOMAR, and MAREMAP, respectively—MIM) shares experiences, knowledge, expertise and technology in order to propose advancement in best practices for geological seabed mapping [163]. The development of a standardized geomorphological classification approach, following previous work published by the British Geological Survey [164], has been one of the key activities of the MIM group. At the European level, the EMODnet-Geology Portal (European Marine Observation and Data Network; [165]) is collecting data from European state members on the geology and geomorphology of European seabeds and harmonizing the geomorphological representation through a vocabulary developed by the Federal Institute for Geosciences and Natural Resources [166] and by the British Geological Survey [167]. In addition to research strictly dealing with geological and geomorphological mapping, further inputs have been also provided by organizations extensively involved in providing seafloor maps for several environmental purposes, but where precise representation and definition of submarine landforms were crucial for their applications. The Integrated Ocean and Coastal Mapping (IOCM) of the National Oceanic and Atmospheric Administration (NOAA) developed the Coastal and Marine Ecological Classification Standard (CMECS), a scheme for classifying benthic habitats that includes a geomorphic level for the description of submarine landforms. At the European level (within the frame of different EU funded projects, such as Mapping European Seabed Habitats (MESH), CoralFish [168] and CoCoNet [169] among others), various classification schemes were proposed [170,171] taking inspiration from the CMECS and/or from the analogue European classification scheme EUNIS [172].

Table 2. List of sensors and techniques cited in the text which are relevant for collecting elevation data in the shallow-water environment (i.e., white ribbon). Active sensors are those sensors that control the projection of acoustic or electromagnetic waves (i.e., sound or light) onto the scene, while the term "passive" refers to those systems or techniques that use ambient light to illuminate the scene or techniques such as Structure from Motion (SfM). MBES: Multi-Beam Echo-Sounder; PDBS: Phase Differencing Bathymetric Systems; MPES: multi-phase echo-sounders; ASV: Autonomous Surface Vessels; USV: Uncrewed surface vehicles; ROV: Remotely Operated Vehicle; UAV: Uncrewed Aerial Vehicle; LiDAR: Light Detecting And Ranging; LLS: Laser Line Scanning; SDB: Satellite-Derived Bathymetry; SfM: Structure from Motion.

Active Sensors				
	Sensor	**Platform**	**Achievable DTM Resolution (Nearshore)**	**Survey Environment**
Acoustic	**MBES (beamforming) PDBS (Interferometers) MPES (Hybrid)**	Vessel ASV/USV ROV	Sub-metric	Water
	PROS: - Accuracy and precision address international standards and can be validated - Dedicated software support accurate data processing and standard requirements		**CONS:** - Time consuming because of unfavorable ratio in coverage/unit of time (i.e., small areas covered in a long period of time) - Sensor must be immerged for data acquisition; very shallow depth measurements (roughly ≤ 2 m) cannot be performed - Not suitable for very shallow and topographically complex seafloor		
Optical	**Bathymetric LiDAR**	Airborne UAV	Sub-metric	Air/water
	PROS: - Favorable ratio in coverage/unit of time (i.e., wide areas covered in a relatively small period of time) - Accuracy and precision address international standards and can be validated		**CONS:** - Data processing is often time consuming - Environmental conditions can pose relevant constrains (water turbidity and atmospheric condition)		
	LLS	Vessel ROV	Sub-metric	Water
	PROS: - Favorable ratio in coverage/unit of time - ROV-based LLS survey can provide high-resolution data in deep and complex environments, which are difficult to cover with traditional ROV-based MBES surveys		**CONS:** - Data processing is often time consuming and results can be difficult to validate - Environmental conditions can pose relevant constrains (water turbidity in particular)		
Passive Sensors and Computational Techniques				
	Name	**Platform**	**DTM Resolution (Nearshore)**	**Survey Environment**
Optical	**SDB**	Satellite	2 m	Air/Water
	PROS: - Highly favorable ratio in coverage/unit of time (i.e., wide areas covered in a relatively small period of time) - Source data are public and available (i.e., Sentinel-2) - Source data also offer photorealistic view of the surveyed scene		**CONS:** - Accuracy and precision are dependent on processing algorithms (they do not rely on sensor performance and are not part of technical specification) - Data processing is often time consuming - Environmental conditions can pose relevant constrains (water turbidity, cloud coverage)		
	SfM	UAV	Sub-metric	Air/Water
	PROS: - Favorable ratio in coverage/unit of time (i.e., wide areas covered in relatively small period of time) - Camera and platforms (i.e., drones) are available also at low costs - Source data also offer photorealistic view of the surveyed scene		**CONS:** - Data processing is often time consuming - There are no available standards and public protocols for data processing to guarantee accuracy of obtained elevation data - Need for ground control points (GCP) to validate results - Environmental conditions can pose relevant constrains (water turbidity, glint, breaking waves)		

A few attempts have recently been made to overcome the critical issue of defining a standard symbology for submarine and integrated land–sea geomorphological mapping, and most of them addressed the issue at the national level. Within the CARG Project (Italian official Geological Cartography), the Institute for the Protection and Environmental Research (ISPRA) promoted the production of Geological and Geothematic Sheets at the scale 1:50,000 of the Italian territory. In particular, in coastal areas, both terrestrial and marine geology and terrestrial and marine geomorphology were represented [173]. Recently, the Working Group on Coastal Morphodynamics of the Italian Association of Physical Geography and Geomorphology—in agreement with ISPRA—revised and updated the legend for the "Geomorphological Map of Italy" [27] with reference to coastal, marine, lagoon and aeolian landforms, processes and deposits [23]. Within the revision process submerged landforms and deposits down to the continental shelf break were included in the new legend. The first applications of this new legend were provided by Brandolini et al. [25], Furlani et al. [26] and Prampolini et al. [6,24]. Brandolini et al. [25] compared multi-temporal maps of anthropogenic landforms in Genoa city (Liguria region, Italy) to analyze the urban development since the Middle Age to provide a tool for geo-hydrological risk assessment. Furlani et al. [26] surveyed and mapped the coastal stretch of the Conero system (Marche region, Italy), focusing on tidal notches as geomorphological markers for analyzing Late-Holocene tectonic stability of the area. Prampolini et al. [6,24] surveyed and mapped coastal and submarine areas of the Maltese archipelago in order to produce the first integrated geomorphological maps of the islands, constituting the basis for geomorphological reconstruction and coastal hazard assessment in an environment highly sensitive to ongoing climate change (Figure 3).

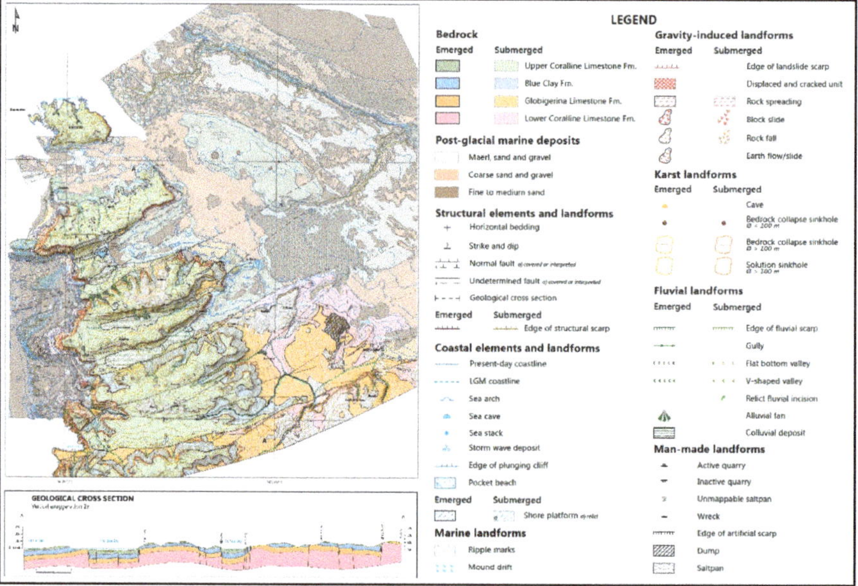

Figure 3. Sketch of the "Integrated geomorphological map of emerged and submerged areas of northern Malta and Comino (central Mediterranean Sea)" at 1:25,000 scale and related legend (modified after Prampolini et al. [6]).

With reference to the coverage of lithological and chronological information, it is possible to produce marine geological maps exploiting data coming from marine geological and geophysical surveys. Examples from the Italian coastal and marine areas are given by ISPRA and Gasparo Morticelli et al. [22]. ISPRA, thanks to the CARG Project, has promoted the mapping of the marine geology of the Adriatic Sea, scale 1:250,000 [174], to represent the main geological structures of the

Adriatic continental shelf. Gasparo Morticelli et al. [22] published the geological map of Marettimo Island and the submerged surroundings (Egadi Islands, central Mediterranean Sea), scale 1:10,000, integrating data from field surveys and previous marine geological and geophysical surveys, carried out within the CARG Project. Moreover, advances are being made in correlating seabed acoustic backscatter with surface sediments to perform automatic ground discrimination and reduce the number of ground truth stations required to produce robust seabed characterization. Recent availability of well-data from the industry (e.g., ViDEPi [175]) and progress in seismic analysis and seismo-stratigraphic correlation, contributed for a wider geological characterization of the subseafloor too. This way, it is possible to get a wider coverage of lithological and chronological information for submarine regions.

2.2. Late-Quaternary Changes of Coastal Landscapes

Scientific literature referring to integrated analyses of terrestrial and marine geospatial datasets related to coastal areas and continental shelves has proved to be critical in outlining Late-Quaternary coastal changes and paleo-landscapes.

Quaternary global sea-level changes have been largely studied worldwide with special attention to the late Pleistocene and Last Glacial Maximum (LGM), when the relative sea level reached its minimum height of ca. 120–130 m below the present sea level (Figure 4) [176,177].

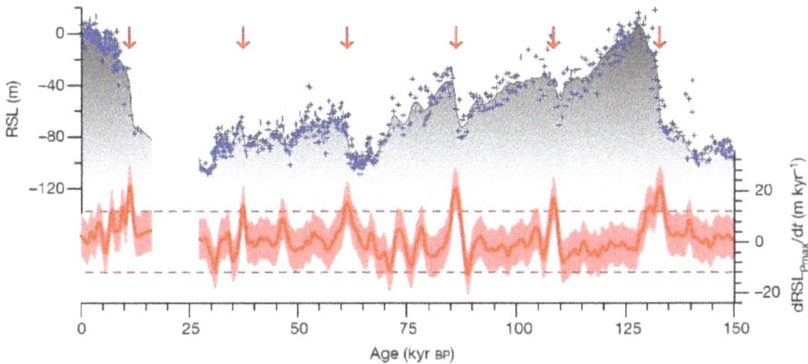

Figure 4. Sea-level changes during the Quaternary (modified after Grant et al. [177]). Relative sea-level (RSL) data (blue crosses) and maximum probability of RSL (grey shading) for the last 150 kyr, with an indication of the rates of sea-level changes (dRSL$_{Pmax}$/dt) expressed as m/kyr.

Consequently, most of the present marine areas down to 130 m of depth, including large portions of the world continental shelves, were emerged during the LGM and modelled by subaerial processes: the paleo-geography of emerged and submerged areas was very different from today, and included land bridges between continental landmasses which do not exist anymore (e.g., the land bridge between Sicily and the Maltese Archipelago—Figure 5—or between Sicily and Egadi Islands of Favignana and Levanzo, in southern Europe [39,42], or the Bering land bridge in the northern Pacific Ocean (Figure 6) [178]).

The post-glacial sea-level rise covered these landscapes, either drowning [33,37] or re-modelling them through marine processes. At some places, the previously emerged areas hosted prehistoric and historic human settlements—as demonstrated by ruins, pollens, bones, or other remains in the sediments, caves, etc. (cf. Section 2.3) [56]. The post-glacial marine transgression led to the disappearance of vast areas, land bridges, human settlements, and a reconfiguration of geographical boundaries. Indeed, the reconstruction of paleo-landscapes and their geomorphological evolution is the basis for further investigations (e.g., on geohazard assessment, identification, and study of geoarchaeological sites).

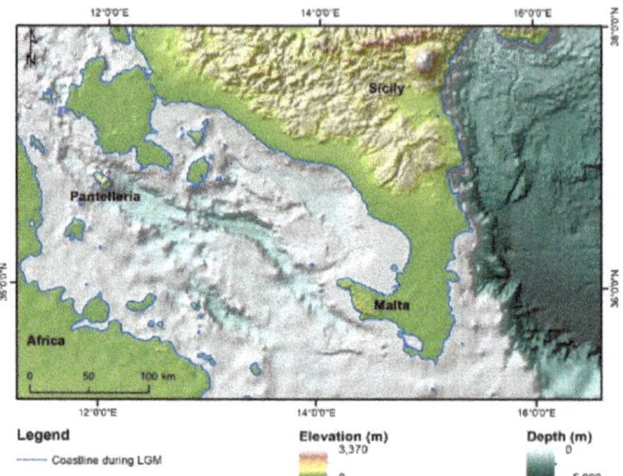

Figure 5. The land bridge between Sicily and Malta (Central Mediterranean Sea) during the LGM (modified after Prampolini et al. [6]).

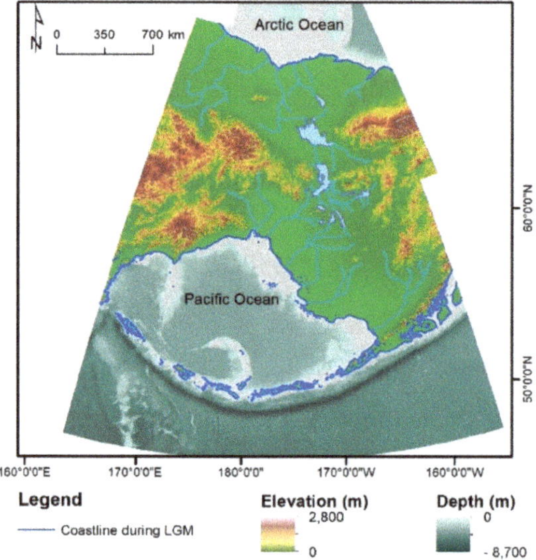

Figure 6. The Bering land bridge during the LGM (modified after Bond [179]).

To reconstruct ancient landscapes located in coastal and shallow-water environments, geomorphological, archaeological, and geophysical observations of both terrestrial and submarine areas are required, together with absolute dating of terrestrial and marine sediments, remains and landforms [54,61]. Hence, technological advances in acquiring data in nearshore areas and in merging terrestrial and marine spatial datasets (e.g., elevation data) are fundamental for the reconstruction of present emerged and submerged topography and to infer paleo-geography.

Numerous examples of terrestrial and marine data integration for the reconstruction of paleo-landscapes are from the Mediterranean Basin [38]. Lo Presti et al. [42] reconstructed the paleo-geography of Egadi Islands and relative sea-level variations from the LGM until today. Furlani and Martin [40] reconstructed the paleo-geography of Faraglioni coast (Ustica Island) that was settlement of a Middle Bronze Age village. They combined geomorphological observation made in nearshore and onshore areas. Miccadei et al. [18,19] reconstructed the Late-Quaternary landscape and geomorphological evolution of Tremiti Islands, located north of Gargano promontory, southern Adriatic Sea. Aucelli et al. [36] carried out a multidisciplinary study of submerged ruins of Roman buildings on the Sorrento Peninsula coast (Gulf of Naples). These archaeological remains enabled the reconstruction of the ancient position of both the sea level and the coastline. Rovere et al. [31] analyzed the submarine geomorphology of the offshore between Finale Ligure and Vado Ligure (western Liguria, NW Mediterranean Sea) for the first time, detecting meaningful submarine geomorphological indicators of former sea levels. Micallef et al. [33] and Foglini et al. [37] reconstructed Late-Quaternary coastal landscape morphology and evolution of the Maltese archipelago, while Furlani et al. [39] focused their research on marine notches of the Maltese Islands that resulted in confirmation of the slowdown of the Late-Holocene marine transgression.

Examples from northern Europe come, among the others, from southern England and northern France [29], and from Northern Ireland [32]. Bridgland et al. [29] analyzed three fluvial sequences, particularly terrace staircases, from southern England and northern France to reconstruct climate fluctuations and paleo-geography of those areas. Westley et al. [32] mapped the continental shelf of northern coast of Ireland and examined the geomorphology for evidence of past sea-level changes, reconstructed the paleo-geography of the area considering sea-level lowstands of −30, −14 and −6 m. This research allowed the identification of ten areas of high archaeological potential.

The combined mapping of emerged and submerged geomorphological features proved to be functional in analyzing the long-term evolution of coastal landslides. Prampolini et al. [6] showed that coastal block slides along the NW coast of Malta prolong below the sea level, reaching a depth of about 40 m, and Soldati et al. [180] demonstrated by means of cosmogenic nuclide dating that they developed in a subaerial environment—when the coastline was much lower than today—having been submerged only later on, during the post-glacial sea-level rise.

2.3. Geoarchaeology

As earlier mentioned, Late-Quaternary sea-level changes have exposed large portion of the present-day continental shelves for long periods of time, resulting in a multitude of archaeological remains lying on the seafloor today [4,45,46,48,49,52]. Therefore, coupling terrestrial and marine datasets can be critical in detecting new archaeological sites in coastal and nearshore areas and for a more comprehensive understanding of already existing ones [60] and references therein. Analyzing coastal archaeological sites can also contribute to the reconstruction of the paleo-geography of ancient landscapes [48], and in particular to infer Late-Holocene relative sea-level oscillations (e.g., [43,51]). Some archaeological remains include functional structures or elements that are unequivocally related to specific elevation of past sea levels, because of their architecture and proximity to the sea. In other cases, an in-depth knowledge of landscape evolution helps inferring about the evolution or the dating of archaeological remains. For example, the proximity of an archaeological site with coastal landforms, whose evolution can be reconstructed, will help in dating and reconstructing the history of the site itself [40] and references therein.

As a matter of fact, early human populations tended to move and expand occupying new territories, in particular during the last glaciation and the early post-glacial period—a period of time characterized by extreme climate fluctuations [56]. In this frame, the most attractive sites for human settlements were coastal lowlands that in some parts of the world, were much more extended than today thanks to sea-level lowstands (e.g., during the LGM, the European land area was 40% wider than presently; [56,181]). During that period, coastal regions were the most densely populated since they profited from more tempered climates that led to enhanced water supplies, and greater ecological diversity. Hence, these areas were sites of prehistoric and historic human settlements, as witnessed by the findings of archaeological remains, pollens, bones, or other ruins in the sediments, caves etc. Then post-glacial sea-level rise led to the flooding of these former territories, redrawing geographical boundaries, and human, plant and animal distributions (cf. Section 2.2) [56,181]. In this context, Harff et al. [56] reported the results achieved within the framework of the SPLASHCOS Project—Submerged Prehistoric Archaeology and Landscapes of the Continental Shelf—(Cooperation in Science and Technology—COST Action TD0902). They succeeded in gathering together experts in geology, archaeology, and climate interested in sea-level changes, paleo-climatology and paleo-geography for reconstructing European submerged landscapes in order to assess their archaeological potential (e.g., [37], cf. Section 2.2).

Geoarchaeological research benefiting from the integration of terrestrial and marine datasets is illustrated with reference to the Mediterranean area by Furlani et al. [50], Aucelli et al. [36,55], Furlani and Martin [40] and Mattei et al. [59]. Furlani et al. [50] studied submerged or partially submerged archaeological structures located along the Maltese coasts providing a first attempt for paleo-environmental reconstruction of the Maltese archipelago from the LGM until today, allowing time and mode of mammal dispersal to the island during the Pleistocene to be inferred. Aucelli et al. [36] analyzed submerged ruins of Roman buildings located along the Sorrento Peninsula coast (Italy) and succeeded in reconstructing sea-level oscillations and coastline changes for the Late-Holocene and tectonic history of the Sorrento Peninsula during the last two millennia. Aucelli et al. [55] explored and mapped the main underwater structures on and below the seabed of the Roman Villa of Marina di Equa (Sorrento Peninsula) and analyzed the geological effects of the 79 A.D. eruption of Vesuvius with the aim of reconstructing the interactions between human and natural events. Furlani and Martin [40] reconstructed the paleo-geography of Ustica Island, focusing on Faraglioni Village, providing clues on the evolution of one of the best-preserved Middle Bronze Age sites in the Mediterranean. Mattei et al. [59] reconstructed the natural and anthropogenic underwater landscape of the submerged Roman harbor of Nisida Island (Gulf of Naples, Italy), the relative sea-level variation in the last 2000 years and outlined the coastal geomorphological evolution of the area.

In northern Europe, Westley et al. [53] exploited the data acquired and analyzed by Westley et al. [32] (cf. Section 2.2) to reconstruct Early–Mid-Holocene paleo-geography of the Ramore Head area (Northern Ireland), hosting evidence of Mesolithic occupation and preserved Early–Mid-Holocene peats both on- and offshore.

Examples of integration of land–sea datasets for geoarchaeological purposes outside the Mediterranean area are provided by Fisher et al. [47] and Cawthra et al. [57] for South Africa, by Bailey et al. [54] for Saudi Arabia, and by Benjamin et al. [58] and Veth et al. [61] for Western Australia. Fisher et al. [47] developed a conceptual tool that enable correlation of the evolution of human behavior within a dynamic model of changes of paleo-environment. Cawthra et al. [57] analyzed paleo-coastal environments, laying on the present continental shelf, offshore of the Pinnacle Point archaeological locality (Mossel Bay, South Africa). During the Pleistocene, these environments were probably settlement of early-modern humans. Bailey et al. [54] analyzed emerged and submerged landscape of SW Saudi Arabia to study human dispersal in Late Pleistocene and Early Holocene. Finally, Benjamin et al. [58] and Veth et al. [61] analyzed a large amount of data (airborne LiDAR, underwater acoustics, cores and scuba dives observations) acquired in Western Australia for archaeological research.

Coupling land–sea data is now more feasible thanks to modern marine research technologies, integration of large databases and proxy data [60] and references therein, allowing further hidden archaeological sites to be discovered and studied in the near future. Recently, this has proved to be successful is the case of the ancient Roman city of Baia located inside the Bay of Pozzuoli and belonging to the Campi Flegrei Archeological Park (Southern Italy). The site is superbly preserved underwater after having been slowly drowned due to bradyseismic movements which characterize this area near the Vesuvius volcano [182].

Geoarchaeological investigations, particularly in coastal and submerged environments are increasing and are taking advantage of new contributions and new approaches in surveying and collecting data using a combination of acoustic and optical remote sensing sources, to recreate a full picture of the present and old landscapes, validated through field surveys observations and absolute dating evidence (e.g., Uncrewed Surface Vehicle simultaneously acquiring geophysical data and images for photogrammetry and drones equipped with cameras).

2.4. Geoheritage and Geodiversity

Integrating terrestrial and marine datasets can be of paramount importance for the assessment of terrestrial and marine sites of geological interest in coastal and shallow-water areas. Both the shore and inner continental shelf show common processes and landforms that should be considered to be a single feature (cf. Sections 2.1 and 2.2) [66].

Geosites—or geodiversity sites (*sensu* Brilha [183])—are places of a certain value due to human perception or exploitation and include geological elements with high scientific, educational, aesthetic, and cultural importance [71]. Geosites and key geodiversity areas are often protected areas thanks to different directives (e.g., EU Habitat Directive, 1992; OSPAR Convention, 1992; EU Marine Strategy Framework Directive, 2008). Indeed, the importance of preserving geodiversity has been acknowledged mainly thanks to the effect that geodiversity has on biodiversity patterns [184]. Hence, the assessment of geosites and key geodiversity areas enhance the identification of areas that need protection (e.g., Marine Protected Areas—MPAs—Geoparks; cf. Sections 2.6 and 2.7).

Geoheritage and geodiversity have been investigated mainly in terrestrial environments [71] and references therein, while only a few studies on this topic refer to underwater environments (cf. [62–64,66]). With reference to the latter, Orrù and Ulzega [62], Orrù et al. [63,64] identified underwater trails for scuba divers in the MPAs of Capo Carbonara and of Capo Caccia (Sardinia, Italy) enhancing the value of the whole underwater environment. Rovere et al. [66] assessed underwater geomorphological heritage in the Bergeggi MPA (Ligurian Sea, Italy) and in the Sigri area (Lesvos Island, Greece); however they considered that a complete approach in studies on geoheritage would take both emerged and submerged landforms of the coastal and nearshore environments into account.

Very few studies deal with the integrated assessment of terrestrial and marine sites of geological interest. This is largely due to technological constraints and, to some extent, to conceptual issues—such as (i) differences in attributes related to geosites in terrestrial and marine environments [185]; (ii) different perception and fruition of abiotic features of the aquatic environment by tourists; and (iii) lack of common schemes and approaches to the identification, assessment, and improvement of submarine geosites. However, attention to these themes is increasing presently.

Coratza et al. [71] identified and assessed the terrestrial and marine geosites of the Portofino Natural Park and MPA (Liguria, Italy), which are internationally known for both terrestrial scenic landforms and quality of the marine ecosystem. They aimed at identifying the most suitable sites for tourist improvement and defining possible connections between terrestrial and marine environments. Finally, they were able to identify a significant number of both terrestrial and marine sites, assessing their scientific value, ecological, cultural, and aesthetic importance, and accessibility, services, and economic potential.

The identification and quantitative assessment of geodiversity in terrestrial areas are already established and several countries have been developing national inventories of geodiversity key areas by means of different methods [183]. Only recently, this discipline has been addressed to the marine environment and its elements of geodiversity. Examples of geodiversity assessment in marine areas come from Scotland [65–70], southeast Brazil [67], Hawaiian and Canarian Islands and the New Zealand subduction zone [68]. Brooks et al. [65] and Gordon et al. [69,70] focused on the contribution of geo-conservation within the Marine Protected Area network for Scottish seawaters. Their work is the first systematic assessment of marine geodiversity key areas comparable to the Geological Conservation Review geo-conservation carried out for the terrestrial geology and geomorphology of Great Britain, which was "a world-first project of its type in the systematic assessment of the whole geological heritage of a country, from first principles" [69]. Mansini Maia and Alencar Castro [67] developed a model for characterizing marine geodiversity at a regional scale in the Vitória–Trindade Volcanic Seamount Ridge and its surroundings (SE Brazil). They aimed at supporting Brazilian marine spatial planning regarding geo-conservation of features related to the geological history of Brazil and the most vulnerable habitats. Finally, Veloo [68] developed a geodiversity index for the seafloor and applied it to three study areas (Hawaii, Canary Islands, and the New Zealand subduction zone), considering abiotic factors such as geomorphology, bathymetry, range of slope angle, and light penetration.

Although there is a growing number of studies on terrestrial and marine geodiversity showing that its understanding is essential for several issues, including geo-conservation [186], still there is no specific literature available on the integrated assessment of geodiversity in coastal and shallow-water areas. However, available knowledge and technological tools call for immediate actions in this field, which would be highly beneficial for holistic management and planning in coastal and nearshore areas.

2.5. Geohazards

Coastal and shallow-water environments are threatened by different kinds of geohazards that can produce significant impacts on (i) the economy, due to possible reduction in tourism and disruption of urbanized areas, (ii) on landscapes, due to possible severe morphological changes both onshore and underwater, and (iii) on ecosystems, due to possible loss of sensitive habitats [187–189]. In this context, the availability of terrestrial and marine spatial datasets is fundamental to get a full picture of coastal geohazards. A combined analysis of terrestrial and marine processes should be considered to be a necessary step in geohazard assessment in coastal environments. In the past few years, there has been an increase of published papers in this field of research. Here we briefly present a review of literature with special reference to tsunami and storm waves, volcanic eruptions, coastal landslides, coastal inundation, and erosion due to sea-level rise.

Recent progress in hydrodynamic modelling and simulation produced considerable results in topics such as tsunami and storm wave hazards in coastal environments, mostly because bathymetric data is a crucial parameter in nearshore wave and hydrodynamic modelling [28,190]. De Jongh and van Opstal [95] proposed an interesting combined analysis of topography and bathymetry in Mozambique to model tsunami and storm surges impacts on land. One of the most common effects to storm and tsunami, after the flooding of low-lying areas, is the detachment of boulders of a variety of sizes (from decametric to metric) from the seafloor. The knowledge of both nearshore and coastal geomorphology is fundamental to develop models reconstructing the height of the wave necessary to produce such a detachment and boulders' possible path on land. Examples of tsunami or storm deposits and models on the waves that caused them come from the Mediterranean Sea, in particular from the Istrian coast (e.g., [116]), the Apulian coasts (e.g., [99] and references therein), the Maltese archipelago (e.g., [104,105]) and the Greek coasts (e.g., [81,113]).

De Gange et al. [88] illustrated the effects of volcanic eruptions on coastal and marine environments, such as spreading of volcanic ashes and pyroclasts, which can affect also terrestrial and marine habitats, earthquakes and landslides (Figure 7).

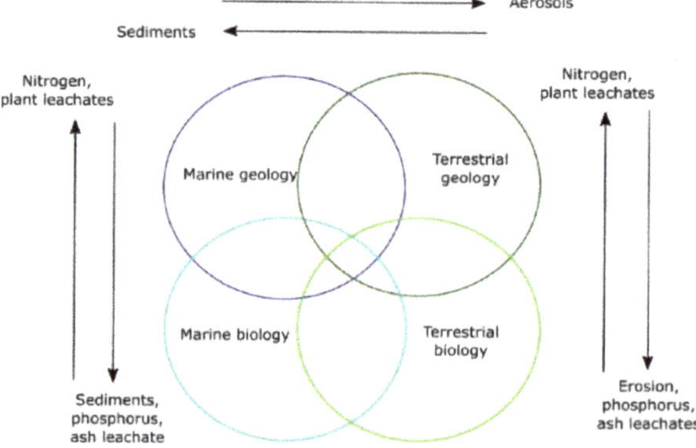

Figure 7. Diagram showing the interplay of marine and terrestrial processes that possibly affect the ecosystem response following a volcanic eruption (modified after De Gange et al. [88]).

Examples of landslides triggered by volcanic eruptions or, in general volcanic setting, are from Mount Etna (Sicily, Italy) and oceanic and insular volcanoes. Urlaub et al. [115] analyzed the deformations of the southeastern flank of Etna volcano that is sliding into the Ionian Sea and carried out the first long-term seafloor displacement monitoring campaign. Oceanic and insular volcanoes commonly experience giant landslides with relevant run-out (i.e., debris avalanches), able to create huge depositional areas in the offshore and even deep domain. Examples come from Canary Islands [76,77,101], Hawaii Islands [72,75,78], Stromboli Island (Figure 8) [83,85,91], Lipari Island [110] etc.

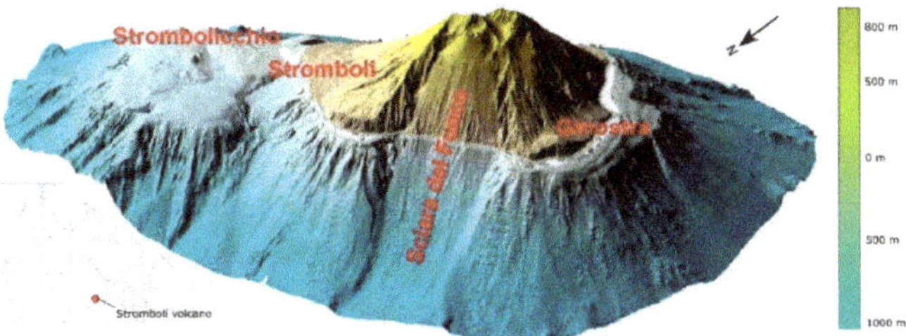

Figure 8. Digital Terrain and Marine Model of Stromboli volcano (Tyrrhenian Sea) showing the Sciara del Fuoco landslide (modified after Chiocci et al. [82]).

As for more common coastal landslides, there are case studies showing how the integration of land and sea datasets can be beneficial for landslide hazard assessment. In the Calabria region, seismic-induced landslides originated on land and reached the seafloor [94,100]. De Blasio and Mazzanti [87] produced a few-centimeters resolution DTLM (Digital Terrain and Lacustrine Model) and a DTMM (Digital Terrain and Marine Model) of two Italian sites in Latium and Calabria affected by coastal rock falls in order to model the falling of material into the water. Casalbore et al. [90] analyzed terrestrial and marine DTMs both pre- and post-hyperpycnal flows at Fiumara (Western Messina Strait, Italy) aiming at detecting the morphological evidence of the event on the seabed and to assess flash flood occurrence a posteriori. Another example of integration of terrestrial and marine datasets for landslide analysis is from the Nice landslide (Ligurian Sea, NW Mediterranean) subsequent to the 1979 catastrophe of the Nice International Airport (NE France) that caused a 2–3 m high tsunami, generated by a landslide that progressively turned into a debris flow and, then, in a turbidity current [73,80]. Several studies analyzed morphology, stratigraphy, geotechnics of the landslide and surrounding terrestrial and marine areas, providing numerical models of the phenomenon [74] and evaluating the possibility of future collapses and related impacts on the environment and human activities [89]. Coastal and marine mass wasting can also be related to past and present sea-level rise due to climate change. This is the case of the Vasto landslide (Abruzzo, Italy), a rotational slide continuing under the sea level, whose geomorphological evolution, and past and historical reactivations have been reconstructed by Della Seta et al. [96]. Another example comes from the Maltese coastal block slides that developed during the last glaciation and were then influenced by the successive sea-level oscillations [180].

Among hazards induced by the ongoing climate change, coastal inundation (especially along stretches of coast affected by subsidence; [108]) and coastal erosion triggered by extreme meteorological events, and sea-level rise are the most reported in the literature. In particular, several papers concern (i) the quantification of sea-level rise (e.g., [2,93,107,111]; Figure 9), (ii) the general impacts of sea-level rise on coastal environments [97,102] and (iii) the assessment of coastal exposure and coastal erosion (e.g., [98,114,117,118]). Terrestrial and marine datasets are differently analyzed and integrated, although bathymetry is often taken into consideration, especially in those works presenting predictive models on sea-level rise and coastal inundation.

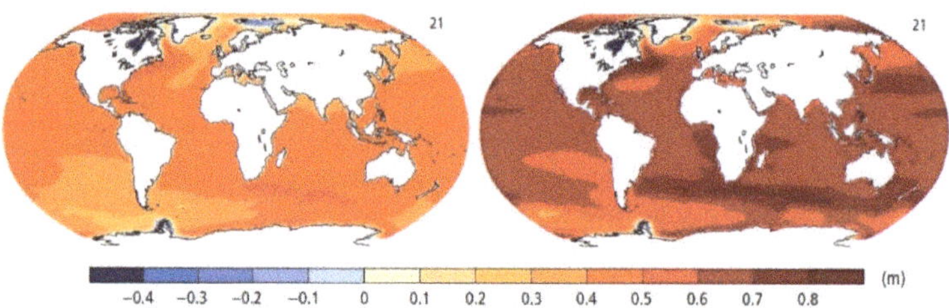

Figure 9. Scenarios for change in sea level for the 2081–2100 period under the RCP2.6 (left) and RCP8.5 (right) (modified after IPCC [2]).

2.6. Marine and Landscape Ecology

In the last few years, attention has been focused on the so-called "mapping from ridge to reef" approach [12] in order to investigate the connectivity between upland watersheds, intertidal zones and shallow coastal areas, including the influence that coastal (or riverine) and submarine morphological features can have on habitats' distribution.

As a matter of fact, habitats can be defined as physically distinct areas characterized by specific physical, chemical and biological properties (and oceanographic properties as well for benthic habitats) and hosting distinct species or communities of species. Among the physical components, seabed morphology and its geomorphological significance can have a remarkable control on ecological processed and associated biota [191,192]. Indeed, it is known that a wide variety of terrain attributes (e.g., orientation, slope angle, roughness), substrate type and chemical and oceanographic variables deeply affect species distribution and in turn biodiversity, providing surrogates used to identify places that deserve protection [68]. In submarine environments, different landforms are usually associated with specific benthic habitats as discussed in Harris and Baker [129] and some species are defined as ecosystem engineers providing themselves typical submarine landforms or geomorphic proxies for habitat detection, even in the deep submarine environments (cf. [193–198]). In tropical coastal environments detailed and accurate representations of topography and bathymetry are essential for habitat mapping [199,200], since they are required for modelling nearshore hydrodynamics, sediment transport and reef evolutionary processes [21] and references therein, [201]. Finally, we must consider that terrestrial and marine ecosystems are linked by freshwater inputs (e.g., rivers discharge) that supply sediments, nutrient exchange and larval transport, and pollutants [121,124].

In this framework, the acquisition of reliable base maps, in terms of elevation data, for both on-land and marine environments constitutes the basis for any studies aiming at analyzing seabed landscapes. Benthic habitat mapping means "plotting the distribution and extent of habitats to create a map with complete coverage of the seabed showing distinct boundaries separating adjacent habitats" as stated within the MESH project [202]. Benthic habitat and, more generally, habitat mapping practices constitute a basic tool for habitats conservation as part of an ecosystem approach [168] (cf. Section 2.7).

The growing interest in seafloor mapping, habitat mapping and development of an integrated management of coastal and marine environment fostered large use of abiotic surrogates to represent biodiversity [126], as witnessed also by the international GeoHab (Marine Geological and Biological Habitat Mapping) community [127,203] and at the European level by the MAREANO program in Norway [204–206], and the MAREMAP in the UK [207,208].

Important contributions including habitat maps generated by coupling terrestrial and underwater geospatial datasets are from Hogrefe et al. [121], McKean et al. [122,123], Vierling et al., [125], Leon et al. [21], Marchese et al. [127] and Prampolini et al. [6,128]. The latter exploited latest technological advancement, among which the LiDAR-derived elevation data of the Maltese coast and of its seafloor down to a depth of ca. 50 m for mapping both geomorphological features and habitats. Leon et al. [21] produced a seamless and high-resolution DEM of the fringing reef system of Lizard Island in northern Great Barrier Reef (Australia), merging multisource 3D models (topographic and bathymetric LiDAR data, passive optical remote sensing data, nautical charts, and single-beam and multi-beam echo-sounder data reaching 30 m b.s.l.). McKean et al. [122,123] tested the high-resolution Experimental Advanced Airborne Research LiDAR (EAARL), a new technology for cross-environment surveys of channels and floodplains, to acquire elevation-depth data of a channel in the Bear Valley Creek (Idaho, USA), and map its landforms and habitats.

Combining terrestrial and bathymetric LiDAR allows reconstructing a 3D view of terrestrial habitats (e.g., St. Joe Woodlands and sagebrush-steppe ecosystem in Idaho (USA) as showed by Vierling et al. [125]). Hogrefe et al. [121] combined depths derived from IKONOS satellite imagery and sonar data to produce a seamless DEM of Tutuila Island (American Samoa) that can be used for evaluating the assessment of human population and land use practices on coral reefs. Finally, worthy of note are the recent studies reporting first applications of photogrammetric technique to UAV imagery to map coral habitats in tropical coastal environments [127,209].

2.7. Coastal Planning and Management

Land and sea interaction is increasingly perceived as relevant in the context of planning and management of terrestrial and sea areas, since most of the activities occurring in the marine environment are also connected with the terrestrial vicinities. The interdependence of land and offshore systems drives the need for integration between terrestrial and marine planning systems, considering driver issues that cross the land/sea boundary [138]. Among others, changes in both landscapes and seascapes due to urbanization and anthropogenic activities represent a key element to consider within any planning processes.

The land–sea interactions and related processes constitute a key element of the Mid-Term Strategy 2016–2021 of UN Environment /MAP adopted with Decision IG. 22/1 [147], and correspond to the first objective of both the Mediterranean Strategy for Sustainable Development (MSSD) 2016–2025, adopted with Decision IG 22/2 [148], and the Sustainable Development Goals 14 and 15 [149,150]. Indeed, the goals of "Life below water" (SDG 14) and "Life on land" (SDG 15) are strictly interconnected [145].

In this context, few diverse approaches facilitate land–sea planning system integration. Among them there are the Integrated Coastal Zone Management (ICZM) and the Maritime Spatial Planning (MSP) tools, coupled with an Ecosystem-Based Approach (EBA) [137].

The ICZM initiatives provide a support to integrated and holistic planning and management of the coastal areas, including both the land (inland limit decided by the countries) and marine (territorial seas) components (Art. 1 of the Protocol on ICZM in the Mediterranean [151]). The importance of considering land and sea space as a whole within the ICZM process is re-affirmed by some of the Protocol's objectives and principles as for example the following: *"Ensure preservation of the integrity of the coastal ecosystems, landscape and geomorphology"* (Art. 5; objective d). Given the definition of the coastal zone provided by the Protocol, this integrity can be preserved only if the land and marine parts of the landscape are considered together.

The "Conceptual Framework for MSP in the Mediterranean" (Barcelona Convention, December 2017 [146]) foster this integration also facilitating the introduction of MSP into ICZM in the framework of the Barcelona Convention Protocols. A step by step methodology for the implementation of the MSP following common principles in the Mediterranean has been designed thanks to the existing guiding documents (Figure 10) [132,134,141–143].

Recent examples of integrated approaches to terrestrial and marine spatial planning occur in the design of a network of MPAs using models such as Marxan ("marine reserve design using spatially explicit annealing"), the most widely used software at global level for conservation planning and designed for solving conservation planning issues in landscapes and seascapes [135].

Following an ecosystem-based approach within the MSP and ICZM leads to an evolution within the different planning and management actions, taking into account the need to embrace multidisciplinary approaches and to advocate cross-realm connectivity [144].

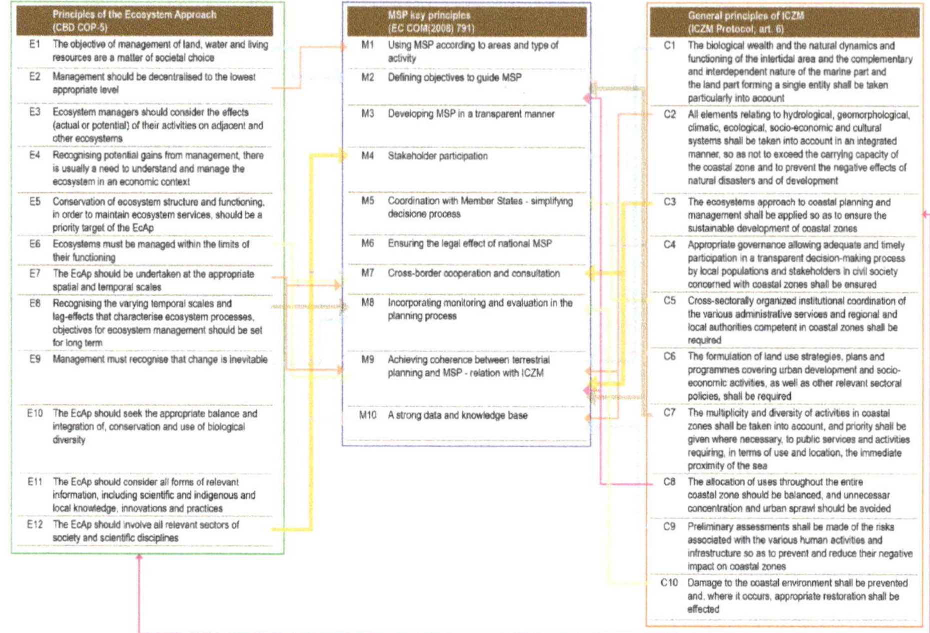

Figure 10. Link between Ecosystem-Based Approach, Maritime Spatial Planning (MSP) and Integrated Coastal Zone Management (ICZM) principles (modified after UNEP-MAP PAP/RAC [146]).

These imply the integration of spatial data stored in a GIS, including relevant Earth Observation services and the characterization of marine habitats and seabed landscapes, especially as premise for representing coastal and maritime space [136].

MSP practice highlights the necessity of having a strong data and knowledge base among its principles [210]. Hence, the integration of terrestrial and marine spatial dataset constitutes a fundamental element to be able to undertake a new interdisciplinary approach for an integrated analysis of marine ecosystems and common maritime space [137].

3. Advances in Data Collection Technology and Data Processing Methodology

The accomplishment of the outcomes reported in the literature and examples listed in previous chapters is mostly due to the widespread recent availability of new technologies and software that enable scientists to acquire geo-environmental spatial data that were unrecoverable before the 1970s in the submarine environments. Their integration with terrestrial data has become feasible especially with the generation of data format and products suitable for implementation into GIS platforms that in turn made possible to handle and analyze complex and heterogeneous datasets from the onshore to the offshore zone, as shown in most of the previously cited works.

The expansion of GIScience can be dated back to the 1990s. As soon as marine datasets became accurately "geo-referenced", thanks to our ability in obtaining geographical positions at sea through the development of GPS, their structure, format and way of representation moved immediately toward the form of geospatially enabling the data to create maps and 3D scenes of the marine environment. Since "marine" GIS has evolved adapting a technology originally designed for land-based applications, the integration of marine and terrestrial datasets has been quite immediate. This especially happened for those marine studies dealing with coastal environments or applications that benefit from the investigation of spatial relationships within and between marine and terrestrial dataset—such as measuring and monitoring the seascape or modelling/predict future scenarios [211]. An important trigger in applying this new approach for coastal and marine data visualization and analysis came undoubtedly from industry, especially in the fields of hydrographic surveying and production of nautical charts and publications.

As soon as marine and terrestrial elevation datasets started to be implemented and harmonized in a continuum dataset, the British Geological Survey introduced the new term "white ribbon" in the hydrographic sector, to designate the information gap of elevation data in the shallow area formed by the intertidal and nearshore zones, meaning the interface between land and sea. Covering the white ribbon with high-resolution bathymetric data became soon a challenge in all first attempts devoted to integrate marine and terrestrial spatial datasets (e.g., [121,124,212–215]), and the scientific community soon realized both the relevance and the issues to be addressed in carrying out topo-bathymetric surveys [15,28,190,216–221]. Most of the difficulties in getting elevation data in the white ribbon are caused by the water depth: it is generally too shallow for traditional bathymetric surveys (because of the draft and the need to submerge the echo-sounders keeping them at a certain distance from the seafloor to obtain an efficient coverage, and because of the unsafe conditions generated by the common occurrence of rocky outcrops and/or waves) and too deep for traditional optical land-based survey methods. Shallow water is in addition more expensive to be surveyed than deeper ones since MBES seafloor coverage is narrowed as water becomes shallower, requiring the vessel to spend excessive time in shallower areas due to the need to run very close survey lines to achieve adequate coverage (Figure 1). Nonetheless, the intertidal and nearshore zones are of extreme importance to investigate for all the reasons listed in the previous sections. This challenge favored the development of new advanced methods and techniques to improve the capability of obtaining shallow-water bathymetric data, and especially:

- The production of new advanced acoustic systems designed for obtaining depth measurements in shallow water;
- The application of cutting-edge visualization technology to images and data collected with optical sensors to obtain elevation data from shallow areas (i.e., underwater photogrammetry, image derived bathymetry, LiDAR, laser scanning).

Finally, given all the technological and economic difficulties mentioned above in mapping the seafloor, both in deep and shallow water, sharing data is increasingly appreciated and encouraged by several research funding programs. The goal "map once, use many times" supports the creation of national (underpinned by governments), regional and international repositories of bathymetric data. Examples of regional repositories are the European EMODnet (www.emodnet-bathymetry.eu) and Baltic Sea Bathymetry Database (http://data.bshc.pro), while at international/global level, GEBCO (https://www.gebco.net/) is a repository of world bathymetry, which is also updated thanks to local portals (e.g., from EMODnet), and currently under update thanks to GEBCO Seabed 2030 Project, aiming at completing mapping of the world ocean by 2030.

3.1. Shallow-Water Acoustic Systems

Most of the interest in surveying shallow water comes from industry (oil and gas companies, port and harbor authorities and maritime engineering among others) and academics and has grown rapidly in recent years. This has pushed manufacturers to both produce MBES systems (i.e., "beamforming system"—[222]) adapted for fast mobilization on smaller vessels (easy-to-use and quick-to-deploy) and explore new innovations in swath bathymetry systems, developing novel swath sonar technology to reach greater seafloor coverage (up to 15 times the depth), such as (i) the interferometric echo-sounders, also known as Phase Differencing Bathymetric Systems (PDBS) [222,223] or (ii) the multi-phase echo-sounders (MPES—[224]).

Technological developments in beamforming system have especially affected the geometry and the performance of the transmit array and sounding frequencies, refining the capability of the systems in offering a wider coverage (up to 7 times the water depth) and narrower acoustic beam (reaching accuracies that have been shown to exceed the IHO-International Hydrographic Organization-Standards for Hydrographic Surveys). The broader coverage (i.e., swath) is obtained using multi-transducer multi-beam products. Two sonar heads (i.e., transducers), for instance, can easily achieve double the coverage, by simply adjusting the angle of the heads.

Developments in shallow-water swath bathymetry systems involved all aspects of the "seafloor mapping system", including all ancillary sensors and software involved in the survey to provide the so-called "integrated survey system", namely the GPS/GNSS positioning systems, the motion sensors, and sound velocity recording sensors. The goal is to simplify installation and calibration procedure and make the shallow-water MBES systems perfect for use on vessels of opportunity, small survey launches, and even Autonomous Surface Vessels (ASVs) or USVs. ASV/USV are vehicles that can navigate and collect data from the surface of the water without a crew. ASV/USV are currently produced to remotely control data acquisition especially in shallow marine water, rivers and channels.

3.2. Optical Sensor for Underwater Imaging and Mapping

Latest technological developments in underwater 3D reconstruction, based on airborne active optical sensor, has given rise to a wide range of new systems and techniques such as the LiDAR systems, Structured Light (SL), Laser Stripe (LS), Laser Line Scanning (LLS), Stereo Vision (SV) and SfM [225–228].

LiDAR systems have substantially improved shallow seafloor mapping in coastal environments. Using infrared laser pulses, topographic LiDAR systems achieve a very high-resolution mapping performance (i.e., meter to sub-meter point spacing) with sub-meter vertical accuracy. Bathymetric LiDAR systems, can even use both infrared and blue-green laser pulses, to simultaneously acquire depth measurements down to ~70 m below Mean Sea Level (MSL), according to water turbidity, typically <40 m is achieved in most applications. Bathymetric LiDAR systems have been the first active optical sensors that provided elevation data from the nearshore areas, allowing surveying shallow seafloor with much more efficiency in terms of coverage and required time (Figure 1). The LiDAR technology can presently be integrated also to terrestrial or surface or underwater platforms, carrying out ROV-based LiDAR inspection surveys, benefiting from increased spatial and temporal resolution, and greater accuracy [229]. LLS can now be used just like a multi-beam although the technology is slightly different. Both subsea LiDAR and Laser scanners generate a relative point cloud (referenced by flow and bearing measurements) with a resolution of even millimeters, i.e., much higher than any acoustics-based system.

The high resolution offered by subsea LiDAR or laser and the need to operate under lighting conditions determine, however, a limited range, strongly regulated by the environmental conditions. The resolution and accuracy typical of LiDAR/submarine laser systems require indeed clear water with good visibility [230]. Thus, they cannot be employed to scan those nearshore areas characterized by high turbidity such as the ones close to river discharge or with sediment/pollutant moved by water movements [231,232].

Hence, starting from the 1990s, active sensors based on underwater acoustic (e.g., multi-beam or interferometric echo-sounders) and light signals LiDAR/LLS systems improved substantially the capacity of obtaining elevation data in underwater environments, despite expensive techniques, especially for small scale surveys. These instruments directly provide a point cloud of bathymetric measurements that can generate a DTM with sub-meter resolution. LiDAR systems can even combine onshore topographic and nearshore bathymetric mapping obtaining detailed emerged and submerged surfaces in a single acquisition (Figure 11) [233].

With the availability of high-resolution images collected by satellite remote sensors (or even by drones), 3D underwater models can now be generated using also passive optical systems, at least for those areas in which visible light can penetrate down to the seafloor. Satellite-Derived Bathymetry (SDB) is indeed the most recently developed method of surveying shallow waters. Different companies developed ad-hoc algorithms since the 1990s to convert the information collected by satellite sensor into bathymetric data. SDB is based on the connection between the seabed reflected energy and the depth of the water [234]. The method, using dedicated computational algorithms, basically exploits, for each pixel of the satellite image where the seafloor is visible, the statistical relationship between the depth of water and the type and intensity of energy detected by the sensor. Since SDB can estimates the water depth of the seafloor up to the extent of light penetration into the water medium (i.e., around 20–30 m under optimal conditions), water transparency is the main limiting factor. Atmospheric absorption, sun glint, high substrate heterogeneity, algal blooms, suspended sediment, or waves, can also all limit SDB performance.

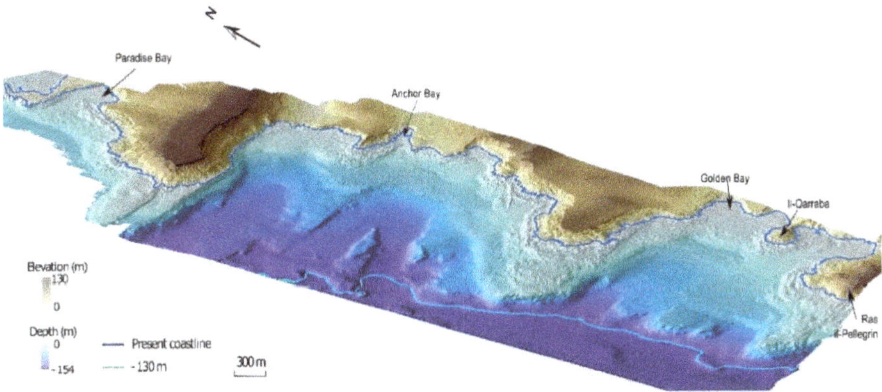

Figure 11. Integration of terrestrial and marine elevation data derived from bathymetric LiDAR: example from the NW coast of Malta, Central Mediterranean Sea (modified after Prampolini et al. [6]).

Finally, with the advent of underwater camera systems, progress in deep-sea robotics, and the increased number of videos and images being captured underwater, researchers began to obtain optical 3D reconstruction of recorded scenes with (sub)centimetric resolution, employing numerous techniques, among which the use of stereo cameras and the principle of photogrammetry even underwater (i.e., SfM technique). Optical underwater imaging is emerging as a key technology for a variety of oceanography applications [227] and [235] among others.

However, very few techniques employing photorealistic seafloor imagery take critically into account the extent to which scattering affects the scenes captured under daylight in shallow water or using active illumination in deep water. In most cases reported in the literature, it is implicitly assumed that light is neither absorbed nor scattered by the medium in which the source and scene are immerged (as it happens in pure air [236]). However, the major challenge facing optical imaging in these applications is the severe degradation of image quality caused by scattering generated by impurities and organisms. SfM has been also applied to UAV-based RGB imagery, on coastal waters [237,238] among others. UAV imagery processed with SfM techniques offers a low-cost alternative to established shallow seafloor mapping techniques providing also important visual information with the generation of an orthomosaic for the surveyed scene [17]. Nevertheless, water refraction introduces severe errors when UAVs imagery is used for bathymetric applications. Although the application of photogrammetric procedures on images captured directly in the water medium (in-water) needs only a thorough calibration to correct the effects of refraction, in instance where the image acquisition occurs through-water (two-media), the sea surface undulations caused by waves [239,240] and the magnitude of refraction that can change at each point of every image, lead to uncertain results [241,242].

Overall, it is clear that no single optical imaging system can meet all the needs of underwater 3D reconstruction. The different systems cover very different spatial scales, resolutions and accuracy, being suitable for different applications. Furthermore, it is important to underline the lack of systematic studies to precisely compare the performance of different sensors in relation to the same scenario and under identical conditions. However, technology and computer vision are definitely on the way of addressing all pitfalls of the mentioned applications, to obtain 3D optical reconstructions more reliable over multiple spatial scales, through innovative sensors and data processing. Finally, it should be emphasized that the possibility of obtaining accurate photorealistic 3D reconstructions, also allows the use of interactive tools for visualization and exploration in 3 dimensions, designed to support the interpretation and analysis of the obtained spatial data. These large, complex and multicomponent spatial datasets can indeed be used to develop innovative learning tools for environmental sciences, presenting new worlds of interactive exploration to a multitude of users [243].

4. Conclusions and Future Perspectives

Most terrestrial landscapes and landforms have always been investigated for several purposes and benefited from high accessibility to surveyed areas and high-resolution data. On the contrary, the submarine environment has struggled in being represented with the same resolution and coverage as its terrestrial counterpart because of its remoteness and/or its limited accessibility and the high costs imposed by the need to use expensive infrastructures and sophisticated technologies (especially research vessels, underwater robotics and a multitude of sensors). Nevertheless, remote sensing tools operating from satellite, aerial platforms and vessels or autonomous vessels and drones have contributed to obtain elevation data for both land and sea areas with a comparable resolution, establishing submarine geomorphology as a field of research that is also remarkably contributing to marine environmental management, with an increase in many associated applicable research.

This paper has pointed out seven good reasons to pursuit such a comprehensive and homogeneous integration of terrestrial and submarine datasets, showing the outputs of relevant research in this field. The interest in producing integrated land–sea geomorphological maps is now at its beginning, even though it would be the basis for further applied research. The integrated assessment of geoheritage and geodiversity in coastal and marine environments has been the subject of a very limited number of papers so far. A much larger number of papers refers instead on the coupling of terrestrial and marine spatial datasets aiming at reconstructing paleo-landscapes in coastal and marine areas and outlining their geomorphological evolution, supporting also the identification and study of archaeological sites. The field of application that has mostly benefited so far from the integration of terrestrial and marine datasets is the integrated assessment of geohazards in coastal and marine areas, with special reference to tsunami and storms, coastal and marine landslides and sea-level-rise-related hazards. Attention to the interrelations between land and sea and their effects on marine habitats is being paid, but still only few works show combined investigation in terrestrial and nearshore environments. Finally, in the last ten years, stakeholders have pointed out the need for an integrated planning and management of marine and terrestrial areas. During this span of time, discussion has extended from MSP alone to ICZM of coastal areas thanks to several directives and plans developed and applied at international level (i.e., European Union, United Nations).

In the near future, it is likely that technological improvements will allow an increasingly easier accessibility to the "white ribbon" and a better integration of terrestrial and marine spatial datasets that strongly relies on the realization of seamless DTMs of land and sea areas. This will enhance further and much wider research in transition environments based on interdisciplinary approaches. It is also desirable that a standardization and/or harmonization of data will be soon achieved, to adopt common terminology and classification schemes for both terrestrial and submarine geomorphological features.

The resulting increased awareness of the interconnection between landscapes of terrestrial and marine areas calls for a holistic approach to better understand environmental changes taking place on Earth and to consequently design appropriate management measures. Scientists and stakeholders typically working on terrestrial and marine areas separately will hopefully understand the benefits of coupling terrestrial and marine investigation and activities, which is of paramount importance for environmental protection and enhancement.

Author Contributions: Conceptualization, F.F.; M.P.; A.S.; M.S.; methodology, F.F.; M.P.; A.S.; M.S.; software, M.P.; investigation, F.F.; M.P.; A.S.; M.S.; data curation, M.P.; writing—original draft preparation, M.P.; writing—review and editing, F.F.; M.P.; A.S.; M.S.; visualization, M.P.; A.S.; supervision, M.S.; project administration, A.S.; M.S.; funding acquisition, A.S.; M.S. All authors have read and agreed to the published version of the manuscript.

Funding: The study was carried out in the frame of the Project "Coastal risk assessment and mapping" funded by the EUR-OPA Major Hazards Agreement of the Council of Europe (2020–2021). Grant Number: GA/2020/06 n° 654503 (Resp. Unimore Unit: Mauro Soldati). This article is also an outcome of the Project MIUR—Dipartimenti di Eccellenza 2018–2022, Department of Earth and Environmental Sciences, University of Milano-Bicocca and of the INTERREG-MED "Actions for Marine Protected Areas—AMAre" 2014–2020. Ref. 649 (Resp. CNR-ISMAR Unit: Federica Foglini).

Acknowledgments: We are thankful to Paola Coratza for her precious suggestions about geoheritage and geodiversity issues, and to the three anonymous reviewers for their constructive comments and suggestions.

Conflicts of Interest: The authors declare no conflict of interest.

References

1. Oreskes, N. The Scientific Consensus on Climate Change. *Science* **2004**, *306*, 1686. [CrossRef] [PubMed]

2. IPCC. *Climate Change 2014–Synthesis Report. Contribution of Working Group I, II, III to the Fifth Assessment Report of the Intergovernmental Panel on Climate Change*; Cambridge University Press: Cambridge, UK; New York, NY, USA, 2014; pp. 1–32.

3. Nicholls, R.J.; Cazenave, A. Sea-level rise and its impact on coastal zones. *Science* **2010**, *328*, 1517–1520. [CrossRef] [PubMed]

4. FitzGerald, D.M.; Fenster, M.S.; Argow, B.A.; Buynevich, I. Coastal Impacts Due to Sea-Level Rise. *Annu. Rev. Earth Planet. Sci.* **2008**, *36*, 601–647. [CrossRef]

5. Kunreuther, H.; Gupta, S.; Bosetti, V.; Cooke, R.; Dutt, V.; Ha-Duong, M.; Held, H.; Llanes-Regueiro, J.; Patt, A.; Shittu, E.; et al. Integrated Risk and Uncertainty Assessment of Climate Change Response Policies. In *Climate Change 2014: Mitigation of Climate Change. Contribution of Working Group III to the Fifth Assessment Report of the Intergovernmental Panel on Climate Change*; Edenhofer, O., Pichs-Madruga, R., Sokona, Y., Farahani, E., Kadner, S., Seyboth, K., Adler, A., Baum, I., Brunner, S., Eickemeier, P., et al., Eds.; Cambridge University Press: Cambridge, UK; New York, NY, USA, 2014.

6. Prampolini, M.; Foglini, F.; Biolchi, S.; Devoto, S.; Angelini, S.; Soldati, M. Geomorphological mapping of terrestrial and marine areas, northern Malta and Comino (central Mediterranean Sea). *J. Maps* **2017**, *13*, 457–469. [CrossRef]

7. Dramis, F.; Guida, D.; Cestari, A. Nature and aims of geomorphological mapping. In *Geomorphological Mapping—Methods and Appplications*; Smith, M.J., Paron, P., Griffiths, J.S., Eds.; Elsevier: Oxford, UK, 2011; Volume 15, pp. 39–73.

8. Paron, P.; Claessens, L. Makers and users of geomorphological maps. In *Developments in Earth Surface Processes*; Smith, M.J., Paron, P., Griffiths, J.S., Eds.; Elsevier: Oxford, UK, 2011; Volume 15, pp. 75–106.

9. Micallef, A.; Krastel, S.; Savini, A. *Submarine Geomorphology*; Springer: Cham, Switzerland, 2018.

10. Sandwell, D.T.; Gille, S.T.; Orcutt, J.; Smith, W.H. Bathymetry from space is now possible. *Eos* **2003**, *84*, 37–44. [CrossRef]

11. Wright, D.J. Introduction. In *Undersea with GIS*; Wright, D.J., Ed.; ESRI Press: Redlands, CA, USA, 2003; pp. xiii–xvi.

12. Wright, D.J.; Heyman, W.D. Introduction to the Special Issue: Marine and Coastal GIS for Geomorphology, Habitat Mapping, and Marine Reserves. *Mar. Geod.* **2008**, *31*, 223–230. [CrossRef]

13. Wölfl, A.-C.; Snaith, H.; Amirebrahimi, S.; Devey, C.W.; Dorschel, B.; Ferrini, V.; Huvenne, V.A.I.; Jakobsson, M.; Jencks, J.; Johnston, G.; et al. Seafloor Mapping–The Challenge of a Truly Global Ocean Bathymetry. *Front. Mar. Sci.* **2019**, *6*, 283. [CrossRef]

14. Mayer, L., Jakobsson, M.; Allen, G.; Dorschel, B.; Falconer, R.; Ferrini, V.; Lamarche, G.; Snaith, H.; Weatherall, P. The Nippon Foundation—GEBCO seabed 2030 project: The quest to see the world's oceans completely mapped by 2030. *Geosciences* **2018**, *8*, 63. [CrossRef]

15. Menandro, P.S.; Bastos, A.C. Seabed Mapping: A Brief History from Meaningful Words. *Geosciences* **2020**, *10*, 273. [CrossRef]

16. Foglini, F.; Grande, V.; Marchese, F.; Bracchi, V.A.; Prampolini, M.; Angeletti, L.; Castellan, G.; Chimienti, G.; Hansen, I.G.; Gudmundsen, M.; et al. Application of Hyperspectral Imaging to Underwater Habitat Mapping, Southern Adriatic Sea. *Sensors* **2019**, *19*, 2261. [CrossRef]

17. Dietrich, J.T. Bathymetric structure-from-motion: Extracting shallow stream bathymetry from multi-view stereo photogrammetry. *Earth Surf. Proc. Land.* **2017**, *42*, 355–364. [CrossRef]

18. Miccadei, E.; Mascioli, F.; Orrù, P.; Piacentini, T.; Puliga, G. Late Quaternary paleolandscape of submerged inner continental shelf areas of Tremiti islands archipelago (northern Puglia). *Geogr. Fis. Dinam. Quat.* **2011**, *34*, 223–234.

19. Miccadei, E.; Mascioli, F.; Piacentini, T. Quaternary geomorphological evolution of the Tremiti Islands (Puglia, Italy). *Quatern. Int.* **2011**, *233*, 3–15. [CrossRef]

20. Miccadei, E.; Orrù, P.; Piacentini, T.; Mascioli, F.; Puliga, G. Geomorphological map of the Tremiti Islands (Puglia, Southern Adriatic Sea, Italy), scale 1: 15,000. *J. Maps* **2012**, *8*, 74–87. [CrossRef]

21. Leon, J.X.; Phinn, S.R.; Hamylton, S.; Saunders, M.I. Filling the 'white ribbon'—A multisource seamless digital elevation model for Lizard Island, northern Great Barrier Reef. *Int. J. Remote Sens.* **2013**, *34*, 6337–6354. [CrossRef]

22. Gasparo Morticelli, M.; Sulli, A.; Agate, M. Sea–land geology of Marettimo (Egadi Islands, central Mediterranean sea). *J. Maps* **2016**, *12*, 1093–1103. [CrossRef]

23. Mastronuzzi, G.; Aringoli, D.; Aucelli, P.P.C.; Baldassarre, M.A.; Bellotti, P.; Bini, M.; Biolchi, S.; Bontempi, S.; Brandolini, P.; Chelli, A.; et al. Geomorphological map of the Italian coast: From a descriptive to a morphodynamic approach. *Geogr. Fis. Dinam. Quat.* **2017**, *40*, 161–195.

24. Prampolini, M.; Gauci, C.; Micallef, A.S.; Selmi, L.; Vandelli, V.; Soldati, M. Geomorphology of the north-eastern coast of Gozo (Malta, Mediterranean Sea). *J. Maps* **2018**, *14*, 402–410. [CrossRef]

25. Brandolini, P.; Faccini, F.; Paliaga, G.; Piana, P. Man-made landforms survey and mapping of an urban historical center in a coastal Mediterranean environment. *Geogr. Fis. Dinam. Quat.* **2018**, *41*, 23–34.

26. Furlani, S.; Piacentini, D.; Troiani, F.; Biolchi, S.; Roccheggiani, M.; Tamburini, A.; Tiricanti, E.; Vaccher, V.; Antonioli, F.; Devoto, S. Tidal Notches (TN) along the western Adriatic coast as markers of coastal stability during Late Holocene. *Geogr. Fis. Dinam. Quat.* **2018**, *41*, 33–46.

27. Campobasso, C.; Carton, A.; Chelli, A.; D'Orefice, M.; Dramis, F.; Graciotti, R.; Guida, D.; Pambianchi, G.; Peduto, F.; Pellegrini, L. Revisione e aggiornamento delle "Linee Guida al Rilevamento della Carta Geomorfologica d'Italia alla scala 1:50.000" e proposta di un modello di cartografia Geomorfologica "a oggetti". *GT&A* **2018**, *2*, 15–27.

28. Genchi, S.A.; Vitale, A.J.; Perillo, G.M.; Seitz, C.; Delrieux, C.A. Mapping topobathymetry in a shallow tidal environment using low-cost technology. *Remote Sens.* **2020**, *12*, 1394. [CrossRef]

29. Bridgland, D.; Maddy, D.; Bates, M. River terrace sequences: Templates for Quaternary geochronology and marine–terrestrial correlation. *J. Quat. Sci.* **2004**, *19*, 203–218. [CrossRef]

30. Pujol, G.L.; Sintes, C.; Lurton, X. High-resolution interferometry for multibeam echosounders. *IEEE Oceans* **2005**, *1*, 345–349.

31. Rovere, A.; Vacchi, M.; Firpo, M.; Carobene, L. Underwater geomorphology of the rocky coastal tracts between Finale Ligure and Vado Ligure (western Liguria, NW Mediterranean Sea). *Quatern. Int.* **2011**, *232*, 187–200. [CrossRef]

32. Westley, K.; Quinn, R.; Forsythe, W.; Plets, R.; Bell, T.; Benetti, S.; McGrath, F.; Robinson, R. Mapping submerged landscapes using multibeam bathymetric data: A case study from the north coast of Ireland. *Int. J. Naut. Archaeol.* **2011**, *40*, 99–112. [CrossRef]

33. Micallef, A.; Foglini, F.; Le Bas, T.; Angeletti, L.; Maselli, V.; Pasuto, A.; Taviani, M. The submerged paleolandscape of the Maltese Islands: Morphology, evolution and relation to Quaternary environmental change. *Mar. Geol.* **2013**, *335*, 129–147. [CrossRef]

34. Kennedy, D.M.; Ierodiconou, D.; Schimel, A. Granitic coastal geomorphology: Applying integrated terrestrial and bathymetric LiDAR with multibeam sonar to examine coastal landscape evolution. *Earth Surf. Process. Landf.* **2014**, *39*, 1663–1674. [CrossRef]

35. Greenwood, S.L.; Clason, C.C.; Mikko, H.; Nyberg, J.; Peterson, G.; Smith, C.A. Integrated use of LiDAR and multibeam bathymetry reveals onset of ice streaming in the northern Bothnian Sea. *GFF* **2015**, *137*, 284–292. [CrossRef]

36. Aucelli, P.; Cinque, A.; Mattei, G.; Pappone, G. Historical sea level changes and effects on the coasts of Sorrento Peninsula (Gulf of Naples): New constraints from recent geoarchaeological investigations. *Palaeogeogr. Palaeocl.* **2016**, *463*, 112–125. [CrossRef]

37. Foglini, F.; Prampolini, M.; Micallef, A.; Angeletti, L.; Vandelli, V.; Deidun, A.; Taviani, M. Late Quaternary Coastal Landscape Morphology and Evolution of the Maltese Islands (Mediterranean Sea) Reconstructed from High-Resolution Seafloor Data. In *Geology and Archaeology: Submerged Landscapes of the Continental Shelf*; Harff, J., Bailey, G., Lüth, L., Eds.; Special Publication; Geological Society: London, UK, 2016; Volume 411, Issue 1, pp. 77–95.

38. Benjamin, J.; Rovere, A.; Fontana, A.; Furlani, S.; Vacchi, M.; Inglis, R.H.; Galili, R.H.; Antonioli, F.; Sivan, D.; Miko, S.; et al. Late Quaternary sea-level changes and early human societies in the central and eastern Mediterranean Basin: An interdisciplinary review. *Quatern. Int.* **2017**, *449*, 29–57. [CrossRef]

39. Furlani, S.; Antonioli, F.; Gambin, T.; Gauci, R.; Ninfo, A.; Zavagno, E.; Micallef, A.; Cucchi, F. Marine notches in the Maltese islands (central Mediterranean Sea). *Quatern. Int.* **2017**, *439*, 158–168. [CrossRef]

40. Furlani, S.; Foresta Martin, F. Headland or stack? Paleogeographic reconstruction of the coast at the Faraglioni Middle Bronze Age Village (Ustica Island, Italy). *Ann. Geophys. Ital.* **2018**, *61*. [CrossRef]

41. De Giosa, F.; Scardino, G.; Vacchi, M.; Piscitelli, A.; Milella, M.; Ciccolella, A.; Mastronuzzi, G. Geomorphological Signature of Late Pleistocene Sea Level Oscillations in Torre Guaceto Marine Protected Area (Adriatic Sea, SE Italy). *Water* **2019**, *11*, 2409. [CrossRef]

42. Lo Presti, V.; Antonioli, F.; Palombo, M.R.; Agnesi, V.; Biolchi, S.; Calcagnile, L.; Di Patti, C.; Donati, S.; Furlani, S.; Merizzi, J.; et al. Palaeogeographical evolution of the Egadi Islands (western Sicily, Italy). Implications for late Pleistocene and early Holocene sea crossings by humans and other mammals in the western Mediterranean. *Earth Sci. Rev.* **2019**, *194*, 160–181. [CrossRef]

43. Antonioli, F.; Anzidei, M.; Lambeck, K.; Auriemma, R.; Gaddi, D.; Furlani, S.; Orrù, P.; Solinas, E.; Gaspari, A.; Karinja, S.; et al. Sea level change during Holocene from Sardinia and northeastern Adriatic (Central Mediterranean Sea) from archaeological and geomorphological data. *Quat. Sci. Rev.* **2007**, *26*, 2463–2496. [CrossRef]

44. Bailey, G.N.; Flemming, N. Archaeology of the continental shelf: Marine resources, submerged landscapes and underwater archaeology. *Quat. Sci. Rev.* **2008**, *27*, 2153–2165. [CrossRef]

45. Harff, J.; Lüth, F. *SINCOS I–Sinking Coasts: Geosphere, Ecosphere and Anthroposphere of the Holocene Southern Baltic Sea*; Rep Roman Germanic Commission: Frankfurt, Germany, 2007; Volume 88, pp. 7–266.

46. Harff, J.; Lüth, F. *SINCOS II–Sinking Coasts: Geosphere, Ecosphere and Anthroposphere of the Holocene Southern Baltic Sea*; R Roman Germanic Commission: Frankfurt, Germany, 2011; Volume 92, pp. 7–380.

47. Fisher, E.C.; Bar-Matthews, M.; Jerardino, A.; Marean, C.W. Middle and Late Pleistocene paleoscape modeling along the southern coast of South Africa. *Quat. Sci. Rev.* **2010**, *29*, 1382–1398. [CrossRef]

48. Benjamin, J.; Bonsall, C.; Pickard, K.; Fischer, A. *Submerged Prehistory*; Oxbow: Oxford, UK, 2011.

49. Bailey, G.; Sakellariou, D.; Members of the SPLASHCOS Network. Submerged prehistoric archaeology and landscapes of the continental shelf. *Antiq. Proj. Gallery* **2012**, *86*. Available online: http://antiquity.ac.uk/projgall/sakellariou334 (accessed on 5 August 2020).

50. Furlani, S.; Antonioli, F.; Biolchi, S.; Gambin, T.; Gauci, R.; Presti, V.L.; Anzidei, M.; Devoto, S.; Palombo, M.; Sulli, A. Holocene sea level change in Malta. *Quatern. Int.* **2013**, *288*, 146–157. [CrossRef]

51. Anzidei, M.; Lambeck, K.; Antonioli, F.; Furlani, S.; Mastronuzzi, G.; Serpelloni, E.; Vannucci, G. Coastal structure, sea-level changes and vertical motion of the land in the Mediterranean. In *Sedimentary Coastal Zones from High to Low Latitudes: Similarities and Differences*; Martini, I.P., Wanless, H.R., Eds.; Special Publications; Geological Society: London, UK, 2014; Volume 388, pp. 453–479. [CrossRef]

52. Evans, A.; Flemming, N.; Flatman, J. *Prehistoric Archaeology of the Continental Shelf: A Global Review*; Springer: New York, NY, USA, 2014.

53. Westley, K.; Plets, R.; Quinn, R. Holocene paleo-geographic reconstructions of the ramore head area, Northern Ireland, using geophysical and geotechnical data: Paleo-landscape mapping and archaeological implications. *Geoarchaeology* **2014**, *29*, 411–430. [CrossRef]

54. Bailey, G.N.; Devès, M.H.; Inglis, R.H.; Meredith-Williams, M.G.; Momber, G.; Sakellariou, D.; Sinclair, A.G.M.; Rousakis, G.; Al Gamdi, A.; Alsharekh, A.M. Blue Arabia: Palaeolithic and underwater survey in SW Saudi Arabia and the role of coasts in Pleistocene dispersals. *Quatern. Int.* **2015**, *382*, 42–57. [CrossRef]

55. Aucelli, P.; Cinque, A.; Giordano, F.; Mattei, G. A geoarchaeological survey of the marine extension of the Roman archaeological site Villa del Pezzolo, Vico Equense, on the Sorrento Peninsula, Italy. *Geoarchaeology* **2016**, *31*, 244–252. [CrossRef]

56. Harff, J.; Bailey, G.; Luth, F. *Geology and Archaeology: Submerged Landscapes of the Continental Shelf*; Geological Society Special Publications 411; Geological Society of London: London, UK, 2016; p. 294, ISBN 978-1-86239-691-3.

57. Cawthra, H.C.; Compton, J.S.; Fisher, E.C.; MacHutchon, M.R.; Marean, C.W. (Eds.) *Submerged Shorelines and Landscape Features Offshore of Mossel Bay, South Africa*; Geological Society Special Publications 411; Geological Society of London: London, UK, 2016; pp. 219–233.

58. Benjamin, J.; O'Leary, M.; Ward, I.; Hacker, J.; Ulm, S.; Veth, P.; Holst, M.; McDonald, J.; Ross, P.J.; Bailey, G. Underwater archaeology and submerged landscapes in Western Australia. *Antiquity* **2018**, *92*, 1–9.

59. Mattei, G.; Troisi, S.; Aucelli, P.; Pappone, G.; Peluso, F.; Stefanile, M. Sensing the Submerged Landscape of Nisida Roman Harbour in the Gulf of Naples from Integrated Measurements on a USV. *Water* **2018**, *10*, 1686. [CrossRef]

60. Sturt, F.; Flemming, N.C.; Carabias, D.; Jöns, H.; Adams, J. The next frontiers in research on submerged prehistoric sites and landscapes on the continental shelf. *Proc. Geol. Assoc.* **2018**, *129*, 654–683. [CrossRef]

61. Veth, P.; McDonald, J.; Ward, I.; O'Leary, M.; Beckett, E.; Benjamin, J.; Ulm, S.; Hacker, J.; Ross, P.J.; Bailey, G. A Strategy for Assessing Continuity in Terrestrial and Maritime Landscapes from Murujuga (Dampier Archipelago), North West Shelf, Australia. *J. Isl. Coast. Archaeol.* **2019**, *27*. [CrossRef]

62. Orrù, P.; Ulzega, A. Rilevamento geomorfologico costiero e sottomarino applicato alla definizione delle risorse ambientali (Golfo di Orosei, Sardegna orientale). *Mem. Soc. Geol. Ital.* **1988**, *37*, 123–131.

63. Orrù, P.; Panizza, V.; Ulzega, A. Submerged geomorphosites in the marine protected areas of Sardinia (Italy): Assessment and improvement. *Ital. J. Quat. Sci.* **2005**, *18*, 167–174.

64. Orrù, P.; Panizza, V. Assessment and Management of Submerged Geomorphosites. A Case Study in Sardinia (Italy). In *Geomorphosites*; Reynard, E., Coratza, P., Regolini-Bissig, G., Eds.; Pfeil: Munchen, Germany, 2009; pp. 201–212.

65. Brooks, A.J.; Kenyon, N.H.; Leslie, A.; Long, D.; Gordon, J.E. *Characterising Scotland's Marine Environment to Define Search Locations for New Marine Protected Areas*; Part 2: The Identification of Key Geodiversity Areas in Scottish Waters (Interim Report July 2011); Commissioned Report No. 430; Scottish Natural Heritage: Inverness, UK, 2011. Available online: http://nora.nerc.ac.uk/id/eprint/16861/1/430.pdf (accessed on 5 August 2020).

66. Rovere, A.; Vacchi, M.; Parravini, V.; Bianchi, C.N.; Zouros, N.; Firpo, M. Bringing geoheritage underwater: Definitions, methods, and application in two Mediterranean marine areas. *Environ. Earth Sci.* **2011**, *64*, 133–142. [CrossRef]

67. Mansini Maia, M.A.; de Alencar Castro, W.J. Methodological proposal for characterization of marine geodiversity in the South Atlantic: Vitória-Trindade Ridge and adjacent areas, Southeast of Brazil. *J. Integr. Coast. Zone Manag.* **2015**, *15*, 293–309.

68. Veloo, F. A New Method to Analyze Seafloor Geodiversity Around the Hawaiian and Canarian Archipelagos and the New Zealand Subduction Cone. Bachelor's Thesis, University of Amsterdam, Amsterdam, The Netherlands, 2017.

69. Gordon, J.E.; Brooks, A.J.; Chaniotis, P.D.; James, B.D.; Kenyon, N.H.; Leslie, A.B.; Long, D.; Rennie, A.F. Progress in marine geoconservation in Scotland's seas: Assessment of key interests and their contribution to Marine Protected Area network planning. *Proc. Geol. Assoc.* **2016**, *127*, 716–737. [CrossRef]

70. Gordon, J.E.; Crofts, R.; Díaz-Martínez, E.; Woo, K.S. Enhancing the role of geoconservation in protected area management and nature conservation. *Geoheritage* **2018**, *10*, 191–203. [CrossRef]

71. Coratza, P.; Vandelli, V.; Fiorentini, L.; Paliaga, G.; Faccini, F. Bridging Terrestrial and Marine Geoheritage: Assessing Geosites in Portofino Natural Park (Italy). *Water* **2019**, *11*, 2112. [CrossRef]

72. Moore, J.G.; Normark, W.R.; Holcomb, R.T. Giant Hawaiian landslides. *Annu. Rev. Earth Planet. Sci.* **1994**, *22*, 119–144. [CrossRef]

73. Mulder, T.; Savoye, B.; Syvitki, J.P. Numerical modelling of a mid-sized gravity flow: The 1979 Nice turbidity current (dynamics, processes, sediment budget and seafloor impact). *Sedimentology* **1997**, *44*, 305–326. [CrossRef]

74. Assier-Rzadkieaicz, S.; Heinrich, P.; Sabatier, P.C.; Savoye, B.; Bourillet, J.F. Numerical modelling of a landslide-generated tsunami: The 1979 Nice event. *Pure Appl. Geophys.* **2000**, *157*, 1707–1727. [CrossRef]

75. Caplan-Auerbach, J.; Fox, C.G.; Duennebier, F.K. Hydroacoustic detection of submarine landslides on Kilauea volcano. *Geophys. Res. Lett.* **2001**, *28*, 1811–1813. [CrossRef]

76. Gee, M.J.; Watts, A.B.; Masson, D.G.; Mitchell, N.C. Landslides and the evolution of El Hierro in the Canary Islands. *Mar. Geol.* **2001**, *177*, 271–293. [CrossRef]

77. Masson, D.G.; Watts, A.B.; Gee, M.J.R.; Urgeles, R.; Mitchell, N.C.; Le Bas, T.P.; Canals, M. Slope failures on the flanks of the western Canary Islands. *Earth Sci. Rev.* **2002**, *57*, 1–35. [CrossRef]

78. McMurtry, G.M.; Watts, P.; Fryer, G.J.; Smith, J.R.; Imamura, F. Giant landslides, mega-tsunamis, and paleo-sea level in the Hawaiian Islands. *Mar. Geol.* **2004**, *203*, 219–233. [CrossRef]

79. Zhang, K.; Douglas, B.C.; Leatherman, S.P. Global warming and coastal erosion. *Clim. Chang.* **2004**, *64*, 41–58. [CrossRef]

80. Dan, G.; Sultan, N.; Savoye, B. The 1979 Nice harbour catastrophe revisited: Trigger mechanism inferred from geotechnical measurements and numerical modelling. *Mar. Geol.* **2007**, *245*, 40–65. [CrossRef]

81. Scheffers, A.; Scheffers, S. Tsunami deposits on the coastline of west Crete (Greece). *Earth Planet Sci. Lett.* **2007**, *259*, 613–624. [CrossRef]

82. Chiocci, F.L.; Romagnoli, C., Tommasi, P.; Bosman, A. The Stromboli 2002 tsunamigenic submarine slide: Characteristics and possible failure mechanisms. *J. Geophys. Res.Solid Earth* **2008**, *113*, B10102. [CrossRef]

83. Baldi, P.; Bosman, A.; Chiocci, F.L.; Marsella, M.; Romagnoli, C.; Sonnessa, A. Integrated subaerial-submarine morphological evolution of the Sciara del Fuoco after the 2002 landslide. In *The Stromboli Volcano: An Integrated Study of the 2002–2003 Eruption*; Calvari, S., Inguaggiato, S., Puglisi, G., Ripepe, M., Rosi, M., Eds.; American Geophysical Union: Washington, DC, USA, 2008; pp. 171–182.

84. Parrott, D.R.; Todd, B.J.; Shaw, J.; Hughes Clarke, J.E.; Griffin, J.; MacGowan, B.; Lamplugh, M.; Webster, T. Integration of multibeam bathymetry and LiDAR surveys of the Bay of Fundy, Canada. In Proceedings of the Canadian Hydrographic Conference and National Surveyors Conference 2008, Victoria, BC, Canada, 5–8 May 2008; Paper 6-2, pp. 1–15.

85. Casalbore, D. Studio di Fenomeni d'Instabilità Gravitativa sui Fondali Marini, con Particolare Riferimento all'Isola di Stromboli. Ph.D. Thesis, Alma Mater Studiorum Università di Bologna, Dottorato di Ricerca in Scienze Della Terra, Bologna, Italy, 2009.

86. Violante, C. Rocky coast: Geological constraints for hazard assessment. In *Geohazard in Rocky Coastal Areas*; Violante, C., Ed.; Geological Society Special Publications 322; Geological Society: London, UK, 2009; pp. 1–31.

87. De Blasio, F.V.; Mazzanti, P. Subaerial and subaqueous dynamics of coastal rockfalls. *Geomorphology* **2010**, *115*, 188–193. [CrossRef]

88. De Gange, A.R.; Vernon Byrd, G.; Walker, L.R.; Waythomas, C.F. Introduction–The Impacts of the 2008 Eruption of Kasatochi Volcano on Terrestrial and Marine Ecosystems in the Aleutian Islands, Alaska. *Arct. Antarct. Alp. Res.* **2012**, *42*, 245–249. [CrossRef]

89. Sultan, N.; Savoye, B.; Jouet, G.; Leynaud, D.; Cochonat, P.; Henry, P.; Stegmann, S.; Kopf, A. Investigation of a possible submarine landslide at the Var delta front (Nice continental slope, southeast France). *Can. Geotech. J.* **2010**, *47*, 486–496. [CrossRef]

90. Casalbore, D.; Romagnoli, C.; Bosman, A.; Chiocci, F.L. Potential tsunamigenic landslides at Stromboli Volcano (Italy): Insights from marine DEM analysis. *Geomorphology* **2011**, *126*, 42–50. [CrossRef]

91. Casalbore, D.; Chiocci, F.L.; Scarascia Mugnozza, G.; Tommasi, P.; Sposato, A. Flash-flood hyperpycnal flows generating shallow-water landslides at Fiumara mouths in Western Messina Straits (Italy). *Mar. Geophys. Res.* **2011**, *32*, 257–271. [CrossRef]

92. Chiocci, F.L.; Ridente, D. Regional-scale seafloor mapping and geohazard assessment. The experience from the Italian project MaGIC (Marine Geohazards along the Italian Coasts). *Mar. Geophys. Res.* **2011**, *32*, 13–23. [CrossRef]

93. Lambeck, K.; Antonioli, F.; Anzidei, M.; Ferranti, L.; Leoni, G.; Scicchitano, G.; Silenzi, S. Sea level change along the Italian coast during the Holocene and projections for the future. *Quat. Int.* **2011**, *232*, 250–257. [CrossRef]

94. Mazzanti, P.; Bozzano, F. Revisiting the February 6th 1783 Scilla (Calabria, Italy) landslide and tsunami by numerical simulation. *Mar. Geophys. Res.* **2011**, *32*, 273–286. [CrossRef]

95. De Jongh, C.; van Opstal, H. *Coast-Map-IO TopoBathy Database. Report Pilot Project*; Leading Partner CARIS BV, 2012; pp. 20–26. Available online: https://www.dhyg.de/images/hn_ausgaben/HN094.pdf (accessed on 5 August 2019).

96. Della Seta, M.; Martino, S.; Scarascia Mugnozza, G. Quaternary sea-level change and slope instability in coastal areas: Insights from the Vasto Landslide (Adriatic coast, central Italy). *Geomorphology* **2013**, *201*, 462–478. [CrossRef]

97. Knight, J.; Harrison, S. The impacts of climate change on terrestrial Earth surface systems. *Nat. Clim. Chang.* **2013**, *3*, 24–29. [CrossRef]

98. Masselink, G.; Russell, P. Impacts of climate change on coastal erosion. *MCCIP Sci Rev.* **2013**, *1*, 71–86.

99. Mastronuzzi, G.; Brückner, H.; De Martini, P.M.; Regnauld, H. Tsunami: From the open sea to the coastal zone and beyond. In *Tsunami: From Fundamentals to Damage Mitigation*; Mambretti, S., Ed.; WIT Press: Southampton, UK, 2013; pp. 1–36.

100. Minelli, L.; Billi, A.; Faccenna, C.; Gervasi, A.; Guerra, I.; Orecchio, B.; Speranza, G. Discovery of a gliding salt-detached megaslide, Calabria, Ionian Sea, Italy. *Geophys. Res. Lett.* **2013**, *40*, 4220–4224. [CrossRef]

101. Carracedo, J.C. Structural Collapses in the Canary Islands. In *Landscapes and Landforms of Spain*; Gutiérrez, F., Gutiérrez, M., Eds.; Springer: Dordrecht, The Netherlands, 2014; pp. 289–306.

102. Cazenave, A.; Cozannet, G.L. Sea level rise and its coastal impacts. *Earth's Future* **2014**, *2*, 15–34. [CrossRef]

103. Mataspaud, A.; Letortu, P.; Costa, S.; Cantat, O.; Héquette, A.; Ruz, M.-H. Conditions météo-marines et facteurs de prédisposition à l'origine de phénomènes de submersion marine: Analyse comparative entre Manche orientale et Mer du Nord méridionale. Presented at International Conference Connaissance et Compréhension des Risques Côtiers: Aléas, Enjeux, Représentations, Gestion. IUEM Brest, Brest, France, 3–4 July 2014; pp. 53–62.

104. Mottershead, D.; Bray, M.; Soar, P.; Farres, P.J. Extreme wave events in the central Mediterranean: Geomorphic evidence of tsunami on the Maltese Islands. *Z. Geomorphol.* **2014**, *58*, 385–411. [CrossRef]

105. Biolchi, S.; Furlani, S.; Antonioli, F.; Baldassini, N.; Deguara, J.C.; Devoto, S.; Di Stefano, A.; Evans, J.; Gambin, T.; Gauci, R.; et al. Boulder accumulations related to extreme wave events on the eastern coast of Malta. *Nat. Hazard. Earth Syst.* **2016**, *16*, 737–756. [CrossRef]

106. Yonggang, J.I.A.; Chaoqi, Z.; Liping, L.; Dong, W. Marine geohazards: Review and future perspective. *Acta Geol. Sin. Engl.* **2016**, *90*, 1455–1470. [CrossRef]

107. Antonioli, F.; Anzidei, M.; Amorosi, A.; Lo Presti, V.; Mastronuzzi, G.; Deiana, G.; De Falco, G.; Fontana, A.; Fontolan, G.; Lisco, S.; et al. Sea-level rise and potential drowning of the Italian coastal plains: Flooding risk scenarios for 2100. *Quat. Sci. Rev.* **2017**, *158*, 29–43. [CrossRef]

108. Aucelli, P.P.C.; Di Paola, G.; Incontri, P.; Rizzo, A.; Vilardo, G.; Benassai, G.; Buonocore, B.; Pappone, G. Coastal inundation risk assessment due to subsidence and sea level rise in a Mediterranean alluvial plain (Volturno coastal plain–southern Italy). *Estuar. Coast. Shelf Sci.* **2017**, *198*, 597–609. [CrossRef]

109. Zaggia, L.; Lorenzetti, G.; Manfé, G.; Scarpa, G.M.; Molinaroli, E.; Parnell, K.E.; Rapaglia, J.P.; Gionta, M.; Soomere, T. Fast shoreline erosion induced by ship wakes in a coastal lagoon: Field evidence and remote sensing analysis. *PLoS ONE* **2017**, *12*, e0187210. [CrossRef] [PubMed]

110. Casalbore, D.; Romagnoli, C.; Bosman, A.; Anzidei, M.; Chiocci, F.L. Coastal hazard due to submarine canyons in active insular volcanoes: Examples from Lipari Island (southern Tyrrhenian Sea). *J. Coast. Conserv.* **2018**, *22*, 989–999. [CrossRef]

111. Di Paola, G.; Alberico, I.; Aucelli, P.P.C.; Matano, F.; Rizzo, A.; Vilardo, G. Coastal subsidence detected by Synthetic Aperture Radar interferometry and its effects coupled with future sea-level rise: The case of the Sele Plain (Southern Italy). *J. Flood Risk Manag.* **2018**, *11*, 191–206. [CrossRef]

112. Moore, R.; Davis, G.; Dabson, O. Applied geomorphology and geohazard assessment for deepwater development. In *Submarine Geomorphology*; Micallef, A., Krastel, S., Savini, A., Eds.; Springer: Cham, Switzerland, 2018; pp. 459–479.

113. Obrocki, L.; Vött, A.; Wilken, D.; Fischer, P.; Willershäuser, T.; Koster, B.; Lang, F.; Papanikolaou, I.; Rabbel, W.; Reicherter, K. Tracing tsunami signatures of the AD 551 and AD 1303 tsunamis at the Gulf of Kyparissia (Peloponnese, Greece) using direct push in situ sensing techniques combined with geophysical studies. *Sedimentology* **2020**, *67*, 1274–1308. [CrossRef]

114. Pennetta, M. Beach Erosion in the Gulf of Castellammare di Stabia in Response to the Trapping of Longshore Drifting Sediments of the Gulf of Napoli (Southern Italy). *Geosciences* **2018**, *8*, 235. [CrossRef]

115. Urlaub, M.; Petersen, F.; Gross, F.; Bonforte, A.; Puglisi, G.; Guglielmino, F.; Krastel, S.; Lange, D.; Kopp, H. Gravitational collapse of Mount Etna's southeastern flank. *Sci. Adv.* **2018**, *4*, eaat9700. [CrossRef]

116. Biolchi, S.; Denamiel, C.; Devoto, S.; Korbar, T.; Macovaz, V.; Scicchitano, G.; Vilibic, I.; Furlani, S. Impact of the October 2018 Storm Vaia on Coastal Boulders in the Northern Adriatic Sea. *Water* **2019**, *11*, 2229. [CrossRef]

117. Buosi, C.; Porta, M.; Trogu, D.; Casti, M.; Ferraro, F.; De Muro, S.; Ibba, A. Data on coastal dunes vulnerability of eleven microtidal wave-dominated beaches of Sardinia (Italy, western Mediterranean). *Data Brief* **2019**, *24*, 103897. [CrossRef] [PubMed]

118. Mucerino, L.; Albarella, M.; Carpi, L.; Besio, G.; Benedetti, A.; Corradi, N.; Firpo, M.; Ferrari, M. Coastal exposure assessment on Bonassola bay. *Ocean. Coast. Manag.* **2019**, *167*, 20–31. [CrossRef]

119. Toker, E.; Sharvit, J.; Fischer, M.; Melzer, Y.; Potchter, O. Archaeological, geomorphological and cartographical evidence of the sea level rise in the southern Levantine Basin in the 19th and 20th centuries. *Quatern. Int.* **2019**, *522*, 55–65. [CrossRef]

120. Rizzo, A.; Vandelli, V.; Buhagiar, G.; Micallef, A.S.; Soldati, M. Coastal vulnerability assessment along the north-eastern sector of Gozo Island (Malta, Mediterranean Sea). *Water* **2020**, *12*, 1405. [CrossRef]

121. Hogrefe, K.R.; Wright, D.J.; Hochberg, E.J. Derivation and Integration of Shallow-Water Bathymetry: Implications for Coastal Terrain Modelling and Subsequent Analyses. *Mar. Geod.* **2008**, *31*, 299–317. [CrossRef]

122. McKean, J.A.; Isaak, D.J.; Wright, C.W. Geomorphic controls on salmon nesting patterns described by a new, narrow-beam terrestrial–aquatic LiDAR. *Front. Ecol. Environ.* **2008**, *6*, 125–130. [CrossRef]

123. McKean, J.; Nagel, D.; Tonina, D.; Bailey, P.; Wright, C.W.; Bohn, C.; Nayegandhi, A. Remote sensing of channels and riparian zones with a narrow-beam aquatic-terrestrial LIDAR. *Remote Sens.* **2009**, *1*, 1065–1096. [CrossRef]

124. Tallis, H.; Ferdana, Z.; Gray, E. Linking terrestrial and marine conservation planning and threats analysis. *Conserv. Biol.* **2008**, *22*, 120–130. [CrossRef]

125. Vierling, K.T.; Vierling, L.A.; Gould, W.A.; Martinuzzi, S.; Clawges, R.M. Lidar: Shedding new light on habitat characterization and modelling. *Front. Ecol. Environ.* **2008**, *6*, 90–98. [CrossRef]

126. Brown, C.J.; Smith, S.J.; Lawton, P.; Anderson, J.T. Benthic habitat mapping: A review of progress towards improved understanding of the spatial ecology of the seafloor using acoustic techniques. *Estuar. Coast. Shelf Sci.* **2011**, *92*, 502–520. [CrossRef]

127. Marchese, F.; Fallati, L.; Corselli, C.; Savini, A. Testing the use of Unmanned Aerial Vehicle and structure from motion technique for acquisition of ultra-shallow water bathymetric data. In Proceedings of the 9th International Conference on Geomorphology (9th ICG), Vigyan Bhawan, New Delhi, India, 6–11 November 2017.

128. Prampolini, M.; Blondel, P.; Foglini, F.; Madricardo, F. Habitat mapping of the Maltese continental shelf using acoustic textures and bathymetric analyses. *Estuar Coast. Shelf Sci.* **2018**, *207*, 483–498. [CrossRef]

129. Harris, P.T.; Baker, E.K. Why map benthic habitats? In *Seafloor Geomorphology as Benthic Habitat–GeoHAB Atlas of Seafloor Geomorphic Features and Benthic Habitats*, 2nd ed.; Harris, P.T., Baker, E.K., Eds.; Elsevier: London, UK, 2020; pp. 3–15.

130. Cicin-Sain, B.; Belfiore, S. Linking marine protected areas to integrate coastal and ocean management: A review of theory and practice. *Ocean. Coast. Manag.* **2005**, *48*, 847–868. [CrossRef]

131. Sardá, R.; Avila, C.; Mora, J. A methodological approach to be used in integrated coastal zone management processes: The case of the Catalan Coast (Catalonia, Spain). *Estuar. Coast. Shelf Sci.* **2005**, *62*, 427–439. [CrossRef]

132. Schultz-Zehden, A.; Gee, K.; Scibior, K. *Handbook on Integrated Maritime Spatial Planning*; Interreg IIIB CADSES PlanCoast Project; April 2008. Available online: https://www.msp-platform.eu/practices/handbook-integrated-maritime-spatial-planning (accessed on 5 August 2019).

133. Cogan, C.B.; Todd, B.J.; Lawton, P.; Noji, T.T. The role of marine habitat mapping in ecosystem-based management. *ICES J. Mar. Sci.* **2009**, *66*, 2033–2042. [CrossRef]

134. Ehler, C.; Douvere, F. *Marine Spatial Planning: A Step-by-Step Approach Toward Ecosystem-Based Management*; (IOC Manual and Guide n. 53, ICAM Dossier n. 6); UNESCO: Paris, France, 2009; p. 99.

135. Watts, M.E.; Ball, I.R.; Stewart, R.S.; Klein, C.J.; Wilson, K.; Steinback, C.; Lourival, R.; Kircher, L.; Possingham, H.P. Marxan with Zones: Software for optimal conservation based land-and sea-use zoning. *Environ. Model. Softw.* **2009**, *24*, 1513–1521. [CrossRef]

136. Meiner, A. Integrated maritime policy for the European Union–consolidating coastal and marine information to support maritime spatial planning. *J. Coast. Conserv.* **2010**, *14*, 1–11. [CrossRef]

137. Schlacke, S.; Maier, N.; Markus, T. Legal implementation of integrated ocean policies: The EU's marine strategy framework directive. *Int. J. Mar. Coast. Law* **2011**, *26*, 59–90. [CrossRef]

138. Smith, D.H.; Maes, F.; Stojanovic, T.A.; Ballinger, R.C. The integration of land and marine spatial planning. *J. Coast. Conserv.* **2011**, *15*, 291–303. [CrossRef]

139. Qiu, W.; Jones, P.J.S. The emerging policy landscape for marine spatial planning in Europe. *Mar. Policy* **2013**, *39*, 182–190. [CrossRef]

140. Kerr, S.; Johnson, K.; Side, J.C. Planning at the edge: Integrating across the land sea divide. *Mar. Policy* **2014**, *47*, 118–125. [CrossRef]

141. Ramieri, E.; Andreoli, E.; Fanelli, A.; Artico, G.; Bertaggia, R. *Methodological Handbook on Maritime Spatial Planning in the Adriatic Sea*; Final Report of Shape Project WP4 "Shipping Towards Maritime Spatial Planning"; February 2014. Available online: http://paprac.org/storage/app/media/Meetings/2019/Sub-regional%20meeting%20Adriatic%20Ionian%20cooperation%20towards%20MSP/Methodological%20Handbook%20on%20MSP%20in%20the%20Adriatic.pdf (accessed on 5 August 2020).

142. Ramieri, E.; Bocci, M.; Markovic, M. SIMWESTMED - Relationship between LSI and ICZM. (R5). *Zenodo* **2019**. [CrossRef]

143. Barbanti, A.; Campostrini, P.; Musco, F.; Sarretta, A.; Gissi, E. *Developing a Maritime Spatial Plan for the Adriatic–Ionian Region*; CNR-ISMAR: Venice, Italy, 2015.

144. Domínguez-Tejo, E.; Metternicht, G.; Johnston, E.; Hedge, L. Marine Spatial Planning advancing the Ecosystem-Based Approach to coastal zone management: A review. *Mar. Policy* **2016**, *72*, 115–130. [CrossRef]

145. UNEP/MAP. *Mediterranean Strategy for Sustainable Development 2016–2025*; Plan Bleu, Regional Activity Centre: Valbonne, France, 2016; p. 84.

146. UNEP-MAP PAP/RAC. Conceptual Framework for Marine Spatial Planning. In Proceedings of the 20th Ordinary Meeting of the Contracting Parties to the Barcelona Convention, Tirana, Albania, 17–20 December 2017.

147. UNEP(DEPI)/MED IG.22/28. Decision IG.22/1: UNEP/MAP Mid-Term Strategy 2016–2021. Available online: https://wedocs.unep.org/rest/bitstreams/8364/retrieve (accessed on 5 August 2020).

148. UNEP(DEPI)/MED IG.22/28. Decision IG.22/2: UNEP/MAP Mid-Term Strategy 2016–2025. Available online: https://wedocs.unep.org/rest/bitstreams/8379/retrieve (accessed on 5 August 2020).

149. Sustainable Development Goals 14 "Conserve and Sustainably Use the Oceans, Seas and Marine Resources for Sustainable Development". Available online: https://unstats.un.org/sdgs/report/2017/Goal-14/ (accessed on 20 July 2020).

150. Sustainable Development Goals 15 "Protect, Restore and Promote Sustainable Use of Terrestrial Ecosystems, Sustainably Managed Forests, Combat Desertification, and Halt and Reverse Land Degradation and Halt Biodiversity Loss". Available online: https://unstats.un.org/sdgs/report/2017/Goal-15/ (accessed on 20 July 2020).

151. UNEP/MAP/PAP: Protocol on Integrated Coastal Zone Management in the Mediterranean. Split, Priority Actions Programme. 2008. Available online: http://iczmplatform.org/ (accessed on 5 August 2020).

152. Smith, M.J.; Paron, P.; Griffiths, J.S. *Geomorphological Mapping: Methods and Applications*; Elsevier: Oxford, UK, 2011.

153. Biolchi, S.; Furlani, S.; Devoto, S.; Gauci, R.; Castaldini, D.; Soldati, M. Geomorphological identification, classification and spatial distribution of coastal landforms of Malta (Mediterranean Sea). *J. Maps* **2016**, *12*, 87–99. [CrossRef]

154. Micallef, A. Marine geomorphology: Geomorphological mapping and the study of submarine landslides. In *Geomorphological Mapping: Methods and Applications*; Smith, M.J., Paron, P., Griffiths, J.S., Eds.; Elsevier: Oxford, UK, 2011; Volume 15, pp. 377–395.

155. Gorini, M.A.V. Physiographic classification of the ocean floor: A multi-scale geomorphometric approach. In *Proceedings of Geomorphometry 2009, Zurich, Switzerland, 31 August–2 September 2009*; Purves, R., Gruber, S., Straumann, R., Hengl, T., Eds.; University of Zurich: Zurich, Switzerland, 2009; pp. 98–105.

156. Rubin, D.M.; McCulloch, D.S. Single and superimposed bedforms: A synthesis of San Francisco Bay and flume observations. *Sediment. Geol.* **1980**, *26*, 207–231. [CrossRef]

157. Ashley, G.M. Classification of large-scale subaqueous bedforms; a new look at an old problem. *J. Sediment. Res.* **1990**, *60*, 160–172.

158. Wynn, R.B.; Stow, D.A. Classification and characterisation of deep-water sediment waves. *Mar. Geol.* **2002**, *192*, 7–22. [CrossRef]

159. Stow, D.A.V.; Hernández-Molina, F.J.; Llave, E.; Sayago, M.; Díaz del Río, V.; Branson, A. Bedform-velocity matrix: The estimation of bottom current velocity from bedform observations. *Geology* **2009**, *37*, 327–330. [CrossRef]

160. Madricardo, F.; Foglini, F.; Kruss, A.; Ferrarin, C.; Pizzeghello, N.M.; Murri, C.; Rossi, M.; Bajo, M.; Bellafiore, D.; Campiani, E.; et al. High resolution multibeam and hydrodynamic datasets of tidal channels and inlets of the Venice Lagoon. *Sci. Data* **2017**, *4*, 170121. [CrossRef]

161. Gavazzi, G.M.; Madricardo, F.; Janowski, L.; Kruss, A.; Blondel, P.; Sigovini, M.; Foglini, F. Evaluation of seabed mapping methods for fine-scale classification of extremely shallow benthic habitats–Application to the Venice Lagoon, Italy. *Estuar. Coast. Shelf Sci.* **2016**, *170*, 45–60. [CrossRef]

162. Selley, R.C.; Cocks, R.; Plimer, I. *Encyclopedia of Geology*; Academic Press: London, UK, 2005.

163. Thorsnes, T.; Bjarnadóttir, L.R.; Jarna, A.; Baeten, N.; Scott, G.; Guinan, J.; Monteys, X.; Dove, D.; Green, S.; Gafeira, J.; et al. National Programmes: Geomorphological Mapping at Multiple Scales for Multiple Purposes. In *Submarine Geomorphology*; Micallef, A., Krastel, S., Savini, A., Eds.; Springer: Cham, Switzerland, 2018; pp. 1–9.

164. Dove, D.; Bradwell, T.; Carter, G.; Cotterill, C.; Gafeira Goncalves, J.; Green, S.; Krabbendam, M.; Mellett, C.; Stevenson, A.; Stewart, H.; et al. *Seabed Geomorphology: A Two-Part Classification System*; British Geological Survey: Edinburgh, UK, 2016; (Unpublished). Available online: http://nora.nerc.ac.uk/id/eprint/514946/ (accessed on 5 August 2020).

165. European Marine Observation and Data Network (EMODnet) Geology. Available online: http://www.emodnet-geology.eu (accessed on 8 April 2020).

166. Asch, K. Interoperability, Standards and EMODnet Geology: Building the Mosaic of European Sea Floor Data. In *Geophysical Research Abstracts*; EGU General Assembly: Munich, Germany, 2019; Volume 21, EGU2019-17389.

167. BGS. *EMODnet Geology III–Work Package 8: Submerged Landscape Data Harmonisation and Confidence Analysis Task Guide*; British Geological Survey: Nottingham, UK, 2018.

168. Grehan, A.J.; Arnaud-Haond, S.; D'Onghia, G.; Savini, A.; Yesson, C. Towards ecosystem based management and monitoring of the deep Mediterranean, North-East Atlantic and Beyond. *Deep-Sea Res. Part II* **2017**, *145*, 1–7. [CrossRef]

169. Boero, F.; Foglini, F.; Fraschetti, S.; Goriup, P.; Macpherson, E.; Planes, S.; Soukissian, T.; The CoCoNet Consortium. CoCoNet: Towards coast to coast networks of marine protected areas (from the shore to the high and deep sea), coupled with sea-based wind energy potential. *Sci. Res. Inform. Technol.* **2016**, *6*, 1–95.

170. Foglini, F.; Angeletti, L.; Campiani, E.; Mercorella, A.; Prampolini, M.; Grande, V.; Savini, A.; Taviani, M.; Tessarolo, C. Habitat mapping for establishing Coast to Coast Network of marine protected areas in the framework of COCONET Project: From coastal area to deep sea in the South Adriatic (Italy). In Proceedings of the GEOHAB—Marine Environment Mapping and Interpretation, Abstract Volume, Lorne, Australia, 4–8 May 2014; p. 32.

171. Davies, J.S.; Guillaumont, B.; Tempera, F.; Vertino, A.; Beuck, L.; Ólafsdóttir, S.H.; Smith, C.; Fosså, J.H.; Van den Beld, I.M.J.; Savini, A.; et al. A new classification scheme of European cold-water coral habitats: Implications for ecosystem-based management of the deep sea. *Deep Sea Res. Part II* **2017**, *145*, 102–109. [CrossRef]

172. European Classification Scheme EUNIS. Available online: https://www.eea.europa.eu/data-and-maps/data/eunis-habitat-classification (accessed on 10 May 2020).

173. ISPRA Geological and Geothematic Sheets at the Scale 1:50,000 of the Italian Territory. Available online: http://www.isprambiente.gov.it/it/cartografia/carte-geologiche-e-geotematiche/carta-geologica-alla-scala-1-a-50000 (accessed on 8 April 2020).

174. Fabbri, A.; Argnani, A.; Bortoluzzi, G.; Correggiari, A.; Gamberi, F.; Ligi, M.; Marani, M.; Penitenti, D.; Roveri, M.; Trincardi, F.; et al. *Carta Geologica dei Mari Italiani alla Scala 1: 250.000. Guida al Rilevamento*; Presidenza del Consiglio dei Ministri, Dipartimento per i Servizi Tecnici Nazionali; Servizio Geologico, Quaderni: Bologna, Italy, 2002; Volume 8, Serie III.

175. Progetto Visibilità dei Dati Afferenti all'Attività di Esplorazione Petrolifera in Italia (ViDEPi). Available online: https://www.videpi.com/videpi/videpi.asp (accessed on 26 May 2020).

176. Waelbroeck, C.; Labeyrie, L.; Michela, E.; Duplessya, J.C.; McManusc, J.F.; Lambeck, K.; Balbona, E.; Labracheriee, M. Sea-level and deep water temperature changes derived from benthic foraminifera isotopic records. *Quat. Sci. Rev.* **2002**, *21*, 295–305. [CrossRef]

177. Grant, K.M.; Rohling, E.J.; Bar-Matthews, M.; Ayalon, A.; Medina-Elizalde, M.; Bronk Ramsey, C.; Satow, C.; Roberts, A.P. Rapid coupling between ice volume and polar temperature over the past 150,000 years. *Nature* **2012**, *491*, 744–747. [CrossRef]

178. Hopkins, D.M. Cenozoic history of the Bering land bridge. *Science* **1959**, *129*, 1519–1528. [CrossRef]

179. Bond, J.D. Paleodrainage Map of Beringia. In *Yukon Geological Survey*; Open File 2019-2; 2019. Available online: http://data.geology.gov.yk.ca/Reference/81642#InfoTab (accessed on 5 August 2020).

180. Soldati, M.; Barrows, T.T.; Prampolini, M.; Fifield, K.L. Cosmogenic exposure dating constraints for coastal landslide evolution on the Island of Malta (Mediterranean Sea). *J. Coast. Conserv.* **2018**, *22*, 831–844. [CrossRef]

181. Marchetti, M.; Soldati, M.; Vandelli, V. The Great Diversity of Italian Landscapes and Landforms: Their Origin and Human Imprint. In *Landscapes and Landforms of Italy*; Soldati, M., Marchetti, M., Eds.; Springer International Publishing AG: Cham, Switzerland, 2017; pp. 7–20.

182. Aucelli, P.P.C.; Brancaccio, L.; Cinque, A. Vesuvius and Campi Flegrei: Volcanic History, Landforms and Impact on Settlements. In *Landscapes and Landforms of Italy*; Soldati, M., Marchetti, M., Eds.; Springer International Publishing AG: Cham, Switzerland, 2017; pp. 389–398.

183. Brilha, J. Inventory and Quantitative Assessment of Geosites and Geodiversity Sites: A Review. *Geoheritage* **2016**, *8*, 119–134. [CrossRef]

184. Zarnetske, P.L.; Read, Q.D.; Record, S.; Gaddis, K.D.; Pau, S.; Hobi, M.L.; Wilson, A.M. Towards connecting biodiversity and geodiversity across scales with satellite remote sensing. *Glob. Ecol. Biogeogr.* **2019**, *28*, 548–556. [CrossRef]

185. Burek, C.V.; Ellis, N.V.; Evans, D.H.; Hart, M.B.; Larwood, J.G. Marine geoconservation in the United Kingdom. *Proc. Geol. Assoc.* **2013**, *124*, 581–592. [CrossRef]

186. Gordon, J.E. Geoconservation principles and protected area management. *Int. J. Geoherit. Parks* **2019**, *7*, 199–210. [CrossRef]

187. Bosello, F.; De Cian, E. Climate change, sea level rise, and coastal disasters. A review of modelling practices. *Energy Econ.* **2014**, *46*, 593–605. [CrossRef]

188. Handmer, J.; Honda, Y.; Kundzewicz, Z.W.; Arnell, N.; Benito, G.; Hatfield, J.; Mohamed, I.F.; Peduzzi, P.; Wu, S.; Sherstyukov, B.; et al. Changes in Impacts of Climate Extremes: Human Systems and Ecosystems. In *Managing the Risks of Extreme Events and Disasters to Advance Climate Change Adaptation*; A Special Report of Working Groups I and II of the Intergovernmental Panel on Climate Change (IPCC); Field, C.B., Barros, V., Stocker, T.F., Qin, D., Dokken, D.J., Ebi, K.L., Mastrandrea, M.D., Mach, K.J., Plattner, K., Allen, S.K., et al., Eds.; Cambridge University Press: Cambridge, UK; New York, NY, USA, 2012; pp. 231–290.

189. Hanson, S.; Nicholls, R.; Ranger, N.; Hallegatte, S.; Corfee-Morlot, J.; Herweijer, C.; Chateau, J. A global ranking of port cities with high exposure to climate extremes. *Clim. Chang.* **2011**, *104*, 89–111. [CrossRef]

190. Quadros, N.D. *What Users Want in Their Bathymetry*; Hydro International: Clevedon, UK, 2012; Volume 18–23.

191. Stallins, J.A. Geomorphology and ecology: Unifying themes for complex systems in biogeomorphology. *Geomorphology* **2006**, *77*, 207–216. [CrossRef]

192. Hopley, D.; Smithers, S.G.; Parnell, K.E. *The Geomorphology of the Great Barrier Reef: Development, Diversity and Change*; Cambridge University Press: Cambridge, UK, 2007.

193. Savini, A.; Vertino, A.; Marchese, F.; Beuck, L.; Freiwald, A. Mapping Cold-Water Coral Habitats at Different Scales within the northern Ionian Sea (central Mediterranean): An assessment of coral coverage and associated vulnerability. *PLoS ONE* **2014**, *9*, e87108. [CrossRef]

194. Bracchi, V.; Savini, A.; Marchese, F.; Palamara, S.; Basso, D.; Corselli, C. Coralligenous habitat in the Mediterranean Sea: A geomorphological description from remote data. *Ital. J. Geosci.* **2015**, *134*, 32–40. [CrossRef]

195. Savini, A.; Marchese, F.; Verdicchio, G.; Vertino, A. Submarine slide topography and the distribution of vulnerable marine ecosystems: A case study in the Ionian Sea (eastern Mediterranean). In *Submarine Mass Movements and Their Consequences, Advances in Natural and Technological Hazards Research*; Lamarche, G., Mountjoy, J., Bull, S., Hubble, T., Krastel, S., Lane, E., Micallef, A., Moscardelli, L., Mueller, C., Pecher, I., et al., Eds.; Springer: Dordrecht, The Netherlands, 2016; Volume 41, pp. 163–170.

196. Bracchi, V.; Basso, D.; Marchese, F.; Corselli, C.; Savini, A. Coralligenous morphotypes on subhorizontal substrate: A new categorization. *Cont. Shelf Res.* **2017**, *144*, 10–20. [CrossRef]

197. Bargain, A.; Marchese, F.; Savini, A.; Taviani, M.; Fabri, M.-C. Santa Maria di Leuca Province (Mediterranean Sea): Identification of suitable mounds for cold-water coral settlement using geomorphometry proxies and Maxent methods. *Front. Mar. Sci. Deep Sea Environ. Ecol.* **2017**, *4*, 338.

198. Lo Iacono, C.; Savini, A.; Basso, D. Cold-Water Carbonate Bioconstructions. In *Submarine Geomorphology*; Micallef, A., Krastel, S., Savini, A., Eds.; Springer: Cham, Switzerland, 2018; pp. 425–455.

199. Riegl, B.; Purkis, S.J. Detection of Shallow Subtidal Corals from IKONOS Satellite and QTC View (50,200 kHz) Single-Beam Sonar Data (Arabian Gulf; Dubai, UAE). *Remote Sens. Environ.* **2005**, *95*, 96–114. [CrossRef]

200. Bejarano, S.; Mumby, P.J.; Hedley, J.D.; Sotheran, I. Combining Optical and Acoustic Data to Enhance the Detection of Caribbean Fore-reef Habitats. *Remote Sens. Environ.* **2010**, *114*, 2768–2778. [CrossRef]

201. Savini, A.; Marchese, F.; Fallati, L.; Corselli, C.; Galli, P. Integrating acoustics and photogrammetry-based 3D point clouds for the generation of a continuous bathymetric model in coral reef environment. In *EGU General Assembly; EGU General Assembly 2020, Online, 4–8 May 2020*; EGU General Assembly: Vienna, Austria, 2020. [CrossRef]

202. Coggan, R.; Populus, J.; White, J.; Sheehan, K.; Fitzpatrick, F.; Piel, S. *Review of Standards and Protocols for Seabed Habitat Mapping*; MESH Mapping European Seabed Habitats, INTERREG European Program; 2007. Available online: www.searchmesh.net/Files/Standards_&_Protocols_2nd (accessed on 5 August 2020).

203. Todd, B.J.; Greene, H.G. *Mapping the Seafloor for Habitat Characterization*; Special Publication 47; Geological Association of Canada: Newfoundland, NF, Canada, 2007.

204. Thorsnes, T. MAREANO–An introduction. *Norw. J. Geol.* **2009**, *89*, 3.

205. Buhl-Mortensen, L.; Buhl-Mortensen, P.; Dolan, M.F.; Holte, B. The MAREANO programme—A full coverage mapping of the Norwegian off-shore benthic environment and fauna. *Mar. Biol. Res.* **2015**, *11*, 4–17. [CrossRef]

206. Buhl-Mortensen, L.; Buhl-Mortensen, P.; Dolan, M.J.F.; Gonzalez-Mirelis, G. Habitat mapping as a tool for conservation and sustainable use of marine resources: Some perspectives from the MAREANO Programme, Norway. *J. Sea Res.* **2015**, *100*, 46–61. [CrossRef]

207. Diesing, M.; Green, S.L.; Stephens, D.; Lark, R.M.; Stewart, H.; Dove, D. Mapping seabed sediments: Comparison of manual, geostatistical, object-based image analysis and machine learning approaches. *Cont. Shelf Res.* **2014**, *84*, 107–119. [CrossRef]

208. Howe, J.A.; Stevenson, A.; Gatliff, R. Seabed mapping for the 21st century—the Marine Environmental Mapping Programme (MAREMAP): Preface. *Earth Environ. Sci. Trans. R. Soc. Edinb.* **2015**, *105*, 239–240. [CrossRef]

209. Castellanos-Galindo, G.A.; Casella, E.; Mejía-Rentería, J.C.; Rovere, A. Habitat mapping of remote coasts: Evaluating the usefulness of lightweight unmanned aerial vehicles for conservation and monitoring. *Biol. Conserv.* **2019**, *239*, 108282. [CrossRef]

210. Commission of the European Communities. *Roadmap for Maritime Spatial Planning: Achieving Common Principles in the EU*; COM (2008) 791 Final; Communication of the European Communities: Brussels, Belgium, 25 November 2008.

211. Ruttenberg, B.I.; Granek, E.I. Bridging the marine–terrestrial disconnect to improve marine coastal zone science and management. *Mar. Ecol. Prog. Ser.* **2011**, *434*, 203–212. [CrossRef]

212. Mason, T.; Rainbow, B.; McVey, S. Colouring the 'White Ribbon–Strategic Coastal Monitoring in the South-East of England. Hydro International. 2006. Available online: http://www.hydro-international.com/issues/articles/id611-Colouring_the_White_Ribbon.html (accessed on 18 September 2019).

213. Kinzel, P.J.; Wright, C.W.; Nelson, J.M.; Burman, A.R. Evaluation of an experimental LiDAR for surveying a shallow, braided, sand-bedded river. *J. Hydraul. Eng.* **2007**, *133*, 838–842. [CrossRef]

214. Boeder, V.; Kersten, T.P.; Hesse, C.; Thies, T.; Sauer, A. Initial experience with the integration of a terrestrial laser scanner into the mobile hydrographic multi sensor system on a ship. In Proceedings of the ISPRS Workshop 2010 on Modeling of optical airborne and spaceborne Sensors, Istanbul, Turkey, 11–13 October 2010; Volume XXXVIII-1/W17, pp. 1–8.

215. Coveney, S.; Monteys, X. Integration potential of INFOMAR airborne LIDAR bathymetry with external onshore LIDAR data sets. *J. Coastal. Res.* **2011**, *62*, 19–29. [CrossRef]

216. Dix, M.; Abd-Elrahman, A.; Dewitt, B.; Nash, L., Jr. Accuracy evaluation of terrestrial LiDAR and multibeam sonar systems mounted on a survey vessel. *J. Surv. Eng.* **2011**, *138*, 203–213. [CrossRef]

217. Stubbing, D.; Smith, K. Surveying from a vessel using a multibeam echosounder and a terrestrial laser scanner in New Zeland. In Proceedings of the Australasian Coasts & Ports Conference 2015: 22nd Australasian Coastal and Ocean Engineering Conference and the 15th Australasian Port and Harbour Conference, Auckland, New Zealand, 15–18 September 2015; Engineers Australia and IPENZ: Auckland, New Zealand, 2015; pp. 860–865.

218. Eakins, B.W.; Taylor, L.A.; Carignan, K.S.; Kenny, M.R. Advances in coastal digital elevation models. *Eos* **2011**, *92*, 149–150. [CrossRef]

219. Eakins, B.W.; Grothe, P.R. Challenges in Building Coastal Digital Elevation Models. *J. Coast. Res.* **2014**, *297*, 942–953. [CrossRef]

220. Biscara, L.; Maspataud, A.; Schmitt, T. Generation of bathymetric digital elevation models along French coasts: Coastal risk assessment. *Hydro Int.* **2016**, *20*, 26–29.

221. Mitchell, P.J.; Aldridge, J.; Diesing, M. Legacy data: How decades of seabed sampling can produce robust predictions and versatile products. *Geosci.* **2019**, *9*, 182. [CrossRef]

222. Lurton, X. Seafloor-mapping sonar systems and Sub-bottom investigations. In *An Introduction to Underwater Acoustics: Principles and Applications*, 2nd ed.; Springer: Berlin, Germany, 2010; pp. 75–114.

223. Maa, K.; Xub, W.; Xu, J. The Comparison between Traditional and Interferometric Multibeam Systems. In Proceedings of the 6th International Conference on Sensor Network and Computer Engineering, Xi'an, China, 8–10 July 2016; Atlantis Press: Xi'an, China, 2016; pp. 261–264.

224. Brisson, L.; Hiller, T. Multiphase Echosounder to Improve Shallow-Water Surveys; Hybrid Approach to Produce Bathymetry and Side Scan Data. *Sea Technol.* **2015**, *56*, 10–14.

225. Massot-Campos, M.; Oliver-Codina, G. Optical Sensors and Methods for Underwater 3D Reconstruction. *Sensors* **2015**, *15*, 31525–31557. [CrossRef]

226. Harpold, A.A.; Marshall, J.A.; Lyon, S.W.; Barnhart, T.B.; Fisher, B.A.; Donovan, M.; Brubaker, K.M.; Crosby, C.J.; Glenn, N.F.; Glennie, C.I.; et al. Laser vision: Lidar as a transformative tool to advance critical zone science. *Hydrol. Earth Syst. Sci.* **2015**, *19*, 2881–2897. [CrossRef]

227. Menna, F.; Agrafiotis, P.; Georgopoulos, A. State of the art and applications in archaeological underwater 3D recording and mapping. *J. Cult. Herit.* **2018**, *33*, 231–248. [CrossRef]

228. Filisetti, A.; Marouchos, A.; Martini, A.; Martin, T.; Collings, S. Developments and applications of underwater LiDAR systems in support of marine science. In Proceedings of the OCEANS 2018 MTS/IEEE, Charleston, SC, USA, 22–25 October 2018; pp. 1–10.

229. Castillón, M.; Palomer, A.; Forest, J.; Ridao, P. State of the Art of Underwater Active Optical 3D Scanners. *Sensors* **2019**, *19*, 5161. [CrossRef]

230. Guenther, G.C.; Brooks, M.W.; LaRocque, P.E. New capabilities of the "SHOALS" airborne lidar bathymeter. *Remote Sens. Environ.* **2000**, *73*, 247–255. [CrossRef]

231. Bailly, J.S.; Le Coarer, Y.; Languille, P.; Stigermark, C.J.; Allouis, T. Geostatistical estimations of bathymetric LiDAR errors on rivers. *Earth Surf. Proc. Land* **2010**, *35*, 1199–1210. [CrossRef]

232. Lague, D.; Brodu, N.; Leroux, J. Accurate 3D comparison of complex topography with terrestrial laser scanner: Application to the Rangitikei canyon (NZ). *ISPRS J. Photogramm.* **2013**, *82*, 10–26. [CrossRef]

233. Fernandez-Diaz, J.C.; Carter, W.E.; Shrestha, R.L.; Leisz, S.J.; Fisher, C.T.; Gonzalez, A.M.; Thompson, D.; Elkins, S. Archaeological prospection of north Eastern Honduras with airborne mapping LiDAR. In *2014 IEEE Geoscience and Remote Sensing Symposium*; IEEE: Quebec City, QC, Canada, 2014; pp. 902–905.

234. Jagalingam, P.; Akshaya, B.J.; Hegde, A.V. Bathymetry mapping using Landsat 8 satellite imagery. *Procedia Eng.* **2015**, *116*, 560–566. [CrossRef]

235. Friedman, A.; Pizarmaaro, O.; Williams, S.B.; Johnson-Roberson, M. Multi-scale measures of rugosity, slope and aspect from benthic stereo image reconstructions. *PLoS ONE* **2012**, *7*, e50440. [CrossRef]
236. Lavest, J.M.; Rives, G.; Lapresté, J.T. Underwater camera calibration. In *European Conference on Computer Vision*; Vernon, D., Ed.; Springer: Berlin/Heidelberg, Germany, 2000; pp. 654–668.
237. Agrafiotis, P.; Skarlatos, D.; Georgopoulos, A.; Karantzalos, K. Shallow water bathymetry mapping from UAV imagery based on machine learning. *Int. Arch. Photogramm. Remote Sens. Spat. Inf. Sci.* **2019**, *XLII-2/W10*, 9–16. [CrossRef]
238. Fallati, L.; Saponari, L.; Savini, A.; Marchese, F.; Corselli, C.; Galli, P. Multi-Temporal UAV Data and Object-Based Image Analysis (OBIA) for Estimation of Substrate Changes in a Post-Bleaching Scenario on a Maldivian Reef. *Remote Sens.* **2020**, *12*, 2093. [CrossRef]
239. Okamoto, A. Wave influences in two-media photogrammetry. *Photogramm. Eng. Remote Sens.* **1982**, *48*, 1487–1499.
240. Fryer, J.G.; Kniest, H.T. Errors in depth determination caused by waves in through-water photogrammetry. *Photogramm. Rec.* **1985**, *11*, 745–753. [CrossRef]
241. Georgopoulos, A.; Agrafiotis, P. Documentation of a submerged monument using improved two media techniques. In Proceedings of the 2012 18th International Conference on Virtual Systems and Multimedia, and VSMM 2012 Virtual Systems in the Information Society, Milano, Italy, 2–5 September 2012; pp. 173–180.
242. Agrafiotis, P.; Georgopoulos, A. Camera constant in the case of two media photogrammetry. *Int. Arch. Photogramm.* **2015**, *40*, 1–6. [CrossRef]
243. Mayer, L.A. Frontiers in sea floor mapping and visualization. *Mar. Geophys. Res.* **2006**, *27*, 7–17. [CrossRef]

Article

Terrestrial and Marine Landforms along the Cilento Coastland (Southern Italy): A Framework for Landslide Hazard Assessment and Environmental Conservation

Domenico Guida [1] and Alessio Valente [2,*]

[1] Department of Civil Engineering, University of Salerno, 84084 Fisciano (SA), Italy; dguida@unisa.it
[2] Department of Sciences and Technologies, University of Sannio, 82100 Benevento, Italy
* Correspondence: valente@unisannio.it; Tel.: +39-0824-305-188

Received: 24 September 2019; Accepted: 30 November 2019; Published: 12 December 2019

Abstract: This study shows the terrestrial and marine landforms present along the Cilento coast in the southern part of the Campania region (Italy). This coast is characterized by the alternation of bays, small beaches, and rocky headlands. In the adjacent submerged areas, there is a slightly inclined platform that has a maximum width of 30 km to the north, while it narrows in the south to approximately 6 km. A wide variety of landforms are preserved in this area, despite the high erodibility of the rocks emerging from the sea and the effects of human activities (construction of structures and infrastructures, fires, etc.). Of these landforms, we focused on those that enabled us to determine Quaternary sea-level variations, and, more specifically, we focused on the correlation between coastal and sea-floor topography in order to trace the geomorphological evolution of this coastal area. For this purpose, the Licosa Cape and the promontory of Ripe Rosse located in northern Cilento were used as reference areas. Methods were used that enabled us to obtain a detailed digital cartography of each area and consequently to apply physical-based coastal evolution models. We believe that this approach would provide a better management of coastal risk mitigation which is likely to become increasingly important in the perspective of climate change.

Keywords: coastal geomorphology; submarine geomorphology; cliffs; sea-level changes; Cilento; southern Italy

1. Introduction

In relatively recent coastal landscapes, such as those of central–western Mediterranean Sea, the events responsible for the landform evolution and the controls they underwent must be sought within the last hundred thousand years. Furthermore, morphogenetic events continue to exert their effect and shape the landscape today, which is complicated by the actions of human beings who built facilities and infrastructures along the coasts to promote tourism or facilitate mobility [1]. These actions are often performed without analyzing landforms and processes carefully, thus causing instability or increasing environmental vulnerability and degradation [2].

This study aims to highlight the main emerged and submerged landforms present along the spectacular coastscape of the National Park of Cilento, in southern Italy. This coastal landscape, with lovely inland areas, received several international awards. In 1997, the entire region was recognized by UNESCO's (United Nations Educational, Scientific and Cultural Organization) Biosphere Reserve with the aim of maintaining a long-term equilibrium between man and his environment by conserving biological diversity, promoting economic development, and preserving cultural values (MAB – Man and Biosphere program), while, in 1998, three sites in the Cilento area (Paestum, Velia, and Padula) were included in the list of UNESCO world heritage sites in the category of "cultural landscapes" of global importance. Finally, in 2010, the area of the National Park was added to the Global Geopark Network of UNESCO, recognized for its rich geological heritage, numerous historical sites, and cultural traditions [3,4]. In this area, many previous studies enable us to reconstruct the morphological evolution of the coast and to determine the consequences of sea-level changes [5,6], where understanding the consequences can help to mitigate the risks affecting some specific sites (e.g., landslides, floods, storm surges, etc.) [7,8]. Furthermore, an estimate of future scenarios, which foresee a global sea-level rise due to global warming, could contribute to the achievement of sustainable planning and sustainable tourism development [9–11].

2. Study Area

The Cilento coastland extends over 100 km along a wide rectangular promontory between the Gulf of Salerno (northwest) and the Gulf of Policastro (southeast) on the southern Tyrrhenian margin of the Italian peninsula. (Figure 1).

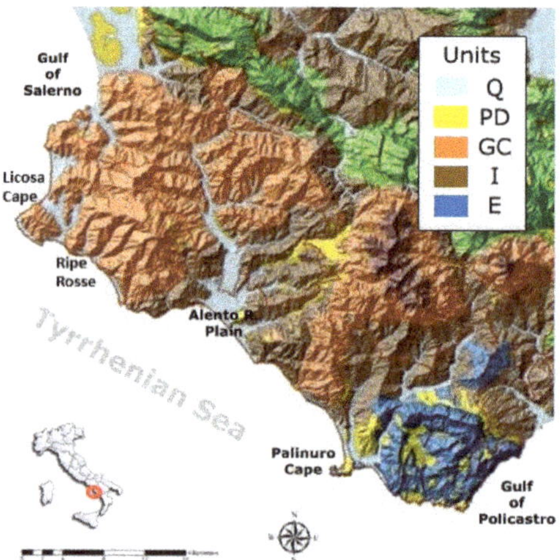

Figure 1. Geological map on DEM (Digital Elevation Model), from Campania Region Technical Cartography at the 1:5000 scale of the. Legend (only for the units shown in the text): Q—Quaternary post-orogenic units; PD—Pliocene deposits; GC—Middle Miocene syn-orogenic units (Cilento Group); I—Lower Tertiary internal units; E—Mesozoic–Lower Tertiary external unit.

Its complex morphology is characterized by mountain reliefs that reach the coast and by narrow floodplains. The causes of this complexity are attributed to the post-orogenic tectonics of the Apennine chain, which occurred from the Early Pliocene to the Middle Pleistocene through extensional faults, which disrupted this sector of the chain. It represents the southern sector of the "fold and thrust belt" formed in the central Tetide area from the late Cretaceous, due to the interaction between the European and African plates, the opening of the Tyrrhenian ocean basin, and the counter-clockwise rotation of the orogenic front [12]. This area has a long and complex lithogenetic history, with various tectono-sedimentary events and orogenic shifts [13], which today enable us to distinguish several lithostratigraphic units outcropping along the coast (Figure 1).

The inner units, comprising principally lower Tertiary deposits (Ligurian complex [14]), are mainly composed of marly and variegated clays, with sedimentary facies belonging to the ocean floor, which are transported upward to calcarenites and calcilutites, often with flint, and then with shales, sandstones, and rare conglomerates formed in a distal turbiditic environment [15]. In outcrops, they generally occupy the lower portion of the sequences, and, in many cases, they represent the lithotypes of the submerged coastal area. In this case, they are partially covered by veils of more recent sediments [16–18].

The external units are mainly composed of Mesozoic–Tertiary carbonates (Bulgheria unit—Middle Liassic to Lower Miocene [13,19]), representative of sedimentary environments ranging from shallow-water carbonates (often back-reef facies) to deep-water carbonates. The outcrops of these units are located on the high and rocky coastline of southern Cilento [18] (Figure 2). On the coastal bottoms, even partially emerged, these rocks are often covered with calcareous algae and animals with calcareous skeletons (sponges, corals, serpulids, bryozoans, mollusks) [17,18,20,21].

Figure 2. An aerial view of a coastal stretch in the calcareous–dolomite successions (External Units).

In disconformity on the previous units, Middle and Upper Miocene syn-orogenic units are present, whose successions are made up mainly of fine to extremely coarse pelitic and calcareous–marly arenaceous turbidites deposited in deep submarine fans (thrust top basin) [22]. Of these sequences, those of the Cilento group (Upper Burdigalian–Upper Tortonian) [23] are the most common along the coast, which are generally found on internal units. In submerged areas, these units are frequently covered by recent sands colonized by fossil organisms and sometimes by seagrass meadows [17,18]. The sand cover usually passes to muds away from the coastal bottoms.

The Quaternary post-orogenic units include all continental sediments, transitional sediments, and marine clastic sediments, deposited after the final emergence of the Apennine Chain, probably beginning in the Lower Pliocene [12,13]. In Cilento, they are represented by exposed aeolian, fluvial, slope, lake, and travertine deposits along the river valleys and on the plains near the coast, as well as by the marine transitional deposits stacked on the emerged and submerged coastal areas. These units may show intercalations of the products of Campania volcanic activity [24–27] (Figure 3).

Figure 3. A cross-section in the last Quaternary alluvial and colluvial deposits in southern Cilento.

The geological and tectonic setting mentioned above led to a prevalent morpho-structural control of the rocky coastal landscape of the Cilento area, sometimes resulting from the retreat and replacement of the previous fault-line scarp, alternated with small, elongated coastal plains (e.g., Alento River plain) [27,28]. These plains were formed by the deepening of the rivers favored by the correspondence with tectonic lineaments and the easy erodibility of the outcropping lithotypes.

The filling of these flared coastal valleys occurred due to over-flooding and marine ingressions. The traces of marine sediments uplifted to different altitudes from terraces, and the transitional sediments on the continental shelf show how sea level variations are superimposed on tectonic events. This is more easily seen along the southern coasts of Cilento composed of external carbonate units. In particular, the coastal profile of Mount Bulgheria shows ancient level surfaces (up to 400 m) with marine sediments from the Lower Pleistocene onward [29,30]. Calcareous cliff faces at sea level are often vertical [3,31].

The rest of the coast is composed of terrigenous deposits of internal and syn-orogenic units that gradually descend toward the sea through stratified escarpments or covered by debris, locally terraced, with generally concave profiles and sometimes composite with different slopes [3,31]. This diversity is due to the presence of marly–clayey levels or pelitic interlayers, which facilitate the occurrence of landslides in continuous evolution. In order to complete this brief geomorphological analysis, it is essential to mention the coastal slopes composed of clastic sediments, such as those represented by steep Pleistocene dune–beach systems of the Pleistocene. The oldest marine abrasion surfaces preserved in soft rock date back to the Upper Pleistocene [30] and are mainly found in the northern coastal section. Lastly, in order to complete the geomorphological scenario, accumulations of debris and sand tongues occurred at the base of the cliffs on the shoreline and close to micro-crags formed by terrigenous and clastic rocks. The former come from the dismantling of the adjacent slope, whereas the latter come from the coastal morphodynamics which transport sand and deposit it in the inlets [3,31].

The analysis of the submerged portion mainly concerns the continental shelf [16,17] (Figure 4) with a variable maximum width of 30 km in the north and a minimum width of 6 km in the south and an edge generally located at a depth of 200m except for the northern stretch of coast, where it is situated at a depth of approximately 230 m (Licosa Cape offshore), and the southeastern stretch of the coastline in the Gulf of Policastro, where it is located in shallower water (<100 m). The average slope varies from 0.3° in the northern sector to 0.8° in the southern sector, in correspondence with the narrowest portion. In this submerged portion, several marine abrasion terraces were identified, which were formed by the action of the sea waves during the Pleistocene paleo-standings of sea level with edges located at various depths [16,17]. Furthermore, in order to confirm that the major structural elements of the emerged part continue beneath the sea surface near the emerged valleys (e.g., Alento River Valley), depressed areas were identified, which are filled with sediments with varying grain size. In geophysical sub-bottom profiles, there is a series of normal and listric faults, oriented northwest to southeast [32]. The latter were caused by the collapse of the Tyrrhenian margin during the Pliocene and the Lower Pleistocene [12,16,17]. This type of fault probably defined the current coastal profile of the Cilento promontory which has the same orientation. At lower depths, sandy plains generally prevail in continuity with the emerged beaches and degrade toward the mudflats offshore. Locally, sands can also be found at greater depths; in this case, they represent ancient relict shorelines which were formed when the sea level was lower than it is today [16–18].

The climate on the Cilento coast is temperate with average annual temperatures of approximately 17 °C (12.6–20.8 °C) and an average annual rainfall that varies from 730 mm in the northern sector to 790 mm in the southern sector. Rainfall is concentrated in spring and late autumn, while, during the summer, there are long periods of drought. This climate is favorable for the development of evergreen forests and Mediterranean scrub along the coast. Of particular interest are the native spontaneous species that grow in the coastal areas, approximately 10% of which are of considerable phytogeographic importance, as they are endemic and/or rare [33,34]. On the beaches, among the sand communities, the increasingly rare sea lily (*Pancratium maritimum*) is still present; phytocoenoses with highly specialized halophytes live on the reefs in direct contact with sea spray and the endemic statice Salerno (*Limonium remotispiculum*) thrives (Figure 5a).

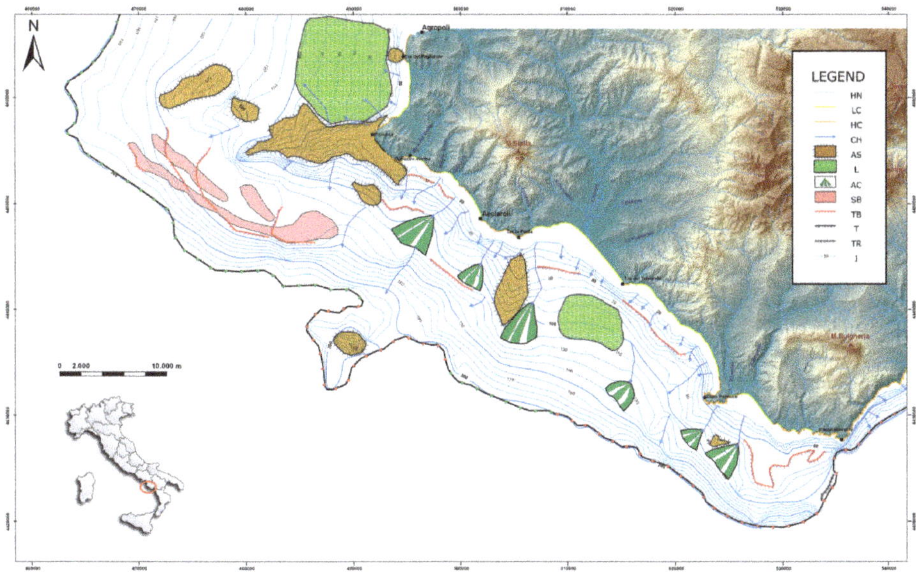

Figure 4. Geomorphological sketch map of Cilento coastal shelf (from [16]). Legend: HN: terrestrial hydrography; LC—low coast; HC—high coast; CH—channels incised in the sea bottom; AS—acoustic substrate rising from the sea bottom; L—depressed areas; AC—stack as ancient mouth complexes; SB—sandy bodies rising from the sea bottom; TB—edges of abrasion terraces; T—morpho-structural terrace; TR—trench; I—isobath.

(a)   (b)

Figure 5. (**a**) Vegetated dune behind Cefalo Beach (southern Cilento); in the background, an inactive calcareous cliff is shown; (**b**) the Palinuro Natural Arch (southern Cilento) with precious *Primula Palinuri* and other rupicolous species.

On the coastal cliffs, the Mediterranean rupicolous species are dotted with precious endemics such as the Primula di Palinuro (*Primula palinuri*), the clove of cliffs (*Dianthus rupicola*), the *Centaurea* (*Centaurea cineraria*), the iberide florida (*Iberis semperflorens*), the Neapolitan *Campanula* (*Campanula fragilis*), and many other flowering plant species that compose a coastal landscape of rare beauty (Figure 5b). In the sunniest and driest areas, we find the ginestra of Cilento (*Cilento genista*), the carob (*Ceratonia siliqua*), the red or Phoenician juniper (*Juniperus phoenicea*), and holm oak and pine woods (*Pinus halepensis*), which seem to be expanding again as they are being reforested.

On these last stretches of coastline, frequent fires and the fact that the roads were widened to reach the homes built on the slopes increased land degradation and reduced slope stability. However, there are still coastal stretches that preserve their original natural condition which are monitored closely by the Cilento, Vallo di Diano, and Alburni National Parks with the aim of mitigating damage and preventing deterioration [3,31]. More recently, the municipalities of Santa Maria di Castellabate in the north and the Costa degli Infreschi and Masseta in the south developed marine conservation and monitoring strategies. The reason for protecting and monitoring these marine areas is because of the richness of their seabeds, which contain biocenoses of great interest, such as pre-coralligenous and coralligenous species, as well as large quantities of *Posidonia* seagrass beds (*Posidonia oceanica*) [17,18].

3. Materials and Methods

Firstly, a review of the existing literature on the geology and geomorphology of Cilento was carried out. Most of these studies were focused on short stretches of coastline that offered particular cues as they were extremely didactic and representative for the development of research (i.e., References [20,25,27]). Previous coastal geomorphological studies did not integrate information on the dynamics and geomorphological evolution of the submerged sectors. This study attempts to fill these research gaps by trying to correlate the emerged and submerged landforms of the northern Cilento sites near Punta Licosa, using an integrated approach (Figure 1).

Integrated geomorphological surveys and analysis were carried out, starting from current terrestrial and submerged landforms. These latter surveys were carried out using sea vision underwater lighting on boats and by performing underwater scuba dives. The results of these surveys were supported by consulting topographic maps of the area. The oldest topographic maps used were the 1956 1:25,000 scale supplied by the IGMI (Istituto Geografico Militare Italiano) and the more recent 2004 1:5000 scale map supplied by the Campania Region. The information on these maps was completed by observing various aerial photogrammetric images obtained from 1943 onward produced by the IGMI, up to those taken in 2012 by the Campania Region. Images found on the web were also analyzed, particularly those taken by Google Earth in 2015 [35], as well as those placed on the National Cartographic Portal of the Italian Ministry of Environment in 2012 [36]. From these images, the LIDAR-derived DEM was extracted for some specific areas.

The new geological cartography created for this area enabled us to highlight the emerged and submerged landforms of the Cilento coast; more specifically, sheets 502 "Agropoli", 519 "Capo Palinuro", and 520 "Sapri" [37–39] represented the basis for defining the nature and genesis of the coastal forms. Subsequently, the availability of a map realized by ISPRA (National Institute for Environmental Protection) for the inclusion of the National Parks of Cilento, Vallo di Diano, and Alburni in the UNESCO Global Geoparks Network provided us with a broader view [18]. In fact, this map not only adds to the information obtained from the sheets mentioned above, but it focuses on some specific aspects, such as the characteristics of protected marine habitats.

The set of information gathered enabled us to highlight the emerged and submerged coastal forms of Cilento in more detail than the existing literature and to qualitatively reconstruct the short- and medium-term geomorphological evolution of various coastal landscapes, such as high cliffs. However, the need to make this information available to planners and administrators for future reference led the authors to develop innovative approaches. Therefore, a cartography was created using the Salerno University geomorphological mapping system (GmIS_UniSa) [40], which is based on a GIS procedure which includes "traditional based on symbol" cartography, as well as polygonal structures, with complete coverage, based on objects and multi-themes of the dataset and the set of rules. This study provides the physical features of simple landforms or composite physical surfaces, by defining elementary polygons or several adjacent polygons and then determining the processes that generated them. Moreover, it enables us to establish the geomorphological model by defining the relationships (geometric, temporal, physical, geological, lithological, and hierarchical) between the different landforms represented [40,41]. Unfortunately, due to our limited knowledge of the seabed, it is not yet possible to create a similar digital map. However, the better representation of the emerged forms emerged with the "object-oriented" cartography and their relationships with the submerged ones led to an improvement of the knowledge in space and time of this coast.

Subsequently, in particular traits, such as Licosa Cape and Ripe Rosse, based on a quantitative restitution of the forms and the role of the coastal processes that generated them in the past, particularly since the late Pleistocene, we tested a physically based numerical model of evolution using SCAPE software with its open-source components and tools [42]. With SCAPE, it was possible to trace the evolution of the basal part of a particularly high coast, where wave action is "almost exclusively" set to continue for the next 500 years. The shape of the coast used for this software was identified by a series of large-scale profiles (1:2000), collected from the same reference line, while, for the basal part of the representative profile modeled by SCAPE, a 1:500 scale was chosen. The execution of the model generated a series of output files with data on the profiles of the rocky cliffs and the beaches below, on the annual flow and transport rate of sediments, and then on the accumulated annual volume. This information, obtained using programs such as Excel and Matlab, enabled us to obtain a graphic representation of the data acquired with the SCAPE program. The results obtained will help us to understand the coastal processes that occur on a particular stretch of coast, and they allow intervention measures and preventing or reducing damage and risks to the environment.

4. Terrestrial and Marine Landforms of Cilento

A detailed description of the terrestrial and marine landforms of the Cilento coast would lose sight of the purpose of this study. In fact, we wish to give emphasis to landforms which are relevant to a better understanding of vulnerable landscapes and to promoting the conservation of the emerged and submerged geomorphological features of the study area [3,18,43]. Of the 100-km-long Cilento coastline, 70% is rocky while 30% includes sand or pebble beaches. Approximately 14 km of coast [44,45] was not considered in these percentages, as they are mainly occupied by anthropic activities [46], and are more concentrated in the port areas (e.g., Agropoli in the north, Casalvelino in Alento River Plain, Marina di Camerota in the south), even if a few were built to protect the eroded sections of the coastline.

The direct survey assisted by aerial photographs, as well as by digital observation systems (LIDAR) on particular stretches, allowed the correlation between the various coastal stretches characterized by rocky outcrops composed of both the calcareous sequences of the external units and the turbidite succession of the internal and syn-orogenic units. Each sequence illustrates a different morphological configuration for geological reasons (lithology and tectonics) and for the erosive–depositional phenomena that influence it [47]. In some cases, these phenomena can be attributed to sea-level changes that occurred during the last hundred thousand years [48]. A further differentiation concerns how these high, rocky coasts are related to the current submerged portion, as the geophysical surveys carried out on the seabed in the last decades detected [16,17], which may be sharp or gradual due to the presence of debris stacks. The combination of these conditions involves a particular morphological evolution that is correlated with each type of rocky coastline [31,47].

Along the Cilento coast, rocks with low erosion resistance (soft rocks) prevail, represented by the sequences in which sandstones and/or calcarenites intercalate at clay levels. These successions are attributable to internal units, and to syn-orogenic ones and post-orogenic deposits. In many cases, the emerged portion is connected to the submerged portion by a broad coastal platform (>200 m) and with a sea-bottom slope that can only exceed 10% locally (Type A in Reference [49]; Type A1 in Reference [47]). The profile is generally convex with an almost uniform gradient (on average 45°), although there may occasionally be concavities in the upper portion of the cliff or gradient differences (Figure 6a).

The evolution of this morphotype takes place due to the parallel retreat of cliffs, which is induced by wave motion that progressively erodes the base of the cliffs, thus causing the collapse of the unstable material of the slopes. Moreover, meteoric degradation occurs on these slopes, which can be decisive when the turbiditic succession presents a high argillaceous fraction. In this case, shortening is also joined to the cliff retreat [47]. Therefore, few landforms created by coastal processes are conserved on these cliffs; however, where wave motion is less forceful (e.g., on a broad, sub-horizontal coastal platform) and there is less degradation (e.g., fewer pelitic intercalations, less extension of the exposed surface), relatively more recent landforms can still be observed today [50–53].

(a)   **(b)**

Figure 6. Sea-cliff morphotypes: (**a**) convex slope (Type A in Reference [39]; (**b**) high coast with shore platform (Type B in Reference [39]).

More specifically, the latest interglacial sediments and landforms (OIS5) are still preserved on coastal stretches with these lithotypes [54]. For example, the age of the sites of Ogliastro Marina and Acciaroli in northern Cilento was determined by analyzing the extent of isoleucine epimerization in protein preserved in molluscan fossils embedded in raised marine deposits outcropping at 4 m (a.s.l.) [55]. They are sandy matrix conglomerates or fossiliferous biocalcarenites containing the fossilized remains of numerous marine species (*Glycimeris glycimeris*, *Astralium* (Bolma) *rugosum*, *Natica* sp., *Venus* sp., *Cardium* sp., *Tapes* sp., *Pecten* sp., *Spondylus gaederopus*, *Cladocora coespitosa*, etc.) without a precise stratigraphic meaning, but certainly indicative of a warm–moderate environment. However, there are rather wide 4–5-m marine abrasion platforms in the northern sectors with slightly cemented sand dunes, which also lie below sea level. These platforms are covered with red or sometimes brown colluviums that may contain the pyroclastic deposits attributed to the Campanian Ignimbrite (39 ka before present (B.P.) [56]). Moreover, at approximately +2 m, a "beach rock" can still be seen in easily erodible soft rocks that could be evidence of one of the last sea transgressions in Late Pleistocene times.

This "2-m bench" reaches a maximum width of approximately 35 m in a few stretches of coastline. It remains uncovered by the sea, yet it is overwashed by storm waves at high tide. It is an almost horizontal platform similar to that described in front of a cliff by Sunamura for Type B [49] (Figure 6b). Its position on the coast north of the promontory of Cape Palinuro means that it was less exposed to the most intense storm surges coming from southeast, as suggested by Reference [2] in similar contexts. However, its presence in other areas (cliffs north of Alento River alluvial plain), even if narrower, shows that they can also be in areas where they are exposed to strong storms. Pools and channels on the platform surface become enlarged and integrated as their protruding edges recede. Cliff recession occurs due to shore platform lowering and flattening, weathering processes, and the removal of weathered material by wave action [50].

On coastal slopes modeled in sequence with lithotype alternations (e.g., turbidite succession), there are widespread landslide phenomena and relative landforms are clearly detectable [57–60]. More than 220 different types of landslides were surveyed by various authorities [61]. Some landslides were caused directly or indirectly by wave action, while others were caused by lithological conditions (e.g., fractured rocks, layering, poorly consolidated sediments) or meteoric degradation (rainfall). The results of the survey show that rotational slides are the most common type of landslide, even though many of these are inactive; falls and complex landslides, such as slide-flows, are also very widespread.

The presence of debris at the base of the cliffs can modify their evolution or accelerate the formation of beaches in coves or bays toward the direction of the current along the coast.

In the southern coastal area of Mount Bulgheria, the cliffs are composed of extremely erosion-resistant limestone (hard rock) (Figure 7a). In many cases, these rocks lie below sea level, as they correspond to structural slopes. The profile is generally vertical or sub-vertical; thus, the action of the waves is drastically reduced. In fact, the depth of the sea at the base of the cliff is greater than the depth of the breakers [47]. Therefore, on these cliffs, defined by Reference [49] as plunging cliffs, subaerial processes can prevail. The most common of these processes is represented by rock falls, generally in correspondence with structural weaknesses [62–65]. Locally, erosional remnants are left on the seabed following cliff retreat, so that the seabed appears articulated, with small terraces, arches, and rocks emerging from sea, as observed on this coastal stretch. However, the retreat rate is lower than the previous morphotype, which allows for the conservation of a great variety of coastal landforms. In particular, at the base of the limestone cliffs, there are tidal notches or fossil biocorrosion grooves, often associated with holes bored by lithophagous species. Caves and hypogean karst cavities formed during the neotectonic period, which developed along the main fractured lines or occasionally along interstatal discontinuities, are almost always remodeled by wave erosion or marine biocorrosion and partially or totally filled with marine and continental sediments [3,29,66] (Figure 7b).

(a)   (b)

Figure 7. (**a**) Sea-cliff morphotypes: high vertical coast (plunging cliff); (**b**) example of caves on limestone cliffs filled with sediments in the south of Cilento (Masseta location).

Marine sediments are generally conglomerates with a coarse, medium-cemented sandy matrix, known as "Panchina", mixed with bioclasts of gastropods and mollusks or coral fragments. They are usually associated with restricted and slightly sloping marine abrasion platforms. In other cases, they are represented by cemented biocalcarenites, such as "beach-rock", coral reefs, or "trottoir", such as "reefs", which are often composed of *Cladocora coespitosa* [20,21].

Continental sediments are almost always associated with low sea-level stands, which are essentially accumulations of pseudo-stratified breccias mainly composed of calcareous elements with sharp or blunt edges, in abundant reddish, colluvial, or pyroclastic matrix. Pre-Tyrrhenian breccias are often well cemented, poor or without a reddish matrix. In other cases, continental deposits are composed of reddish sands of colluvial or wind origin. There are occasionally karst speleothemes and concretionary accumulations in situ. The presence of pyroclastites (fine ash) is of particular importance as they are excellent chronostratigraphic markers [20,29]. Brown and immature soils settle on both breccias and colluvial deposits [20,29].

Unlike soft rocks, it is quite common to observe a series of landforms created during the oldest paleo sea-level stand on limestone and dolomite in southern Cilento. In fact, five marine terraces are located in this sector between 170/180 m and 40/50 m, and at lower altitudes such as +8/8.5 m, 3.5/5 m, and 2 m [20,21] (Figure 8).

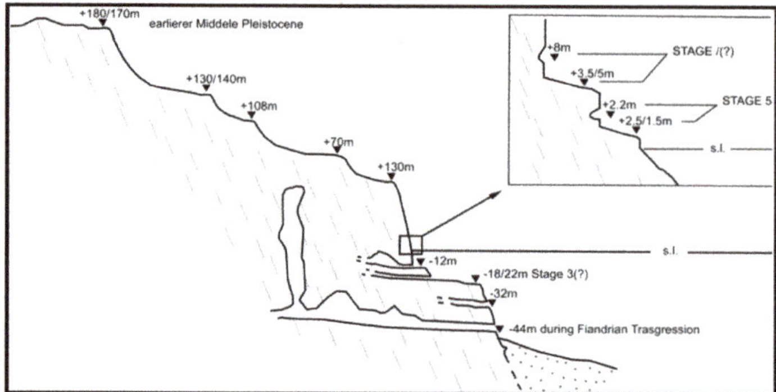

Figure 8. Schematic cross-section summarizing the evidence of paleo sea-level stands recognized in the subaerial and submarine sectors of Palinuro Cape [20].

The highest of the heights, which occasionally can present marine deposits, are attributable to the Middle Pleistocene for physical continuity with similar forms [30]. On the other hand, lower wave-cut terraces, represented by sea-notches and bioconstruction, are correlated to OIS5 [29,30]. The differences in position derive from the tectonic uplift this relief underwent during the Pleistocene [29,54]. According to the estimates carried out on the Middle Pleistocene marine terraces, the uplift rate should have reached 0.2 mm/year during the last 700 ka B.P. period [20,54], although the uplift rate may be significantly lower considering the traces of the Upper Pleistocene sections. In the submerged portion, evidence of several paleo-sea level stands were found, which are mainly represented by wave-cut terraces and sea-notches outcropping along the underwater cliff, and occasionally by marine conglomerates with *Lithophaga* burrows, which can be divided into four main groups located at depths of −44/46, −18/24, −12/14, and −7/8 m below sea level (Figure 8).

Particular morphotypes observed along the Cilento coastline are known as "slope-over-wall cliffs", which are generally composed of soft rocks [2] and have vegetated slopes (typically with a gradient of 20°–30° but locally up to 45°) that descend down a sub-vertical rocky cliff face to the sea. The upper part of slope may have an almost uniform gradient (especially where it follows stratification by immersion toward the sea, cleavage, joint or fault planes), but, more often, it is convex in shape like a hog's back and, occasionally, it can be concave, where the lower slopes of the deposit that covers it are preserved. Their genesis is generally attributable to the rise in sea levels during the Holocene after the last glaciation period. Because of glacio-eustatic sea-level changes, this morphotype underwent alterations due to wave actions during interglacial periods and sub-aerial (therefore, not marine) modifications during glacial periods (Figure 9).

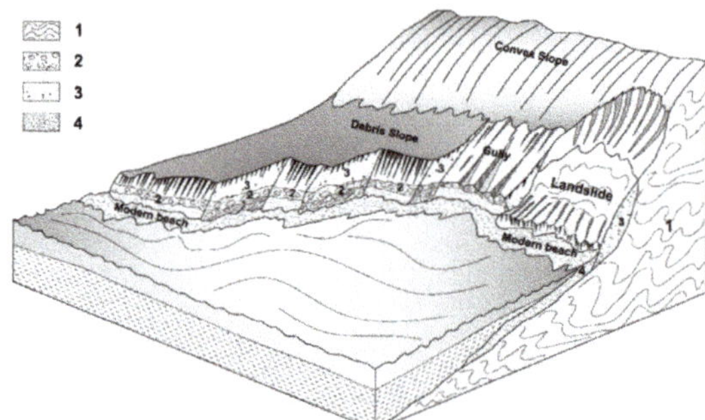

Figure 9. Evolution of "slope-over-wall" profile in a schematic cross-section. Legend: 1. deformed substratum; 2. solifluction deposits; 3. debris slope deposits; 4. modern beach sediments.

It can, therefore, be deduced that climate change played a fundamental role in their evolution, thus determining a climate-induced of erosion alternation of erosive conditions, sometimes attributed to sea processes and sometimes land processes, which occurred in various ways [48,53,67]. This is particularly evident on the stretch of coastline called Ripe Rosse in northern Cilento, and on the coastal stretch called Marina di Pisciotta in the south, adjacent to Palinuro Cape (Figure 1).

Even if the south of Cilento is not well known, due to its morphological conditions, there are some lovely beaches, which are popular seaside destinations. They are mainly situated at the mouths of incisions in valleys or in small bays. Long beaches can only be found in Santa Maria di Castellabate, between Casalvelino and Marina di Ascea, to the north and south of Palinuro [44,45] (Figure 10a). Only the "central" stretch develops in the small coastal alluvial plain crossed by the Alento River (Figure 10b). This river lies in a Pleistocene morphotectonic depression that lowers the succession of the Cilento group toward the sea [27].

(a)   (b)

Figure 10. (**a**) Cala del Cefalo beach at south of Palinuro Cape, including wide dune with endemic species; (**b**) Alento River Coastal Plain: in the foreground, a stretch of the Casalvelino–Ascea marine coast with the port of Casalvelino and a series of defense works parallel to the coast; in the background, a stretch of low coast (beach–dune system).

The plain is dominated by a large sub-horizontal surface of a terrace composed of fluvial sediments of various grain sizes, and it contains fragments of building bricks, which partly cover the ancient Magna Graecia remains of the port of the city of Elea. This city was the seat of the famous philosophical school where Zenone and Parmenides settled, which was later seized by the Romans and given the name of Velia. The archaeological excavations carried out there today are an important tourist and cultural attraction. The aforementioned ancient marine terrace is responsible for the retreat of the coast over 500 m to the west. The area became a marshy area following the silting of the Greek port in the first century anno Domini (A.D.) and was definitively abandoned in the ninth century. Today, the Alento river and its tributaries are engraved on the terrace for 1–2 m. It is difficult to link this coastal variation to historic variations in sea-level rise; the alluvial progression appears to be related to climate changes that may have generated greater sedimentary deposits during the High Middle Ages and caused greater slope degradation [68]. In fact, the slopes that dominate the terrace are covered with a thick eluvio-colluvial cover composed of reddish clays and silt that form part of the foundations of the ancient Greek city and contain archaeological remains. The pedogenized deposits of the dune to the west of Velia lie on the terraced deposits and the historical colluvial sediments [69]. The submerged area nearest the emerged area is characterized by a gently sloping sandy bottom covered with current ripple marks formed by waves except for a few stretches [18]. It is generally colonized by fossil organisms (e.g., *Donax* spp., *Chamelea gallina*, *Callista chione*) that are able to resist wave and current action. Offshore, beyond 20/25 m, the muddy fraction contains fossil organisms such as mollusks of the Veneridi family, worms, and crustaceans. Large areas of the sandy plain are covered with meadows of phanerogams (*Posidonia oceanica, Cymodocea nodosa*), while the muddy coastal plain is characterized by "fields" of soft corals (Pennatulacei), particularly in the area in front of the Alento estuary [18].

On the sea bottom near the high coast characterized by a "slope-over-wall profile", small banks of gravel and coarse organic sand can be seen. These are low-relief seabeds with almost horizontal surfaces, due to erosion caused by low-sea-level stands following the Upper Pleistocene [18]. They are adjacent to the emerged part of the northernmost stretch of the study area in front of Mount Tresino and between Acciaroli and Pioppi, while it is more detached at a depth of 5 m in front of Ripe Rosse and Marina di Pisciotta (close to Palinuro Cape).

The submerged landscape which lies in front of Licosa Cape is of particular interest. Along this stretch, the continental shelf reaches a maximum length of approximately 23 km with a border that slopes gradually down to the ocean floor. In the profile, various edges of sub-horizontal surfaces modeled by wave action were recognized up to 150 m. According to Reference [16], the progradation of this platform toward the sea occurred until the last glacial expansion (18 ka B.P.), while the sub-flat surfaces were formed during the last sea level rise. In fact, the acoustic profiles, surveyed in this area, show a truncation of the prograding bodies near an erosion surface, covered by a thin drape of Holocene sediments. In order to confirm the sedimentary characteristics of these prograding bodies, a core sample was collected from the deepest part of the shelf at −149 m. At approximately 73 cm from the bottom of the core sample, there are coarse sands containing numerous whole or fragmented mollusk shells, including *Arctica islandica*, a cold-water species of the Pleistocene [16], which survived in the Mediterranean until the end of the Würm.

The channels identified on the continental shelf by the geophysical analysis were probably formed during the same period, near rivers and streams [17]. They represent the relict forms of a hydrographic network of subaerial origin when the sea retreated to the isobath of 110–120 m, while the sediments that cover them date back to the subsequent sea-level rise [70]. Therefore, these channels would have been formed when the continental shelf emerged from the sea during the last glaciation (18 ka B.P.) [16]. Some of these channels also show sedimentary bodies in their termini located at approximately −90 m, which can be interpreted as mouth bar complexes.

Finally, it should be noted that, according to References [16,17], the continental shelf has three terraces located at depths of 54 m, 86 m, and 107 m, modeled on the rocky bottom (acoustic substrate). Such a bottom has a limited extension and cannot be easily followed. Other terraces with irregular surface morphology were recognized by Reference [17], and depressed areas full of different size sediment grains were identified during the last study (e.g., north of Licosa Cape and in front of the Alento River mouth), which may be due to distensive tectonic lineaments activated during the Pliocene and Pleistocene. Among these, those that border the Alento River Plain continue in front of the seabed [17].

5. Future Scenarios

Understanding the evolution of a coastline is important for its conservation and enhancement. Firstly, we focused on a site in the Cilento that protects landforms from sea-level changes both on land and on the seabed, where a series of geomorphological processes took place from the Pleistocene epoch until today. Secondly, we evaluated the risks induced by geomorphic processes that occurred over time on a coastal cliff and how the knowledge acquired could be used for developing mitigation strategies. These interventions lower the degree of vulnerability and, consequently, the risk of losing structures and infrastructures present on the coastal landscape. These considerations also take into account future climate predictions [71–73], which indicate an increase in temperatures, which would significantly increase sea levels. According to the Intergovernmental Panel on Climate Change (IPCC) hypothesis, the sea level could rise by more than a meter by 2100 if there is a global temperature increase of 1.5 °C [74], which would affect coastal processes and seriously change the Cilento coastscape. This landscape attracts tourists from all over the world and helps the economy to thrive. For this reason, we try to predict what will happen in the future by reconstructing the geomorphological evolution of the coastal landforms [75,76].

Licosa Cape promontory is a site of Cilento that needs to be protected and enhanced (Figure 11a). In fact, both the emerged and the submerged areas are recognized as priority areas for protection. In fact, they are included in the National Park of Cilento, Vallo di Diano, and Alburni and in the marine protected area of Santa Maria di Castellabate. It represents a high morphological, northwest (NW)–southeast (SE)-oriented area characterized by rounded ridges with regular, moderately steep slopes, or less frequently with concave–convex profiles; transversely, the slope shows triangular-shaped facets. The promontory consists entirely of the basal turbiditic succession of the Cilento Group (Pollica formation: Upper Burdigalian–Langhian [23]. This arenaceous–pelitic succession, composed of thin/thick tabular layers, emerges along the outer edge of the promontory, and its height varies from 4 to 10 m (Figure 11b). On the eastern edge, the slope is joined by thick and polycyclic colluvial taluses (Late Pleistocene) and alluvial fans (Middle Pleistocene) [37,77]. The former are mainly composed of angular arenaceous clasts in a yellow to yellowish-brown and reddish-brown matrix that varies from loamy sand to sandy loam, while the latter have sub-rounded clasts and positive or inverse gradation. Both taluses and fans are dissected by minor canals and incisions, generally V-shaped, which are filled with alluvial deposits.

(a)   (b)

Figure 11. Licosa Cape: (**a**) in the foreground, the islet of Licosa in front of the sub-horizontal promontory; (**b**) the profile of the marine abrasion terrace at +4 m of the Upper Pleistocene (in the background) cut in the turbidite succession of the Cilento group.

Along the north and southwest flanks of the Monte Licosa ridge, the basal debris deposits gradually adapt to the terraced surfaces of the promontory, close to several marine abrasion platforms [25]. The highest and largest terrace (20–25 m a.s.l.) with a surface area of 500 m^2 in the southwestern area, probably gives the promontory of Licosa its almost quadrangular shape. In addition to this terrace, there are three other orders of marine terraces suspended at different heights above sea level with relative organogenic and pyroclastic deposits (Table 1).

These characteristics indicate that the terraces were formed between the Middle and Late Pleistocene and, therefore, demonstrate the exact sequence of eustatic events and tectonic movements. More precisely, the chronological reconstruction of the terraces was based on (i) epimerization of isoleucine and U/Th dating methods on biogenic samples [78,79]; (ii) presence of Paleolithic pre-Mousterian industries [25,80]; (iii) presence of a pyroclastic marker layer, widespread along the southern Tyrrhenian coastal areas, which dates back to the OIS 3–2 transition [26,81]; and (iv) stratigraphic correlations on a regional scale [78,82].

Table 1. Synthetic table of morphological indicators of paleo sea levels at Licosa promontory; a.s.l.—above sea level.

Measured Heights (m a.s.l.)	Geomorphological Markers/Other Indicators	Age/Chronostratigraphy	References
20–25 m	Wave-cut terrace/tephra layer + lithic industry	No dating/reliable a correlation to OIS7 (for stratigraphic position, tephra surveyed and characters of the stony artifacts)	[25,26,80]
8–10.5 m	Wave-cut terrace/tephra layer	110 ka (^{230}Th/^{234}U)/correlated to OIS5e	[25,26,78]
3–5 m	Wave-cut terrace/tephra layer	102 ± 4 ka (U series dating)/correlated to OIS5c	[25,26,79]
1.5–2 m	Wave-cut terrace/tephra layer	25.3 ± 0.3 Ka/correlated to OIS5a	[25,26]

A full-coverage object-oriented mapping was performed in order to provide a complete representation of the promontory of Licosa Cape (features and evolution processes) at different scales. All the surface features identified by field surveys and aero-photogrammetry analysis were automatically identified, hierarchically organized, and mapped using the GmIS_UniSa procedure [40] (Figure 12). Special attention was given to the objective recognition, classification, and mapping of present-day land forming processes (incised channels, rock cliffs, and shallow landslides) superimposed on stadial Pleistocene landforms (terraced alluvial fans, marine terraces, talus slopes, colluvial hollows in headwaters) (Figure 12).

Figure 12. Geomorphological map of Licosa Cape made with the GmIS_UniSa procedure [40].

This was not the case for the submerged area in front of Cape Licosa, for which the submerged landscape map was developed by Reference [18] for the Cilento National Park, Vallo di Diano, and Alburni (Figure 13). As previously mentioned, the submerged landscape is extremely interesting for the topographic features that are visible on the sea floor and for those that can be highlighted by the acoustic profiles that were realized in the area. In particular, close to the shoreline, the rocky bottom corresponds with the sea floor except for a light veil of silty/sandy sediments covered by hydroids and stooling silicones. This rocky bottom emerges at a short distance forming a little island with an almost flat-topped summit. Offshore, at the depth of about 25 m, there is another sub-horizontal surface that slopes gently seaward, which is composed of organogenic sands and gravels produced by the fragmentation of coralligenous bioconstructions. According to the survey carried out on this terrace, the surface is characterized by sediment waves (megaripples). Also, in this case, its shape reveals phenomena that occurred when the sea level was lower during the upper Pleistocene–Holocene period or during the sea-level rise after the last glaciation as suggested by various authors [16,17,77].

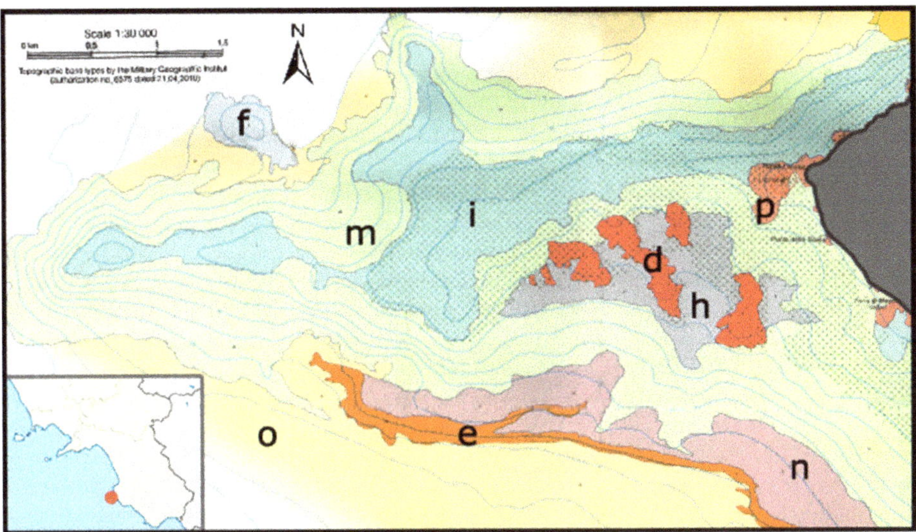

Figure 13. Submerged landscape map of Licosa Cape (extracted from Reference [17]). Legend: d—spur with coralligenous bioconstruction; e—rocky bedding planes covered by bioturbated mud; f—bank with coarse organogenic cover; h—depositional terrace composed of organogenic sand and gravel; i—wave-cut terrace with mixed organogenic cover; m—slope with mixed organogenic sediments; n—deep terrace with muddy bioclastic cover; o—shelf muddy plain; p—rock; the dotted cover indicates the phanerogam plants.

Other depositional bodies are found at greater depths and run parallel to the edge of the platform. They are characterized by a type of echo with an indistinct background without reflections in the substrate [16], which indicates the presence of more reflective sandy deposits than pelitic deposits. The upper part is sometimes covered with a thin layer of Holocene sediments. According to Reference [16], the characteristics of the sandy deposits are attributable to a beach environment when the sea was at its lowest level (18 ka B.P.), which are useful for carrying out beach nourishment interventions along the coasts.

The emerged and submerged landforms detected close to the Licosa promontory suggest that the polyphasic and polycyclic evolution that occurred during the Quaternary was affected by climatic variations. In the emerged portion, the debris deposits at the foot of Mount Licosa could be due to the cold phases of the Upper Pleistocene, when there was little or no forest cover and the land was covered with semiarid vegetation such as grasses and shrubs [83,84]. These phases favored the fragmentation of the rocks (cryoclastic processes due to freezing/thawing cycles) when large amounts of debris were produced on the slopes. At that time, sea levels were low, and the emerged area reached its maximum extension, as confirmed by the acoustic recordings and the beach sediments found in the previously mentioned core sample. Moreover, during the interglacial or interstadial–stadial periods, the relatively finer parts of the upper and steeper parts of the slopes were removed by different transport processes (gravity and/or water) [82,85]. This material, which was distributed on the wide coastal plains during the coldest periods, accumulated close to the coast in the warmest periods. On the basis of these characteristics, at least two generations of debris deposits were identified, including the oldest glacial/interglacial stages (OIS9(?) 7–6) and the last interglacial/glacial cycles (OIS5 to 4–2). Under these conditions, with the rising sea level, semi-submerged terraced surfaces with relative deposits were modeled, similar to those identified in this site.

The highest terrace may have reached its present position between 20 and 25 m due to a tectonic uplift, which probably occurred in the Upper Pleistocene. According to Reference [77], it was modeled in the Middle Pleistocene (OIS7), corresponding to a time range between 245,000 and 190,000 years before the present. However, the overlapping of fossil-rich sandstone deposits associated with this terrace, on dark-red deposits belonging to continental dunes, could make the older traces recede to a previous colder stage (OIS8). The terraces at lower altitudes are not easily recognizable, except for those that can be observed at approximately 4 m along the whole promontory. This may be due to the worsening of erosion phenomena along their escarpments which occurred during warm periods. With regard to the best represented terrace, organogenic deposits are associated with thicknesses of approximately 50 cm with a pyroclastic level. Using the data obtained from these elements, it was possible to trace the formation of these terraces back to the Upper Pleistocene (OIS5c [79]). The Licosa Cape promontory is currently covered by typical Mediterranean woodlands even if they appear to be quite degraded [34], which is due to repeated deforestation carried out until the middle of the 20th century for agricultural purposes. More recently, the innermost area was used for grazing animals, while the area nearest the coastline was placed under protection. In fact, these areas were left to a slow and spontaneous re-naturalization. The man-induced degradation of the landscape probably increased the geomorphic instability of some escarpments, especially in the piedmont area. Moreover, a hypothetical sea level rise could accelerate erosion and reshape this landscape, as occurred in the past. In a future scenario, the emerged and submerged landforms described will be less visible. However, the documentation for the valorization of the site may prove useful for promoting the geomorphological evolution of this stretch of known coast.

The other stretch of coastline analyzed in detail was Ripe Rosse, which lies to the south of Licosa Cape (Figure 14). It shows how a better understanding of the coastal geomorphological evolution can be useful for mitigation and protection. On this high rocky stretch of coastline, the risk of landslide increased significantly, which may be due to the anthropogenic changes of the upper slope caused by the construction of an important road for tourism facilities and commercial activities in the Cilento and by a particular geomorphological evolution, as occurred on other coastal stretches of Cilento.

Figure 14. An example of "slope-over-wall" profile at Ripe Rosse in the northern Cilento; note that plants on the detritus cover the slope and the gravel/pebbly beach at the foot of the cliff; along the cliff, thin and fine turbidite outcrops can be seen.

Ripe Rosse is reachable from the beaches that surround it. There is a rather narrow strip (2 to 4 m wide) where debris of all sizes accumulated, which indicates the numerous rockfalls that make up the cliff. It is an outcrop with a large turbidite succession greater than 150 m thick, belonging to basal formation of the Cilento group (Pollica formation: Upper Burdigalian–Langhian [23]. In particular, this succession is composed of coarse-grained and medium-grained sandstone layers, generally with clear bases, which pass upward to finer sand, silt, and mud. The sandstone layers are sometimes replaced by conglomerates with erosive and concave bases. Laterally adjacent to these coarser deposits, there are finer grain sizes and thinner turbidite layers and chaotic intervals interlayered with these turbidites, interpreted as submarine landslides, in a basin floor fan [86,87]. The plants (e.g., *Genista cilentina, Ceratonia siliqua, Juniperus phoenicea, Pinus halepensis*) that cover the slope belong to the Mediterranean scrub, whereas, in the adjacent submerged areas, seagrass meadows (*Posidonia oceanica, Cymodocea nodosa*) are widely diffused. This coast represents a key biodiversity asset, as it performs important ecological functions that are highly considered by the UNESCO Man and Biosphere program since 1997 [4].

As previously mentioned, "Ripe Rosse" has a slope-over-wall profile, which is composed of a convex upper part and a vertical lower part, which was formed by the sea-level rise following the last glaciation [31]. This is confirmed by the presence of a small wave-cut platform covered with coarse organogenic sands and gravels at depths of 5 m to 25 m from the sea bottom in front of the cliff, which was formed during the last sea-level drop [17,18]. Moreover, as revealed by Reference [88], the gradient of the shallowest part of the coastal shelf is very low and has an irregular topography with small scarps and other positive morphologies up to the terraces at −43 m, which does not allow precise sea-level estimation. Therefore, it is believed that this was due to the climatic oscillation occurred in the last Pleistocene and Holocene and, consequently, the processes influenced by it, which influenced its geomorphological evolution and led to the current condition of the cliff. Moreover, this evolution could also be decisive in the future, when a sea-level rise is expected.

To this aim, a SCAPE numerical model was used, which gave promising results on coastal risks and mitigation methods. This model was preceded by geomorphological analysis including field surveys, elaboration of maps (1956 and 2004) and aerial photos (1943, 2012, and 2015). The multi-temporal processing was completed using available multi-temporal Google Earth (GE) images from 2015 [35] and images placed on the National Cartographic Portal by the Italian Ministry of the Environment [36]. The DEM obtained by LIDAR was extracted from this website.

This detailed analysis enabled us to determine the geomorphological features of "Ripe Rosse", as well as to reconstruct its short- and medium-term geomorphological evolution, as we obtained information on the processes that occurred in the past, especially since the late Pleistocene. The spatio-temporal information of the area was obtained and digitalized on a geomorphological map using the GmIS_UniSa procedure [40], which proved useful for the numerical model but is not reported in this paper. The model, which was calculated on several profiles of the coastline, includes their geometrical features, input data, as well as files, describing wave conditions, tidal levels, average sea level, annual sediment flow, sediment transport, and accumulated annual volume.

The profile of "Ripe Rosse" mainly consists of an upper portion with a moderate slope (mean 40°) that descends toward a vertical basal cliff (mean 80°) into the sea with a slightly inclined coastal platform up to 200 m in width. More specifically, Ripe Rosse has a convex, colluvial, debris flow slope on the remnants of a buried, uplifted marine platform, covered with rounded, gravelly marine deposits, hanging onto the cliffed bedrock slope. The original, longer convex–concave profile was connected to a lower sea level during the last glacial age. The cliffed slope was progressively modeled by a slope retreat mechanism due to the post-glacial sea level rising until the present day. A threshold behavior of the entire coastal slope profile, with a general gravitational collapse, was identified after the complete disruption of the buried marine platform [67].

Such peculiar profiles could be formed on coasts where cliffs of relatively resistant rock are degraded by periglacial freeze-and-thaw processes resulting in solifluction, and they form coastal slopes that are then undercut by marine erosion. This process is still active on high-latitude coasts, but it was more widespread during Pleistocene times, when coasts that are now temperate were subjected to the down-slope movement of frost-shattered rubble during cold phases with low sea levels. The Pleistocene cliffs became slopes covered with solifluction deposits composed of angular gravel. This slope apron deposit extended out onto what is now the sea floor. Late in Pleistocene times, the climate became milder and vegetation grew on these coastal slopes. The sea level rose, and the slopes were undercut by erosion.

This evolution was simulated to predict future climatic conditions, since climate tropicalization will be the most popular topic for the next few hundred years. Starting from its current state and bearing the sea-level rise in mind, the effect of the marine processes at the foot of the cliff was reconstructed (Figure 15).

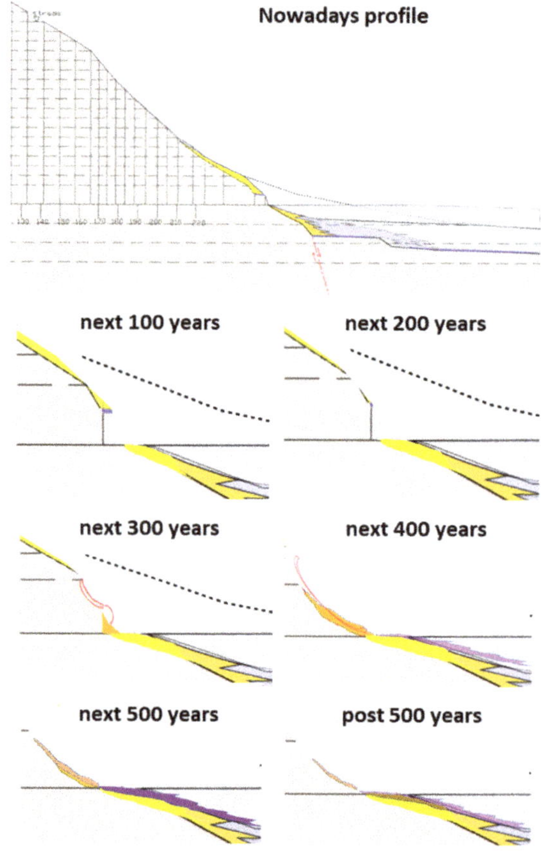

Figure 15. Qualitative reconstruction (step by step) of the geomorphological evolution for the next 500 years. The dotted line indicates the topographic surface at −15,000 years from the present with the sea level at −130 m from the current position. In yellow, the detrital material covering the slope and then deposited at the base of the cliff is shown; in orange, the material dismantled from the wall is shown, which determines the general collapse of the cliff, once the threshold is exceeded.

The removal of the material collapsed from the slope and the formation of a large platform of coastal erosion was also considered. The formation of a vast beach at the foot of the cliff, made of sediments transported from the adjacent coast in erosion or by piles of rocks that fell from the slope above, is the result of erosion processes which change the profile of the cliff. The morphological expression of this change in the coastal platform is represented by the increase in its gradient and the decrease in its height, which accelerates the recession rate. The simulation realized with SCAPE software was tested for 500 years starting from the current conditions and considering the hypothesis of a sea-level rise of 1 mm/year on a 10-m-high cliff. The result was a 140-m cliff retreat, represented graphically by the Excel and Matlab programs along the modeled section and in a representative profile (Figure 16).

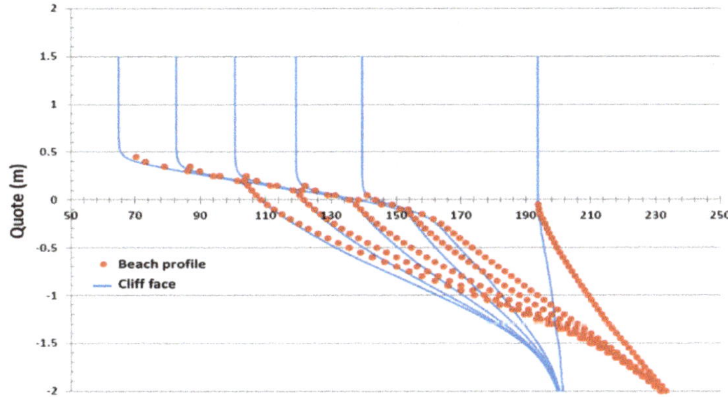

Figure 16. Parallel retreatment of the "wall" and landward shifting of the beach profile in the next 500 years simulated by SCAPE software.

The simulation showed clearly that the vertical basal part of the coastal slope recedes parallel to itself with uniform denudation intensity if the slope processes are constant and/or the rock resistances are uniform. It is important to note that the recession is facilitated by the continuous removal of debris from the base of the slope and the formation of a partially submerged accumulation.

Unfortunately, it is not yet possible to simulate the entire slope above the wall; however, the progressive retreat of the wall should intercept the threshold of the slope portion with the detrital material, which would accelerate the evolution of the entire coastal slope as confirmed by the geomorphological reconstruction. If this were to happen in hot and humid climatic conditions or under high anthropogenic pressure (slope cuts and wildfires), there would be an emphasis on the subaerial processes extended to the entire slope with a consequent evolution of the "substitution" of the slope shape. This evolution could lead to the consumption of the top portion and shorten the coastal slope, which would increase the risks to which the road would be subjected, which is the only road leading to the coastal resorts located southward.

The coastal slope of Marina di Pisciotta, northward to Palinuro Cape, is in a similar situation. In this case, a slow-moving landslide [89] occurred on the cliff escarpment, which affected both the roads and the railway line that connect northern and southern Italy (Salerno–Reggio Calabria line). Also, in this case, it is a high coast with a "slope-over-wall" profile. Gaining knowledge of the geomorphological evolution of this type of coast would enable us to implement appropriate risk mitigation strategies, which would prevent roads from being damaged and improve mobility and the economy.

6. Conclusions

The coastal landscape of Cilento (southern Italy) has a great variety of terrestrial and marine landforms. Despite the continuous degradation of rocks with different degrees of erodibility and the negative effects of mankind on the territory, these forms are able to maintain their morphological characteristics. These characteristics make the landscape attractive to tourists, who choose the Cilento coastal areas for their holidays, but they also capture the interest of researchers and experts in coastal geomorphology [3,18]. For this reason, the Cilento territory and the contiguous marine areas are protected, both at the national and at the international level. However, even if it is possible to protect the environment and ensure sustainable development inland, it can be difficult in coastal areas, due to both anthropic pressure and climate change effects. As previously mentioned, the Cilento coastline was already affected by erosion processes that led to coastal erosion and shoreline retreat [44,45] and by numerous landslides that occurred on the cliffs or the slopes behind the beaches [61]. Seas and oceans are under considerable anthropic pressure due to structures and infrastructures built close to the coast to the detriment of the conservation of the environment, and the ports and coastal defenses are not entirely adequate for the context. On the other hand, a sea-level rise caused by an increase in temperatures would have further effects on the coastline that cannot be fully controlled. These impacts would be greater where the adjacent beaches and structures cannot be effectively protected, and greater still on soft rock cliffs, like those found in the study area.

With regard to Licosa Cape, where anthropic pressure is not so high, climate change effects should be considered for the conservation of the landforms. Due the presence of a wide coastal platform, the estimated rise in sea level would probably not have significant short- and medium-term effects on the area close to the terrace. However, there could be an intensification of landslide phenomena along the slopes of the Monte Licosa ridge and swamping in the terrace area, which already occurred in the past in warm periods. The case of the "slope-over-profile" profile would be completely different as verified by the application; once the threshold represented by the wall is exceeded, there would be a huge earth flow followed by the complete collapse of the slope and the destruction of the structures/infrastructures built on it. In Cilento, there are numerous infrastructures such as roads, but risk mitigation in order to conserve these landforms would entail huge economic costs. Zoning regulations could help to protect the area, as the result of a detailed knowledge of the landscape and its space–time evolution [90]. To this end, efforts should be made to adopt multidisciplinary approaches that use innovative topographic and geo-morphometric analyses that enable us to develop a detailed digital geomorphological map and enhance our spatio-temporal knowledge.

This paper provides useful information on the landforms for planners and operators working in the area. Meanwhile, for the site of Ripe Rosse and other places located in areas prone to landslides, a proposal was put forward to establish the "prototypal moving geosites" within the Geopark Network in order to emphasize their scientific, educational, and social relevance [91]. To this aim, we wish to invite researchers to monitor these particular geosites as students strive to understand the forms and processes related to them. Mankind should implement activities that do not damage directly and indirectly our geological and geomorphological heritage in order to conserve all terrestrial land and marine landforms.

Author Contributions: This research was conducted in collaboration between the authors. Both authors, G.D. and A.V. were involved in the conceptualization of the problem, in the identification of the data and in the preparation of the manuscript. While A.V. mainly conducted data integration and contributed to the drafting of the document, G.D. set up the methods to develop future scenarios. Discussions and analyzes were conducted by both authors. Both authors contributed to the revision of the manuscript.

Funding: This research received no external funding.

Acknowledgments: The authors would like to thank the three anonymous reviewers for their fruitful comments and the guest editor for suggestions. The paper was funded by C.U.G.R.I. (InterUniversity Consortium), recognized by MIUR (Italian Minister for University and Research), as the Interuniversity Center for the Prevision and Prevention of Great Risk and the Department of Civil Engineering of Salerno University (Italy).

Conflicts of Interest: The authors declare no conflicts of interest.

References

1. Prampolini, M.; Foglini, F.; Biolchi, S.; Devoto, S.; Angelini, S.; Soldati, M. Geomorphological mapping of terrestrial and marine areas, northern Malta and Comino (central Mediterranean Sea). *J. Maps* **2017**, *13*, 457–469. [CrossRef]

2. Griggs, G.B.; Trenhaile, A.S. Coastal cliffs and platforms. In *Coastal Evolution: Late Quaternary Shoreline Morphodynamics*; Carter, R.W.G., Woodroffe, C.D., Eds.; Cambridge University Press: Cambridge, UK, 1994; pp. 425–476.

3. Valente, A.; Magliulo, P.; Russo, F. The coastal landscape of Cilento (southern Italy): A challenge for protection and tourism valorization. In *Landscapes and Landforms of Italy*; Soldati, M., Marchetti, M., Eds.; Springer International Publishing: Manhatton, NY, USA, 2017; Volume 35, pp. 409–419.

4. De Vita, A.; Aloia, A.; Catino, N.; Marsicano, A. Cilento and Vallo di Diano Geopark. In *Italian Geoparks*; Aloia, A., Burlando, M., Eds.; Forum Nazionale Geoparchi Italiani: Salerno, Italia, 2013; pp. 46–51.

5. Antonioli, F. Sea level change in Western-Central Mediterranean since 300 Kyr: Comparing global sea level curves with observed data. *Alp. Mediterr. Quat.* **2012**, *25*, 15–23.

6. Antonioli, F.; Lo Presti, V.; Rovere, A.; Ferranti, L.; Anzidei, M.; Furlani, S.; Mastronuzzi, G.; Orru, P.E.; Scicchitano, G.; Sannino, G.; et al. Tidal notches in Mediterranean Sea: A comprehensive analysis. *Quat. Sci. Rev.* **2015**, *119*, 66–84. [CrossRef]

7. Walker, H.; McGraw, M. Geomorphology and coastal hazards. In *Geomorphological Hazards and Disaster Prevention*; Alcántara-Ayala, I., Goudie, A., Eds.; Cambridge University Press: Cambridge, UK, 2010; pp. 129–144.

8. Finkl, C.W. *Coastal Hazards*; Finkl, C.W., Ed.; Springer: Manhattan, NY, USA, 2013; p. 840. [CrossRef]

9. Brandolini, P.; Faccini, F.; Piccazzo, M. Geomorphological hazard and tourist vulnerability along Portofino Park Trails (Italy). *Nat. Haz. Earth Syst. Sci.* **2006**, *6*, 563–571. [CrossRef]

10. May, V.J. Integrating the geomorphology environment, cultural heritage, tourism and coastal hazards in practice. *Geogr. Fis. Din. Quat.* **2008**, *31*, 187–194.

11. Vogiatzakis, I.N.; Griffiths, G.H.; Cassar, L.F.; Morse, S. *Mediterranean Coastal Landscape: Management Practices, Typology and Sustainability, Final Report*; University of Reading: Reading, UK, 2005.

12. Patacca, E.; Sartori, R.; Scandone, P. Tyrrhenian basin and Apenninic arcs: Kinematic relations since Late Tortonian times. *Mem. Soc. Geol. Ital.* **1990**, *45*, 425–451.

13. Patacca, E.; Scandone, P. Geology of southern Apennines. Results of the CROP Project, Sub-Project CROP-04. *Boll. Soc. Geol. Ital.* **2007**, 75–119.

14. Bonardi, G.; Amore, F.O.; Ciampo, G.; De Capoa, P.; Miconnet, P.; Perrone, V. Il Complesso Liguride Auct.: Stato delle conoscenze e problemi aperti sulla sua evoluzione pre-appenninica ed i suoi rapporti con I'Arco Calabro. *Mem. Soc. Geol. Ital.* **1988**, *41*, 7–35, (In Italian with English abstract).

15. Ciarcia, S.; Mazzoli, S.; Vitale, S.; Zattin, M. On the tectonic evolution of the Ligurian accretionary complex in Southern Italy. *Geol. Soc. Am. Bull.* **2012**, *124*, 463–483. [CrossRef]

16. Ferraro, L.; Pescatore, T.; Russo, B.; Senatore, M.R.; Vecchione, C.; Coppa, M.G.; Di Tuoro, A. Studi di geologia marina del margine tirrenico: La piattaforma continentale tra Punta Licosa e Capo Palinuro (Tirreno meridionale). *Boll. Soc. Geol. Ital.* **1997**, *116*, 473–485, (In Italian with English abstract).

17. Pennetta, M.; Bifulco, A.; Savini, A. Ricerca di depositi di sabbia sottomarina relitta sulla piattaforma continentale del Cilento (SA) utilizzabile per interventi di ripascimento artificiale dei litorali. *Geol. Dell Ambiente* **2013**, *1*, 1–22.

18. ISPRA. *Geological Map and Submerged Landscape Map of the National Park of Cilento, Vallo di Diano and Alburni*; ISPRA: Rome, Italy, 2013.

19. Scandone, P.; Sgrosso, I.; Bruno, F. Appunti di geologia sul Monte Bulgheria (SA). *Boll. Soc. Natur. Napoli* **1964**, *72*, 19–26.

20. Antonioli, F.; Cinque, A.; Ferranti, L.; Romano, P. Emerged and submerged quaternary marine terraces of Palinuro Cape (southern Italy). *Mem. Descr. Carta Geol. D Ital.* **1994**, *52*, 237–260.

21. Antonioli, F.; Puglisi, C.; Silenzi, S. Rilevamento morfostratigrafico della costa emersa e sommersa del promontorio di Capo Palinuro. *Mem. Descr. Carta Geol. D Ital.* **1994**, *52*, 225–236, (In Italian with English abstract).

22. Cavuoto, G.; Martelli, L.; Nardi, G.; Valente, A. Depositional systems and architecture of Oligo-Miocene turbidite successions in Cilento (southern Apennines). *Geoacta* **2004**, *3*, 129–147.

23. Cammarosano, A.; Cavuoto, G.; Danna, M.; De Capoa, P.; De Rienzo, F.; Di Staso, A.; Giardino, S.; Martelli, L.; Nardi, G.; Sgrosso, A.; et al. Nuovi dati e nuove interpretazioni sui flysch terrigeni del Cilento (Appennino meridionale, Italy). *Boll. Soc. Geol. Ital.* **2004**, *123*, 253–273, (In Italian with English abstract).

24. Brancaccio, L.; Cinque, A.; Romano, P.; Rosskopf, C.M.; Russo, F.; Santangelo, N.; Santo, A. Geomorphology and neotectonic evolution of a sector of the Tyrrhenian flank of the Southern Apennines (Region of Naples, Italy). *Zeit Geomorph. N. F.* **1991**, *82*, 47–58.

25. Cinque, A.; Romano, P.; Rosskopf, C.; Santangelo, N.; Santo, A. Morfologie costiere e depositi quaternari tra Agropoli e Ogliastro Marina (Cilento, Italia meridionale). *Il Quat.* **1994**, *7*, 3–16, (In Italian with English abstract).

26. Marciano, R.; Munno, R.; Petrosino, P.; Santangelo, N.; Santo, A.; Villa, I. Late quaternary tephra layers along the Cilento coastline (southern Italy). *J. Volcan Geotherm. Res.* **2008**, *177*, 227–243. [CrossRef]

27. Cinque, A.; Rosskopf, C.; Barra, D.; Campajola, L.; Paolillo, G.; Romano, M. Nuovi dati stratigrafici e cronologici sull'evoluzione recente della piana del fiume Alento. *Il Quat.* **1995**, *8*, 323–338, (In Italian with English abstract).

28. Ascione, A.; Romano, P. Vertical movements on the eastern margin of the Tyrrhenian extensional basin. New data from Mt Bulgheria (Southern Appenines, Italy). *Tectonophysics* **1999**, *315*, 337–358. [CrossRef]

29. Esposito, C.; Filocamo, F.; Marciano, R.; Romano, P.; Santangelo, N.; Scarciglia, F.; Tuccimei, P. Late Quaternary shorelines in Southern Cilento (Mt. Bulgheria): Morphostratigraphy and chronology. *Il Quat. It J. Quater Sci.* **2003**, *16*, 3–14.

30. Romano, P. La distribuzione dei depositi marini pleistocenici lungo le coste della Campania. Stato delle conoscenze e prospettive di ricerca. *Studi Geol. Camerti N. Spec.* **1992**, *1*, 265–269, (In Italian with English abstract).

31. Guida, D.; Aloia, A.; Valente, A. Classification of the Cilento, Vallo di Diano and Alburni National Park—European Geopark Coastland. In Latest trends in Energy, Environment and Development. In Proceedings of the 7th International Conference on Environmental and Geological Science and Engineering, Salerno, Italy, 3–5 June 2014; pp. 121–126.

32. Bartole, R.; Savelli, C.; Tramontana, M.; Wezel, F.C. Structural and sedimentary features in the Tyrrhenian margin off Campania, Southern Italy. *Mar. Geol.* **1984**, *55*, 163–180. [CrossRef]

33. Blasi, C.; Capotorti, G.; Copiz, R.; Guida, D.; Mollo, B.; Smiraglia, D.; Zavattaro, L. Classification and mapping of ecoregions of Italy. *Plant Biosyst.* **2014**, *148*, 1255–1345. [CrossRef]

34. Corbetta, F.; Pirone, G.; Frattaroli, A.R.; Ciaschetti, G. Lineamenti vegetazionali del Parco Nazionale del Cilento e Vallo di Diano. *Braun Blanquetia* **2004**, *36*, 1–61, (In Italian with English abstract).

35. Google Earth. 2015. Available online: https://google-earth-pro.com (accessed on 20 September 2019).

36. National Geoportal of the Italian Ministry of Environment. 2012. Available online: http://www.pcn.minambiente.it/ (accessed on 20 September 2019).

37. ISPRA. Geological Maps of Italy, Scale 1: 50,000. Sheets 502 "Agropoli". 2016. Available online: http://www.isprambiente.gov.it/Media/carg/502_AGROPOLI/Foglio.html (accessed on 20 September 2019).

38. ISPRA. Geological Maps of Italy, Scale 1: 50,000. Sheets 519 "Capo Palinuro". 2016. Available online: http://www.isprambiente.gov.it/Media/carg/519_CAPO_PALINURO/Foglio.html (accessed on 20 September 2019).

39. ISPRA. Geological Maps of Italy, Scale 1: 50,000. Sheets 520 "Sapri". 2016. Available online: http://www.isprambiente.gov.it/Media/carg/520_SAPRI/Foglio.html (accessed on 20 September 2019).

40. Dramis, F.; Guida, D.; Cestari, A. Nature and aims of geomorphological mapping. In *Geomorphological Mapping: Methods and Applications*; Smith, M., Paron, P., Griffiths, J.S., Eds.; Elsevier: Amsterdam, The Netherlands, 2011; pp. 39–73.

41. Bishop, M.P.; Allan James, L.; Shroder, J.F., Jr.; Walsh, S.J. Geospatial technologies and digital geomorphological mapping: Concepts, issues and research. *Geomorphology* **2012**, *137*, 5–26. [CrossRef]

42. SCAPE (Scalable Preservation Environments) Open Source Software. Available online: https://scape-project.eu/software/scape-open-source-software (accessed on 20 September 2019).

43. Aloia, A.; De Vita, A.; Guida, D.; Valente, A.; Troiano, A. The geological heritage of Cilento and Vallo di Diano Geopak as key in the evolution of the central Mediterranean in the last 200 MY. In Proceedings of the 10th European Geopark Conference, Langesund, Norway, 16–20 september 2011; Rangnes, K., Ed.; European Geoparks Network: Porsgrum, Norway, 2011; pp. 32–41.

44. Regione Campania, Piano Stralcio per la Difesa Costiera (P.S.E.C.). 2007. Available online: http://www.adbsxsele.it (accessed on 20 September 2019).

45. GNRAC (Gruppo Nazionale per la Ricerca sull'Ambiente Costiero). Lo stato dei litorali italiani. *Studi Costieri.* **2006**, *10*, 3–113.

46. De Pippo, T.; Donadio, C.; Pennetta, M.; Petrosino, C.; Terlizzi, F.; Valente, A. Coastal hazard assessment and mapping in Northern Campania, Italy. *Geomorphology* **2008**, *97*, 451–466. [CrossRef]

47. De Pippo, T.; Pennetta, M.; Terlizzi, F.; Valente, A. Principali tipi di falesia nella Penisola Sorrentina e nell'Isola di Capri: Caratteri e lineamenti morfo-evolutivi. *Boll. Soc. Geol. Ital.* **2007**, *126*, 181–189, (In Italian with English abstract).

48. Trenhaile, A.S. The effect of Holocene changes in relative sea level on the morphology of rocky coast. *Geomorphology* **2010**, *114*, 30–41. [CrossRef]

49. Sunamura, S. *Geomorphology of Rocky Coasts*; John Wiley: New York, NY, USA, 1992; p. 302.

50. Trenhaile, A.S. Rock coasts, with particular emphasis on shore platforms. *Geomorphology* **2002**, *48*, 7–22. [CrossRef]

51. Carpenter, N.E.; Dickson, M.E.; Walkden, M.J.A.; Nicholls, R.J.; Powrie, W. Effects of varied lithology on soft-cliff recession rates. *Mar. Geol.* **2014**, *354*, 40–52. [CrossRef]

52. Ashton, A.; Walkden, M.; Dickson, M. Equilibrium responses of cliffed coasts to changes in the rate of sea level rise. *Mar. Geol.* **2011**, *284*, 217–229. [CrossRef]

53. Sunamura, T. Rocky coast processes: With special reference to the recession of soft rock cliffs. *Proc. Jpn. Acad. Ser. B Phys. Biol. Sci.* **2015**, *91*, 481–500. [CrossRef]

54. Ferranti, L.; Antonioli, F.; Mauz, B.; Amorosi, A.; Dai Pra, G.; Mastronuzzi, G.; Monaco, C.; Orru, P.; Pappalardo, M.; Radtke, U.; et al. Markers of the last interglacial sea level high stand along the coast of Italy: Tectonic implications. *Quat. Int.* **2006**, *145–145*, 30–54. [CrossRef]

55. Hearty, P.J.; Miller, G.H.; Stearns, C.S.; Szabo, B.J. Aminostratigraphy of Quaternary shorelines in the Mediterranean basin. *Geol. Soc. Am. Bull.* **1986**, *97*, 850–858. [CrossRef]

56. De Vivo, B.; Rolandi, G.; Gans, P.B.; Calvert, A.; Bohrson, W.A.; Spera, F.J.; Belkin, H.E. New constraints on the pyroclastic eruptive history of the Campanian volcanic Plain (Italy). *Miner. Pet.* **2001**, *73*, 47–65. [CrossRef]

57. Iadanza, C.; Trigila, A.; Vittori, E.; Serva, L. Landslides in coastal areas of Italy. In *Geohazard in Rocky Coastal Areas*; Violante, C., Ed.; Geological Society: London, UK, 2009; Volume 322, pp. 121–141.

58. Chelli, A.; Pappalardo, M.; Llopis, I.A.; Federici, P.R. The relative influence of lithology and weathering in shaping shore platforms along the coastline of the Gulf of La Spezia (NW Italy) as revealed by rock strength. *Geomorphology* **2010**, *118*, 93–104. [CrossRef]

59. Piacentini, D.; Devoto, S.; Mantovani, M.; Pasuto, A.; Prampolini, M.; Soldati, M. Landslide susceptibility modeling assisted by Persistent Scatters Interferometry (PSI): An example from the northwestern coast of Malta. *Nat. Haz.* **2015**, *78*, 681–697. [CrossRef]

60. Budetta, P.; Galletta, G.; Santo, A. A methodology for the study of the relation between coastal cliff erosion and the mechanical strength of soils and rock masses. *Eng. Geol.* **2000**, *56*, 243–256. [CrossRef]

61. Budetta, P.; Santo, A.; Vivenzio, F. Landslide hazard mapping along the coastline of Cilento region (Italy) by means of a GIS-based parameter rating approach. *Geomorphology* **2008**, *94*, 340–352. [CrossRef]

62. Kogure, T.; Aoki, H.; Maekado, A.; Hirose, T.; Matsukura, Y. Effect of the development of notches and tension cracks on instability of limestone coastal cliffs in the Ryukyus, Japan. *Geomorphology* **2006**, *80*, 236–244. [CrossRef]

63. Lim, M.; Rosser, N.J.; Allison, R.J.; Petley, D.N. Erosional processes in the hard rock coastal cliffs at Staithes, North Yorkshire. *Geomorphology* **2010**, *114*, 12–21. [CrossRef]

64. Budetta, P.; De Luca, C.; Santo, A. Recurrent rockfall phenomena affecting the seacliffs of the Campania shoreline. *Rend. Online Soc. Geol. Ital.* **2015**, *35*, 42–45.

65. Biolchi, S.; Furlani, S.; Covelli, S.; Busetti, M.; Cucchi, F. Morphoneotectonics and lithology of the eastern sector of the Gulf of Trieste (NE Italy). *J. Maps* **2016**, *12*, 936–946. [CrossRef]

66. Esposito, C.; Filocamo, F.; Marciano, R.; Romano, P.; Santangelo, N.; Santo, A. Genesi, evoluzione e paleogeografia delle grotte costiere di Marina di Camerota (Parco Nazionale del Cilento e Vallo di Diano, Italia Meridionale). *Thalass. Salentina* **2003**, *26*, 165–174.

67. Bird, E. *Coastal cliffs: Morphology and Management*; Springer: Basel, Switzerland, 2016.

68. Lippmann-Provansal, M. L'Apennin Campanien Meridional (Italie). Etude Geomorphologique. Ph.D. Thesis, Universite d'Aix—Marseille II, Aix en Provence, France, 1987.

69. Ortolani, F.; Pagliuca, S.; Toccaceli, R.M. Osservazioni sull'evoluzione geomorfologica olocenica della piana costiera di Velia (Cilento, Campania) sulla base di nuovi rinvenimenti archeologici. *Geogr. Fis. Dinam. Quat.* **1991**, *14*, 163–169, (In Italian with English abstract).

70. Pennetta, M. Margine tirrenico meridionale: Morfologia e sedimentazione tardo pleistocenica – olocenica del sistema di piattaforma-scarpata. *Boll. Soc. Geol. Ital.* **1996**, *115*, 339–354, (In Italian with English abstract).

71. Clark, A.R.; Fort, D.S.; Davis, G.M. The strategy, management and investigation of coastal landslides at Lyme Regis, Dorset. In *Landslides in Research, Theory and Practice*; Bromhead, E., Dixon, N., Ibsen, M.L., Eds.; Thomas Telford: London, UK, 2000; pp. 279–286.

72. Moore, R.; Davis, G. Cliff instability and erosion management in England and Wales. *J. Coast. Conserv.* **2015**, *19*, 771–784. [CrossRef]

73. Barton, M. Climate change, sea level rise and coastal landslides. In *Engineering Geology for Society and Territory. Climate Change and Engineering Geology*; Lollino, G., Manconi, A., Clague, J.J., Shan, W., Chiarle, M., Eds.; Springer: Basel, Switzerland; Cham (CH), Switzerland, 2015; Volume 1, pp. 415–418.

74. IPCC. *Global Warming of 1.5 °C*; Special Report, Intergovernmental Panel on Climate Change; IPCC: Rome, Italy, 2018.

75. Lambeck, K.; Antonioli, F.; Anzidei, M.; Ferranti, L.; Leoni, S.; Scicchitano, G.; Silenzi, S. Sea level change along the Italian coast during the Holocene and projections for the future. *Quat. Int.* **2011**, *232*, 250–257. [CrossRef]

76. Walkden, M.; Dickson, M. Equilibrium erosion of soft rock shores with a shallow or absent beach under increased sea level rise. *Mar. Geol.* **2008**, *251*, 75–84. [CrossRef]

77. Martelli, L.; Nardi, G.; Cammarosano, A.; Cavuoto, G.; Aiello, G.; D'Argenio, B.; Marsella, E. Note illustrative della Carta Geologica d'Italia (scala 1:50.000), Foglio 502 "Agropoli". Servizio Geologico d'Italia, ISPRA. 2016. Available online: http://www.isprambiente.gov.it/Media/carg/campania.html (accessed on 20 September 2019)(In Italian with English extended abstract).

78. Brancaccio, L.; Cinque, A.; Russo, F.; Belluomini, G.; Branca, M.; Delitala, L. Segnalazione e datazione di depositi marini tirreniani sulla costa campana. *Boll. Soc. Geol. Ital.* **1990**, *109*, 259–265, (In Italian with English abstract).

79. Iannace, A.; Romano, P.; Santangelo, N.; Santo, A.; Tuccimei, P. The OIS 5c along Licosa Cape promontory (Campania region, Southern Italy): Morphostratigraphy and U/Th dating. *Zeit Geomorph. N. F.* **2001**, *45*, 307–319.

80. Gambassini, P.; Martini, F.; Palma di Cesnola, A.; Peretto, C.; Piperno, M.; Ronchitelli, A.M.; Sarti, L. Il Paleolitico dell'Italia centro-meridionale. In *Guide Archeologiche di Preistoria e Protostoria 1*; ABACO Edizioni: Forlì, Italy, 1996.

81. Lirer, L.; Pescatore, T.; Scandone, P. Livello di piroclastici nei depositi continentali post-Tirreniani del litorale sud-tirrenico. *Atti Accad. Gioenia Sci. Nat. Catania* **1967**, *18*, 85–115.

82. Scarciglia, F.; Terribile, F.; Colombo, C.; Cinque, A. Late Quaternary climatic changes in Northern Cilento (South Italy): An integrated geomorphological and paleopedological studies. *Quat. Int.* **2003**, *106–107*, 141–158. [CrossRef]

83. Huntley, B.; Watts, W.; Allen, J.; Zolitschka, B. Paleoclimate, chronology and vegetation history of the Weichselian Lateglacial: Comparative analysis of data from three cores at Lago Grande di Monticchio, southern Italy. *Quat. Sci. Rev.* **1998**, *18*, 945–960. [CrossRef]

84. Russo Ermolli, E.; di Pasquale, G. Vegetation dynamics of southweastern Italy in the last 28000 yr inferred from pollen analysis of a Tyrrhenian Sea core. *Veg. Hist. Archaeobotany* **2002**, *11*, 211–220. [CrossRef]

85. Coltorti, M.; Dramis, F. The significance of stratified slopewaste deposits in the Quaternary of Umbria-Marche Apennines, Central Italy. *Zeit Geomorph. N. F.* **1988**, *71*, 59–70.

86. Cocco, E.; De Pippo, T.; Valente, A. Sedimentologia del Flysch del Cilento: Le arenarie di Pollica. *Geol. Rom.* **1986**, *25*, 25–32, (In Italian with English abstract).

87. Cavuoto, G.; Valente, A.; Nardi, G.; Martelli, L.; Cammarosano, A. A prograding Miocene Turbidite System. In *Atlas of Deep-Water Outcrops*; Nilsen, T.H., Shew, R.D., Steffens, G.D., Studlick, J.R.J., Eds.; AAPG Studies in Geology; AAPG and Shell Exploration & Production: Tulsa, OK, USA, 2008; Volume 56, p. 54.

88. Savini, A.; Basso, D.; Bracchi, V.A.; Corselli, C.; Pennetta, M. Maerl-bed mapping and carbonate quantification on submerged terraces offshore the Cilento peninsula (Tyrrhenian Sea, Italy). *Geodiversitas* **2012**, *34*, 77–98. [CrossRef]

89. De Vita, P.; Carratù, M.T.; La Barbera, G.; Santoro, S. Kinematics and geological constraints of the slow-moving Pisciotta rock slide (southern Italy). *Geomorphology* **2013**, *201*, 415–429. [CrossRef]
90. Valiante, M.; Bozzano, F.; Guida, D. The Sant'Andrea-Molinello Landslide system (Mt. Pruno, Roscigno, Italy). *Rend. Online Soc. Geol. Ital.* **2016**, *41*, 214–217. [CrossRef]
91. Calcaterra, D.; Aloia, A.; Budetta, P.; De Vita, A.; De Vita, P.; Guida, D.; Zampelli, S. Moving geosites: How landslides can become focal points in a Geopark. In Proceedings of the 6th International UNESCO Conference on Global Geoparks 2014—Stonehammer Geopark, Fundy Bay, NB, Canada, 19–22 september 2014; pp. 12–13.

Article

Geomorphology of Canyon Outlets in Zrmanja River Estuary and Its Effect on the Holocene Flooding of Semi-enclosed Basins (the Novigrad and Karin Seas, Eastern Adriatic)

Ozren Hasan [1,*], Slobodan Miko [1], Dea Brunović [1], George Papatheodorou [2], Dimitris Christodolou [2], Nikolina Ilijanić [1] and Maria Geraga [2]

[1]   Croatian Geological Survey, Sachsova 2, 10000 Zagreb, Croatia; smiko@hgi-cgs.hr (S.M.); dbrunovic@hgi-cgs.hr (D.B.); nilijanic@hgi-cgs.hr (N.I.)

[2]   Laboratory of Marine Geology and Physical Oceanography, Department of Geology, University of Patras, 26504 Patras, Greece; gpapathe@upatras.gr (G.P.); dchristo@upatras.gr (D.C.); mgeraga@upatras.gr (M.G.)

*   Correspondence: ohasan@hgi-cgs.hr; Tel.: +385-91-207236

Received: 28 August 2020; Accepted: 1 October 2020; Published: 10 October 2020

Abstract: Detailed multi-beam bathymetry, sub-bottom acoustic, and side-scan sonar observations of submerged canyons with tufa barriers were used to characterize the Zrmanja River karst estuary on the eastern Adriatic coast, Croatia. This unique karst environment consists of two submerged karst basins (Novigrad Sea and Karin Sea) that are connected with river canyons named Novsko Ždrilo and Karinsko Ždrilo. The combined use of high-resolution geophysical data with legacy topography and bathymetry data in a GIS environment allowed for the description and interpretation of this geomorphological setting in relation to the Holocene sea-level rise. The tufa barriers had a predominant influence on the Holocene flooding dynamics of the canyons and karst basins. Here, we describe the possible river pathways from the basins during the lowstand and the formation of a lengthening estuary during the Holocene sea-level rise. Based on the analyzed morphologies and the relative sea-level curve for the Adriatic Sea, the flooding of the Novsko Ždrilo occurred 9200 years before present (BP) and Karinsko Ždrilo was flooded after 8400 years BP. The combination of high-resolution geophysical methods gave an accurate representation of the karst estuarine seafloor and the flooding of semi-isolated basins due to sea-level rise.

Keywords: eastern Adriatic coast; estuary; sea-level rise; tufa; multi-beam; sub-bottom profiler; holocene

1. Introduction

The rapid development of swath acoustic techniques has enabled seabed mapping at high spatial resolutions and accuracies. The results of the high-resolution acoustic technologies, such as multibeam echosounder (MBES) bathymetry, MBES backscattering, sub-bottom profiling (SBP), side-scan sonar (SSS), and their derivatives, represent an excellent platform for geomorphological and geological classifications of the seabed [1–3]. Despite the problems regarding the acquisition of the data in shallow water or problems related to the incident angle due to steep slopes [4], MBES bathymetry data and quantitative terrain indices, such as the slope, curvature, and roughness, prove to be very useful as a tool for seabed characterization (best summarized by [5,6]). Backscatter data of the acoustic intensity scattered by the seabed collected during MBES surveys [7] gives us valuable information about bottom-type sediment characteristics [1,8]. GIS-based classification techniques and packages for MBES bathymetry and MBES backscattering data have become numerous and available, and have undergone significant development and improvements in the past decade [7,9–12]. Sub-bottom profiles, in conjunction with SSS and bathymetry data and its derivatives, can give insight into depositional dynamics, therefore enabling geomorphological reconstructions [3,13].

This study investigated a unique karst geomorphological setting that was recognized along the Croatian karst on the eastern Adriatic coast; it consists of two canyon-type outlets (Novsko Ždrilo and Karinsko Ždrilo) connecting two semi-isolated basins (Novigrad and Karin Seas) to each other and the open sea. We used the above-mentioned acoustic techniques and the ArcGIS tools available in its Spatial Analyst extension (slope, curvature, maximum likelihood classification (ML), interactive unsupervised classification (uISO), and segment mean shift classification (SMS)), as well as the ArcGIS package Benthic Terrain Modeller (BTM) [12] in order to estimate the timing of the Holocene marine flooding of the semi-isolated sedimentary basins and to make paleoenvironmental reconstructions. During the last glacial maximum (LGM) lowstand, the Novigrad and Karin Seas acted as karst poljes, i.e., interior valleys [14]. After the post-LGM transgression [15–17], the present-day landscape was formed by creating a submerged karst landscape with drowned canyons (including Zrmanja River) and karst poljes [14]. Geomorphological conditions that could prevent marine flooding of the area in the form of submerged barriers within the narrow channels (called ždrila) had to be determined before the reconstruction was possible. Prior to the present study, the existence of submerged tufa barriers was described by scientist divers [18]. The calcareous tufa barriers are unique karst geomorphological features that are common in the karstic region of Croatia [19–22]. They are formed as rheophilic algae and mosses are encrusted by carbonate in waters with a high concentration of dissolved $CaCO_3$, forming porous sediment that grows laterally and vertically, creating a series of obstacles, lakes, and waterfalls [22,23]. Here, we estimated the timing of the flooding of the Novigrad and Karin Seas based on the available relative sea-level curves [15–17] and multibeam-derived elevations of barriers within the drowned Novsko Ždrilo and Karinsko Ždrilo. The submerged canyon, Novigrad Sea, and the Zrmanja River are the three major elements that influenced the formation of the Holocene Zrmanja estuary and its geomorphology during the sea-level rise. An extensive literature review has shown very few examples of similar semi-enclosed marine environments or lakes, with most studies focusing on biota and habitat mapping [23–26]. Furthermore, these studies were performed in non-karstic environments.

We hypothesized that the geomorphological features present in this area had a predominant influence on the Holocene sea flooding. Furthermore, MBES bathymetry, MBES backscattering, SBP, and SSS data and their derivatives proved to be complementary, enabling both surface and subsurface characterizations of the sediments. Therefore, the present study benefited from the use of various remote sensing techniques combined with GIS classification tools in order to obtain a comprehensive overview of the seafloor's physical and geomorphological properties. This type of study can be applicable to other estuarine environments, as well as other coastal environments.

2. General Geological and Geomorphological Features

The Novigrad and Karin Seas are two small semi-enclosed marine bays located in northern Dalmatia on the eastern Adriatic coast, Croatia (Figure 1a). The bays are interconnected via a submerged river canyon called Karinsko Ždrilo (KZD) (Figure 1b). The Novigrad Sea is on the northern side, connected to Velebit Channel and open sea via another canyon-like passage called Novsko Ždrilo (NZD). The bays have a flat bottom with average depths of 26 to 30 m in the Novigrad Sea and 12 to 13 m in the Karin Sea [27]. Both canyons are oriented in the NNW–SSE direction (Figure 1b). Novsko Ždrilo is 3750 m long and 150 m to 250 m wide at sea level. Its steep slopes rise to 150 m above sea level (a.s.l.). The water depth within the NZD channel is 25 to 46 m. Karinsko Ždrilo has a less dramatic morphology with slopes rising 20 to 40 m a.s.l., whereas the canyon width is 115 m to 250 m at sea level. The water depth in KZD ranges from 15 to 25 m. Due to the region's geographic importance in the connection of the interior with northern Dalmatia, two bridges cross NZD. The larger highway bridge is located at the central part of NZD, while the smaller local Maslenica Bridge is at the southern end of the channel (Figure 1b).

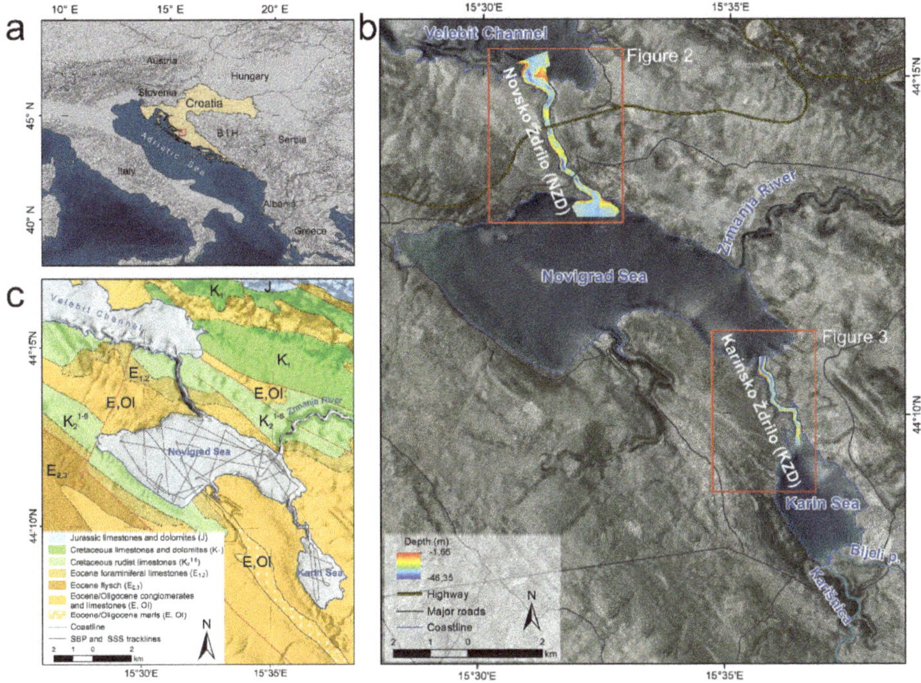

Figure 1. (**a**) General study site location in the red square. (**b**) Multibeam echosounder (MBES) bathymetry data for Novsko Ždrilo (NZD) and Karinsko Ždrilo (KZD) draped over a shaded relief and an orthophoto map [27]. The study areas are highlighted with red squares and presented in detail in Figures 2 and 3. (**c**) Geological map with the plotted sub-bottom profiling (SBP) track lines.

Freshwater enters the Karin Sea via several periodical torrential rivers, but mainly through Karišnica and Bijeli Potok. The karst river Zrmanja, with a total length of 69 km, feeds the Novigrad Sea through a river canyon. It forms an estuary, which extends from Novsko Ždrilo up to 14 km inland until it reaches the tufa waterfall Jankovića Buk [28]. The estuary is highly stratified most of the year [29,30]. The river is characterized by many tufa barriers that were formed as autogenous calcite deposits on macrophytes and microphytes [31]. Tufa barriers across the Zrmanja River were previously studied from biological, geochemical, fossil evolution, and chronological points of view [21,31–33]. Tufa growth forms barriers, barrages, or waterfalls across the river valley. When carbonate-rich water falls vertically, vertical cascade tufas result. If water flows over steep slopes, the tufa occurs in the shape of fans, cones, or mounds [19,23,32]. Often, tufa barrier growth forms dams and lakes. Some barriers in the Zrmanja River are still active and some are fossilized. Available studies from the Dinaric karst and the Zrmanja River show that the majority of them are of the Holocene age, with the most intensive growth in the last 7000 years [31,33,34]. The existence of tufa barriers within Novsko Ždrilo has already been confirmed by scientist divers [18], who claimed barrier heights of 10–20 m. The river's complex karst hydrological characteristics are well described by Bonacci [35]. Despite the numerous monitoring measurements, it was not possible to determine the exact karst aquifer (catchment) extent, but it is suggested that the Zrmanja River is connected with the neighboring Krka River through complex karst underground flows [35]. There are also several periodical rivers feeding the Novigrad Sea from the west and south. The composition of sedimentary records has been analyzed from a geochemical perspective on short cores from the Novigrad Sea and Zrmanja River in order to determine the distribution of trace elements and differentiate the anthropogenic impacts from the natural background values [36–38].

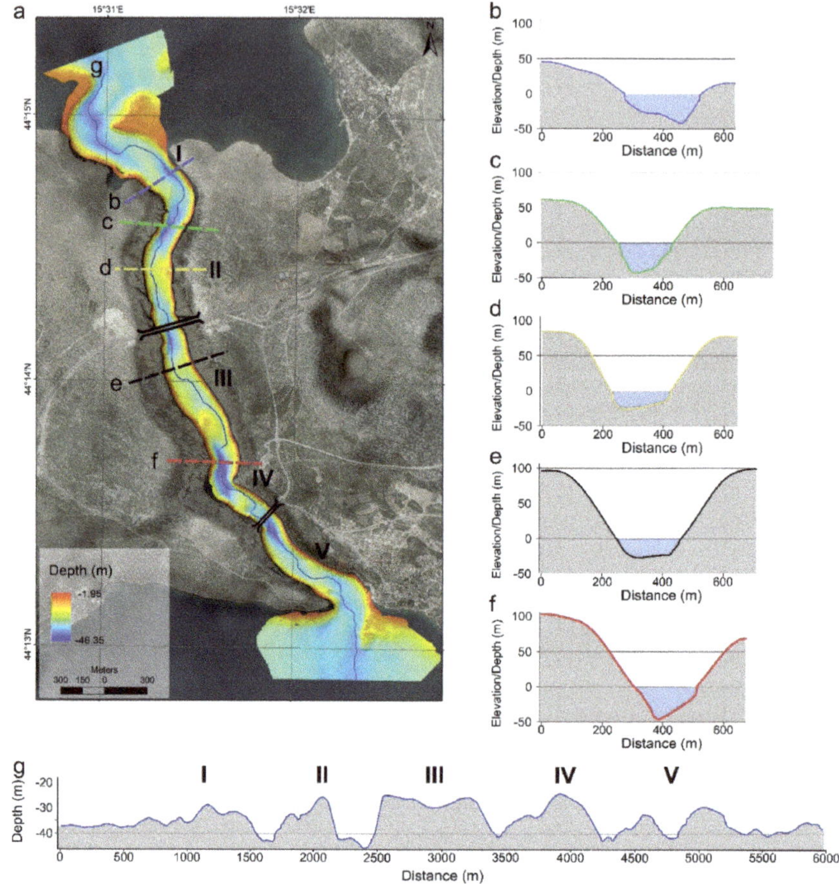

Figure 2. (**a**) Bathymetry image of NZD draped over the DTM and the orthophoto image [27]. Prominent barriers are marked with numbers I–V. (**b–f**) Profiles showing the barriers and deepest parts of the NZD canyon with steep and high sides with depths of over 45 m and elevations of up to 150 m a.s.l. (**g**) The lowest profile through NZD showing a possible path of the water flow (out of and into the Novigrad Sea) during the lowstand and sea-level rise. The same barriers from (**a**) are marked with numbers I–V. The two bridges across NZD are illustrated as references.

As a result of their isolated geomorphological location, sea currents and waves have a slight influence on the bays [39]. Conversely, the Zrmanja River brings 2–3 times more freshwater annually than the total volume of the Novigrad Sea [36]. Together with the freshwater flowing into the Karin Sea by Karišnica and Bijeli Potok, this amount of water can cause strong outflow currents in narrow channels, such as NZD and KZD.

The study area is a part of the Croatian karst Dinarides and consists of a thick carbonate succession deposited from the Late Palaeozoic to the Eocene. During the period from the Mesozoic to the Cenozoic, the area was a part of the large Adriatic–Dinaric Carbonate Platform (ADCP, [40,41]). The disintegration of the ADCP started in the Late Cretaceous and is characterized by the development of flysch basins and carbonate deposition on the margins. The transition from the Cretaceous to the Paleogene was marked by the regional emergence of the entire platform, followed by dynamic tectonics in the Paleogene. The final uplift of the entire Dinaric area culminated in the Oligocene and the Miocene as a result of the collision of Adriatic and Dinaric segments of the Platform [40,41]. The geological background of the studied area consists mainly of Mesozoic, Paleogene, and Neogene carbonates and clastic deposits [42–46] (Figure 1c), with *terra rossa* soils and cambisols on limestone as the dominant soil types. The Mesozoic comprises Jurassic limestones and dolomites at the base, with a succession of Cretaceous limestones, dolomites, and carbonate breccias. Eocene limestones; dolomites and flysch; Oligocene limestones, conglomerates, and marls transgressively overlie Mesozoic rocks [42–46] (Figure 1c). Occurrences and deposits of bauxites can be found in the study area and the wider region, as well as a disused bauxite processing factory in the city of Obrovac [42,47,48]. Prior to the rapid Pleistocene–Holocene transgression, the present-day Novigrad and Karin Seas acted as karst poljes. They were subsequently submerged, creating a typical drowned karst landscape together with the drowned canyon of the Zrmanja River [14].

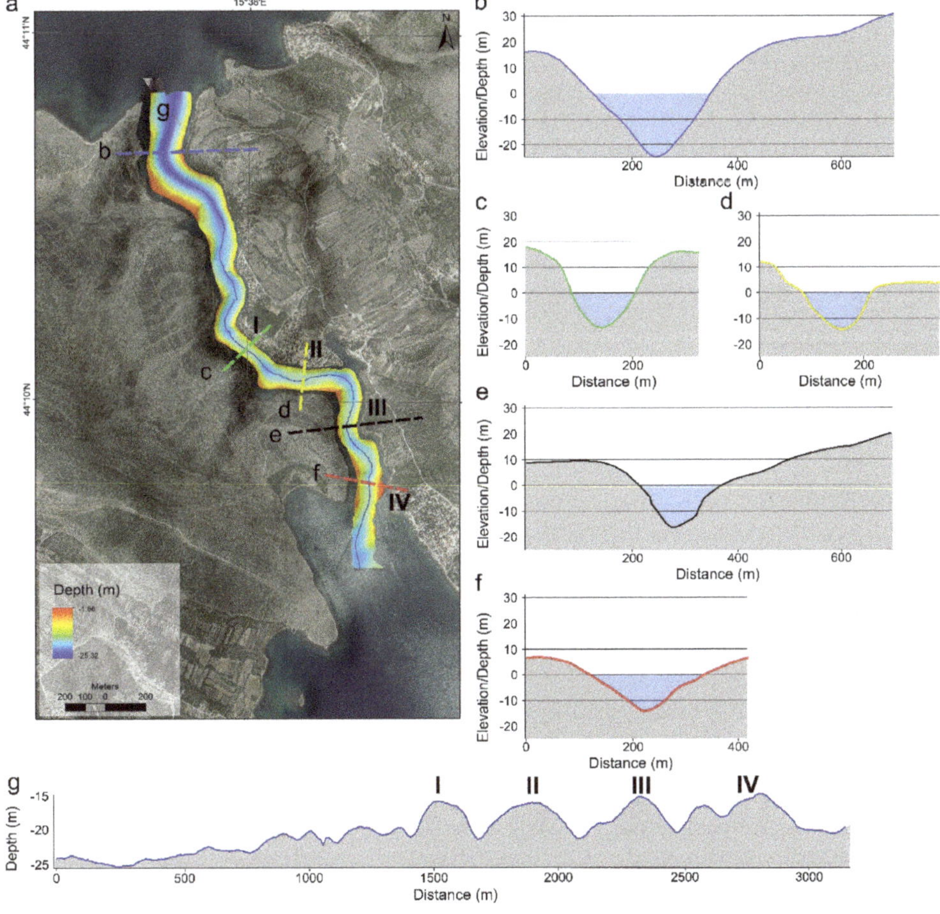

Figure 3. (**a**) MBES bathymetry image of KZD draped over the DTM and orthophoto image [27]. Prominent barriers are marked with numbers I-IV. (**b–f**) Profiles showing the barriers and deepest parts of KZD canyon with depths of 25 m and elevations up to 40 m a.s.l. (**g**) The lowest profile through KZD showing a possible path of water flow (out of and into the Novigrad Sea) during the lowstand and sea-level rise. The barriers from (**a**) are marked with numbers I-IV.

3. Materials and Methods

To study the morphology of the submarine canyons, we used pinger profiles and shipborne multibeam bathymetric data collected during the two surveys conducted in 2015 and 2019. A 2015 campaign comprised SBP and SSS surveys. We used a 3.5 kHz pinger (ORE), GeoAcoustics Ltd. (Great Yarmouth, UK) GeoPulse Transmitter model 5430A, and a GeoAcoustics Ltd. (Great Yarmouth, UK) GeoPulse Reciever model 5210A. SBP data was logged using a Triton SB-Logger (v 7.3, Triton Imaging Inc., Capitola, CA, USA). The signal penetration was limited by the water depth and shallow limestone bedrock and never exceeded 23 ms. Assuming a sound velocity of 1500 m/s, the vertical signal penetration was up to 17 m. A towfish EG&G 272 TD TVG (EG&G Inc., Gaithersburg, MD, USA) was towed 3 m below the sea surface and SSS data were recorded with an EdgeTech 4100P Topside Processor unit (EdgeTech Inc., Escondido, CA, USA). The positioning was obtained usinga Hemisphere DGPS (Hemisphere GNSS, Inc., Scottsdale, AZ, USA). The equipment was mounted on a 6 m long vessel moving at an average speed of 3.5 knots. Afterward, the SBP data were exported in a SEG-Y format (Society of Exploration Geophysicists Exchange Tape Format) and further interpreted in the Geosuite Allworks software (version 2.6.7., Geo Marine Survey System, Rotterdam, Netherlands).

The second campaign dataset was taken in 2019 comprised MBES mapping of the canyons. For this purpose, we used a WASSP S3 MBES (Furuno ENL, Auckland, New Zealand), which is capable of recording multibeam and backscatter data. It was side-mounted on a vessel moving at an average speed of 3.5 knots. The used MBES system works at a typical frequency of 160 kHz with an effective beam width (angular coverage) of 120 degrees using 224 beams. The beam width is 4.4 × 3.2 (PS/FA) with a transmitting voltage response of 155 dB and a receiving voltage response of -194 dB. The vessel motion was corrected for with a WASSP Sensorbox (Furuno ENL, Auckland, New Zealand) inertial measuring unit (IMU), which makes corrections for the pitch, roll, and heave. The IMU worked in conjunction with the Hemisphere Vector V103 DGPS compass antenna (Hemisphere GNSS, Inc., Scottsdale, AZ, USA) used for positioning. The data were acquired with WASSP CDX software (version 3.9, Furuno ENL, Auckland, New Zealand). Cleaning, processing, and validation of the MBES data were performed with the hydrographic software BeamworX Autoclean (version 2020.1.1.0., BeamworX BV, Utrecht, Netherlands).

Morphometric Analyses

The final MBES bathymetry and MBES backscatter grids were exported from BeamworX as a 1 m pixel ASCII grid for further analysis in ArcGIS (version 10.2.1, ESRI inc., Redlands, CA, USA). To create a more meaningful base for the GIS analyses, we gridded together the MBES bathymetry data with the available onshore and bathymetry data digitized from topographic maps 1:25,000 [27] as line and point data into a georeferenced digital terrain model (DTM) with a 1 m pixel size.

Following an extensive literature review [1,3,5,6,12,49], we calculated a range of secondary features to classify and interpret the collected MBES bathymetry, MBES backscatter, and SSS data. ArcGIS with the Spatial Analyst extension to do the DTM, shaded relief, and slope analyses. We used the joined DTM to do the shaded relief and slope analysis, while other morphometric analyses were applied only for the MBES bathymetry/MBES backscatter data. A multidirectional hillshade was created to highlight the morphological features of the terrain, including channels and the bottom morphology. A slope analysis, which is relevant in a geomorphological context linked to the acceleration of currents, the stability of sediments, and erosion [6], was calculated using the standard ArcGIS algorithm proposed by [50]. To describe the heterogeneity of the studied canyons, we used a vector ruggedness measure (VRM). It was calculated using a Benthic Terrain Modeler ArcGIS tool package (version 3.0) [12]. The calculated values are dimensionless and range from 0 (no variation) to 1 (complete variation). Typical values are small (up to 0.4) in natural data [12]. Variations in the range were better observed when calculated for the MBES bathymetry data resampled to a 10 m cell size.

Curvature (the second derivative of the bathymetric surface, or the first derivative of the slope) was calculated using the standard ArcGIS tools according to the method proposed by [51].

The curvature can be calculated parallel to the slope (profile curvature), where it describes the acceleration or deceleration of the flow, or perpendicular to the slope (plan or planiform curvature), which describes the convergence or divergence of the flow. The planiform curvature can be useful when defining ridges, valleys, and slopes along the side of the features [5]. While values close to 0 indicate that the surface is flat, moderate relief values vary from −0.5 to 0.5 and extreme relief values vary between −4 and 4 or more. Physically, the calculated attributes can affect the marine flow, internal waves, and current channeling [12]. Variations in the curvature were also better observed when calculated for the MBES bathymetry data that was resampled to a 10 m cell size.

The morphometric analyses included a combination of the fill DTM, flow direction, and flow accumulation needed to determine the flowline in the channels. The flowline presents the lowest possible pathway for water flowing out of the poljes during the sea-level lowstand, and a pathway for the sea to enter the poljes during the sea-level rise. By applying this methodology, it is possible to determine a correct relative sea level, and consequently, the timing of the Novigrad and Karin Sea drownings.

We made several attempts to classify the backscatter data using ArcGIS tools Spatial analyst tools (version 10.2.1, ESRI inc., Redlands, CA, USA). The best results of the unsupervised backscatter acoustic classifications were achieved using the ML uISO, and SMS. In ML, the mean vector and the covariance matrix characterize each class. A statistical probability can be calculated for each class based on these two cell values. This leads to the determination of the membership of the cell to a specific class [49]. The procedure is based on Bayes' theory of joint probabilities, which accounts for marginal distributions of datasets and their respective internal correlations under the assumption of multivariate normality in N-dimensional Euclidean space [52]. uISO provides a quantitative unsupervised clustering using the functionalities of the Iso Cluster and Maximum Likelihood Classification tools. SMS determines clusters in the MBES backscatter raster by grouping adjacent pixels with similar spectral characteristics. The mean shift algorithm is a non-parametric clustering method for image segmentation. After applying the function, all convergence points are found and clusters are built from them. All convergence points closer than the range defined in the spatial domain are grouped. The number of significant clusters present in the feature space is automatically determined by the number of significant modes detected [53].

4. Results

The use of available high-resolution bathymetry data (bathymetry, seismic, and side-scan sonar data) incorporated with the already available topographic data enabled us to undertake spatial and morphometric analyses of the Novsko Ždrilo and Karinsko Ždrilo channels.

4.1. Bathymetry and Morphometric Analyses

Both studied channels were characterized by elongated geometries and steep slopes (Figure 1). The MBES bathymetry results showed a very distinct seabed within the channels, with multiple barriers, which is typical for a karst morphology (Figures 2 and 3). This is well depicted in the profile lines (Figures 2d and 3d), where multiple pronounced barriers are visible. The water depth at the northern entrance to Novsko Ždrilo was 39 m, whereas, on the SSE end of the channel, the water depth was 37 m. There was an S-shaped bend at the northern entrance to NZD, where the first barrier in NZD could be observed (marked as I in Figure 2). After the bend toward the south, there was a deep part of the canyon with depths of over 40 m below sea level (b.s.l.) extending to the next barrier, which rose to 24.5 m b.s.l. in the lowest part (marked as II in Figure 2a,d). The central part of the canyon was shallower, with two joining barriers at depths of 25–30 m b.s.l. in the lowest part of the crown (marked as III in Figure 2a,d). To the south, this shallow part deepened steeply to the deepest part of the canyon below 45 m b.s.l., then rose steeply again to 26 m b.s.l. (marked as IV in Figure 2). This was the most pronounced barrier as the channel deepened beyond this barrier toward the south, below 40 m b.s.l. The bottom elevated to form two minor barriers near the exit (marked as V in Figure 2), then flattened toward the southern end.

The northern part of KZD was the deepest, with the depth at the northern end reaching 24 m b.s.l. (Figure 3a,d). The bottom gradually elevated in the middle part of the KZD canyon, where the first barrier could be observed (marked as I in Figure 3a,d). Altogether, there were four barriers at the southern part of the canyon (marked as I-IV in Figure 2a,d), all with similar heights (14–16 m b.s.l.) and declining to similar depths (21–22 m b.s.l.). The depth at the southern end of the channel was also the deepest part of the Karin Sea, with a depth of 20.6 m b.s.l.

The slope analysis of the broader area surrounding the channels revealed what was already described: steep slopes rose to 150 m a.s.l. and the continuation of these slopes underwater, which reached almost to the bottom of the channels, where they flattened. The sidewalls in NZD had a maximum inclination of up to 44 degrees on the western side of the canyon close to barrier II (Figures 2a and 4a). In the rest of the canyon, the slopes were still steep and inclined at 28–35°, flattening further at both channel ends.

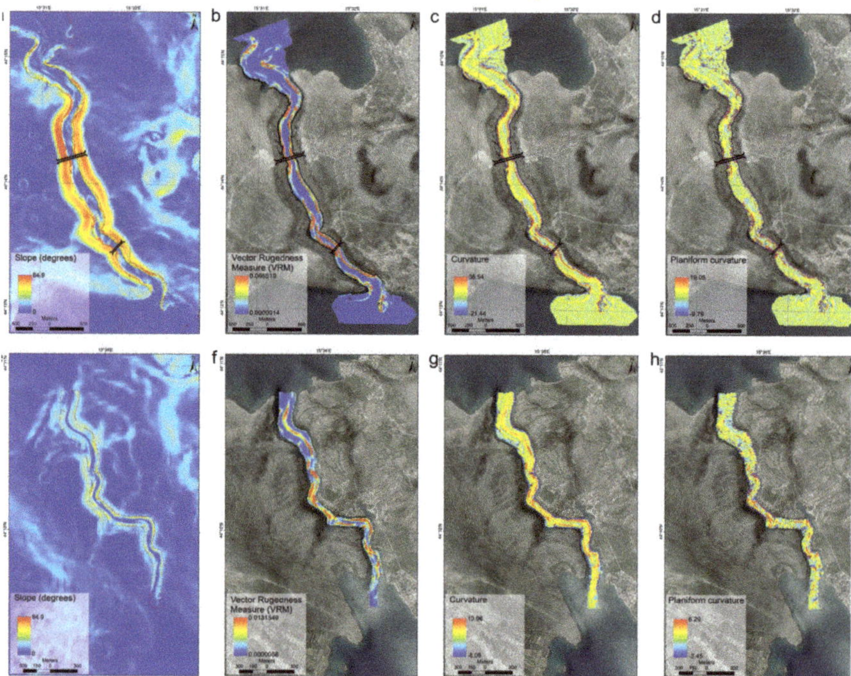

Figure 4. (**a**) Slope analysis of NZD and its surroundings. (**b**) Vector ruggedness measure of NZD. (**c**) Curvature analysis of the NZD area. (**d**) Planiform curvature calculated for the NZD area; the two bridges across NZD are illustrated as references. (**e**) Slope analysis of KZD and its surroundings. (**f**) Vector ruggedness measure of the KZD. (**g**) Curvature analysis of the KZD area. (**h**) Planiform curvature calculated for the KZD area.

The steepest slopes in KZD barely reached the inclinations observed with the NZD sidewalls, with a maximum inclination of 27 degrees, while the average inclinations were 15 to 20°. The steepest parts of the KZD were canyon slopes on both sides of the central-to-southern part of this channel (Figure 4e).

We used the BTM VRM to present the surface roughness, where higher values should represent rockier surfaces (Figure 4b,f). In the analysis of the generalized MBES bathymetry data with a 10 m cell size, the higher surface roughness was clearly visible on the sides in most of the NZD channel. In some areas in the southern and northern parts of the channel (red colours in Figure 4b), high roughness areas extend throughout almost the whole width of the channel. The northern part of KZD, as well as the farthest southern flat-bottom parts of the KZD channel, exhibited low roughness values (Figure 4f). The central part of the channel had high roughness values through its whole width (Figure 4f).

Curvature analysis of the NZD 10 m MBES bathymetry data produced high values (blue and red colors, Figure 4c) near the Maslenica bridge and 500 m to NW, under the highway bridge, and N toward the canyon exit. The curvature of the rest of the channel was low. The planiform curvature (which is meant to emphasize convex or concave forms in the relief) especially emphasized the area around the Maslenica bridge. The curvature analysis of the KZD exhibited elevated positive or negative values in the central and southern parts of the channel (Figure 4g). The values in the central part were somewhat higher. The planiform curvature highlighted the central part of the channel as a part that had elevated values (Figure 4h).

4.2. Acoustic Backscatter Characteristics and Its Derivatives

The backscatter intensity ranged from −10 db to −45 dB for 99.9% of the data in NZD, and from −16 dB to −38 dB in KZD (Figure 5a,e). The backscatter physiography of the channels consisted of low acoustic backscatter surfaces at the canyon ends. Within the channels, the backscatter intensity increased, especially from the steep canyon sides.

A segmentation classification was created with 10 classes, out of which, 5 classes with values higher than 202 were relevant to NZD (Figure 5b). The classification that was derived from the backscatter intensity in NZD resulted in diversification of the central part of the channel, while the NW and SE channel ends had lower values. Areas under the bridges had the highest values. Elevated values extended toward the north and south of the Maslenica bridge and north of the highway bridge. uISO created seven classes, where based on the created dendrogram, it was possible to further reduce the number of classes (Figures 5c and 6). Classes with distances lower than 1 were merged, namely, classes 3 and 4 and classes 5 and 6, creating a classification with five classes (Figures 5c and 6). The derived layer showed three classes within the channel, with a majority of the channel covered by class 6, while class 7 covered the areas near the bridges and north of them. Classes 3 and 4 were limited to the entrance/exit areas of the channel. A raster classification using ML derived seven classes. The classification showed that three classes were dominant within the channel. A majority of the channel was covered with class 7, while class 8 covered the areas near both bridges and north of them. Classes 3, 4, and 5 were limited to the entrance/exit areas of the channel. ML composed very similar visual results to uISO.

The segmentation classification for KZD created 10 classes. Classes with lower values (176–199) were dominant at both channel ends. At the NW end of KZD, classes with values 176–199 reached well within the channel, approximately 800 m from the NW end (Figure 5). Going southward, classes with green and light-yellow colors dominated (values 199–210), while the highest values could be detected in the central part and on the eastern channel sides along most of the channel length. The uISO for KZD consisted of seven classes. A reduction to five classes based on the dendrogram (Figure 6) produced a similar result. Classes 1, 2, and 3 were dominant at the channel ends, and were more pronounced in the NW area. Classes 6 and 7 dominated the rest of the channel. Class 8 appeared on the eastern steep sides of most of KZD and on the western sides of the central part of the channel. ML derived seven classes. The classification showed that classes 1 and 3 were dominant at the channel ends. At the NW end of KZD, classes 1 and 3 reached 800 m within the channel from the NW end. Classes 6 and 7 covered the rest of the channel, and class 8 covered the channel sides.

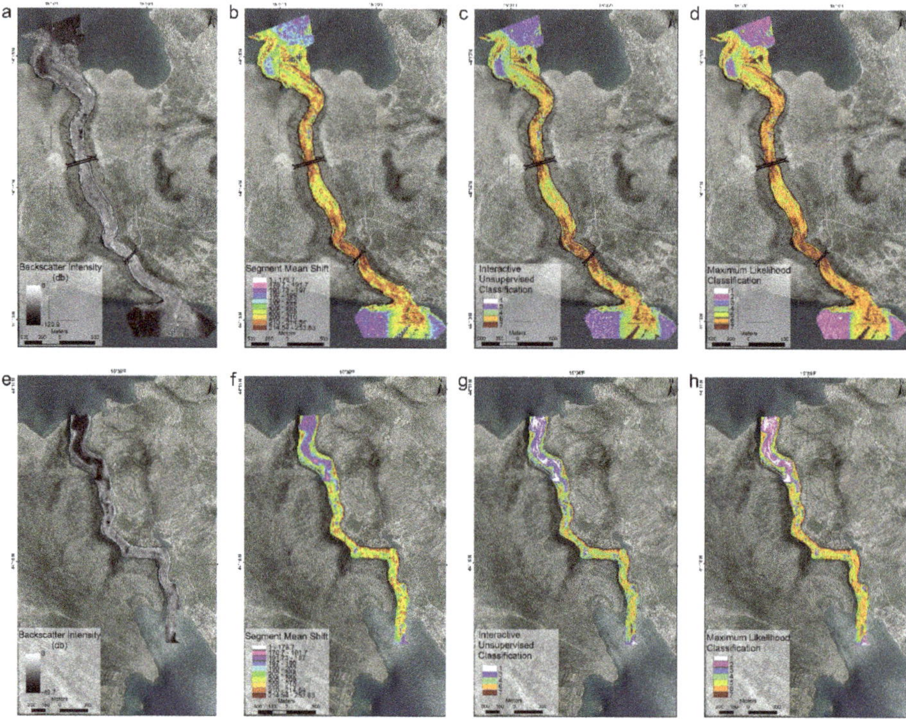

Figure 5. (**a**) Backscatter data of the NZD channel. (**b**) Segment mean shift classification of the NZD area. (**c**) Interactive unsupervised classification (uISO) of the NZD area. (**d**) Maximum likelihood classification of the NZD area; the two bridges across NZD are illustrated as references. (**e**) Backscatter data of the KZD channel. (**f**) Segment mean shift classification of KZD. (**g**) uISO classification of KZD. (**h**) Maximum likelihood classification of KZD.

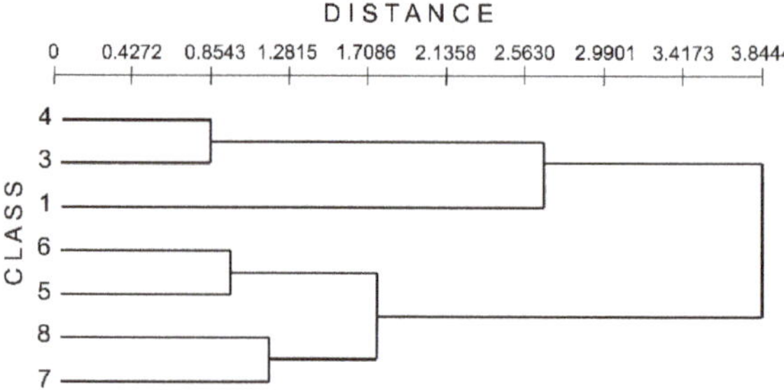

Figure 6. Correlation of determination (R^2) for the linear Pearson correlation measure between seven variables created using uISO for NZD and KZD.

4.3. Side-Scan Sonar

A visual analysis of the SSS mosaic revealed several morphological characteristics that helped to interpret the MBES bathymetry, MBES backscatter, and SBP data.

The central-bottom part of the NZD canyon entrance and the deeper central parts of the channel exhibited low reflectivity of the planar surface in the SSS mosaic. The sides of the channel showed high reflectivity throughout the whole length of the channel. The eastern steep sides appeared to have more exposed boulders and rocky outcrops on the slopes (Figure 7). This was especially highlighted on the sharp bends within the canyon. There was a collapsed steel structure visible in the southern part of the canyon, spreading across the whole width of the channel (Figure 7). This underwater construction was the remains of a bridge demolished during the War of Independence in 1991 [54]. It is located below the present steel bridge that was constructed in 2005 [54]. After the bridge demolition, a floating bridge was constructed at the southern entrance to the canyon. Its concrete supports were still well visible on the SSS mosaic image (Figure 7b). The shallows at the southern end of the SSS mosaic had rocky outcrops and were partially covered with sediments with visible waveforms (Figure 7b).

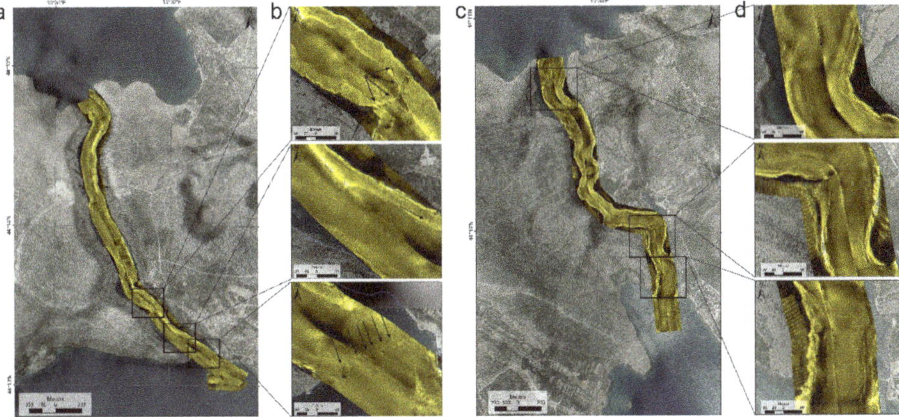

Figure 7. Side-scan sonar mosaic data showing the morphologic characteristics and anthropogenic impact in (**a**) the NZD channel; (**b**) enlarged details of the SSS mosaic showing the collapsed bridge construction (marked with arrows), the flat bottom with the pipeline crossing the channel, and the remains of the concrete slabs used to secure a floating bridge; (**c**) the KZD channel; (**d**) northern end with a flat bottom and rocky sides, a sharp bend with rocky outcrops, and the southern end of the channel with a flat bottom and several waveforms.

Similar to NZD, KZD had a flat bottom-central part of the channel with a lower reflectivity on the SSS mosaic, while steep sides exhibited a higher reflectivity (Figure 7). A low reflectivity was the most pronounced at the northern part of the canyon entrance (for approximately 800 m southward). Rocky outcrops and solitary boulders were exposed, especially on the steep sides of the sharp bends (Figure 7d).

The effort to classify the SSS mosaic using ArcGIS tools failed to provide useful results, creating only a small number of classes.

4.4. Sub-Bottom Profiler

The penetration of the SBP seismic signal was limited due to a very thin sediment cover over the limestone bedrock and shallow water depth causing the occurrence of multiples. Three acoustic units could be determined at the SE end of NZD on the profile perpendicular to the channel (Figure 8b). The uppermost seismic unit (unit 1, Figure 8) was acoustically semi-transparent. Some internal parallel reflectors with weak amplitudes were visible at the base of this unit. The lower sediment unit (unit 2) was characterized by high acoustic amplitudes. The acoustic signal penetrated for less than 10 ms through this unit, indicating coarse sediments. Units 1 and 2 were separated by a high amplitude unconformity with an irregular surface. The acoustic basement (unit 3) was interpreted as bedrock. Due to its relative position in the stratigraphic succession and surrounding surface geology, it is safe to presume that it was bedrock constituted of karstified limestones. The acoustic profile through NZD showed a very dynamic morphology of the upper surface with many barriers in the form of ridges, where some were covered with sediment. Side-echo refractions, caused by the steep adjacent barriers, were observed throughout the profile. A thin overlay of acoustically semi-transparent sediments was visible in the central to northern part of the profile (Figure 8a). It was separated from the bedrock by a very weak amplitude reflector. Due to the water depth, multiples could also be observed throughout the whole SBP profile.

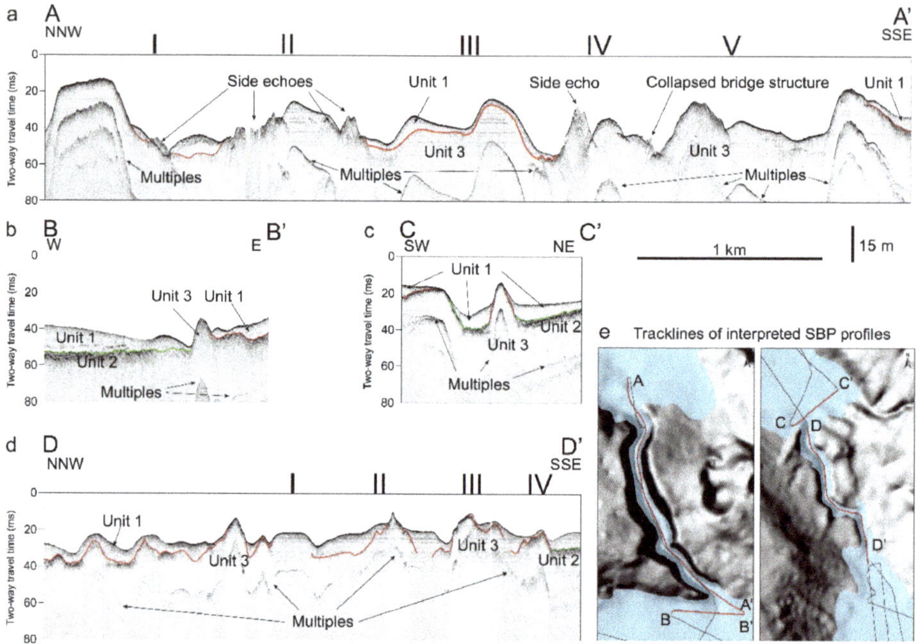

Figure 8. High-resolution seismic profiles surveyed in the canyons. (**a**) Profile A–A′. Only a thin sediment cover was visible over the karstified NZD seabed with multiple side echoes. A thicker sediment succession was visible at the channel ends and in the central part. (**b**) Profile B–B′. A thicker sedimentary succession with two distinct units was visible at the southern NZD end. (**c**) Profile C–C′. The northern end of the KZD with a thicker sediment succession. (**d**) Profile D–D′. The northern part of KZD was covered with sediment (unit 1); in the southern half, rocky barriers stood out with sediment infill between them. (**e**) The SBP track lines of the interpreted profiles are marked with red lines on the maps. The barriers from Figures 2 and 3 are marked with Roman letters.

The same three acoustic units could be determined at the SE end of KZD on the profile perpendicular to the KZD channel (Figure 8c). The uppermost unit was acoustically semi-transparent (unit 1) with some weak reflectors on the bottom of the unit. The underlying unit 2 had high acoustic amplitudes and attenuated the signal penetration. Unit 3 exhibited sharp and steep ridges in the middle of the profile and along the base of the western side of the profile, which were draped by units 1 and 2.

An SBP profile along KZD indicated a very dynamic bathymetry. Several steep carbonate ridges (unit 3) penetrated through the sediment cover to the seabed surface (Figure 8d). Most of the bedrock was covered with the acoustically semi-transparent sediments of unit 1. On the NW third of the profile, the bedrock was draped with the thicker sediment succession of unit 2. Southward, several carbonate ridges pointed out to the surface, while the space in between was partially filled with acoustically semi-transparent sediments. A thicker unit 2 succession was determined at the SE end of KZD.

5. Discussion

Integration of the acoustic and morpho-bathymetric surveys enabled us to reconstruct a detailed bottom morphology of the two narrow karst canyons that connect two semi-enclosed bays, the Novigrad and Karin Seas. Merging the available high-resolution hydroacoustic data-bathymetric, high-resolution seismic, and side-scan sonar data with the already available topographic data enabled us to make spatial and morphometric analyses and create maps to describe the unique environment that acted as a river discharge passage during the sea lowstand, as well as an inlet of the sea into the basins and estuary. Steep slopes and a pronounced bottom morphology characterized the canyons.

5.1. Morphology of the Canyons

Analysis of the MBES bathymetry measurements in NZD revealed six barriers extending along the channels to their steep sides. The steepness of the channel sides was highlighted in the slope and curvature analyses that provided typical values for extreme relief [55]. Three barriers rose to a depth of 25 m b.s.l., with a height difference of 15–20 m, extending to a maximum depth of 45 m b.s.l. The bottom morphology of NZD was very irregular, as depicted on the bathymetric profile and as highlighted by many side-echo refractions visible on the SBP profile (Figure 8a). Adjacent steep rocky outcrops or steep sidewalls cause side-echo refractions [13]. The sediment thickness was higher in the bays, as evidenced by the SBP profile perpendicular to the NZD channel. Within the channel, the sediment overlay was thin or non-existent on the most barriers, and a significant sediment build-up was only noticed on barriers 2 and 3 in the central part of the channel. Thin sediment cover was emphasized by the increased surface roughness of the steep channel sides, depressions, and some barriers (Figure 4). This lack of sediment cover was most likely caused by strong present-day currents in the narrow channels. Strong sea-bottom currents can be caused by a significant input of freshwater into two bays, as the Zrmanja River alone brings 2–3 times more water annually than the total volume of the Novigrad Sea [37]. To this volume, we must add the contribution of the rivers Karišnica and Bijeli Potok flowing into the Karin Sea, as well as karst underground flows ending as submarine springs in the Novigrad Sea [35]. Tidal currents have a negligible effect on the estuary, as tides are rather weak, with M2 amplitudes below 10 cm [28]. As the MBES backscatter signal differs due to the bottom type and its physical characteristics, namely, its hardness or softness [1,8]. Thus, it was possible to classify the MBES backscatter data. A thicker sediment succession at the ends of the NZD channel, which was visible on the SBP profiles, was well delineated in the MBES backscatter derivatives (Figure 5) due to different characteristics compared to sediments within the channel. Within the NZD channel, the MBES backscatter intensity increased, pointing to a rockier surface with a high acoustic backscatter. Sediments in the deepest areas or depressions were also well defined as a class with different sediment characteristics.

The bathymetry data for KZD showed the deeper and more even bottom of the northern part, while the southern half of the channel revealed five barriers. The barriers were equally deep and had similar heights rising to a depth of 14–16 m b.s.l. The SBP data pointed to the fact that most of the channel was covered with at least several meters of sediment, with only peaks of the barriers in the southern part of the KZD comprising a thin sediment overlay (Figure 8). The northern part of the channel bedrock was covered with a thicker sediment succession that increased toward the Novigrad Sea. Higher surface roughness in the central part of KZD, as well as toward the southern part, highlighted the more uneven morphology of the southern part of the channel. The diversification in the MBES backscatter signal derivatives due to the difference in the physical characteristics of the sediment [1,8] was most pronounced in the northern part of KZD, where a thicker sediment succession was visible on the seismic profiles. Similar characteristics could be observed on the southern end of the channel.

Despite the similarity in their appearance, it seems as if the KZD was a "reduced" version of NZD in many ways. The steep sides of NZD rose 150 m a.s.l., with depths below 40 m, while the KZD slopes rose to 40 m a.s.l. and the channel was only up to 20 m deep. The pattern was similar in the case of the bottom morphology, which was more prominent in NZD. One of the reasons for the milder morphology was the thicker sediment cover in KZD. There were several reasons for the preservation of the thicker succession of sediments within KZD: the currents were not as strong as in NZD due to the reduced freshwater influx that only came from the short periodical rivers Karišnica and Bijeli Potok, which is in contrast to NZD, where the volume of the freshwater influx was significantly higher as a result of rivers flowing into the Karin and Novigrad Seas, including the Zrmanja River [37]. Another reason can be found in the easily erodible flysch sediments abundant in the Karišnica and Bijeli Potok watersheds [45] (Figure 1c).

It is clear that the majority of submerged barriers within the canyons were karstic limestone forms. What is still unclear is whether all the submerged barriers in the canyons are made of tufa. The reason for the assumption of tufa barriers in the canyon is the existence of many relicts and recent tufa barriers in the Zrmanja River [21,31–33]. Therefore, favorable conditions for tufa growth in the studied canyons also existed during the lowstand. Ultimately, the morphology of tufa deposits is controlled by the topography and water flow regime [32]. The growth and calcification of rheophilic algae and mosses produce porous hardened substrates and results in a lateral displacement that extends across the river, forming dams with lakes behind them [22]. The tufas in Zrmanja River are described as waterfall and barrage tufas, with some waterfalls being more than 8 m high [21,32]. Tufa barriers higher than 10 m have been detected in the NZD by scientist divers, with one barrier reaching 20 m high with a crest at a depth of 26 m b.s.l. [18]. Five barriers could be detected in the presented data with crests at the highest depths of 16–30 m b.s.l. in NZD, and four barriers with crowns at the highest depths of 14–16 m b.s.l. in KZD (Figure 9; see also Figures 2, 3 and 8).

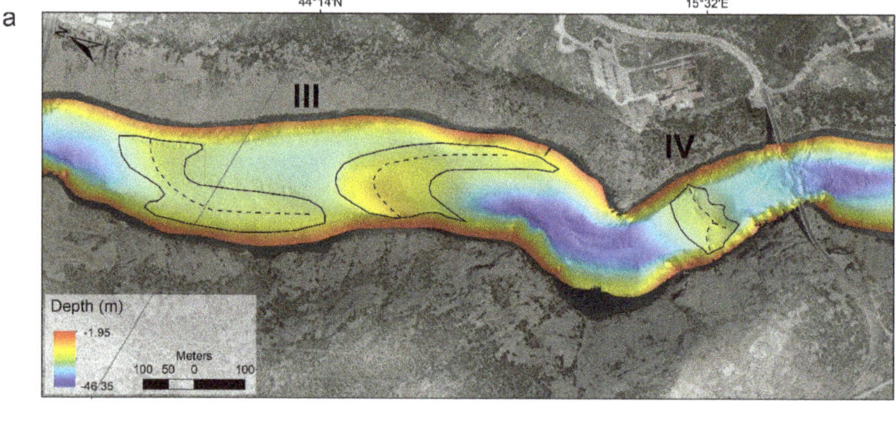

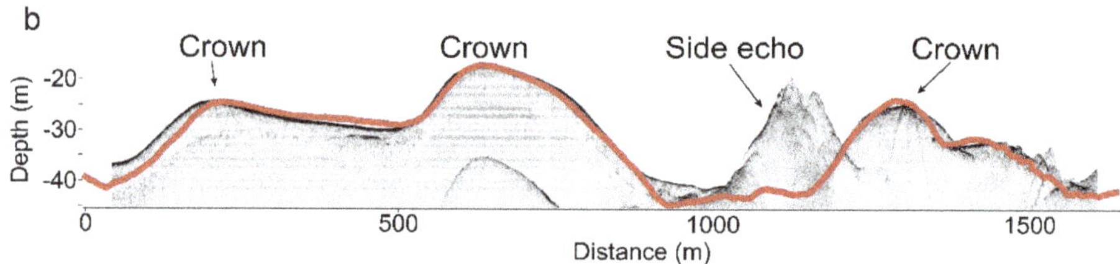

Figure 9. (**a**) Detailed MBES bathymetry map of barriers III and IV in NZD, with the outlined barriers and their crests (dashed line). (**b**) SBP profile of barriers III and IV that are overlain with the actual bathymetry profile over the central part of the canyon (red line).

The SSS mosaic proved to be very useful for determining rocky areas, as well as single boulders collapsed from the steep canyon sides. The anthropogenic effect on the NZD bottom was best recorded on the SSS mosaic, documenting the remains of a collapsed metal bridge construction, which comes from past war efforts during the 1990s War of Independence [54]. The anthropogenic effect was also visible in the form of concrete blocks delineating the path of a temporary floating bridge that was constructed at the southern entrance of the NZD due to the collapse of the pre-1990 bridge [54].

5.2. The Role of the Channels and the Barriers in the Holocene Flooding of the Novigrad and Karin Seas

There are many definitions of an estuary, where many include not only its present state under the influence of the river and the sea but also its morphogenetic origin [56,57]. In this way, Dalrymple et al. [58] define an estuary as "the seaward portion of a drowned river valley system which receives sediment from both fluvial and marine sources and which contains facies influenced by tide, wave and fluvial processes." Evans and Prego [56] conclude that estuaries were produced by a relative rise in sea level and drowning of a previous erosional depression produced via fluvial erosion. Due to the rapid late Pleistocene–Holocene transgression, the river canyons and the poljes in the present-day Novigrad and Karin Seas were submerged [14]. Based on the data gathered in this study and the available data on the relative sea-level rise, we can make hypotheses regarding the evolution of the lower part of the Zrmanja River estuary during the Holocene sea-level rise (Figure 10). During that period the sea level rose 65 m [15,16,59], and eventually formed today's Zrmanja River estuary along with the Novigrad and Karin Seas. The tufa barriers presented a significant factor for the flooding of poljes, as they created 10-20 m high barrages that prevented water from flowing directly (Figure 10). The similarity with the present Zrmanja River is evident, whose present estuary ends at Jankovića Buk, a tufa waterfall that creates a border between an estuary and a river [30,35]. The present river flow is intermittent with plenty of active and fossil tufa barriers [1,31]. The created flowline, as the lowest possible water path in NZD and KZD, allowed us to determine the pathway through the canyon and the relative sea level during the flooding of the Novigrad and Karin Seas. The sea level reached the lowest part of the crest of the barriers in NZD at the present depth of −24.5 to −25 m, and afterward, flooded the Novigrad Sea (Figure 10a–c). Flooding of the polje in the Novigrad Sea area enabled the formation of the Zrmanja River underwater fan, as the sea-level rise caused the deposition of river sediment at the exit of the canyon (Figure 10a). When the sea level rose to the depths of −14 m to −16 m, it reached the crest of the barriers in KZD (Figure 10a,b,d). As the sea level continued to rise, it flooded the Karin Sea until it reached the present level (Figure 10a,b,d). Based on the relative sea-level curve [16] and the heights of the barriers, the estimated time of the flooding of NZD, and consequently Novigrad Sea, was after 9200 BP, while the time of the flooding of KZD, and consequently Karin Sea, can be estimated as having occurred after 8400 BP (Figure 10b).

Further investigations of the sediments deposited in these basins (including sediment cores and high-resolution acoustic methods) will provide more conclusive answers about the timing of the Holocene sea-level rise.

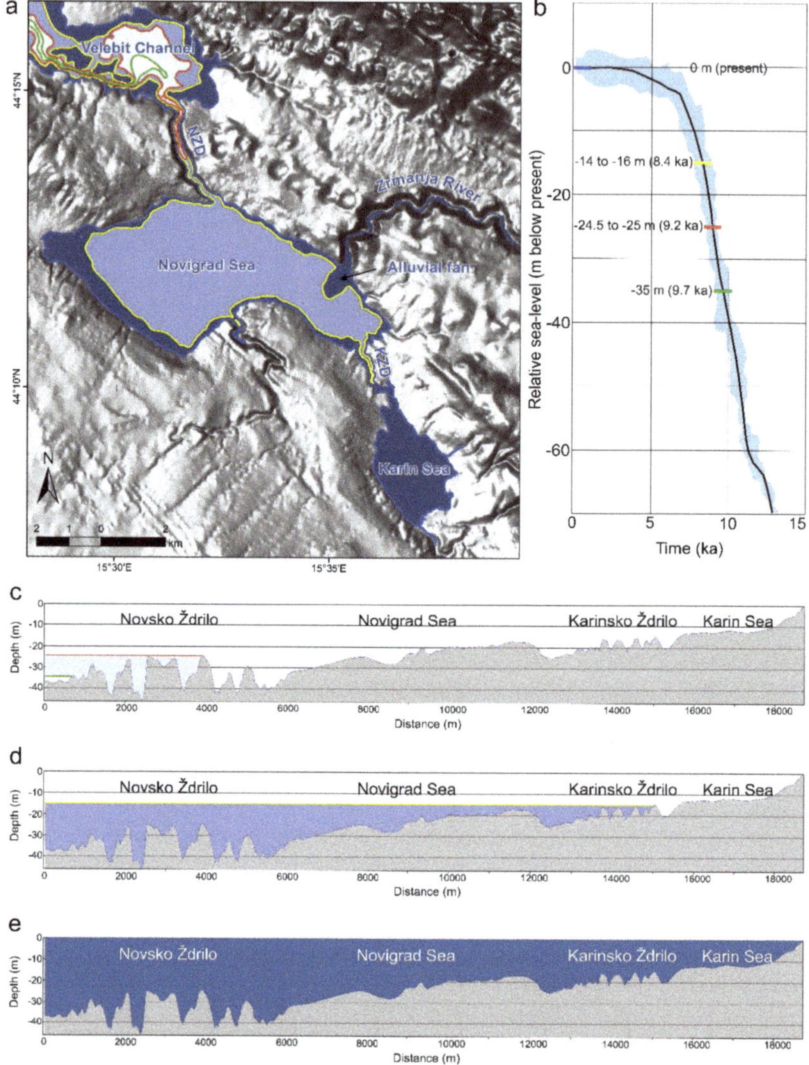

Figure 10. (**a**) Schematic of the coastline position in the study area at sea level at −35 m b.s.l. before the canyons were flooded (white polygon with a green outline), at −25 m b.s.l. (light blue polygon with a red outline), at −15 m b.s.l. (medium blue polygon with a yellow outline), and at the present sea level (dark blue polygon with a blue outline) based on the recent bathymetry data and the Holocene sea-level rise [59]. (**b**) Sea-level curve (modified from [59]) with relative depths of the lowest position of barrier crowns in NZD (red line) and KZD (yellow line) that prevented flooding into the Novigrad and Karin Seas, respectively, as well as the sea level at −35 m b.s.l. before the flooding (green line) and the present sea level (blue line). (**c**) Profile based on the available bathymetry data with sea levels before the NZD canyon was flooded at −35 m b.s.l. (white polygon with a green outline) and at sea level at −25 m b.s.l. (light blue polygon with a red outline). (**d**) Profile based on available bathymetry data with the sea level at −15 m b.s.l. after both canyons were flooded (medium blue polygon with a yellow outline). (**e**) Profile based on the available bathymetry data with the present sea level (dark blue polygon with a blue outline).

6. Conclusions

High–resolution MBES bathymetry, MBES backscatter, SBP, and SSS measurements were carried out in two canyons to provide the first insights into the contemporary physiography of this unique environment consisting of two semi-isolated basins (the Novigrad and Karin Seas) connected with narrow steep channels Novsko Ždrilo and Karinsko Ždrilo.

NZD connects the Novigrad Sea with the open sea. It is also a part of the 15-km-long estuary of the Zrmanja River flowing into the Novigrad Sea and brings large volumes of freshwater. NZD canyon comprised steep and high side slopes extending up to 150 m a.s.l. Six barriers were detected within NZD, extending from the top of the barriers at −25 m b.s.l. to the bottom at −45 m b.s.l. The barriers were most probably all made of tufa, as detected by divers and as an analog with past and present tufa growth in the Zrmanja River. Those barriers prevented marine flooding during the Holocene sea–level rise since during the lowstand, the semi–isolated Novigrad and Karin Seas acted as karst poljes. Strong outflow currents prevented significant sediment build-up. A thicker sediment succession was detected at the ends of the channel extending to the open sea and bay. This was well depicted in SBP profiles and delineated in the MBES backscatter intensity and its derivatives. KZD connects the Karin Sea with the Novigrad Sea. The KZD canyon characteristics were less prominent compared to NZD, with lower and less steep sides. The five barriers detected in KZD were mainly covered in sediment and extended from 14 to 16 m b.s.l.

The post-LGM sea-level rise drowned the coastal karstic landscapes in the Eastern Adriatic Coast, including the Zrmanja River canyon and two poljes, the present-day Novigrad and Karin Seas. Determination of the lowest possible water path in NZD and KZD allowed us to determine the elevation of the relative sea-level rise. The sea level reached the top of the barriers in NZD at a present depth of −24.5 m to −25 m, and in KZD from −14 m to −16 m. The timing of the flooding of the channels and the bays was estimated based on the relative sea–level curve for the Adriatic Sea after 9200 BP in NZD and after 8400 BP in KZD.

SSS proved useful for determining the anthropogenic effect on the NZD bottom, where the remains of metal construction of the collapsed bridge, as well as concrete supports of the floating bridge, were detected.

Knowledge of the geomorphology of the aforementioned karst features is the most important for the relative sea-level reconstruction. When merged with additional investigations of the sediments deposited in the studied basins, which include sediment cores and high-resolution acoustic methods, these results will provide new insights into the timing of the rapid Holocene relative sea-level rise in the eastern Adriatic coast, as well as in other Mediterranean coastal areas [60].

Author Contributions: Conceptualization, O.H. and S.M.; methodology, O.H., S.M. and D.B.; software, O.H.; validation, O.H., S.M. and D.B.; formal analysis, O.H.; investigation, O.H., S.M., G.P., D.C., M.G., D.B. and N.I.; resources, S.M.; writing—original draft preparation, O.H., S.M. and D.B.; writing—review and editing, O.H., D.B., S.M., N.I., G.P., D.C. and M.G.; visualization, O.H., S.M. and D.B.; supervision, S.M.; project administration, S.M. and N.I.; funding acquisition, S.M., O.H. and N.I. All authors have read and agreed to the published version of the manuscript.

Funding: This research was funded by the Croatian Science Foundation (HRZZ) under the project "Lost Lake Landscapes of the Eastern Adriatic Shelf" (LoLADRIA), grant number HRZZ-IP-2013- 11-9419, and the EMODNet Geology project, grant number EASME/EMFF/2018/1.3.1.8/Lot1/SI2.811048.

Acknowledgments: The authors would like to thank Nikos Georgiou, Xenophon Dimas, George Ferentinos, and Margarita Iatrou from the University of Patras for their support during the seismic survey and during the interpretation of the seismic data. The authors would like to acknowledge the anonymous reviewers and editors for their valuable comments, which helped to improve the quality of this manuscript.

Conflicts of Interest: The authors declare no conflict of interest. The funders had no role in the design of the study; in the collection, analyses, or interpretation of data; in the writing of the manuscript, or in the decision to publish the results.

References

1. Bellec, V.K.; Bøe, R.; Rise, L.; Lepland, A.; Thorsnes, T.; Rún Bjarnadóttir, L. Seabed sediments (grain size) of Nordland VI, offshore north Norway. *J. Maps* **2017**, *13*, 608–620. [CrossRef]
2. Buhl-Mortensen, L.; Buhl-Mortensen, P.; Dolan, M.F.J.; Holte, B. The MAREANO programme—A full coverage mapping of the Norwegian off-shore benthic environment and fauna. *J. Mar. Res.* **2015**, *1*, 4–17. [CrossRef]
3. Deiana, G.; Holon, F.; Meleddu, A.; Navone, A.; Orrù, P.E.; Paliaga, E.M. Geomorphology of the continental shelf of Tavolara Island (Marine Protected Area 'Tavolara-Punta Coda Cavallo'—Sardinia NE). *J. Maps* **2019**, *15*, 19–27. [CrossRef]
4. Amiri-Simkooei, A.R.; Snellen, M.; Simons, D.G. Riverbed sediment classification using multi-beam echo-sounder backscatter data. *J. Acoust. Soc. Am.* **2009**, *126*, 1724–1738. [CrossRef]
5. Wilson, F.J.; O'Connell, B.; Brown, C.; Guinan, J.C.; Grehan, A.J. Multiscale Terrain Analysis of Multibeam Bathymetry Data for Habitat Mapping on the Continental Slope. *Mar. Geod.* **2007**, *30*, 3–35. [CrossRef]
6. Dolan, M.F.J. *Calculation of Slope Angle from Bathymetry Data Using GIS—Effects of Computation Algorithms, Data Resolution and Analysis Scale*; NGU-Report 2012.041; Geological Survey of Norway: Trondheim, Norway, 2012; p. 44.
7. Che Hasan, R.; Ierodiaconou, D.; Laurenson, L.; Schimel, A. Integrating Multibeam Backscatter Angular Response, Mosaic and Bathymetry Data for Benthic Habitat Mapping. *PLoS ONE* **2014**, *9*, e97339. [CrossRef]
8. Lurton, X.; Lamarche, G. (Eds.) *Backscatter Measurements by Seafloor-Mapping Sonars*; Guidelines and Recommendations; GeoHab Group: Venice, Italy, 2015; p. 192.
9. Dartnell, P.; Gardner, J.V. Predicting seafloor facies from multibeam bathymetry and backscatter data. *Photogramm. Eng. Remote Sens.* **2004**, *9*, 1081–1091. [CrossRef]
10. Erdey-Heydorn, M.D. An ArcGIS Seabed Characterization Toolbox Developed for Investigating Benthic Habitats. *Mar. Geod.* **2008**, *31*, 318–358. [CrossRef]
11. Le Bas, T. RSOBIA—A new OBIA Toolbar and Toolbox in ArcMap 10.x for Segmentation and Classification. *Photogramm. Eng. Remote Sens.* **2016**, *70*, 1081–1091. [CrossRef]
12. Walbridge, S.; Slocum, N.; Pobuda, M.; Wright, D.J. Unified Geomorphological Analysis Workflows with Benthic Terrain Modeler. *Geosciences* **2018**, *8*, 94. [CrossRef]
13. Trottier, A.-P.; Lajeunesse, P.; Gagnon-Poiré, A.; Francus, P. Morphological signatures of deglaciation and postglacial sedimentary processes in a deep fjord lake (Grand Lake, Labrador). *Earth Surf. Process. Landf.* **2020**, *45*, 928–947. [CrossRef]
14. Pikelj, K.; Juračić, M. Eastern Adriatic Coast (EAC): Geomorphology and Coastal Vulnerability of a Karstic Coast. *J. Coast. Res.* **2013**, *29*, 4. [CrossRef]
15. Correggiari, A.; Roveri, M.; Trincardi, F. Late Pleistocene and Holocene Evolution of the North Adriatic Sea. *II Quat.* **1996**, *9*, 697–704.
16. Lambeck, K.; Antonioli, F.; Anzidei, M.; Ferranti, L.; Leoni, G.; Scicchitano, G.; Silenzi, S. Sea level change along the Italian coast during the Holocene and projections for the future. *Quat. Int.* **2011**, *232*, 250–257. [CrossRef]
17. Benjamin, J.; Rovere, A.; Fontana, A.; Furlani, S.; Vacchi, M.; Inglis, R.H.; Galili, E.; Antonioli, F.; Sivan, D.; Miko, S.; et al. Late Quaternary sea-level changes and early human societies in the central and eastern Mediterranean Basin: An interdisciplinary review. *Quat. Int.* **2017**, *449*, 29–57. [CrossRef]
18. Bakran-Petricioli, T.; Petricioli, D. Habitats in Submerged Karst of Eastern Adriatic Coast—Croatian Natural Heritage. *Croat. Med. J.* **2008**, *49*, 455–458. [CrossRef] [PubMed]
19. Viles, H.A.; Goudie, A.S. Tufas, travertines and allied carbonate deposits. *Prog. Phys. Geogr. Earth Environ.* **1990**, *14*, 19–41. [CrossRef]
20. Primc-Habdija, B.; Habdija, I.; Plenković-Moraj, A. Tufa deposition and periphyton overgrowth as factors affecting the ciliate community on travertine barriers in different current velocity conditions. *Hydrobiologia* **2001**, *457*, 87–96. [CrossRef]
21. Pavlović, G.; Prohić, E.; Miko, S.; Tibljaš, D. Geochemical and Petrographic Evidence of Meteoric Diagenesis in Tufa Deposits in Northern Dalmatia (Zrmanja and Krupa Rivers, Croatia). *Facies* **2002**, *46*, 27–34. [CrossRef]
22. Golubić, S.; Violante, V.; Plenković-Moraj, A.; Grgasović, T. Travertines and calcareous tufa deposits: An insight into diagenesis. *Geol. Croat.* **2008**, *61*, 363–378. [CrossRef]

23. Peña, J.L.; Sancho, C.; Lozano, M.V. Climatic and tectonic significance of late Pleistocene and Holocene tufa deposits in the Mijares river canyon, eastern Iberian Range, northeast Spain. *Earth Surfa. Proc. Land.* **2000**, *25*, 1403–1417. [CrossRef]

24. Ferentinos, G.; Papatheodorou, G.; Geraga, M.; Iatrou, M.; Fakiris, E.; Christodoulou, D.; Dimitriou, E.; Koutsikopoulos, C. Fjord water circulation patterns and dysoxic/anoxic conditions in a Mediterranean semi-enclosed embayment in the Amvrakikos Gulf, Greece. *Estuar. Coast. Shelf Sci.* **2010**, *88*, 473–481. [CrossRef]

25. Cukur, D.; Krastel, S.; Çağatay, M.N.; Damcı, E.; Meydan, A.F.; Kim, S.P. Evidence of extensive carbonate mounds and sublacustrine channels in shallow waters of Lake Van, eastern Turkey, based on high-resolution chirp subbottom profiler and multibeam echosounder data. *Geo-Mar. Lett.* **2015**, *35*, 329–340. [CrossRef]

26. Manoutsoglou, E.; Hasiotis, T.; Kyriakoudi, D.; Velegrakis, A.; Lowag, J. Puzzling micro-relief (mounds) of a soft-bottomed, semi-enclosed shallow marine environment. *Geo-Mar. Lett.* **2018**, *38*, 359–370. [CrossRef]

27. *Topographic Map scale 1:25000 (TK25)*; Državna Geodetska Uprava: Zagreb, Croatia, 2010.

28. Viličić, D.; Terzić, S.; Ahel, M.; Burić, Z.; Jasprica, N.; Carić, M.; Caput Mihalić, K.; Olujić, G. Phytoplankton abundance and pigment biomarkers in the oligotrophic, eastern Adriatic estuary. *Environ. Monit. Assess* **2008**, *142*, 199–218. [CrossRef] [PubMed]

29. Burić, Z.; Cetinić, I.; Viličić, D.; Caput Mihalić, K.; Carić, M.; Olujić, G. Spatial and temporal distribution of phytoplankton in a highly stratified estuary (Zrmanja, Adriatic Sea). *Mar. Ecol.* **2007**, *28*, 169–177. [CrossRef]

30. Viličić, D. Ecological characteristics of the Zrmanja estuary. *Hrvat. Vode* **2011**, *19*, 201–214.

31. Horvatinčić, N.; Krajcar Bronić, I.; Obelić, B. Differences in the 14C age, δ13C and δ18O of Holocene tufa and speleothem in the Dinaric Karst. *Palaeogeogr. Palaeocimatol. Palaeoecol.* **2003**, *193*, 139–157. [CrossRef]

32. Pedley, M. Tufas and travertines of the Mediterranean region: A testing ground for freshwater carbonate concepts and developments. *Sedimentology* **2009**, *56*, 221–246. [CrossRef]

33. Veverec, I.; Faivre, S.; Barešić, J.; Buzjak, N.; Horvatinčić, N. Evolucija fosilne sedrene barijere Gazin kuk u rijeci Zrmanji (Hrvatska). In *Zbornik Sažetaka Završne Radionice*; projekt HRZZ-IP-2013-11-1623 Reconstruction of the Quaternary environment in Croatia using isotope methods—REQUENCRIM, 2014–2018; Institut Ruđer Bošković: Zagreb, Croatia, 2018; p. 33. (In Croatian)

34. Horvatinčić, N.; Geyh, M.A. Interglacial growth of tufa in Croatia. *Quat. Res.* **2000**, *53*, 185–195. [CrossRef]

35. Bonacci, O. Water circulation in karst and determination of catchment areas: Example of the River Zrmanja. *Hydrol. Sci. J.* **2009**, *44*, 373–386. [CrossRef]

36. Fiket, Ž.; Mikac, N.; Kniewald, G. Sedimentary records of the Zrmanja River estuary, eastern Adriatic coast—Natural vs. anthropogenic impacts. *J. Soils Sediments* **2017**, *17*, 1905–1916. [CrossRef]

37. Fiket, Ž.; Pikelj, K.; Ivanić, M.; Barišić, D.; Vdović, N.; Dautović, J.; Žigovečki Gobac, Ž.; Mikac, N.; Bermanec, V.; Sondi, I.; et al. Origin and composition of sediments in a highly stratified karstic estuary: An example of the Zrmanja River estuary (eastern Adriatic). *Reg. Stud. Mar. Sci.* **2017**, *16*, 67–78. [CrossRef]

38. Fiket, Ž.; Ivanić, M.; Furdek Turk, M.; Mikac, N.; Kniewald, G. Distribution of Trace Elements in Waters of the Zrmanja River Estuary (Eastern Adriatic Coast, Croatia). *Croat. Chem. Acta* **2018**, *91*, 29–41. [CrossRef]

39. Juračić, M.; Crmarić, R. Holocene sediments and sedimentation in estuaries eastern Adriatic coast. In Proceedings of the 3rd Croatian Conference on Water, Osijek, Croatia, 28–31 May 2003; pp. 227–233.

40. Vlahović, I.; Tišljar, J.; Velić, I.; Matičec, D. Evolution of the Adriatic Carbonate Platform: Palaeogeography, main events and depositional dynamics. *Palaeogeogr. Palaeoclimatol. Palaeoecol.* **2005**, *220*, 333–360. [CrossRef]

41. Korbar, T. Orogenic evolution of the External Dinarides in the NE Adriatic region: A model constrained by tectonostratigraphy of Upper Cretaceous to Paleogene carbonates. *Earth-Sci. Rev.* **2009**, *96*, 296–312. [CrossRef]

42. Ivanović, A.; Sakač, K.; Marković, B.; Sokač, B.; Šušnjar, M.; Nikler, L.; Šušnjara, A. *Basic Geologcal Map SFRY 1:100 000. Sheet Obrovac L33-140*; Institute of Geology: Zagreb, Croatia, 1967. (In Croatian)

43. Ivanović, A.; Sakač, K.; Sokač, B.; Vrsalović-Carević, I.; Zupanić, J. *Basic Geological Map SFRY 1:100 000. Geology of Sheet Obrovac*; Institute of Geology: Zagreb, Croatia, 1967. (In Croatian)

44. Majcen, Ž.; Korolija, B. *Basic Geologcal Map SFRY 1:100 000. Geology of Sheet Zadar*; Institute of Geology: Zagreb, Croatia, 1973. (In Croatian)

45. Majcen, Ž.; Korolija, B.; Sokač, B.; Nikler, L. *Basic Geologcal Map SFRY 1:100 000. Sheet Zadar L33-139*; Institute of Geology: Zagreb, Croatia, 1973.

46. *Geological Map of the Republic of Croatia scale 1:300,000*; Croatian Geological Survey: Zagreb, Croatia, 2009; 1 sheet.

47. Kovačević Galović, E.; Ilijanić, N.; Peh, Z.; Miko, S.; Hasan, O. Geochemical discrimination of Early Palaeogene bauxites in Croatia. *Geol. Croat.* **2012**, *65*, 53–65. [CrossRef]

48. Hasan, O.; Miko, S.; Ilijanić, N.; Brunović, D.; Dedić, Ž.; Šparica, M.M.; Peh, Z. Discrimination of topsoil environments in a karst landscape: An outcome of a geochemical mapping campaign. *Geochem. Trans.* **2020**, *21*, 22. [CrossRef]

49. Montereale Gavazzi, G.; Madricardo, F.; Janowski, L.; Kruss, A.; Blondel, P.; Sigovini, P.; Foglini, F. Evaluation of seabed mapping methods for fine-scale classification of extremely shallow benthic habitats—Application to the Venice Lagoon, Italy. *Estuar. Coast. Shelf Sci.* **2016**, *170*, 45–60. [CrossRef]

50. Horn, B.K.P. Hill shading and the reflectance map. *Proc. IEEE* **1981**, *69*, 14. [CrossRef]

51. Zevenbergen, L.W.; Thorne, C.R. Quantitative analysis of land surface topography. *Earth Surf. Process. Landf.* **1987**, *12*, 47–56. [CrossRef]

52. Conese, C.; Maselli, F. Use of error matrices to improve area estimates with maximum likelihood classification procedures. *Remote Sens. Environ.* **1992**, *40*, 113–124. [CrossRef]

53. Demirović, D. An Implementation of the Mean Shift Algorithm. *Image Process. Line* **2019**, *9*, 251–268. [CrossRef]

54. Nadilo, B. Reconstruction of the old Maslenica bridge. *Građevinar* **2004**, *56*, 755–762.

55. Zahra, T.; Paudel, U.; Hayakawa, Y.S.; Oguchi, T. Knickzone Extraction Tool (KET)—A new ArcGIS toolset for automatic extraction of knickzones from a DEM based on multi-scale stream gradients. *Open Geosci.* **2017**, *9*, 73–88. [CrossRef]

56. Evans, G.G.; Prego, R. Rias, estuaries and incised valleys: Is a ria an estuary? *Mar. Geol.* **2003**, *196*, 171–175. [CrossRef]

57. Perillo, G.M.E. Definitions and geomorphologic classifications of estuaries. In *Geomorphology and Sedimentology of Estuaries. Developments in Sedimentology*; Elsevier: Amsterdam, The Netherlands, 1995; Volume 53, pp. 17–47.

58. Dalrymple, R.W.; Zaitlin, B.A.; Boyd, R. A conceptual model of estuarine sedimentation. *J. Sediment. Petrol.* **1992**, *62*, 1130–1146. [CrossRef]

59. Lambeck, K.; Rouby, H.; Purcell, A.; Sun, Y.; Sambridge, M. Sea level and ice volume since the glacial maximum. *Proc. Natl. Acad. Sci.* **2014**, *111*, 15296–15303. [CrossRef]

60. Prampolini, M.; Savini, A.; Foglini, F.; Soldati, M. Seven Good Reasons for Integrating Terrestrial and Marine Spatial Datasets in Changing Environments. *Water* **2020**, *12*, 2221. [CrossRef]

 *water*

Article

Bridging Terrestrial and Marine Geoheritage: Assessing Geosites in Portofino Natural Park (Italy)

Paola Coratza [1], Vittoria Vandelli [1,*], Lara Fiorentini [2], Guido Paliaga [3] and Francesco Faccini [3,4]

[1] Dipartimento di Scienze Chimiche e Geologiche, Università degli Studi di Modena e Reggio Emilia, Via Campi 103, 41125 Modena, Italy; paola.coratza@unimore.it

[2] Settore Politiche delle Aree interne, Antincendio, Forestazione, Parchi e Biodiversità, Regione Liguria, Via D'Annunzio 111, 16121 Genova, Italy; lara.fiorentini@regione.liguria.it

[3] CNR Istituto di Ricerca per la Protezione Idrogeologica, sede di Torino, Strada delle Cacce 73, 10135 Torino, Italy; guido.paliaga@irpi.cnr.it (G.P.); faccini@unige.it (F.F.)

[4] Dipartimento di Scienze della Terra, dell'Ambiente e della Vita, Università degli Studi di Genova, Corso Europa 26, 16132 Genova, Italy

* Correspondence: vittoria.vandelli@unimore.it

Received: 28 August 2019; Accepted: 8 October 2019; Published: 11 October 2019

Abstract: Interest in geoheritage research has grown over the past 25 years and several countries have issued laws to encourage improvement and conservation. Investigations on geosites are prevalently carried out on land environments, although the study of underwater marine environments is also of paramount scientific importance. Nevertheless, due to the constraints of underwater environments, these sites have been little explored, also on account of the higher costs and difficulties of surveying. This research has identified and assessed the terrestrial and marine geosites of the Portofino Natural Park and Protected Marine Area, which are internationally famous owing to both the land scenic features and the quality of the marine ecosystem. The goal was to pinpoint the most suitable sites for tourist improvement and fruition and identify possible connections between the two environments. In all, 28 terrestrial sites and 27 marine sites have been identified and their scientific value as well as their ecological, cultural, and aesthetic importance has been assessed. In addition, accessibility, services, and economic potential of geosites has also been taken into account. Both the updated database of terrestrial and marine geosites in the Portofino protected areas and the assessment procedure adopted can become useful tools for the managers of these sites and provide decision-makers with possible strategies for tourist development.

Keywords: underwater geoheritage; geosites; geomorphological survey; geotourism; Portofino Park; Italy

1. Introduction

Geoheritage and geosite studies have assumed growing scientific importance in the past 25 years, and territorial legislative initiatives have emerged all around the world. Geoheritage studies have usually been carried out in terrestrial environments: Mountain areas (e.g., [1–6]), coastal areas (e.g., [7–11]), karst areas (e.g., [12–16]), fluvial areas (e.g., [17–19]), and volcanic areas ([20–23]). Recently, a great deal of interest has concerned also geoheritage in urban areas (e.g., [24–30]).

For what concerns the definition of geosites and their different types of values, they have been much debated within the scientific community (cfr., [31,32] and reference therein). Up to now, two main approaches can be distinguished for defining what geosites are: A restrictive and a broader definition.

According to the restrictive definition, geosites are in situ elements with high scientific value [33], i.e., sites "having particular importance for the comprehension of the history of the Earth and of its present and future evolution" [34,35]. According the broader definition, geosites—or geodiversity sites (*sensu* [33])—are defined as geological elements that present a certain value due to human perception or exploitation, e.g., elements with high scientific, educational, aesthetic, and cultural value. Often, geosites are included in protected areas even if their institution is, in most countries, related to the biological aspects more than the geological ones. In fact, geology has often been inadequately accounted for in parks creation, planning and management. Nevertheless, after decades of focus on the protection of biological heritage, a great deal of progress has been made in the last 20 years (cf., [36] and reference therein). In this respect, particularly notable is the UNESCO Global Programme, which intends to "promote a global network of geoparks safeguarding and developing selected areas having significant geological features" [31,37,38]. Moreover, natural disasters and their tangible evidence in landscape may be important geosites, ideal to promote geological education [39] and geotourism [40,41].

In Italy as elsewhere, the nature conservation in coastal and marine environment is provided by marine protected areas whose nature conservation policy primarily addresses the biodiversity, often underestimating or nearly neglecting abiotic features. Among the European legislative framework worthy of note are the EU Birds Directive (1979), the Habitats Directive (1992), the OSPAR Convention (1992), and the EU Marine Strategy Framework Directive (2008), which have focused the attention towards the marine environment.

Concerning underwater geoheritage and marine geosites [42], despite their importance, only few studies have been developed; this is particularly true when compared with studies on marine biotic heritage ([43] and references therein). This is mainly due to the physical constraints of the marine environment that influence the high costs of underwater surveys and the difficulty of investigating near shore areas, where navigation is not possible. In addition, as highlighted by Burek et al. [44], there are general differences in attributes related to sites of geological and geomorphological interest in terrestrial and marine environments. In a marine environment, geological heritage is largely invisible, except in clear and shallow water, and hardly accessible. These characteristics have reduced the opportunities for promotion, education, and interpretation activities for the public, but at the same time, they have also reduced vulnerability to man-made damage. Furthermore, the different perception and enjoyment of abiotic features of the aquatic environment by tourists has led to a delay in developing common schemes and approaches to the identification, assessment, and improvement of submarine geosites [43].

While many studies have dealt with emerged shorelines [45–47], geoheritage research in underwater environments still lacks common investigation schemes and approaches, again especially in comparison to studies on marine biotic features [48,49]. Specific studies on submerged geoheritage are few and were developed mainly by Italian researchers [43,50–55]. In particular, in Orrù et al. [52], the selection of sites of geomorphological interest was carried out by considering several significant valences as: (i) Model of geomorphological evolution; (ii) exemplarity; (iii) paleo-geomorphological testimonial; and (iv) ecological valence. In the same work, the geosite assessment was carried out considering their scientific interest and other types of interest such as cultural, educational, and historical interests. Similar to this approach was the one used by Rovere et al. [43]; in fact, they evaluated underwater geomorphological heritage in two Mediterranean marine areas by considering two sets of values, that were the scientific and the additional values. The two sets were further divided in subcategories inspired by those proposed for terrestrial environment (e.g., [47,56,57]). Recently, Flores-de la Hoya et al. [58] prosed a method to rapidly assess coastal underwater spots to be used as recreational scuba diving sites. In the latter work, the assessment was based on several criteria inspired by the methodology provided by Ramos [59] for the evaluation of diving site attractiveness in the Algarve region.

As regards marine geoconservation, a growing interest has been recently observed, especially in the UK, where geoheritage has started to be integrated in the management of protected areas ([44,60,61]) and a methodology to assess geodiversity key areas on the seabed has been developed.

From a geoheritage viewpoint, submerged areas are particularly interesting for several reasons:

- Relict landforms, testifying past geological and geomorphological events or paleo-environments, and direct consequences of human interaction are usually better preserved than in a continental environment [52]. In particular, research on climate change occurring in the past 22 ka BP has allowed sea-level fluctuations to be identified up to about −120 m with respect to present levels. The numerous marine markers identifiable in the submerged strip of present-day Mediterranean coasts constitute exceptional archives of long-term paleo-environmental change, with particular reference to climate and sea-level changes (e.g., [62–64]).

- Abiotic heritage has strong interconnections with human life and marine biodiversity, since it plays an important role in providing benefits through the functioning of ecosystems (cf., [65]). The benefits include ecological benefits, such as habitat provision and improvement of fish stocks, social and cultural benefits related to nature appreciation, economic benefits of tourism, and recreational enjoyment of the marine environment.

- Submerged areas are often tourist destinations, with a potential for geotourist popularization of their geological and geomorphological heritage. Enjoyment of the underwater environment focuses mainly on biological attractions, such as marine biota and habitats [51] or cultural l.s elements, such as archeological remnants (e.g., [66,67]) or shipwrecks (e.g., [68,69]) whilst the importance of natural abiotic features is often underestimated [55]. The submerged environments are also used for tourist activities, especially for cultural, historical, and religious purposes. Examples of links between submerged cultural heritage and submerged geoheritage in the Mediterranean have been developed in marine protected areas in Liguria [43,53–55], in the Greek Islands [47], in Sardinia [50,52], and in Malta [70].

According to Rovere et al. ([43] and references therein), a complete approach in the studies of geoheritage in coastal zones should necessarily include the description of both the shore and inner continental shelf, according to the fact that two environments showing common processes and landforms must be considered as a single feature [71]. The need for integrating terrestrial and submerged datasets in geomorphological studies is not new. Examples of studies coupling land and sea data available in literature have considered several aspects, such as: (i) Archaeological investigations (e.g., for the northern coast of Ireland by Westley et al. [72] and Harff et al., [73]); (ii) paleo-environmental reconstruction (e.g., Quaternary geomorphological evolution of the Tremiti Islands, southern Italy [74,75])—especially in fluvial environments; (iii) marine spatial planning [76]; (iv) coastal hazards assessment and risk reduction (e.g., mitigation of the risk due to tropical cyclones, tsunamis, floods, and sea-level rise along the Mozambique coasts [77,78]); (v) integrated geomorphological mapping of emerged and submerged areas (e.g., [79] in the Netherlands; [80] in the Tremiti Islands, southern Italy; [70,81] in the Maltese Archipelago).

The goal of this study is to identify and assess terrestrial and marine geosites—intended, in a broad sense, as component of the cultural heritage of a territory [82,83]—in the Portofino Natural Park (Liguria Region, northern Italy), in order to select sites more suitable for a geotourism exploitation, pinpointing a potential morphogenetic bridge between terrestrial and marine features. These latter are poorly known by the general public especially from geological and geomorphological perspectives. The Portofino Natural Park, which comprises a terrestrial protected area, established in 1935, and a marine protected area, established in 1999, is well known at an international level thanks to its landscape, environmental, and cultural characteristics (Figure 1). Over 1 million people a year visit the sea hamlets of Portofino and Camogli, as well as the coast between Rapallo and Portofino, whereas the 80 km long footpath network is trodden throughout the year by over 100,000 hikers [11]. In recent times, scuba diving activities, managed by the protected marine area administration, have significantly increased. Scuba divers arrive at properly chosen buoys starting from the diving centers of Santa Margherita, Camogli, and San Michele di Pagana (located between Santa Margerita Ligure and Rapallo). The remarkable environmental and cultural features conserved both in terrestrial and marine areas of the Portofino Natural Park led the study area to become an ideal site for the development of geotourism, defined according to the broader approach of the National Geographic in the United States as "tourism that sustains or enhances the geographical character of the place being visited, including its environment, culture, aesthetics, heritage and the well-being of its residents" [84].

Figure 1. Location map of the study area (modified from [11]).

2. Geographical Setting

The Promontory of Portofino breaks the continuity of the coastline between Genoa and La Spezia, along a perimeter of 13 km and an area of 18 km^2. The orography is characterized by rather high peaks, considering the short distance from the sea [85]. There is a WNW–ESE oriented relief, the culmination of which corresponds to the Mount of Portofino (610 m). Hydrographic catchments are less than 1 km^2 wide, with channels of the second order at the most [86]. Among the most important catchments of the southern slope, the following can be quoted: Cala dell'Oro catchment, located west of San Fruttuoso bay; San Fruttuoso catchment; Ruffinale and Vessinaro catchments, both located between San Fruttuoso Bay and Portofino promontory. Whereas, on the eastern side, the Rio del Fondaco at Portofino and Fosso dell'Acquaviva at Paraggi are found [87,88].

Due to the torrential regime, the flow rates of watercourses are substantially nil for most of the year; in the case of heavy rainfall of short duration (not infrequent in the area), the maximum flow rates, for return times of 200 years, range between 20 and 40 m^3/sec (flow rate unit contribution of 40 m^3/sec/km^2 for catchment areas of less than 1 km^2).

The Portofino Park protects the area of the promontory bearing the same name, which is located less than 20 km away to the east of Genoa. To date, the protected area is 1056.26 ha, out of which 58.61 ha make up the integral reserve, 597.31 ha are the general reserve area, and 362.50 ha are the protection area. The remaining 37.84 ha belong to the economic promotion area [11]. The contiguous territory adds an extra 932 ha to the Park (Figure 2).

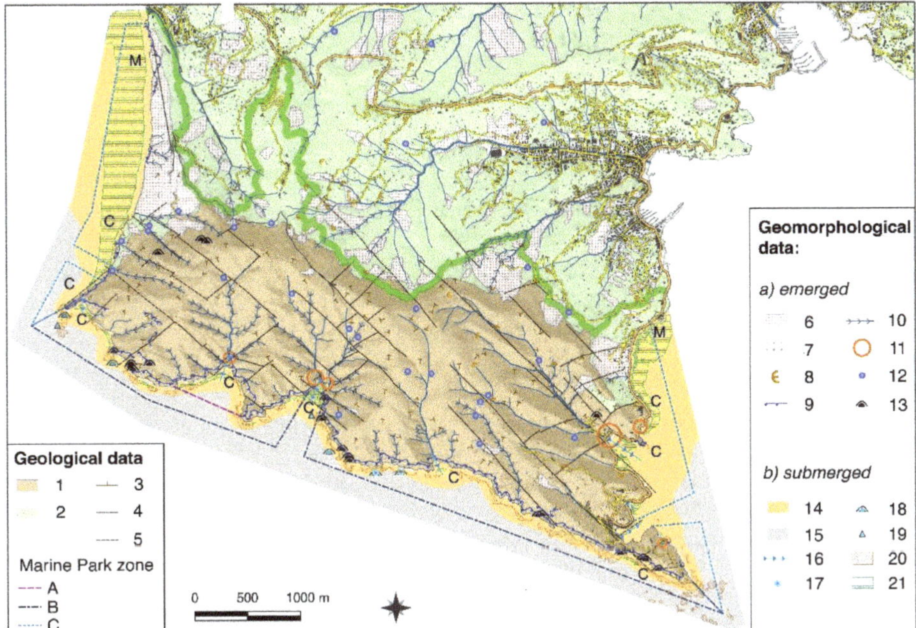

Figure 2. Geological and geomorphological sketch of the study area (modified from [11,86–88]):
(**1**) Conglomerate; (**2**) marly limestone flysch; (**3**) bed attitude; (**4**) fault; (**5**) tectonic lineament; (**6**) landslide
and debris covers; (**7**) anthropic fill; (**8**) soil slip; (**9**) edge of sea cliff scarp; (**10**) incising channel;
(**11**) pocket beach; (**12**) spring; (**13**) cave; (**14**) submerged cliff with boulders, sands in the small bays
(San Fruttuoso, Portofino and Paraggi); (**15**) sea bottom with sands and muddy sands; (**16**) submerged
channel; (**17**) submerged spring; (**18**) submerged cave; (**19**) submerged peak; (**20**) coralligenous reef;
(**21**) seagrass meadow.

The area of the Park stretches over the municipal territories of Camogli, Portofino, and Santa
Margherita Ligure, whereas the contiguous area is part of the municipal territory of Rapallo. The residing
population of the Park is around 750 inhabitants. The presence of tourists is high throughout the year:
At the village of Portofino, there are over 1 million tourists per year, whereas at San Fruttuoso, tourist
boats carry some 400,000 tourists/year around the Gulfs of Tigullio and Paradiso [11,86]. Apart from
seaside tourism, there is also a high presence of hikers along the over 80 km long footpaths: Just the
stretch from Portofino Vetta to Pietre Strette is trodden by over 70,000 hikers per year.

Thanks to its landscape, natural, and cultural values [89,90], the Promontory of Portofino has
been protected since 1935 by Italian Law no. 1251 (Establishment of the local authority of Mount of
Portofino). Since 1995, it has been managed as 'Ente Parco', established by Ligurian Regional Law
no. 12/95 (Reorganization of protected areas), which redefined the borders of the protected area with
Regional Law no. 29/2001 (Identification of the perimeter of the Portofino Natural Regional Park).

The Marine Protected Area of Portofino, established by the Italian Ministry for the Environment,
was added to the Park with the Decree of 26/04/1999, which implemented the Italian Law no. 979/1982
(Measures for the Sea Protection). The marine area is subdivided into three zones of safeguard (A, B,
and C), in which free navigation, hunting or catching of fauna, underwater fishing, and diving are
forbidden. In addition, all underwater activities that require contact with the seabed are forbidden,
as well as the anchoring of any boat [91]. Zone A (Integral Reserve) comprises the sea area of Cala
dell'Oro bay (west of San Fruttuoso bay). Access to this area is permitted only for emergency rescue
and authorized scientific research. Zone B (General Reserve) stretches from the Portofino lighthouse

point to Punta Chiappa, excluding the access corridor to the harbor of San Fruttuoso. Less restraining issues characterize this zone: Authorized sport fishing is allowed for residents, scuba diving is allowed for diving centers and authorized private subjects, whereas bathing is free. This marine area is very popular among scuba divers, who are attracted by the considerable natural beauty of the seabed and, in particular, by the great number of violescent sea-whips (*Paramuricea clavata*) and the richness of sea fauna. Zone C (Partial Reserve) stretches between the two sides of the Promontory of Portofino and owes its fame to the vast prairies of Mediterranean tapeweed (*Posidonia oceanica*). Bathing, scuba diving, and sport fishing are allowed. On the whole, over 70,000 scuba divers per year plunge into the water of the Portofino Protected Marine Area [92].

Recently, in December 2017, the Portofino National Park was established. It comprises both the terrestrial area and the protected marine area. By the end of 2019, the Italian Ministry for the Environment will establish the new borders of this National Park with a specific law.

3. Geological, Geomorphological, and Hydrogeological Setting

The geology of the Portofino National Park is known at an international level owing to the presence of Portofino Conglomerate, whose lithological nature and geological and geomorphological significance have in fact been widely studied (e.g., [93–95] and references therein) as well as its geomechanical behavior (e.g., [88,96]). The conglomerate forms the trapezoid-shaped promontory between Punta Chiappa to the West and the Portofino lighthouse to the East. The geological root of the Portofino Mount, between Camogli and Rapallo, is characterized by a marly-calcareous flysch (Mt. Antola Flysch). The boundary between these two geological formations (pudding stone and flysch) is partially ascribable to tectonic causes and shows a WNW–ESE trend (Figure 2). The Promontory morphology is derived from a structure bounded by normal faults, typical of a continental margin subject to disjunctive tectonics [97,98].

The Portofino Conglomerate is made up of marly-calcareous clasts and, to a lesser degree, sandstones, ranging in size from centimeters to meters, arranged in several-meter thick layers with rare sandstone intervals, often accompanied by thin coal layers. Ophiolite, limestone, cherts, and gneiss clasts are also found, although less frequently. This conglomerate, which lacks a fossil record, was dated doubtfully to the Oligocene due to the scarcity of biostratigraphic records [97,98].

On the whole, the structural setting of the Conglomerate shows a SE to SW dip, with a less than 20° inclination. The rock mass is affected by various joint systems, easily identifiable at a meso- and macro-scale. The NW–SE and NE–SW oriented systems, which are ascribable to normal faults, are the most important. At a slope scale, the intersection between the various joint systems produces the subdivision of the conglomerate into several decameter-thick blocks [99].

Mt. Antola Flysch, dating to the Cretaceous, is made up of calcareous marls and marly limestones, marls with argillite levels, siltites, and calcarenites. The structural setting of the flysch is constrained by diverse deformation phases, both ductile and brittle, which affected this rock mass. An isoclinal-fold arrangement was identified in this formation; it shows a SSW vergence with a WNW–ESE oriented axis [95].

Landforms in the study area are controlled by geological-tectonic setting and conditioned by meteo-climate conditions [87,88]. Rocky cliffs up to 200 m high, the highest of the Mediterranean coast, characterize the southern slope of the Promontory of Portofino [93]. The average inclination of the slope is 45° to 65°, although many are the coastal stretches characterized by vertical cliffs [94]. The action due to swell is important and is determined by both SE wind ('Scirocco', dominant wind), and SW wind ('Libeccio', prevailing wind). Sea storms are rather frequent, with wave heights exceeding 5 m; they can cause serious damage to buildings and infrastructures, as in the event of 27–29 October 2018, which affected the Promontory eastern coast, between Rapallo and Portofino.

The profile of the emerged cliff continues underwater up to a depth of some 70 m. Up to the margin of the shelf, some 140 m deep, the inclination of the seabed is rather homogeneous and gentle.

The margin of the shelf, which is not influenced by the presence of the promontory, is found at a distance of 3.5 to 4 km from the coast [100].

The base of the narrow continental slope is found at a depth of 0.6 to 1 km, in correspondence with a furrow named 'Canyon della Riviera di Levante', which stretches in an E–W direction, with the confluence of a small canyon formed in the West front of the Promontory of Portofino [85,100].

The morphologically significant tectonic alignments, which contour Mount Portofino with landforms such as saddles, towers, and triangular facets, continue in the submerged portion. Some of the faults are considered active, since they disrupt the seabed in their underwater part.

In the high conglomerate cliffs of the southern slope, there are often rock falls, even along very steep fluvial channels, mostly of the first order, as in the case of the torrents Ruffinale and Vessinaro [96]. On the western slope, the cliff has been prevalently modelled in Mt. Antola Flysch, attaining heights exceeding 100 m. This stretch of coast is subject to a SW swell, which is one of the main causes for occurrence of rapid slope movements, such as debris/mud flows and rock avalanches, which often have a high destructive power [94]. There are also slow slope movements with surface of rupture in the marly-calcareous bedrock. In this case, numerous morphotectonic clues suggest a process of the mountain slope deformation type [101]. Along the boundary between the conglomerate and flysch, there are landslides of diverse origin and state of activity owing to the contrast of resistance and deformability between adjacent rock masses [11]. Among the landslide bodies surveyed, worthy of note is the accumulation found at Sotto Le Gave, on the eastern slope, which is partially due to mountain slope deformation and has affected buildings and infrastructures even in the recent past.

In the submerged area comprised within 200 m from the coastline, morphological rises linked to neotectonic modelling are found, as South of Punta Chiappa (*Secca dell'Isuela*), E of San Fruttuoso (*Secca Gonzatti*), and SE of Punta Portofino. This portion of the seabed reveals exceptional biodiversity [91], also resulting from geomorphological features. The widespread coralline biocoenosis and tapeweed prairies, which characterize most of the seabed near the coast, are developed on large rock blocks (>1 m).

The meteo-climatic characteristics of this area are linked to the cyclogenesis of the Gulf of Genoa, which causes events of short but intense precipitation (less than 6 h, with rain peaks exceeding 50 mm/h) between mid-summer and mid-autumn [93,102,103]. Consequently, the most common effects at ground level are flash floods, hyper-concentrated fluxes, and debris/mud flows. Among the most significant and destructive events in living memory, those of 1915, 1961, and 1995/1996 should be mentioned. Also, in the 2000–2018 period, many extreme hydro-meteorological events occurred on Portofino Promontory, causing important effects at ground level with considerable damage to buildings and infrastructures: The average, on a historical basis, is over one event per year [11].

The Ligurian Speleological Registry lists 20 caves in the Portofino Conglomerate [104]. Their origin is prevalently tectonic although, to a much lesser extent, is due to chemical–physical dissolution or processes linked to the sea wave action. In addition, several natural caves have been surveyed in the submerged portion of the cliff, up to a depth of 60 to 70 m; also, their genesis is a result of tectonic modelling.

The intense joint network of the conglomerate, the contrast of hydraulic conductivity with the marly-calcareous flysch, and the climate characteristics of the territory cause significant effective infiltration with widespread presence of groundwater and springs [99]. Effective infiltration ranges from 350 mm/y at sea level up to over 500 mm/y at higher elevations. The water springs are located either along the contact between the Conglomerate and the marly-limestone Flysch, or in the Conglomerate rock mass, along tectonic lineation, or along the interface with the sandy interlayers. Underground aquifers are extremely fragmented, with annual intermittent flow rates ranging from less than 1 L/min in dry summer to over 10 L/s in late autumn. Some of these springs have been used for a long time and today still feed local water-supply systems [93]. There are also significant springs underwater, along the submerged cliffs, and at the connection with the shelf. The latter is an important morphological element indicating the position of the sea-level at the end of the Würm regression. Furthermore, these features bear witness to the neotectonic activity taking place in the Plio-Quaternary, with uplifting and lowering phenomena affecting the seabed.

Among anthropic forms, drywall slope terracing is a very common farming technique, which dates back to ancient times. Terracing has deeply modified the geomorphological, vegetation, and dwelling landscape at a slope scale. Well-preserved examples of slope terracing are found in the Valloni di Paraggi, Portofino, and San Fruttuoso; they make up an important cultural and landscape asset.

4. Materials and Methods

The increasing interest in the promotion of geotourism requires the selection and assessment of geosites in order to determine priorities in site management and geoconservation strategies. Based on these premises, a research program for the identification and assessment of geosites at the Portofino Natural Park has been developed.

4.1. Geosites Identification

Research on geosites at Portofino Natural Park has taken advantage of the numerous thematic maps and scientific publications on the geology and geomorphology of the study area, concerning both emerged and submerged areas of the Park, as well as tourist maps and guidebooks. Some milestone publications have been particularly significant for the aims of this study, such as Ristori [105] on the Conglomerate and groundwater regime at Mount Portofino and Pellati [85] on the geomorphological characteristics of the Promontory of Portofino. As for geological and petrographic features of the conglomerate, the contributions by Giammarino et al. [97] and Giammarino and Messiga [98] should be mentioned. As concerns geomorphological features, in recent times there have been contributions on geomorphological hazard and tourist vulnerability along the Park footpaths [86], on the landslides of the western slope of Mount Portofino [94], and on geomorphological mapping of San Fruttuoso and Portofino [87,88]. In addition, other publications have been taken into account: The debris flows along the coast [93], the hydrogeology of the Caselle springs [99], and the terracing of the Park considered as a cultural asset [90]. Salmona and Varardi [91] discuss the socioeconomic aspects of the protected marine area, whereas other contributions deal with underwater tourism and related impact on the ecosystem. Cerrano et al. [92] stress the importance of volunteer scuba divers for scientific activities aiming at the conservation of Mediterranean natural resources and [106] describe the success of scuba diving in the Portofino protected marine area. Furthermore, Lucrezi et al. [107] illustrate the contribution of scuba divers in the management of protected marine areas and, again, Lucrezi et al. [107] pinpoints the correct balance between scuba diving activities and environmental sustainability. Saayman and Saayman [108] discuss the economic benefits resulting from scuba diving in protected marine areas, and Di Carro [109] describes an approach for assessing human impact on the Portofino protected marine area. Finally, Markantonatou et al. [110] develops a study on social networks and the flow of information for responsible and sustainable planning in the Portofino protected marine area.

For the selection of terrestrial geosites (Table 1 and Figure 3), this study took advantage of the inventory developed by Faccini et al. [11] where geosites have been selected and classified according to their main scientific relevance in: Geological, geomorphological, mineralogical-petrographic, hydrogeological geosites, and viewpoints (*sensu* [111]).

A geoheritage inventory for the underwater part of the area investigated was lacking. Therefore, marine geosites were selected (Table 2) by combining geological and geomorphological data in strict collaboration with park managers. The sites were classified according to their main scientific relevance as geomorphological, speleological, and hydrogeological geosites (Tables 1 and 2 and Figure 3). Since geosite assessment is important for the promotion of the area from a geotourism perspective, two marine sites of cultural interest and great tourist potential have been included (*Cristo degli Abissi*—ID 21S; and Mowak Deer shipwreck—ID 13S). These sites show a complex relationship between the natural and/or human heritage of the Portofino Park [11]. The underwater geosites have been classified into two categories according to the skills of the visitors: (i) Snorkeling sites (more or less available to everybody) and (ii) sites equipped for qualified scuba divers. Snorkeling sites have been classified based on direct observations and comprise practically all the free bathing sites of the Portofino Protected Marine Area. They are pocket beaches, often fed by annual beach nourishments, apart from the Punta Chiappa site, which is a rocky cliff. The scuba diving sites are managed by the Portofino protected marine area and are identified by 21 signaling buoys, where one or two boats can be moored. The diving sites have been classified into three categories, according to their technical difficulties. The scientific data reported in the cards of each diving point, which have been elaborated by the Portofino protected marine area [112], have been updated with new original observations on each diving point up to a maximum depth of 45 m.

Table 1. Terrestrial geosites.

Nr	Name/Location	Scientific Interest	Features/Description
1T	Pietre Strette	Geological	Conglomerate
2T	St George Church	Geological	Conglomerate
3T	St Rocco	Geological	Marly limestone
4T	P.ta Chiappa	Geological	Conglomerate
5T	P.ta Pedale	Geological	Marly limestone
6T	Pietre Strette	Geomorphological	Boulders
7T	Vitrale	Geomorphological	High cliffs
8T	P.ta Cervara	Geomorphological	Sea stack
9T	Mt. Campana	Geomorphological	Mass movement (lateral spread)
10T	P.ta Budego	Geomorphological	High cliffs
11T	Cala dell'Oro	Geomorphological	Inlet
12T	Pietre Strette	Minero-Petrographical	Anagenite
13T	Cala dell'Oro	Minero-Petrographical	Coal interlayers
14T	St Rocco	Minero-Petrographical	Abandoned quarry
15T	Rio Gentile	Minero-Petrographical	Abandoned quarry
16T	Coppelli	Hydrogeological	Natural springs
17T	Acquaviva	Hydrogeological	Natural springs
18T	Caselle	Hydrogeological	Natural springs
19T	Vegia	Hydrogeological	Natural springs
20T	St Rocco	Viewpoints	Viewpoints
21T	Batterie	Viewpoints	Viewpoints
22T	Toca saddle	Viewpoints	Viewpoints
23T	Castelletto	Viewpoints	Viewpoints
24T	Rocca del Falco	Viewpoints	Viewpoints
25T	Base O	Viewpoints	Viewpoints
26T	Mt Campana	Viewpoints	Viewpoints
27T	Semaforo Nuovo	Viewpoints	Viewpoints
28T	Sotto le Gave	Viewpoints	Viewpoints

4.2. Geosite Assessment

Evaluation of geosites has been developing since the 1990s (cf., [32,113–115]). In spite of many published methods about the assessment of sites, the scientific literature reveals that there is still a great debate concerning values and criteria to be used in the geosite assessment process (see [31,116] and reference therein) and there is no general accepted method. One of the most popular approaches for geosite assessment is the comparative analysis of geosites within a given area, by applying numerical evaluation of their values, based on several criteria and respective indicators (e.g., [33,56,117–121]). The aim of a quantitative assessment is to reduce subjectivity [122] associated with any evaluation procedure, since the intrinsic value of these environmental elements cannot really be measured. Indeed, the scientific quality of a geosite is a purely indicative numerical quantity, which can be subject to variations determined by the subjectivity of the operators and the general characteristics of the area under examination.

Figure 3. Examples of geosites in the Portofino Natural protected areas: (**1**) Grotta dell'Eremita (44.31797 N 9.15120 E, marine geosites 3S and cave on the sea cliff); (**2**) Cristo degli Abissi (44.314038 N 9.174979 E, marine geosites 21S); (**3**) Mt. Campana (44.31973 N 9.15529 E, terrestrial geosites 26T); (**4**) High cliffs at Vitrale (44.30389 N 9.20164 E, terrestrial geosites 7T); (**5**) Punta Cervara stack (44.31355 N 9.21291 E, 'lo scoglio della Carega', terrestrial geosites 8T); (**6**) Rock fall boulders at Castello di Paraggi (44.31112 N 9.21241 E, marine geosites 26D). Image 1 and 2 from Portofino Marine Protected Area archive [100].

Table 2. Marine geosites (minimum and maximum depth are expressed in meters). The eighth column refers to the qualitative level of difficulty to reach a certain submerged geosite through scuba diving (source: [112]).

Nr	Name/Location	Scientific Interest	Features/Description	Protection Zone	Min Depth	Max Depth	Difficulty
1S	Punta Chiappa di Levante	Speleological	Cave	B	10	40	high
2S	Punta della Targhetta	Geomorphological	Submerged cliff	B	8	20	low
3S	Grotta dellEremita	Geomorphological	Cave	B	5	40	low
4S	Punta della Torretta	Geomorphological	Submerged cliff	B	10	35	high
5S	Punta dell'Indiano	Geomorphological	Submerged cliff	B	16	45	high

Table 2. *Cont.*

Nr	Name/Location	Scientific Interest	Features/Description	Protection Zone	Min Depth	Max Depth	Difficulty
6S	Il Dragone	Geomorphological	Landslide blocks, cliff	B	5	40	high
7S	La Colombara	Speleological	Cave	B	10	30	medium
8S	Secca Gonzatti (Secca Carega)	Geomorphological	Submarine relief, shoal	B	5	30	medium
9S	Targa Gonzatti	Geomorphological	Landslide blocks	B	12	33	high
10S	Scoglio del Raviolo (Andrea Ghisotti)	Hydrogeological/ Geomorphological	Cave and submarine spring	B	10	35	medium
11S	Testa del Leone	Hydrogeological/ Geomorphological	Cave and submarine spring	B	8	35	medium
12S	Scoglio del Diamante	Geomorphological	Landslide blocks	B	10	30	low
13S	Mohawk Deer	Geomorphological/ Cultural	Landslide blocks and scarp	B	10	40	high
14S	Punta Vessinaro	Geomorphological	Landslide blocks, submarine cliff	B	10	35	high
15S	Casa del Sindaco - Vitrale	Geomorphological	Submarine cliff, landslide blocks	B	20	40	high
16S	Chiesa di San Giorgio (La Liscia)	Geomorphological	Landslide blocks, cave	B	10	40	medium
17S	Punta del Faro	Geomorphological	Landslide blocks, cliff	B	16	40	medium
18S	Secca dell'Isuela	Geomorphological	Submarine relief	B	14	40	high
19S	Punta dell'Altare	Geomorphological	Submarine cliff	B	10	35	medium
20S	Punta Chiappa ponente	Geomorphological	Cliff, debris covered sea bottom	B	5	25	low
21S	Cristo degli Abissi	Geomorphological/ Cultural	Debris covered sea bottom, rocky blocks	B	12	30	low
22D	Punta Chiappa	Geomorphological/ Geological	Cliff, submarine scarp	B/C	0	5	low
23D	Baia di San Fruttuoso	Geomorphological/ Ecological	Seabed, submarine scarp, Posidonia meadows	B	0	3	low
24D	Baia dell'Olivetta	Geomorphological	Boulders, submarine cliff	C	0	3	low
25D	Baia di Paraggi	Geomorphological/ Ecological	Seabed, submarine cliff, Posidonia meadows	C	0	3	low
26D	Castello di Paraggi	Geomorphological	High cliff, rock fall boulders	C	0	5	low
27D	Punta Cervara-Punta Pedale	Geomorphological/ Geological	Landslide boulders, Posidonia meadows	C	0	5	low

Some methods for quantitative assessment of geosites are based on combined numerical indices to obtain a final score, often named Q-value or global value (e.g., [57,123,124]). This index corresponds to the combination of three sets of criteria relevant to: (i) intrinsic characteristic of a geosite (e.g., degree of scientific knowledge), (ii) potential for use, (iii) need for protection.

Other authors preferred methodologies based on independent criteria (e.g., Brilha's [33] methodology) without the determination of a final score, but considering the results of each set of criteria relevant to a given site. This is because the criteria considered are independent of each other and because the independent numerical evaluation for each criterion enables the individual analysis of each geosite. Specific geosite assessment procedures vary in terms of both the number and type of criteria considered as well as weighing individual parameters and indicators. The criteria generally used for geosite and geomorphosite assessment can be classified into five categories as follows [120,125]:

1. Scientific/intrinsic (scientific merit) values;
2. Exemplarity and educational potential of the site;
3. Accessibility to the site and presence of tourist infrastructures;

4. Existing threats and risks;
5. Added values.

The criteria are preferably adapted to the geological and geomorphological context of the study area.

In the present study, the recognized terrestrial and marine geosites have been quantitatively assessed by applying a methodology that has been specifically set up on the basis of previous works ([47,56,57,119,126]), which concerned the evaluation of terrestrial sites of geomorphological interest and which were applied also to underwater geosites [11]. Although the coastline marks the boundary between terrestrial and marine environments, there is a continuity of geological and geomorphological features across this boundary [61]. In order to fulfil this continuity between land and sea the same assessment methodology for terrestrial and marine geosites was applied. This methodology is based on three sets of values relevant to scientific, additional, and potential for use (Tables 3–5) and the evaluation process builds on bibliographical data and on the detailed and well consolidated knowledge of the geological and geomorphological features of the study area acquired by authors. Scientific value was divided into four sub-criteria (Table 3): Integrity (INT), representativeness (REP), rareness (RAR), and paleogeographic model (PAL). Additional value was divided into three sub-criteria (Table 4): Ecological (ECOL), aesthetical (AEST), and cultural (CULT). Potential for use value was divided into three sub-criteria (Table 5): Accessibility (ACC), services (SER), and economic potential (ECON).

A score between 1 and 5 was attributed to each sub-criterion. For each geosite, the total scientific/additional/potential for use value (*Totval*) was estimated by summing the score of each sub-criterion (a_i) and dividing by the number of sub-criteria (n_a) for each set of values (cf. Equation (1)):

$$Tot\ val = \frac{\sum_i a_i}{n_a}. \tag{1}$$

The aesthetic value is the most subjective one and for the definition of criteria and its assessment, research on landscape perception (see e.g., [34,127] for a review) has been taken into account. Table 4 specifies which features are to be considered in order to assess the aesthetic value of a given geosite. According to Coratza et al. [119], these features are: (i) panoramic quality, (ii) colour diversity, (iii) vertical development, iv) naturalness. The cultural value is the more heterogeneous sub-criterion (Reynard et al., 2007). Therefore, in order to guide the assessment procedure, the features considered to estimate the cultural value for a given geosite are specified in Table 4. Considering the tourist vocation of the area, attention was devoted to the assessment of the potential for use value. In particular, for the estimation of accessibility (ACC), two sets of sub-criteria were taken into account for terrestrial and marine geosites, as shown in Table 5. In order to estimate the economic potential of a site, the number of visitors per year has been taken into account. In fact, it can be assumed that the greater the number of visitors, the more the economic income. In particular, for the study area, two different sets of thresholds, regarding the number of visitors per year, were considered in the assessment of the economic value of terrestrial and marine geosites, respectively. An exception was made for estimating the economic value of marine geosites accessible via snorkeling. For these, the same visitor thresholds as the terrestrial geosites have been here considered, since they are comparable to terrestrial geosites in terms of number of visitors. The data on visitor influx along the footpath network across the present geosites, which have been given by the Park authority, are an indispensable element for judging the economic potential of the area. For this purpose, eco-counters aiming to monitor hikers have been installed. As for the number of visitors to marine geosites accessible with scuba diving equipment, the management is ruled by an agreement between diving centers and the administration board of the protected marine area, which has also provided us with attendance data concerning each diving point.

Table 3. Sub-criteria used for the numerical assessment of geosite scientific value.

Scientific Value	1	2	3	4	5
Integrity (INT): State of conservation of a landform	Poor conservation due to natural causes (after [43])	Poor conservation due to inadequate management inadequate management	Damage may occur in some parts of landform but landscape integrity is preserved	Good conservation due to proper management	Good conservation due to natural conditions
Representativeness (REP): exemplarity with respect to a reference space [119]	No exemplarity (after [43])	Poor example of process or landform	Fair example of process or landform	Good example of process or landform	Reference site (in scientific literature) for the description of process or landform
Rareness (RAR): rarity of the site with respect to a reference space [57]	Very common	Rare at a local scale[1]	Rare at a regional scale	Rare at a national scale	Rare at an international scale
Paleogeographical model (PAL): Importance of a site in defining processes or environments characterizing the Earth history (modified after [43])	No paleogeographic value (after [43])	Scarce paleogeographic significance	Good representation of a paleoprocess	Good representation of a paleoenvironment	Good representation of a paleoprocess and a paleoenvironment

[1] Rare at the scale of Portofino Natural Protected area.

Table 4. Sub-criteria used for the numerical assessment of geosite additional value.

Additional Value		1	2	3	4	5
Ecological value (ECOL): presence of ecotypes and level of the site protection for its natural features [128]		No ecotypes and no site protection	Presence of ecotypes without any protection	Presence of rare ecotypes and protection at a local level	Presence of rare ecotypes and protection at a regional level	Presence of rare ecotypes and protection at a national level
Aesthetic value (AEST): [119]	**Panoramic quality**	Site not visible from any viewpoint	Site visible from one viewpoint	Site visible from more than one viewpoint	Site visible at 360° but within a close distance	Site visible from many viewpoints also at a great distance
	Color diversity	No color diversity	Low color diversity	Moderate color diversity	High color diversity	Very high color diversity
	Vertical development	Same level as the surrounding ground	Slightly emerging from the surrounding ground	Moderately emerging from the surrounding ground	Significantly emerging from the surrounding ground	Imposing feature in the landscape
	Naturalness	Completely modified by human intervention	Strongly affected by human intervention but some natural features are still preserved	Moderately affected by human intervention but most of the natural features are preserved	Slightly affected by human intervention	No traces of human intervention
Cultural value (CULT): [119]	**Religious importance**	No religious importance	Religious importance but no connection to geological and geomorphological features of the site	Religious importance with connection to geological or geomorphological features of the site	Local religious importance with connection to geological and geomorphological features of the site	National religious importance with connection to geological and geomorphological features of the site
	Historical importance	No historical importance	Historical importance but no connection to geological and geomorphological features of the site	Historical importance with connection to geological or geomorphological features of the site	Local historical importance with both connections to geological and geomorphological features of the site	National historical importance with both connections to geological and geomorphological features of the site
	Artistic and/or literature importance	No artistic and literature importance	Artistic and/or literature importance but no connection to geological and geomorphological features of the site	Artistic and/or literature importance with connection to geological or geomorphological features of the site	Local artistic and/or literature importance with connections to both geological and geomorphological features of the site	National artistic and/or literature importance with connections to both geological and geomorphological features of the site

Table 5. Sub-criteria used for the numerical assessment of geosite potential for use.

Potential for Use Value		1	2	3	4	5
Accessibility (ACC): level of accessibility	Land	No access	Accessible only by experts with specific technical skills (e.g., climbers, speleologists)	Accessible by experts but no specific technical skills are required	Accessible by people with normal movement capacity	Accessible by people with limited movement capacity
	Sea [43]	No accessibility or only with indirect methods (e.g., ROV, submersible)	Accessible to expert professional divers or speleology divers	Accessible to 2nd level SCUBA divers (max depth 40 m)	Accessible to 1st level SCUBA divers (max depth 18 m)	Accessible to snorkeling
Services	Land: presence of equipment and support services in the nearby [119]	No services	Support services within a walkable distance but subject to seasonal availability	Equipment available but subject to seasonal availability	Equipment and services in the near proximity of the site subject to seasonal availability	Equipment and support services in the near proximity of the site, available 7/24 all year round
(SER)	Sea: distance from the nearest boarding dock [58]	distance from the boarding dock > 10 km	distance from the boarding dock between 10 and 7 km	distance from the boarding dock between 7 and 5 km	distance from the boarding dock between 5 and 2 km	distance from the boarding dock < 2 km
Economic potential (ECON): number of visitors per year	land	Visitors < 5000	5000 < visitors ≤ 20,000	20,000 < visitors ≤ 50,000	50,000 < visitors ≤ 70,000	visitors > 70,000
	sea	Visitors ≤ 100	100 < visitors ≤ 400	400 < visitors ≤ 700	700 < visitors ≤ 1000	Visitors > 1000

5. Results

Twenty-eight terrestrial geosites and 27 marine geosites were identified and assessed (Figure 4 and Tables 1 and 2). The terrestrial geosites are mainly sites of geomorphological interest of tectonic origin or gravity-induced slope landforms or even coastal landforms, all strictly linked to each other in terms of origin and geomorphological evolution. The footpath network follows the distribution of the terrestrial geosites, which are widespread all over Portofino Park. Marine geosites are mainly concentrated between Punta Chiappa di Levante and Punta Portofino, which is the largest outcrop area of the Portofino Conglomerate.

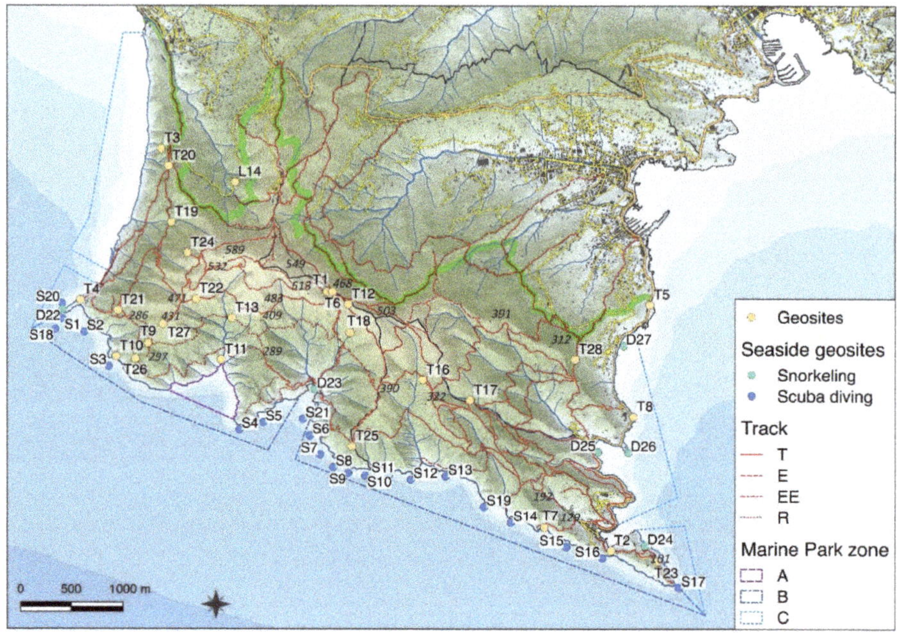

Figure 4. Location of terrestrial and marine geosites.

The geosites selected were evaluated considering their scientific value. Moreover, the research assessed numerically geosite additional value in terms of ecological and eventually cultural importance of the sites as well as their aesthetic quality. Additionally, the potential for use was estimated taking into account the visit conditions, proximity, and availability of services and economic potential of each site. If we compare the average scientific, additional, and potential for use values in both marine and terrestrial geosites (Figure 5), it can be noticed that these values are similar and comparable to each other and no significant variation was identified. Notwithstanding the adoption of the same assessment methodology for terrestrial and marine geosites, some sub-criteria (accessibility, services, and economic potential) have been adapted considering the different characteristics between terrestrial and marine sites. This approach has allowed a balanced evaluation of geosites between the terrestrial and marine area. Moreover, the relationships between the two environments have been better defined.

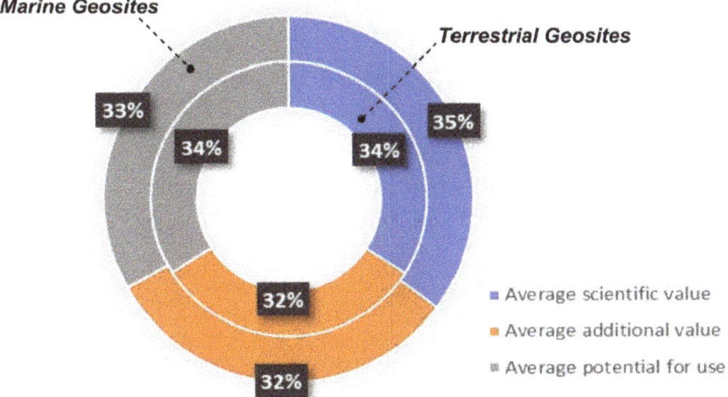

Figure 5. Compared graphs referring to the average of the scientific, additional, and potential for use total values (cf. Section 4.2), calculated considering all the assessed terrestrial (inner circle) and marine geosites (outer circle), respectively.

At the same time, the use of independent assessment criteria permitted the individual analysis of the geosites and the identification of opportunities, weaknesses, and restrictions on tourism development. Indeed, these are fundamental steps for the enforcement of Park management strategies [129].

Multivariate representation (Figures 6 and 7) allows one to compare geosites to each other. Different colors indicate the main scientific interest (geomorphological, speleological, or hydrogeological) of each geosite and the form of the cartograms indicates groups of geosites (e.g., geosites with high scientific value but low potential for use value).

Regarding the scientific value, the assessment of the sub-criteria has revealed that the majority of geosites both terrestrial and marine are well preserved. This confirms the success of conservation strategies applied in the area by the Park authorities. Moreover, most of the geosites are fair to good examples of geological/geomorphological processes and landforms, in terms of level of representativeness (REP). Both terrestrial and marine geosites can be considered as rare at a regional scale, whereas the *Cristo degli Abissi* site (ID 21S) is exceptional at an international scale. High cliffs, more than 150 m a.s.l., set up along active normal faults, continue below sea-level and evolve due to retrogressive erosion, as witnessed by submerged rock fall deposits located on the sea bottom at different distances from the cliffs. These rock fall deposits are good examples of retrogressive processes favored by intense faulting and originated by gravity-induced processes. In addition, good examples of paleo-processes are offered by submerged caverns of structural origin, which are the continuation of land caves. Terrestrial conglomerate outcrops and submarine reliefs are good representatives of past paleo-environments. In particular, submarine reliefs (e.g., the *Secca Gonzatti* geosite—ID 8S) and saddles are ascribable to deep-seated gravitational slope deformations. Many terrestrial springs gushing in the conglomerate or at the boundary between conglomerate and marly-calcareous flysch, are found also below sea-level and witness the different uplift rate due to neotectonic activity between the emerged and submerged areas of the Park.

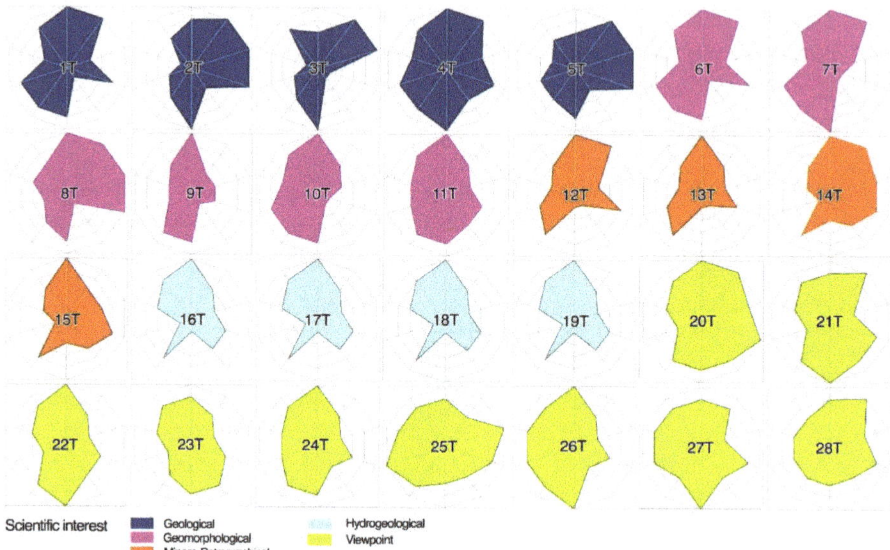

Figure 6. Multivariate representation of the results of terrestrial geosite assessment. Along the axes, the quantitative results of the assessment of the scientific, additional, and potential for use values and their sub-criteria are shown for each terrestrial geosite (for axis legend, see Figure 7). The colors represent different scientific interests, as specified in the figure legend.

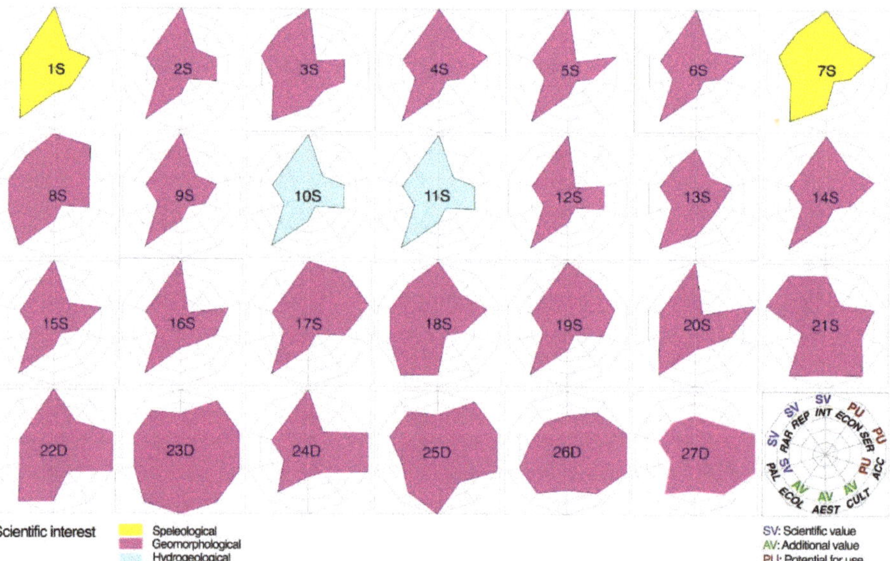

Figure 7. Multivariate representation of the results of marine geosite assessment. Along the axes, the quantitative results of the assessment of the scientific, additional, and potential for use values and their sub-criteria are shown for each marine geosite. The colors represent different scientific interests, as specified in the figure legend.

The majority of terrestrial geosites is protected at a regional level, since it is included within the Portofino Regional Natural Park. The great majority of marine geosites is characterized by high ecological value; most of them are included in the protected marine area of type B, while the others are located within the type C protected area. Marine sites have a high ecological importance owing to the presence of coralline biocoenosis and prairies of *Posidonia oceanica* both on sandy and rocky seabeds. Furthermore, the presence of rocky blocks resulting from rock falls and topples originating in the overhanging conglomerate and, to a lesser extent, marly-calcareous flysch cliffs, favors exceptional biodiversity recognized at an international level.

Viewpoints generally have the greatest additional value due to their scenic quality and color diversity (aesthetic value, AEST). Among marine geosites, the *Baia di San Fruttuoso* (ID 23D), *Baia di Paraggi* (ID 25D), and the *Cristo degli Abissi* site (ID 21S) have the highest additional value due to their aesthetic quality and cultural relevance. For instance, the *Baia di San Fruttuoso* hosts the abbey bearing the same name, which dates back to the 10th century CE, while the internationally famous *Cristo degli Abissi* statue was placed in 1954 and has now assumed historical relevance.

The assessment of the potential for use is crucial in choosing management strategies aimed at the promotion of geotourism in the area; therefore, particular attention has been devoted to choosing the most suitable sub-criteria for this evaluation. The latter revealed that terrestrial geosites are generally easily accessible except from the cliffs, of course, which require climbing skills for their fruition, while panoramic points are generally the easiest accessible terrestrial geosites. The most accessible marine geosites are the ones reachable by snorkeling, while all the other sites are accessible only to second-level certified scuba divers.

Low values are recorded for services in the proximity of terrestrial geosites. In fact, despite the support services being generally located at a walkable distance, they are subject to seasonal availability (mainly in spring and summer). For marine geosites, the services—i.e., boarding docks—are mostly located at a distance of 4 to 7 km. Every year, tens of thousands of visitors choose to hike along the footpaths of the Portofino Natural Park (from 20,000 to more than 70,000 visitors per year), making the economic value of terrestrial geosites very high. As for marine geosites, the number of visitors, in terms of scuba dives per year, is two orders of magnitude lower than the terrestrial geosite visitors, except for snorkeling sites, which are attended as much as land ones. It should be mentioned that data on scuba diving are underestimated due to the presence of illegal non-registered scuba divers.

6. Conclusions

The Portofino Natural Park boast some of the most impressive sceneries of the Mediterranean area, displaying a large variety of geological landscapes as well as unique ecological systems, both in terrestrial and marine environment.

This research has led to the identification and assessment of 28 terrestrial and 27 marine geosites of the Portofino Natural Park and protected marine area, pinpointing the most suitable sites for geotourism promotion, for both their contribution to the understanding of the geological processes acting through time on landscapes as well as their aesthetic importance.

In fact, the area is a well-known seaside resort and the present economy is almost exclusively based on onshore and offshore tourism. Nevertheless, tourism activities focus mainly on the rich marine biota and habitats and for recreational purposes while the geological and geomorphological features are usually neglected. Instead, these features, including submarine ones, could play a relevant role in developing a sustainable and safe tourism fruition, thanks to a deeper understanding of the complex geological and geomorphological contexts.

Moreover, for the first time, geosite assessment has been performed by applying a common methodology to both terrestrial and marine geosites. Some sub-criteria (accessibility, services, and economic potential) have been adapted considering the different characteristics between terrestrial and marine sites. This approach has allowed us to emphasize the relationships between the terrestrial and marine environments. The selected geosite network is meant to show common processes and landforms between these environments creating the ground for diversified but common and more efficient management and conservation actions and policies.

Author Contributions: Conceptualization, P.C. and V.V.; methodology, P.C. and V.V.; formal analysis, P.C. and V.V.; investigation, L.F., G.P., and F.F.; data curation, L.F., G.P., and F.F.; writing—original draft preparation, P.C., V.V., and F.F.; writing—review and editing, P.C., V.V., L.F., G.P., and F.F.; visualization, P.C., V.V., L.F., G.P., and F.F.; supervision, P.C. and F.F.

Funding: This research received no external funding.

Acknowledgments: The authors would like to thank Giorgio Fanciulli and Alberto Girani for all the material provided and for the profitable discussions on the geosites of the Portofino Park and Portofino Marine Protected Area.

Conflicts of Interest: The authors declare no conflict of interest.

References

1. Garavaglia, V.; Pelfini, M.; Bollati, I. The influence of climate change on glacier geomorphosites: The case of two Italian glaciers (Miage Glacier, Forni Glacier) investigated through dendrochronology. *Géomorphologie* **2010**, *2*, 153–164. [CrossRef]
2. Bosson, J.B.; Reynard, E. Geomorphological heritage, conservation and promotion in high-alpine protected areas. *eco. mont* **2012**, *4*, 13–22. [CrossRef]
3. Ravanel, L.; Bodin, X.; Deline, P. Using terrestrial laser scanning for the recognition and promotion of high-alpine geomorphosites. *Geoheritage* **2014**, *6-2*, 129–140. [CrossRef]
4. Reynard, E.; Coratza, P. The importance of mountain geomorphosites for environmental education. Examples from the Italian Dolomites and the Swiss Alps. *Acta Geogr. Slov.* **2016**, *56-2*, 291–303. [CrossRef]
5. Bollati, I.; Pellegrini, M.; Reynard, E.; Pelfini, M. Water driven processes and landforms evolution rates in mountain geomorphosites: Examples from Swiss Alps. *Catena* **2017**, *158*, 321–339. [CrossRef]
6. Bollati, I.; Coratza, P.; Panizza, V.; Pelfini, M. Lithological and structural control on Italian mountain geoheritage: Opportunities for tourism, outdoor and educational activities. *Quaest. Geogr.* **2018**, *37*, 53–73. [CrossRef]
7. Van den Ancker, H.J.A.M.; Jungerius, P.D. Geodiversity, geoheritage and geoconservation along the Dutch coast. *Ned. Geogr. Stud.* **2004**, *325*, 63–72.
8. Brocx, M.; Semeniuk, V. Coastal geoheritage: Encompassing physical, chemical, and biological processes, landforms, and other geological features in the coastal zone. *J. R. Soc. West Aust.* **2009**, *92*, 243–260.
9. Brocx, M.; Semeniuk, V. Coastal geoheritage: A hierarchical approach to classifying coastal types as a basis for identifying geodiversity and sites of significance in Western Australia. *J. R. Soc. West Aust.* **2010**, *93*, 81–113.
10. Coratza, P.; Gauci, R.; Schembri, J.; Soldati, M.; Tonelli, C. Bridging Natural and Cultural Values of Sites with Outstanding Scenery: Evidence from Gozo, Maltese Islands. *Geoheritage* **2016**, *8*, 91–103. [CrossRef]
11. Faccini, F.; Gabellieri, N.; Paliaga, G.; Piana, P.; Angelini, S.; Coratza, P. Geoheritage map of the Portofino Natural Park (Italy). *J. Maps* **2018**, *14*, 87–96. [CrossRef]
12. Martín-Duque, J.F.; Caballero García, J.; Carcavilla Urquí, L. Geoheritage information for Geoconservation and Geotourism Through the Categorization of Landforms in a Karstic Landscape. A Case Study from Covalagua and Las Tuerces (Palencia, Spain). *Geoheritage* **2012**, *4*, 93–108. [CrossRef]
13. Huo, S.J.; Sun, J.H.; Sun, K.Q. An Analysis on Resource Management of Geoheritage: A Case Study of South China Karst. *Adv. Mater. Res.* **2012**, *518*, 5909–5920. [CrossRef]
14. Hoblea, F.; Delannoy, J.J.; Jaillet, S.; Ployon, E.; Sadier, B. Digital Tools for Managing and Promoting Karst Geosites in Southeast France. *Geoheritage* **2014**, *6*, 113–127. [CrossRef]
15. Ballesteros, D.; Jiménez-Sánchez, M.; Domínguez-Cuesta, M.J.; García-Sansegundo, J.; Meléndez-Asensio, M. Geoheritage and Geodiversity Evaluation of Endokarst Landscapes: The Picos de Europa National Park, North Spain. In *Hydrogeological and Environmental Investigations in Karst Systems*; Andreo, B., Carrasco, F., Durán, J., Jiménez, P., LaMoreaux, J., Eds.; Springer: Berlin/Heidelberg, Germany, 2015; pp. 619–627.
16. Antić, A.; Tomić, N.; Marković, S. Karst geoheritage and geotourism potential in the Pek River lower basin (Eastern Serbia). *Geogr. Pann.* **2019**, *23*, 32–46. [CrossRef]
17. Bollati, I.; Pelfini, M.; Pellegrini, L. A geomorphosites selection method for educational purposes: A case study in Trebbia Valley (Emilia Romagna, Italy). *Geogr. Fis. Din. Quat.* **2012**, *35*, 23–35.

18. Álvarez-Vázquez, M.Á.; De Uña-Álvarez, E. Inventory and Assessment of Fluvial Potholes to Promote Geoheritage Sustainability (Miño River, NW Spain). *Geoheritage* **2017**, *9*, 549–560. [CrossRef]

19. Clivaz, M.; Reynard, E. How to Integrate Invisible Geomorphosites in an Inventory: A Case Study in the Rhone River Valley (Switzerland). *Geoheritage* **2018**, *10*, 527–541. [CrossRef]

20. Migoń, P.; Pijet-Migoń, E. Overlooked geomorphological component of volcanic geoheritage—Diversity and perspectives for tourism industry, Pogórze Kaczawskie region, SW Poland. *Geoheritage* **2016**, *8*, 333–350. [CrossRef]

21. Nemeth, K.; Moufti, M.R. Geoheritage Values of a Mature Monogenetic Volcanic Field in Intra-continental Settings: Harrat Khaybar, Kingdom of Saudi Arabia. *Geoheritage* **2017**, *9*, 311–328. [CrossRef]

22. Nemeth, K.; Casadevall, T.; Moufti, M.R.; Marti, J. Volcanic Geoheritage. *Geoheritage* **2017**, *9*, 251–254. [CrossRef]

23. Zacek, V.; Hradecky, P.; Kycl, P.; Sevcik, J.; Novotny, R.; Baron, I. The Somoto Grand Canyon (Nicaragua)-a Volcanic Geoheritage Site One Decade After Discovery: From Field Geological Mapping to the Promotion of a Geopark. *Geoheritage* **2017**, *9*, 299–309. [CrossRef]

24. Del Monte, M.; Fredi, P.; Pica, A.; Vergari, F. Geosites within Rome City center (Italy): A mixture of cultural and geomorphological heritage. *Geogr. Fis. Din. Quat.* **2013**, *36*, 241–257. [CrossRef]

25. Palacio-Prieto, J.L. Geoheritage within cities: Urban geosites in Mexico City. *Geoheritage* **2015**, *7*, 65–373. [CrossRef]

26. Chan, M.A.; Godsey, H.S. Lake Bonneville Geosites in the Urban Landscape: Potential Loss of Geological Heritage. In *Developments in Earth Surface Processes*; Oviatt, C.G., Shroder, J.F., Eds.; Elsevier: Amsterdam, The Netherlands, 2016; pp. 617–633. [CrossRef]

27. De Wever, P.; Baudin, F.; Pereira, D.; Cornée, A.; Egoroff, G.; Page, K. The Importance of Geosites and Heritage Stones in Cities—A Review. *Geoheritage* **2017**, *9*, 561–575. [CrossRef]

28. Reynard, E.; Coratza, P.; Pica, A. Urban Geomorphological Heritage. An Overview. *Quaest. Geogr.* **2017**, *36*, 7–20. [CrossRef]

29. Portal, C.; Kerguillec, R. The Shape of a City: Geomorphological Landscapes, Abiotic Urban Environment, and Geoheritage in the Western World: The Example of Parks and Gardens. *Geoheritage* **2018**, *10*, 67–78. [CrossRef]

30. Sacchini, A.; Ponaro, M.I.; Paliaga, G.; Piana, P.; Faccini, F.; Coratza, P. Geological landscape and stone heritage of the Genoa walls Urban park and surrounding area (Italy). *J. Maps* **2018**, *14*, 528–541. [CrossRef]

31. Brilha, J. Geoheritage: Inventories and evaluation. In *Geoheritage: Assessment, Protection and Management*; Reynard, E., Brilha, J., Eds.; Elsevier: Amsterdam, The Netherlands, 2018; pp. 69–86.

32. Reynard, E.; Coratza, P.; Regolini-Bissig, G. *Geomorphosites*; Verlag Dr. Friedrich Pfeil: Munich, Germany, 2009.

33. Brilha, J. Inventory and Quantitative Assessment of Geosites and Geodiversity Sites: A Review. *Geoheritage* **2016**, *8*, 119–134. [CrossRef]

34. Grandgirard, V. Géomorphologie, protection de la nature et gestion du paysage. Ph.D. Thesis, Doctorat-Faculté des Sciences, Université de Fribourg, 1997.

35. Grandgirard, V. L'évaluation des géotopes. *Geol. Insubrica* **1999**, *4*, 59–66.

36. Gordon, J.E.; Crofts, R.; Díaz-Martínez, E.; Woo, K.S. Enhancing the role of geoconservation in protected area management and nature conservation. *Geoheritage* **2018**, *10*, 191–203. [CrossRef]

37. Patzak, M.; Eder, W. UNESCO Geopark. A new programme—A new UNESCO label. *Geol. Balc.* **1998**, *28*, 33–35.

38. UNESCO. Geoparks Programme: A new initiative to promote a global network of geoparks safeguarding and developing selected areas having significant geological features. In Proceedings of the 156th UNESCO Executive Board, Paris, France, 4 March 1999.

39. Coratza, P.; De Waele, J. Geomorphosites and natural hazards: Teaching the importance of geomorphology in society. *Geoheritage* **2012**, *4*, 195–203. [CrossRef]

40. Wu, J.H.; Lin, W.K.; Hu, H.T. Assessing the impacts of a large slope failure using 3DEC: The Chiu-fen-erh-shan residual slope. *Comput. Geotech.* **2017**, *88*, 32–45. [CrossRef]

41. Migoń, P.; Pijet-Migoń, E. Natural disasters, geotourism, and geo-interpretation. *Geoheritage* **2019**, *11*, 629–640. [CrossRef]

42. Kotilainen, A. Marine Geotope Protection. In *Encyclopedia of Marine Geosciences*; Harff, J., Meschede, M., Petersen, S., Thiede, J., Eds.; Springer: Berlin/Heidelberg, Germany, 2014; pp. 448–449. [CrossRef]

43. Rovere, A.; Vacchi, M.; Parravicini, V.; Bianchi, C.N.; Zouros, N.; Firpo, M. Bringing geoheritage underwater: Definitions, methods, and application in two Mediterranean marine areas. *Environ. Earth Sci.* **2011**, *64*, 133–142. [CrossRef]

44. Burek, C.V.; Ellis, N.V.; Evans, D.H.; Hart, M.B.; Larwood, J.G. Marine geoconservation in the United Kingdom. *Proc. Geol. Assoc.* **2013**, *124*, 581–592. [CrossRef]

45. Arisci, A.; De Waele, J.; Di Gregorio, F.; Ferrucci, I.; Follesa, R. Integrated, sustainable touristic development of the karstic coastline of SW Sardinia. *J. Coast. Conserv.* **2003**, *9*, 81–90. [CrossRef]

46. Carobene, L.; Firpo, M. Conservazione e valorizzazione dei geositi costieri in Liguria: L'esempio del tratto di costa tra Varazze e Cogoleto. In *La Valorizzazione Dello Spazio Fisico Come via Alla Salvaguardia Ambientale*; Terranova, R., Brandolini, P., Firpo, M., Eds.; Patron Editore: Bologna, Italy, 2005; pp. 87–102.

47. Zouros, N. Geomorphosite assessment and management in protected areas of Greece. The case of the Lesvos island coastal geomorphosites. *Geogr. Helv.* **2007**, *62*, 169–180. [CrossRef]

48. Bianchi, C.N. From bionomic mapping to territorial cartography, or from knowledge to management of marine protected areas. *Biol. Mar. Mediterr.* **2007**, *14*, 22–51.

49. Bianchi, C.N.; Morri, C.; Navone, A. Classificazione degli ambienti sommersi e cartografia tematica. In *Tavolara: Nature at Work … Working in Nature*; Navone, A., Trainito, E., Eds.; Carlo Delfino Editore: Sassari, Italy, 2008; pp. 145–165.

50. Orrù, P.; Ulzega, A. Rilevamento geomorfologico costiero e sottomarino applicato alla definizione delle risorse ambientali (Golfo di Orosei, Sardegna orientale). *Mem. Soc. Geol. Ital.* **1988**, *37*, 123–131.

51. Orrù, P.; Panizza, V. Assessment and management of submerged geomorphosites. A case study in Sardinia (Italy). In *Geomorphosites*; Reynard, E., Coratza, P., Regolini-Bissig, G., Eds.; Pfeil: Munchen, Germany, 2009; pp. 201–212.

52. Orrù, P.; Panizza, V.; Ulzega, A. Submerged geomorphosites in the marine protected areas of Sardinia (Italy): Assessment and improvement. *Ital. J. Quat. Sci.* **2005**, *18*, 167–174.

53. Rovere, A.; Parravicini, V.; Donato, M.; Riva, C.; Diviacco, G.; Coppo, S.; Firpo, M.; Bianchi, C.N. Surveys of the Punta Manara shoals: An ecotipological approach. *Biol. Mar. Mediterr.* **2006**, *13*, 210–211.

54. Rovere, A.; Carobene, L.; Firpo, M. Geoheritage conservation in coastal and submerged landscapes: Mapping methods, GIS approach and management perspectives from the Bergeggi area. In Proceedings of the 33rd International Geological Conference, Oslo, Norway, 6–14 August 2008.

55. Rovere, A.; Vacchi, M.; Parravicini, V.; Morri, C.; Bianchi, C.N.; Firpo, M. Bringing geoheritage underwater: methodological approaches to evaluation and mapping. *Géovisions* **2010**, *35*, 65–80.

56. Pereira, P.; Pereira, D.; Caetano Alves, M.I. Geomorphosite assessment in Montesinho Natural Park (Portugal). *Geogr. Helv.* **2007**, *62*, 159–168. [CrossRef]

57. Reynard, E.; Fontana, G.; Kozlik, L.; Scapozza, C. A method for assessing scientific and additional values of geomorphosites. Geographica Helvetica. *Geogr. Helv.* **2007**, *62*, 148–158. [CrossRef]

58. Flores-de la Hoya, A.; Godínez-Domínguez, E.; González-Sansón, G. Rapid assessment of coastal underwater spots for their use as recreational scuba diving sites. *Ocean Coast. Manag.* **2018**, *152*, 1–13. [CrossRef]

59. Ramos, J.; Santos, M.N.; Whitmarsh, D.; Monteiro, C.C. The usefulness of the analytic hierarchy process for understanding reef diving choices: A case study. *Bull. Mar. Sci.* **2006**, *78*, 213–219.

60. Brooks, A.J.; Kenyon, N.H.; Leslie, A.; Long, D.; Gordon, J.E. *Characterising Scotland's Marine Environment to Define Search Locations for New Marine Protected Areas. Part 2: The Identification of Key Geodiversity Areas in Scottish Waters (Interim Report July 2011)*; Commissioned Report No. 430; Scottish Natural Heritage: UK, 2011. Available online: http://nora.nerc.ac.uk/id/eprint/16861/1/430.pdf (accessed on 11 October 2019).

61. Gordon, J.E.; Brooks, A.J.; Chaniotis, P.D.; James, B.D.; Kenyon, N.H.; Leslie, A.B.; Long, D.; Rennie, A.F. Progress in marine geoconservation in Scotland's seas: Assessment of key interests and their contribution to marine protected area network planning. *Proc. Geol. Assoc.* **2016**, *127*, 716–737. [CrossRef]

62. Silenzi, S.; Devoti, S.; Gabellini, M.; Magaletti, E.; Nisi, M.F.; Pisapia, M.; Angelelli, F.; Antonioli, F.; Zarattini, A. Le variazioni del clima nel Quaternario. *Geo-Archeol.* **2004**, *1*, 15–50.

63. Lambeck, K.; Antonioli, F.; Purcell, A.; Silenzi, S. Sea-level change along the Italian coast for the past 10,000 yr. *Quat. Sci. Rev.* **2004**, *23*, 1567–1598. [CrossRef]

64. Anzidei, M.; Antonioli, F.; Benini, A.; Lambeck, K.; Sivan, D.; Serpelloni, E.; Stocchi, P. Sea level change and vertical land movements since the last two millennia along the coasts of southwestern Turkey and Israel. *Quat. Int.* **2011**, *232*, 13–20. [CrossRef]

65. Gray, M. Geodiversity: The Backbone of Geoheritage and Geoconservation. In *Geoheritage*; Reynard, E., Brilha, J., Eds.; Elsevier: Amsterdam, The Netherlands, 2018; pp. 13–25. [CrossRef]

66. Raban, A. Archaeological park for divers at Sebastos and other submerged remnants in Caesarea Maritima, Israel. *Int. J. Naut. Archaeol.* **1992**, *21*, 27–35. [CrossRef]

67. Abd-el-Maguid, M.M. Underwater Archaeology in Egypt and the Protection of its Underwater Cultural Heritage. *J. Marit. Arch.* **2012**, *7*, 193–207. [CrossRef]

68. Jeffery, B. Realising the cultural tourism potential of South Australian shipwrecks. *Hist. Environ.* **1990**, *7*, 72.

69. Edney, J.; Howard, J. Review 1: Wreck diving, Scuba Diving Tourism. In *Contemporary Geographies of Leisure, Tourism and Mobility*; Musa, G., Dimmock, K., Eds.; Routledge: Abington, UK, 2013; pp. 52–56.

70. Prampolini, M.; Foglini, F.; Biolchi, S.; Devoto, S.; Angelini, S.; Soldati, M. Geomorphological mapping of terrestrial and marine areas, northern Malta and Comino (Central Mediterranean sea). *J. Maps* **2017**, *13*, 457–469. [CrossRef]

71. Bird, E.C.F. *Coastal Geomorphology*, 2nd ed.; John and Wiley and Sons: Hoboken, NJ, USA, 2008; p. 436.

72. Westley, K.; Quinn, R.; Forsythe, W.; Plets, R. Mapping submerged landscapes using multibeam bathymetric data: A case study from the north coast of Ireland. *Int. J. Naut. Archaeol.* **2011**, *40*, 99–112. [CrossRef]

73. Harff, J.; Bailey, G.; Luth, F. *Geology and Archaeology: Submerged Landscapes of the Continental Shelf*; Geological Society: London, UK, 2016; p. 294.

74. Miccadei, E.; Mascioli, F.; Piacentini, T. Quaternary geomorphological evolution of the Tremiti Islands (Puglia, Italy). *Quatern. Int.* **2011**, *233*, 3–15. [CrossRef]

75. Miccadei, E.; Mascioli, F.; Orrù, P.; Piacentini, T.; Puliga, G. Late Quaternary palaeolandscape of submerged inner continental shelf areas of Tremiti Islands archipelago (northern Puglia). *Geogr. Fis. Din. Quat.* **2011**, *34*, 223–234.

76. Kerr, S.; Johnson, K.; Side, J.C. Planning at the edge: Integrating across the land sea divide. *Mar. Policy* **2014**, *47*, 118–125. [CrossRef]

77. Parrott, D.R.; Todd, B.J.; Shaw, J.; Hughes Clarke, J.E.; Griffin, J.; MacGowan, B.; Lamplugh, M.; Webster, T. Integration of multibeam bathymetry and LiDAR surveys of the Bay of Fundy, Canada. In Proceedings of the Canadian Hydrographic Conference and National Surveyors Conference, Victoria, BC, Canada, 2008; pp. 1–15.

78. De Jongh, C.; van Opstal, H. *Coast-Map-IO TopoBathy Database*. Report Pilot Project, Leading Partner CARIS BV. 2012, pp. 20–26. Available online: https://www.dhyg.de/images/hn_ausgaben/HN094.pdf (accessed on 11 October 2019).

79. Van Alphen, J.S.L.J.; Damoiseaux, M.A. A geomorphological map of the Dutch shoreface and adjacent part of the continental shelf. *Geol. Mijnb.* **1989**, *68*, 433–443.

80. Miccadei, E.; Orrù, P.; Piacentini, T.; Mascioli, F.; Puliga, G. Geomorphological map of the Tremiti Islands (Puglia, southern Adriatic Sea, Italy), scale 1:15,000. *J. Maps* **2012**, *8*, 74–87. [CrossRef]

81. Prampolini, M.; Gauci, C.; Micallef, A.S.; Selmi, L.; Vandelli, V.; Soldati, M. Geomorphology of the north-eastern coast of Gozo (Malta, Mediterranean Sea). *J. Maps* **2018**, *14*, 402–410. [CrossRef]

82. Panizza, M. Geomorphosites: Concepts, methods and examples of geomorphological survey. *Chin. Sci. Bull.* **2001**, *46*, 4–6. [CrossRef]

83. Panizza, M.; Piacente, S. Geomorphological asset evaluation. *Z. Geomorph.* **1993**, *87*, 13–18.

84. Stokes, A.M.; Cook, S.D.; Drew, D. *Geotourism: The New Trend in Travel*; Travel Industry America and National Geographic Traveler: Washington, DC, USA, 2003.

85. Pellati, A. La Penisola di Portofino, note geomorfologiche. *Riv. Sci. Nat.* **1934**, *25*, 13–34.

86. Brandolini, P.; Faccini, F.; Piccazzo, M. Geomorphological hazard and tourist vulnerability along Portofino Park trails (Italy). *Nat. Hazard. Earth. Syst. Sci.* **2006**, *6*, 563–571. [CrossRef]

87. Faccini, F.; Piccazzo, M.; Robbiano, A. Environmental Geological Maps of San Fruttuoso Bay (Portofino Park, Italy). *J. Maps* **2008**, *4*, 431–443. [CrossRef]

88. Faccini, F.; Piccazzo, M.; Robbiano, A.; Roccati, A. Applied Geomorphological Map of the Portofino municipal territory (Italy). *J. Maps* **2008**, *4*, 451–462. [CrossRef]

89. Cimmino, F.; Faccini, F.; Robbiano, A. Stones and coloured marbles of Liguria in historical monuments. *Per. Mineral.* **2004**, *73*, 71–84.

90. Paliaga, G.; Giostrella, P.; Faccini, F. Terraced landscape as cultural and environmental heritage at risk: An example from Portofino Park (Italy). *Annal. Ser. Hist. Sociol.* **2016**, *26*, 1–10.

91. Salmona, P.; Verardi, D. The marine protected area of Portofino, Italy: A difficult balance. *Ocean. Coast. Manag.* **2011**, *44*, 39–60. [CrossRef]

92. Cerrano, C.; Milanese, M.; Ponti, M. Diving for science-science for diving: Volunteer scuba divers support science and conservation in the Mediterranean Sea. *Aquat. Conserv.* **2017**, *27*, 303–323. [CrossRef]

93. Faccini, F.; Piccazzo, M.; Robbiano, A. Natural hazards in San Fruttuoso of Camogli (Portofino Park, Italy): A case study of a debris flow in a coastal environment. *Boll. Soc. Geol. Ital.* **2009**, *128*, 641–654.

94. Brandolini, P.; Faccini, F.; Robbiano, A.; Terranova, R. Geomorphological hazard and monitoring activity along the western rocky coast of the Portofino Promontory (Italy). *Quatern. Int.* **2007**, *171*, 131–142. [CrossRef]

95. Corsi, B.; Elter, F.M.; Giammarino, S. Structural fabric of the Antola unit (Riviera di Levante, Italy) and implications for its Alpine versus Apennine origin. *Ophiolites* **2001**, *26*, 1–8.

96. Cevasco, A.; Faccini, F.; Nosengo, S.; Olivari, F.; Robbiano, A. Valutazioni sull'uso delle classificazioni geomeccaniche nell'analisi della stabilità dei versanti rocciosi: Il caso del 'Promontorio di Portofino (Provincia di Genova). *GEAM* **2004**, *111*, 31–38.

97. Giammarino, S.; Nosengo, S.; Vannucci, G. Risultanze geologiche-paleontologiche sul Conglomerato di Portofino (Liguria Orientale). *Atti Istituto di Geologia dell'Università di Genova* **1969**, *7*, 305–363.

98. Giammarino, S.; Messiga, B. Clasti di meta-ofioliti a paragenesi di alta pressione nel Conglomerato di Portofino: Implicazioni paleogeografiche e strutturali. *Ofioliti* **1979**, *4*, 25–41.

99. Bonaria, V.; Faccini, F.; Galiano, I.C.; Sacchini, A. Hydrogeology of conglomerate fractured-rock aquifers: An example from the Portofino's Promontory (Italy). *Rendiconti Online Soc. Geol. Ital.* **2016**, *41*, 22–25. [CrossRef]

100. Area Marina Protetta Portofino. Available online: http://www.portofinoamp.it/ (accessed on 6 May 2019).

101. Sacchini, A.; Faccini, F.; Ferraris, F.; Firpo, M.; Angelini, S. Large-scale landslide and Deep-Seated Gravitational Slope Deformation of the Upper Scrivia Valley (Northern Apennine, Italy). *J. Maps* **2016**, *12*, 344–358. [CrossRef]

102. Faccini, F.; Paliaga, G.; Piana, P.; Sacchini, A.; Watkins, C. The Bisagno stream catchment (Genoa, Italy) and its major floods: Geomorphic and land use variations in the last three centuries. *Geomorphology* **2016**, *273*, 14–27. [CrossRef]

103. Brunetti, M.; Bertolini, A.; Soldati, M.; Maugeri, M. High-resolution analysis of 1-day extreme precipitation in a wet area centered over eastern Liguria, Italy. *Theor. Appl. Climatol.* **2019**, *135*, 341–353. [CrossRef]

104. Ligurian Cave Inventory. Available online: https://www.catastogrotte.net/ (accessed on 6 May 2019).

105. Ristori, G. Il Conglomerato miocenico e il regime sotterraneo delle acque nel Promontorio e Monte di Portofino. *Atti Università di Pisa* **1901**, *18*, 49–67.

106. Du Plessis, E.; Saayman, M. What makes scuba diving operations successful: The case of Portofino, Italy. *EJTR* **2017**, *17*, 164–176.

107. Lucrezi, S.; Milanese, M.; Sarà, A.; Palma, M.; Saayman, M.; Cerrano, C. Profiling scuba divers to assess their potential for the management of temperate marine protected areas: A conceptual model. *Tourism. Mar. Environ.* **2018**, *13*, 85–108. [CrossRef]

108. Saayman, M.; Saayman, A. Are there economic benefits from marine protected areas? An analysis of scuba diver expenditure. *Eur. J. Tour. Res.* **2018**, *19*, 23–39.

109. Di Carro, M.; Magi, E.; Massa, F.; Castellano, M.; Mirasole, C.; Tanwar, S.; Olivari, E.; Povero, P. Untargeted approach for the evaluation of anthropic impact on the sheltered marine area of Portofino (Italy). *Mar. Pollut. Bull.* **2018**, *131*, 87–94. [CrossRef]

110. Markantonatou, V.; Noguera-Méndez, P.; Semitiel-García, M.; Hogg, K.; Sano, M. Social networks and information flow: Building the ground for collaborative marine conservation planning in Portofino Marine Protected Area (MPA). *Ocean. Coast. Manag.* **2016**, *120*, 29–38. [CrossRef]

111. Migoń, P.; Pijet-Migoń, E. Viewpoint geosites—Values, conservation and management issues. *Proc. Geol. Assoc.* **2017**, *128*, 511–522. [CrossRef]

112. Scuba Diving Sites of the MPA of Portofino. Available online: http://www.portofinoamp.it/ubacquea/i-siti-di-immersione-dellarea-marina-protetta (accessed on 6 May 2019).

113. Carton, A.; Cavallin, A.; Francavilla, F.; Mantovani, F.; Panizza, M.; Pellegrini, G.B.; Tellini, C.; Bini, A.; Castaldini, D.; Giorgi, G.; et al. Ricerche ambientali per l'individuazione e la valutazione dei beni geomorfologici. Metodi ed esempi. *Ital. J. Quat. Sci.* **1994**, *7*, 365–372.

114. Rivas, V.; Rix, K.; Frances, A.; Cendrero, A.; Brunsden, D. Geomorphological indicators for environmental impact assessment: Consumable and non-consumable geomorphological resources. *Geomorphology* **1997**, *18*, 169–182. [CrossRef]

115. Vujicic, M.D.; Vasiljevic, D.E.; Markovic, S.B.; Hose, T.A.; Lukic, T.; Hadzic, O.; Janicevic, S. Slankamen villages preliminary geosite assessment model (GAM) and its application on Fruska Gora Mountain, potential geotourism destination of Serbia. *Acta Geogr. Slov.* **2011**, *51*, 361–377. [CrossRef]

116. Reynard, E.; Coratza, P. Scientific research on geomorphosites. A review of the activities of the IAG working group on geomorphosites over the last twelve years. *Geogr. Fis. Dinam. Quat.* **2013**, *36*, 159–168.

117. Bruschi, V.M.; Cendrero, A. Direct and parametric methods for the assessment of geosites and geomorphosites. In *Geomorphosites*; Reynard, E., Coratza, P., Regolini-Bissig, G., Eds.; Pfeil: München, Germany, 2009; pp. 73–88.

118. Reynard, E. The assessment of geomorphosites. In *Geomorphosites*; Reynard, E., Coratza, P., Regolini-Bissig, G., Eds.; Pfeil: Munchen, Germany, 2009; pp. 63–71.

119. Coratza, P.; Galve, J.P.; Soldati, M.; Tonelli, C. Recognition and assessment of sinkholes as geosites: Lessons from the Island of Gozo (Malta). *Quaest. Geogr.* **2012**, *31*, 25–35. [CrossRef]

120. Kubaliková, L.; Kirchner, K. Geosite and Geomorphosite Assessment as a Tool for Geoconservation and Geotourism Purposes: A Case Study from Vizovická vrchovina Highland (Eastern Part of the Czech Republic). *Geoheritage* **2016**, *8*, 5–14. [CrossRef]

121. Reynard, E.; Perret, A.; Bussard, J.; Grangier, L.; Martin, S. Integrated approach for the inventory and management of geomorphological heritage at the regional scale. *Geoheritage* **2016**, *8*, 43–60. [CrossRef]

122. Bruschi, V.M.; Cendrero, A. Geosite evaluation; Can we measure intangible values? *Ital. J. Quat. Sci.* **2005**, *18*, 293–306.

123. Brilha, J. *Património Geológico e Geoconservação: A Conservação da Natureza na sua Vertente Geológica*; Palimage Editores: Viseu, Portugal, 2005; pp. 1–190.

124. Coratza, P.; Giusti, C. Methodological proposal for the assessment of the scientific quality of geomorphosites. *Ital. J. Quat. Sci.* **2005**, *18*, 303–313.

125. Kubalíková, L. Geomorphosite assessment for geotourism purposes. *Czech. J. Tour.* **2013**, *2*, 80–104. [CrossRef]

126. Cappadonia, C.; Coratza, P.; Agnesi, V.; Soldati, M. Malta and Sicily joined by geoheritage enhancement and geotourism within the framework of land management and development. *Geosciences* **2018**, *8*, 253. [CrossRef]

127. Droz, Y.; Miéville-Ott, V. *La Polyphonie du Paysage*; Presses Polytechniques et Universitaires Romandes: Lausanne, Switzerland, 2005; p. 236.

128. Štrba, L.; Rybár, P.; Baláž, B.; Molokáč, M.; Hvizdák, L.; Kršák, B.; Lukáč, M.; Muchová, L.; Tometzová, D.; Ferenčíková, J. Geosite assessments: Comparison of methods and results. *Curr. Issues Tour.* **2015**, *18*, 496–510. [CrossRef]

129. Coratza, P.; Vandelli, V.; Soldati, M. Environmental rehabilitation linking natural and industrial heritage: A Master Plan for dismissed quarry areas in the Emilia Apennines (Italy). *Environ. Earth Sci.* **2018**, *77*, 455. [CrossRef]

Article

Geodiversity Evaluation and Water Resources in the Sesia Val Grande UNESCO Geopark (Italy)

Luigi Perotti *, Gilda Carraro, Marco Giardino, Domenico Antonio De Luca and Manuela Lasagna

Earth Sciences Department, University of Torino, 10125 Turin, Italy; gilda.carraro@edu.unito.it (G.C.); marco.giardino@unito.it (M.G.); domenico.deluca@unito.it (D.A.D.L.); manuela.lasagna@unito.it (M.L.)
* Correspondence: luigi.perotti@unito.it

Received: 16 September 2019; Accepted: 4 October 2019; Published: 9 October 2019

Abstract: This paper aims at systemizing knowledge related to geodiversity assessment for water resources and its evaluation. The novel aspect connected to geodiversity of this paper is the analysis of the components of hydrological system, both at the superficial and underground level, in the territory of the Sesia Val Grande United Nations educational, scientific, and cultural organization (UNESCO) Global Geopark (Northwest Italy). More specifically, the research establishes a conceptual model and a specific procedure for the evaluation of geodiversity connected to water resources on a regional scale, by means of a qualitative-quantitative geographic information system (GIS) process, renamed here as hydro-geodiversity assessment. For these purposes, a targeted ecosystem approach is applied to consider the assets of the Geopark territory that has been derived from the interaction between water and other components of geodiversity, i.e., the hydro-geosystemic services. The element selection and processing operations led to the identification of areas characterized by greater values of hydrological geodiversity, in which the link between surface and underground hydrodynamics became closer and intense. The single geodiversity factor maps that were obtained from partial data aggregations were added together in map algebra operations, then subjected to weighing to formulate the hydro-geodiversity map of the Sesia Val Grande UNESCO Global Geopark. The results of the present study strengthen the strategic management of geological, geomorphological, and hydrological heritages of the study area by identifying different landscapes and local peculiarities determined by mutual influences between geology and hydrological dynamics.

Keywords: water resources; geodiversity assessment; geosystem services; geoheritage; hydro-geodiversity; Sesia Val Grande UNESCO Global Geopark

---

1. Introduction

The term "geodiversity" has no intrinsic value; its importance relies on the quality of the relationships built between the systems or spheres of which it is composed, i.e., the Earth system science (atmosphere, lithosphere, hydrosphere, and biosphere; Figure 1) [1] and those specifically addressed to describe surface processes, landforms, and materials, such as pedosphere and anthroposphere. Interactions between these different spheres constitute the variety of geological and geomorphological phenomena and landscapes [2] to which human beings attribute several values. Therefore, we agree with Sharples [3] in considering geodiversity as "the quality we are trying to conserve," and geoconservation as "the endeavor to conserve it".

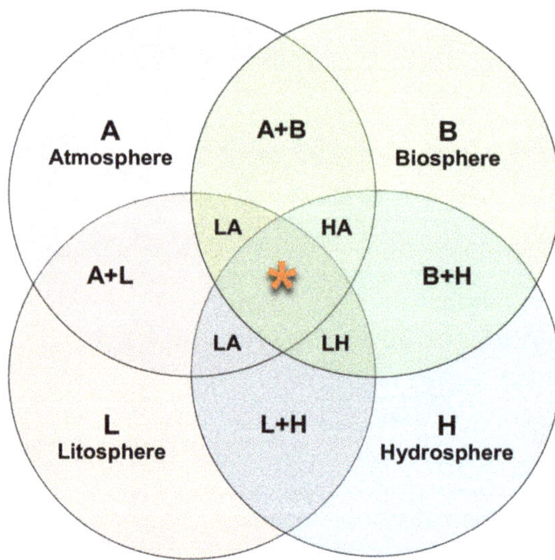

Figure 1. Conceptual scheme of the relationships between the main spheres considered by the Earth systems science [1]. Asterisk indicates the area of relationships with pedosphere and anthroposphere.

According to Gray's (2013) definition, geodiversity is not just a matter of different Earth features [1] but also of their assemblages, structures, systems, and contribution to landscapes. The complexity of geodiversity is a challenge for its assessment. A system of norms and modes of action, as well as a model of interaction of all the related variables, has to be created in order to achieve significant geodiversity assessment, in term of both biotic and abiotic ecosystem services. Such a comprehensive assessment can offer relevant contributions to both geoconservation and sustainable use of georesources [4].

This work is focused on systematizing the relationships between geodiversity and water resources by analyzing parts of the hydrological cycle that interact with geological features, satisfy essential needs, and for allow biological and human evolution. The first research question we want to address is: What value should be attributed to water resources in the definition of geodiversity?

Despite the broad interests and various focuses of contemporary geodiversity methods and tools, the related literature considers hydrology as a fundamental component of geodiversity assessment [1,5]. However, relationships between geology and water resources have been frequently analyzed in a sectorial way:

1) by strictly considering the hydrogeological aspects with a purely quantitative method in order to establish precise numerical values about productivity of an aquifer [6];
2) by qualitatively interpreting hydrological details, either in a landscape analysis perspective [7], or by applying a morphographic approach to hydro-geodiversity issues [8].

Although the methods used in the two aforementioned studies have opposite aims and results (quantitative versus qualitative), their main tendency is to perform partial analyses of the interaction between water and geodiversity, i.e., by focusing on specific geological or geomorphological phenomena (in these last two cases: landforms and river dynamics), thus neglecting other relevant elements of the hydrological system.

On the other hand, some authors [9,10] posed a greater effort in a systematic attempt to define the structure of the hydrogeological interactions, including factors and variables useful for creating a map of water resources. In the study by Arajuo and Pereira, such a factor map represents a fundamental part of the final geodiversity map of the State of Cearà, in Brazil [9].

Despite the undoubted inclusion of hydrological processes in the geodiversity equation adopted by Brazilian and Polish assessments [9,10], the real crux of the matter is that, to date, there is no uniform choice of essential variables to be considered for classification of hydrological elements relevant for geodiversity assessment. This is particularly relevant for regional geodiversity studies, such as large-scale surveys and assessments, where a theoretical framework has to be carefully discussed for achieving successful applications [11].

To overcome the problem, we sought to define and test a set of relevant variables for a qualitative-quantitative assessment of hydrological-geodiversity in the Alpine territory of the Sesia Val Grande UNESCO Global Geopark (UGG). In a preliminary phase, it was necessary to devise a relational model that could provide support for the collection and management of information to be processed. This important and preliminary research operation included the definition of a conceptual model for geodiversity assessment in relation to water resources, based on the specific territorial context, and the adopted regional scale of analysis.

2. Study Area

The Sesia Val Grande UGG [12] study area covers about 2000 km^2 and is located in the Piemonte Region (NW Italy) (Figure 2). The area includes 106 municipalities across 4 provinces: Verbania, Vercelli, Novara, and Biella. Elevation of the territory varies from 190 m a.s.l. at the lower alpine piedmont area to 4554 m a.s.l. at the top of Monte Rosa (Pennine Alps), the second highest massif of the European Alps. Indeed, the area is mainly mountainous, including high plains and large floodplains, as well as a portion of the Maggiore Lake.

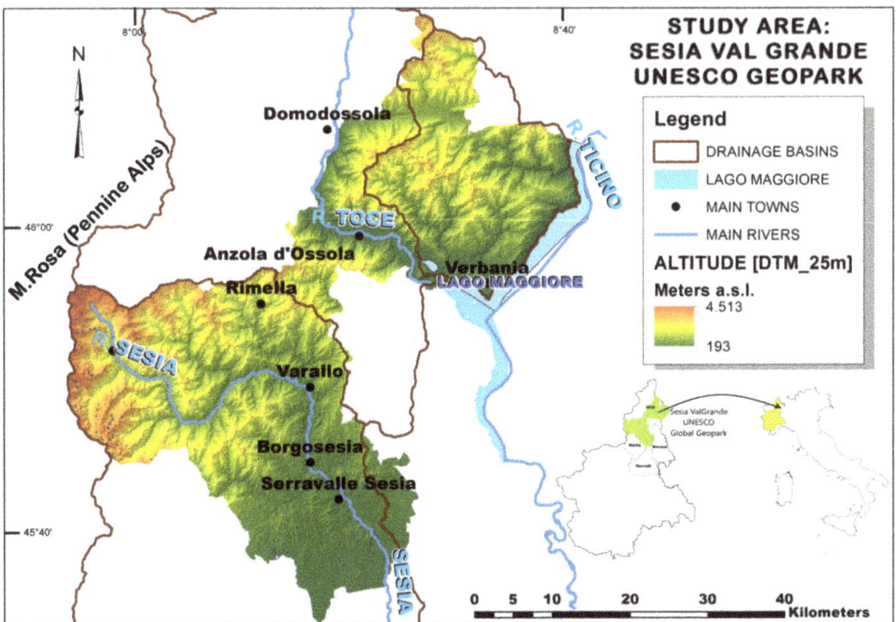

Figure 2. The study area of the Sesia Val Grande United Nations educational, scientific, and cultural organization (UNESCO) Global Geopark (UGG).

The study area comprises three important hydrographic sub-basins of the Po drainage basin (Figure 2): those of the Sesia river (3079 sqkm area, 138 km length), Toce river (1785 sqkm area, 57 km length), and Ticino river (6033 sqkm area, 284 km length) [13]. Many factors and competing morphodynamic processes contributed to the shaping of the mountain relief and the valleys system, i.e., the litho-structural and tectonic conditions posed by alpine orogeny and the morphoclimatic variations, such as the Pleistocene glacial/interglacial phases and later Holocene stages. These "regional" factors are related to long-term processes, giving the basic shape of the Alpine mountains and valleys, then followed by important Holocene "local" morphogenetic processes of fluvial/torrential, as well as of gravitational origin. Currently, the dominant geomorphic agency in the valleys of Sesia Val Grande Geopark is the fluvial-torrential one, which is accompanied by consistent gravitational instabilities, where the slopes are steeper. [14]. A large portion of the territory shows fluvial/torrential landforms along steep slopes (16% or more), such as deep river incision, mainly in bedrock, with abundant debris deposits, often forming debris fans up to km-size. On the other hand, in valley floor and high plain areas, rivers created terraces at the valley sides, and multiple to single channels had a tendency to meander at the valley mouth [15].

At higher elevations, glaciers have been the most important morphogenetic agent during Holocene. Indeed, the upper areas of the high valleys are dominated by glacial and periglacial processes [16]. Despite the ongoing climate warming and the predominant southern exposure of the slope of the Monte Rosa Massif, seven glaciers are still present [17], whose hydro-geosystem value is undeniable: they constitute the sources of the Sesia river, as well as of a beautiful landscape, even if they are extremely sensitive and endangered by climate change.

Concerning the tectonic setting, Figure 3 shows the main structural units and geological complexes, in which the geopark is located. Units of the Southern Alps are aligned along a northeast-southwest direction, juxtaposed to the Austroalpine units along the Insubric Line (towards W) and to the lower Pennidic units along the Sempione-Centovalli Line (towards N). In turn, the Austroalpine is in tectonic contact by means of complex polyphasic deformation with the Pennidic domain, herein represented mainly by oceanic units [18].

The lithological geodiversity constrains the Geopark's hydrological features. The whole area (Figure 3) is mainly characterized by crystalline basement units, i.e., lithologies whose permeability depends on fractures density. A little portion of conglomerates, limestones, and marbles outcrops in the Lower Sesia Valley, while fluvial and fluvioglacial deposits characterize the valley floors and the piedmont. These last units represent the reloading areas of the high productivity aquifers present in the territory.

The various geo-lithological complexes, combined with their structural settings, contribute to the large geological diversity of the area and, at the same time, define highly diversified hydrographic network and rich hydrogeological structures [19], as follows:

1. Alpine and pre-Alpine magmatic and metamorphic rocks (a large diversity including granite, gabbro, diorite, peridotite, quartzite, gneiss, micashist, amphibolite, granulite, etc.) from main lithotectonic units of both internal (Southern Alps: Ivrea-Verbano, Serie dei Laghi, and periadriatic magmatic units) and axial alpine belt (Sesia-Lanzo, Monte Rosa, and Liguria-Piemonte oceanic units). Generally, groundwater circulation is absent or limited to surface fracture systems and major faults. The prevalent permeability varies from low to very low but along the most fractured zones, the degree of permeability can also vary from medium to high.

2. Dolostones and breccias of Triassic-Jurassic sedimentary cover the Southern Alps and metamorphic carbonatic rocks (marble) within the mentioned alpine units. They are characterized by remarkable water circulation due to superficial and deep karst phenomena. The prevalent permeability (for fracturing and karst phenomena) results from high to medium.

3. Sands (Pliocene Asti Sand of marine origin): the prevalent permeability for porosity has a medium degree. The coarser terms of this complex represent aquifers of good productivity.

4. Lacustrine, marsh, and fluvial sediments ("Villafranchian" deposits): the prevalent permeability for porosity is of medium degree, even if they are characterized by a high heterogeneity, depending on the depositional environment.

5. Pleistocene to present-day glacial deposits. The prevalent permeability for porosity varies from medium to low.

6. Alluvial deposits: these sediments are located in the valleys bottom and in the piedmont. Due to their porosity they show prevalent permeability from high to medium and contain a shallow unconfined aquifer in connection with surface rivers.

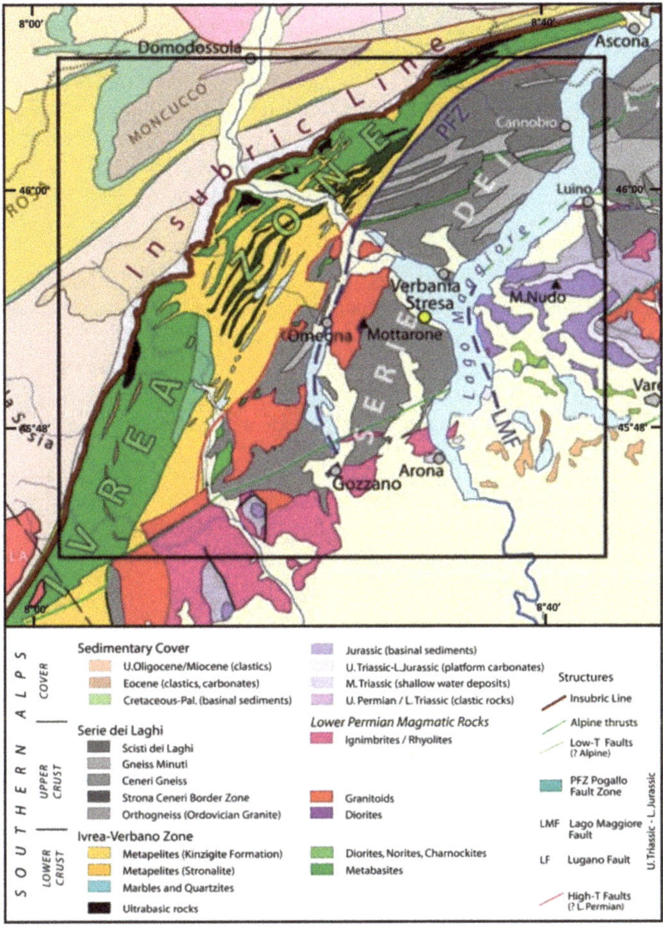

Figure 3. Map of the geological and structural units of the Sesia Val Grande UNESCO Geopark. (modified from Brak et al. 2010) [18].

3. Materials and Methods

In the following paragraph, we describe the adopted methodological approach, the selected parameters, and the methodology chosen for carrying out the evaluation process that led to the final map of hydro-geodiversity of the Sesia Val Grande UGG.

3.1. Methodology

Before defining the input data for hydro-geodiversity assessment, it is necessary to define the conceptual structure of the assessment. In this specific evaluation, the intention is to consider values and services that the territory and community derived from the abiotic components in connection with the dynamics of the water and the formation of aquifers.

Gray's (2013) model of geosystemic services identifies five different categories of geoservices [2] and was adapted in a hydrogeological overview and applied to the local context of Sesia Val Grande UNESCO Global Geoparks area (UGGp). This kind of approach, usually part of qualitative methods to assess geodiversity [20], is human-centered. Our analyses identified hydro-geosystem services, which represented hydrogeological features capable of offering a range of specific services and goods.

The conceptual process guided the assessment of hydro-geosystem services and is described in Figure 4. Starting from the analyses of relationships between geodiversity and water, we proposed a framework for hydro-geodiversity. Since it represents the part of geodiversity concerning the hydrosphere, it includes hydrogeological phenomena that interacts with geolithological features, the component of geomorphological landscape, and the way in which human societies manage them.

According to the conceptual scheme of geosystem services by Gray (2013), interlinked categories have been found and the intensity and types of relationships are described in Figure 4:

- regulating dynamics (atmospheric, geological, geomorphological, and anthropogenic processes);
- provisioning (of drinking water, water for industry, agriculture, or energy production);
- cultural processes (related to the development of the spiritual, religious, and collective identity of local communities and to the maintenance of psycho-physical health);
- knowledge processes (which reconstruct the evolutionary history of terrestrial cycles, deal with monitoring quality, and presence of water in glaciers, canals, aquifers, which develop strategies for the management of hydrogeological risk in a context of climate change).

These interactions determine a range of hydro-geoservices that forge the structure of the hydro-geosystem, which corresponds to a certain degree of hydro-geodiversity. The conceptual definition of the hydro-geosystem services in the territory under study was essential to understand and define the input data to consider the hydro-geodiversity map of Sesia Val Grande UNESCO Geopark.

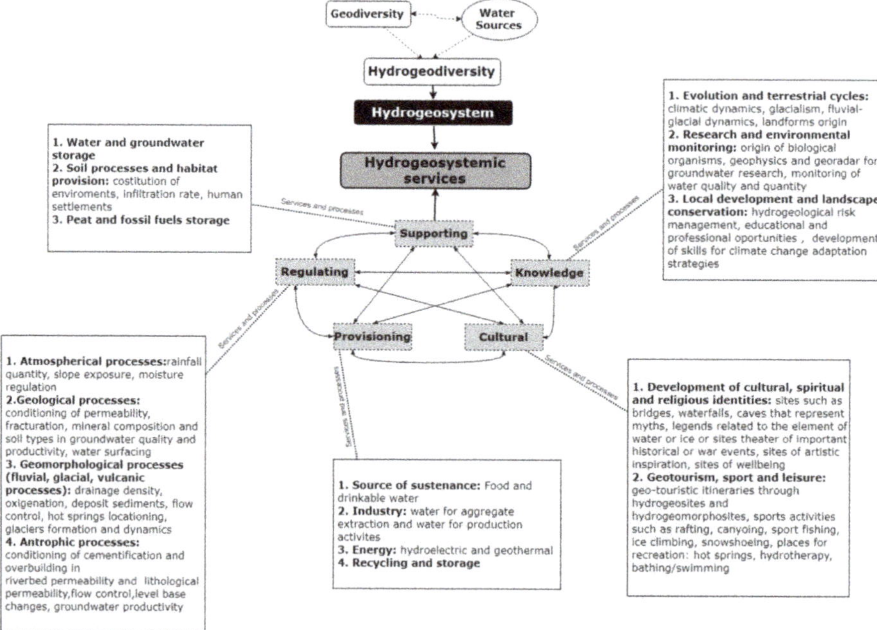

Figure 4. Hydrogeoservices from Gray (2013) [2].

3.2. Hydro-Geodiversity Assessment

Once analyzed, the characteristics of the territorial context defined the conceptual setup of the research, it was possible to proceed with the definition of the parameters of the hydro-geodiversity assessment. The specific parameters are described in detail in Table 1. The operational purpose is to identify areas characterized by high hydro-geodiversity using a qualitative-quantitative evaluation technique.

Table 1. Chosen parameters for hydro-geodiversity assessment in the case study area (modified from Zwolinski, Najwer, and Giardino, 2018) [20].

Chosen Parameters for Hydro-Geodiversity Assessment in the Sesia Val Grande UNESCO Global Geopark		
Purpose	1° = COGNITIVE	1°: Define a conceptual structure of geodiversity linked to water resources
	2° = OPERATIONAL	2°: Identified areas characterized by high hydrological geodiversity
Data Source	INDIRECT	Cnr-Regione Piemonte [21] Siri - Regione Piemonte [22] PPR piemonte [23] Arpa Piemonte [24] Autorità di bacino po [25] Corine land cover [26]
Subject	SELECTIVE APPROACH	Choice of a set of components of the natural abiotic environment
Spatial Scale	REGIONAL	Analysis Scale 1:100.000
Time Scale	CURRENT	Most updated data
Evaluation Criterion	RELATIVE	Hydro-geosystem services, human-centred
Evaluation Technique	MIXED = QUANTITATIVE-QUALITATIVE	Expert and automatic classification
Representation of the results of evaluation	CARTOGRAPHIC	ESRI ArcGis

The hydro-geodiversity assessment procedure is typically quantitative, based on the pioneering work of Serrano and Ruiz [11]. It is the kind of approach based on the construction of map algebra indexes and techniques, using geographic information system (GIS) software to process information. To achieve this practical purpose, we performed a GIS analysis by using ArcGis 10.5 software (developed by ESRI Redlands, USA) on a complete set of georeferenced spatial data. On the basis of the available data and the survey scale, a hydro-geodiversity equation was established, whose variables corresponded to the factor maps that were added together using weighing techniques in the map algebra phase.

Due to the large study area, the chosen scale for the evaluation was 1:100,000. Indeed, several relevant features for geodiversity assessment are at a nominal scale of 1:10,000. In order to obtain a final representation that was compatible with the finale factor maps and consistent with the chosen scale of representation (1:100,000), we did a semi-automatic data generalization.

Based on the selected scale of analyses, a geodatabase was constructed by collecting the public data provided by regional and territorial agencies, as described in the data source field described in Table 1.

The main steps for the hydrological geodiversity assessment were:

1) Construction of a georeferenced database in GIS environment;
2) analysis and interpretation of the information retrieved based on the guiding model of hydro-geosystem services created previously and considering the significant factors for local communities;

3)   define factor maps and variables using an iterative approach;
4)   combine factor maps, attributing weights in map algebra operations;
5)   choose and create the final hydro-geodiversity map for the Sesia Val Grande UNESCO Global Geopark;
6)   identify hydro-geodiversity landscapes and promote their conservation.

Data selected in the initial phase have undergone changes due to scale compatibility or type of input data. An iterative process was adopted [27]. Available data were collected and analyzed, and we observed the results and determined which data could be used and how.

Four main factors were chosen for the evaluation of hydro-geodiversity:

- Basement rocks and deposits permeability, integrated with fracturing index (tP), for the factor map of total permeability;
- land use integrated with the slope instability index (tLU) for the factor map of total land use;
- springs and wells location (SWD) for the factor map of springs and wells density;
- Hydrography, glaciers location, glacial cirques, landslides, and alluvial fan location (MR) for the factor map of morphogenetic relevance.

These factors represent the variables of the hydro-geodiversity (HGD) equation, which can be summarized as:

$$HGD = tP + tLU + SWD + MR \tag{1}$$

From the vector data and the expert classification of the elements, the data has been transformed into a raster format. This allows us to assign a value to each identified class and to add the obtained images with a final resolution of 25 m.

Figure 5 briefly describes the methodology adopted in the evaluation assessment, as well as highlighting the relationships between the fur factor maps in creating the hydro-geodiversity map.

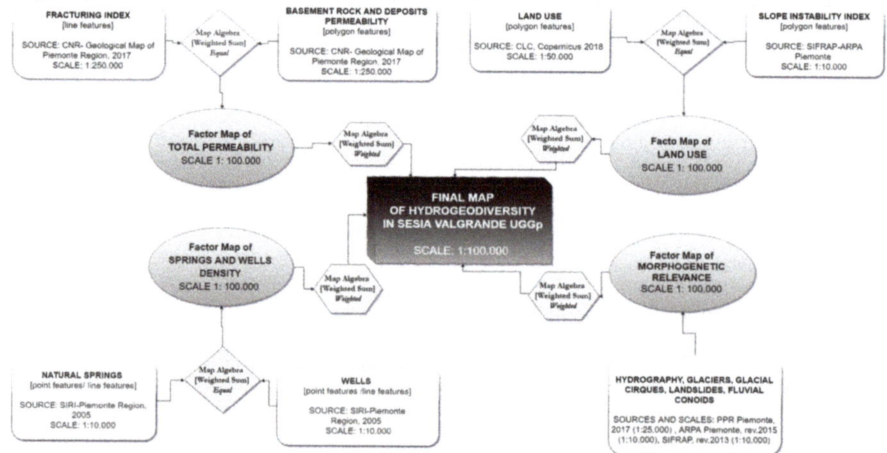

Figure 5. Flow diagram for the creation of the final map of hydro-geodiversity in the Sesia Val Grande UNESCO Global Geopark.

Once the four factor maps have been obtained, the next step includes the processing of partial maps via attributing weights through map algebra. By varying the weight of the individual partial maps in the map algebra phase, it was possible to evaluate and compare the results of the weights assigned to the individual factor maps. To find the best weight proportions used to obtain a result that identifies sufficiently homogeneous areas of hydrological geodiversity, the AHP (Analytic Hierarchy Process) method [28] was used, which is a multi-criteria decision support technique.

Finally, seven hypotheses of hydro-geodiversity assessment were formulated. Only one was chosen as a representative for the Sesia Val Grande UGG hydro-geodiversity. It was then reclassified into three distinct classes using the natural break method.

Consequently, an interpretation of the results were made, leading to the definition, within the geopark, of a certain number of landscapes characterized by high hydro-geodiversity.

4. Results

4.1. Factor Map of Total Permeability

The geological lithology [21] was divided by type and degree of permeability. More specifically, the basement rocks, quaternary, and pre-quaternary deposits were classified with values from 1 to 5 based on the hypothetic degree of permeability, which is directly related to the constitution and the productivity of aquifers that hide from lithological formations (Table 2). Both the deposits and rock basements were classified by hypothetical permeability, which underpins their predisposition to constitute aquifers.

Table 2. Lithology of rock basement and pre-quaternary deposits in the Sesia Val Grande UGG.

Rock Basement and Pre-Quaternary Deposits	Value
Amphibolite, Diorite, Metabasite, Gneiss, Granite, Granodiorite, Peridotite, Serpentinite	1
Micaschist, Calc-schist, Paragneiss, Phillite, Mylonite	2
Conglomerate, Andesite, Pyroclastic rock	3
Marble, Limestone, Dolomite, Sandstone	4
Villafranchiano, Asti Sand	5

The highest values were assigned to quaternary and pre-quaternary gravelly and sandy deposits. Intermediate values were instead assigned to lithologies such as marble, limestone, and sandstone with mixed deposits or debris flow. Ultimately, very low values were assigned to most coherent lithologies or glacial deposits (Table 3).

Table 3. Classification of quaternary deposits in the Sesia ValGrande UGG.

Quaternary Deposits	Value
Glacial deposits and rock glacier deposits (active and inactive)	2
Mixed deposits	3
Alluvial terraces and debris flow (fl2)	4
Fluvial Deposits (fl1) and fluvioglacial deposits	5

A particular procedure was followed with regard to lacustrine and marsh deposits, since these deposits contain clay and are characterized by low permeability. However, theses deposits constitute layers of protection for aquifers and represent superficial environments dominated by the water dynamics that constitute extremely precious biotopes (e.g., peat bogs, marshes, ponds). Thus, classifying these deposits on the basis of permeability means giving them low values; this is not in line with the objective of the present study, which seeks to enhance the centrality of the water element in its interaction with geodiversity. This led to the choice of not considering these deposits in the present classification, but to evaluate them with high value in the factor map of total land use. To complete the factor map of total permeability, the state of fracturing of the substrate and influencing the degree of permeability of the rock was considered. Indeed, the lithologies of the basement, such as crystalline rocks, have a wide permeability range depending on the level of incidence of tectonic structures. In the study area faults, fault systems and ductile shear systems were distinguished [21]. Moreover, structures responsible for ductile, ductile-fragile, or brittle-ductile deformation were distinguished, because in ductile-fragile areas in the tectonic action produce a greater incidence of fracturing.

In order to create a map of the fracturing index, several hypotheses of classification were advanced. The final decision was to create manually areas of fracturing relevance, attributing more value to fragile deformations as compared to ductile ones. These areas were added to the map of deposits and rock basement permeability, in order to obtain a map of total permeability (Figure 6).

In the map, it is possible to observe how rocks with low degree of lithological permeability assumed maximum values, i.e., the Insubric Line, the Cossato-Mergozzo-Brissago, and the Pogallo Lines.

4.2. Factor Map of Land Use

Land use is an important factor to consider in the equation of hydro-geodiversity because it explains the human impact on natural environments. In the hydro-geodiversity assessment, it is important to highlight all variables that seal or pollute the ground. Thus, the factor map of land use collects all types of land uses identified by the corine land cover satellite tracking system, which integrates them with wetlands, marsh areas, and lakes (Figure 7).

The identified elements were classified based on their possible effect on ground permeability, as well as the possibility to create underground reserves and water resource pollution.

Because of this, the factor map of land use collects all types of land uses identified by the corine land cover satellite tracking system. The elements identified are then classified based on the possible conditioning of the ground or riverbed permeability, as well as the quality of the underground reserves.

Table 4 summarizes the considered variables, classifying them from the lowest hydro-geodiversity value (1) to the highest (5). Regarding the lake data, provided by the regional landscape plan (PPR), only the water elements with a surface greater than 100 m × 100 m were selected. This measure corresponds to the minimum "cartographic" resolution at the 1:100,000 scale.

Table 4. Expert classification of Land Use types in Sesia Val Grande UGGp.

Land Use	Value
Continuous Urban, Fabric, Industrial or Commercial Units, Road and Rail Networks and Associated Land, Bare Rocks	1
Discontinuous Urban, Fabric, Sport and Leisure Facilities, Non-Irrigated Arable Land, Complex Cultivation Patterns, Sparsely Vegetated Areas	2
Pastures, Land Principally Occupied by Agriculture, With Significant Areas of Natural Vegetation, Coniferous Forest, Moors and Heathland	3
Broad-Leaved Forest, Stable Meadows	4
Beaches, Dunes, Sands, Glaciers and perpetual Snow, Inland Marshes, Lacustrine Deposits and Peats, Water Bodies, Lakes, Mineral Extraction Sites, Rice Fields	5

Considering the number of areal landslides, the map of land use has been integrated with the map of slope instability. Landslide phenomena were considered on the basis of their density; for reasons of scale adaptation, only landslides with a surface area greater than 100 m × 100 m were selected. The slope instability index was obtained by analyzing the density of the area landslides converted to point format. The result of the kernel density analysis was then reclassified into three classes using the natural breaks method (values from 0 to 2).

4.3. Factor Map of Springs and Wells Density

The location of springs and wells were mapped separately. Then, in order to obtain a final representation compatible with the other factor maps and consistent with the chosen scale of representation, areas with a higher density of springs and wells were identified (Figure 8).

Natural springs and wells were subjected to a kernel density with a radius of 1000 m [29]. The raster file obtained was then reclassified into four classes (0, 3, 4, 5) with manual classification, turning the areas characterized by low density into a value of 0. For this classification, high values were used to stress the importance of these factors.

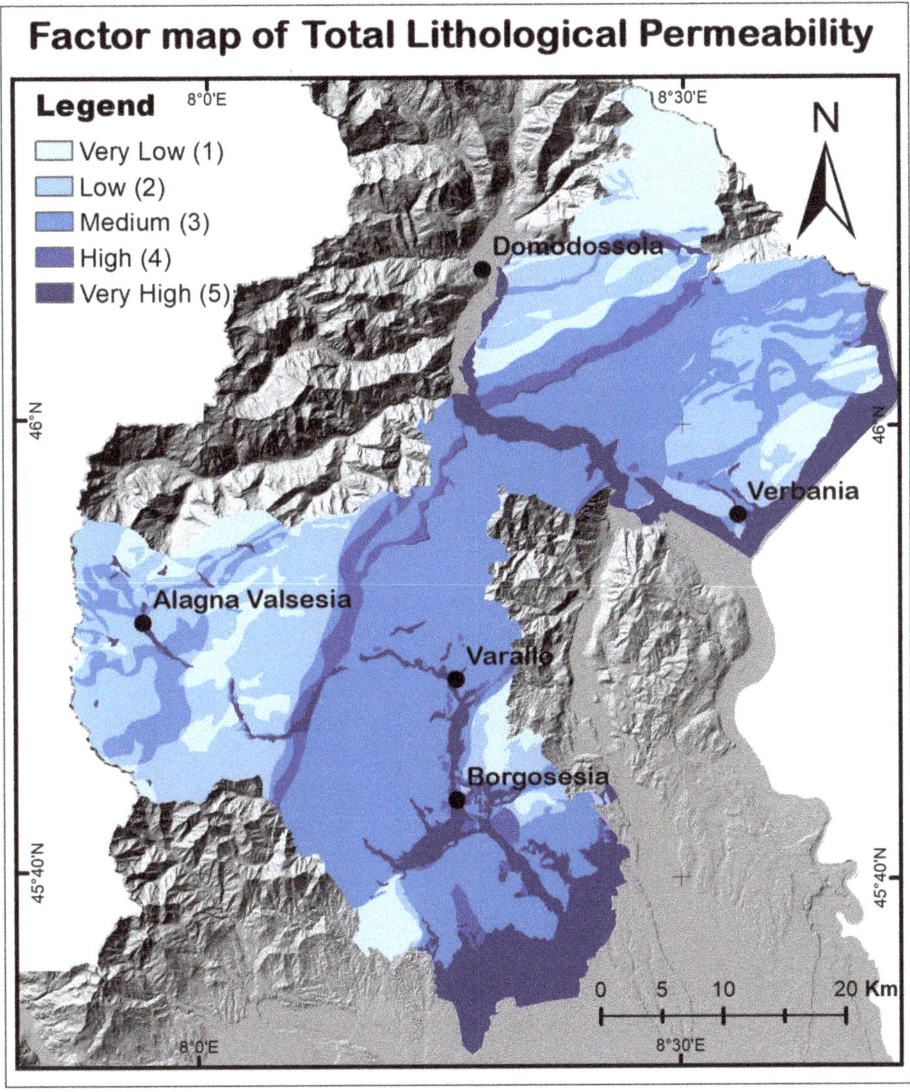

Figure 6. Factor map of lithological permeability integrated with fracturing index. (Esri ArcGis 10.5, [30].

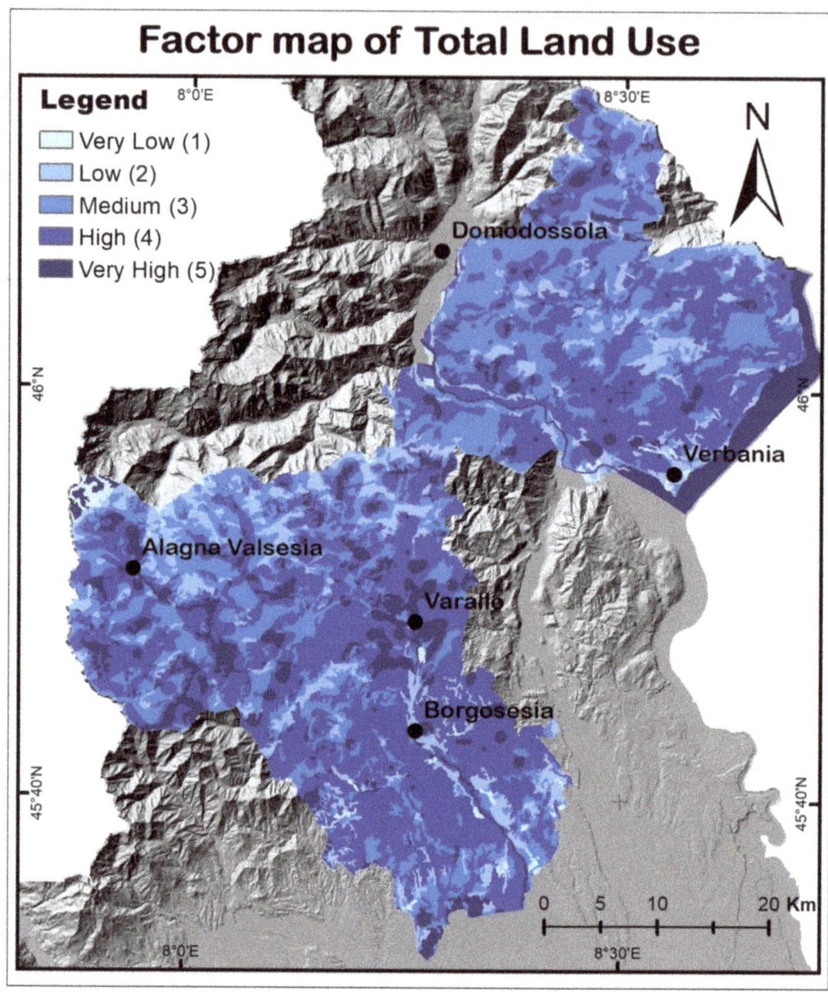

Figure 7. Factor map of land use integrated with landslides density index [30].

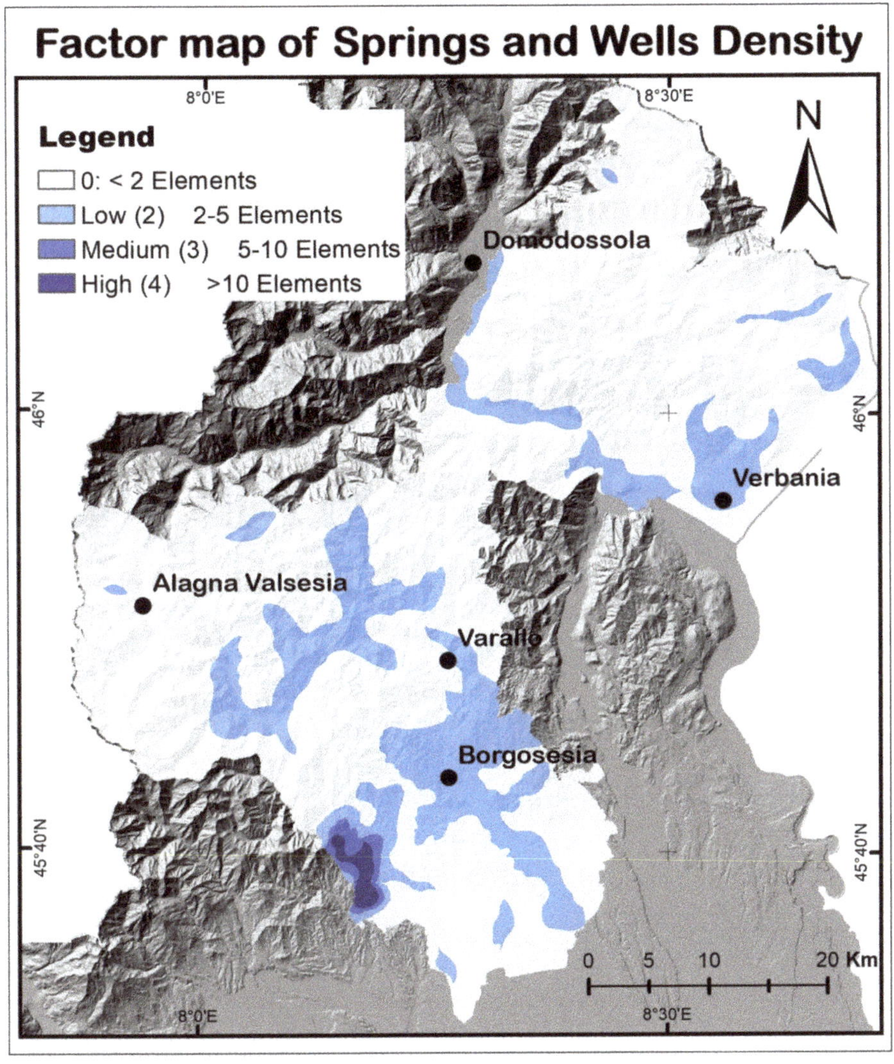

Figure 8. Factor map of springs and wells density [30].

4.4. Factor Map of Morphogenetic Relevance

The factor map of morphogenetic relevance (Figure 9) is used to consider the predominant geomorphological factors that characterize the study area, as well as the dynamics and genetic processes that are the basis of morphological conformation. The territory of the Sesia Val Grande UGG was therefore divided into three areas of morphogenetic relevance: glacial, fluvial, and gravitational. Geomorphological elements taken into consideration are glaciers and glacial cirques for glacial relevance, hydrographic network, alluvial fan, lakes for fluvial relevance, and areal landslides for gravitational relevance. These areas were expertly classified with values from 3 to 5. For the areas dominated by the glacial processes was given the maximum value (5). Glacial modelling is indeed a central factor in the geomorphology of alpine areas.

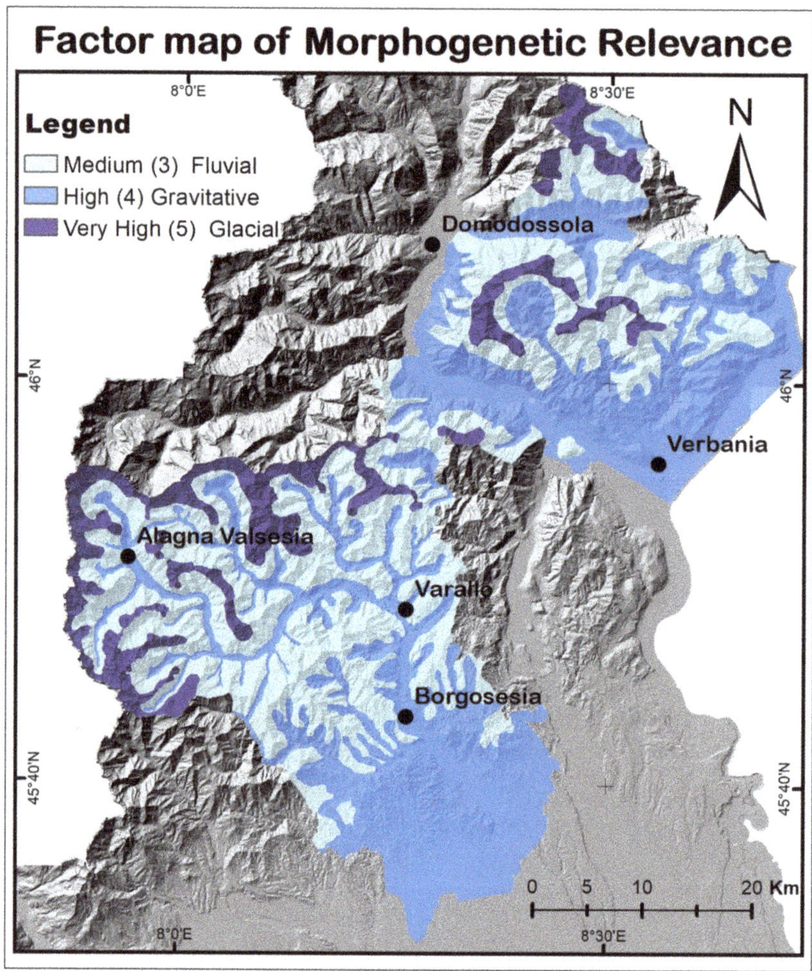

Figure 9. Factor map of morphogenetic relevance [30].

Glacial cirques often host lakes and mountain pastures, which constitute rare habitats and areas used for anthropic purposes for grazing. At the same time, glaciers constitute a reserve for drinkable water and important element of river flow regulation.

The river and lake elements constitute fluvial relevance, which represents a high value of hydrological geodiversity. Lakes and rivers are reservoirs of water. Moreover, rivers can be used for energy production (e.g., dams, hydroelectric power plants) and provide aggregates (e.g., gravel, sand, silt, peat) for various uses. Lastly, the lowest value (3) was attributed to the areas of gravitational relevance. This value has a moderate to high estimation since the landslide processes are firmly interrelated with the water dynamics. They can, in fact, activate and be activated by water processes.

Once the final factor maps were obtained (Figures 6–9), they were added together in a map algebra operation, their sums weighted with GIS, and put through AHP.

This criterion allowed us to create measures that judged consistency, derived priorities between criteria that allowed for comparisons, and established a hierarchy of priorities among the elements. The weighing method adopted made it possible to elaborate many hydro-geo-assessment solutions (from HG_A_1 to HG_A_7 in Table 5). Each time, a greater weight was assigned to one of the four factors, as illustrated in Table 5.

Table 5. Hydro-geodiversity assessment solutions. The underline values indicate the group of factors that has more weight in the map algebra process.

		Same Weight	Method Priority Calculation with the Analytic Hierarchy Process (AHP)					
	Factor Maps	**HG_A_1**	**HG_A_2**	**HG_A_3**	**HG_A_4**	**HG_A_5**	**HG_A_6**	**HG_A_7**
1	Lithological Permeability	1	0,243	0,157	0,157	0,298	0,197	0,175
2	Land Use	1	0,197	0,319	0,281	0,27	0,379	0,409
3	Springs and Wells Density	1	0,182	0,281	0,319	0,246	0, 182	0,175
4	Morphogenetic Relevance	1	0,379	0,243	0,243	0,246	0,243	0,241
	Consistency rate		4,30%	4,30%	4,30%	2,20%	4,30%	5,70%

The last two solutions, HG_A_6 and HG_A_7, are the result of a reasoning that considers a more rigorous approach, which was adopted in the present work and the objectives set. A greater weight was used in the land use factor map, containing the elements interacting with the human dimension. While the HG_A_6 shows an increasing value of factors, the HG_A_7 shows that lower weights are equal. In the HG_A_7 solution, land use weight is equal to 40% of the total weight. Moreover, the springs and wells density is equal to 25%. In fact, since these factors are connected to human activities, these results indicates systems of provisioning and pumping of water resources, as well as a strong point for monitoring the quality and quantity of water in deep and shallow aquifers. To the natural abiotic factors, like permeability and areas of morphogenetic relevance, a weight of 17% was assigned to both levels. It was therefore considered that the HG_A_7 solution was the best solution for the hydro-geodiversity assessment.

In order to obtain more homogeneous areas, the raster file was reclassified into three classes using the natural breaks method (presented in Figure 10 with specific areas and hydrogeosites).

5. Discussions and Conclusions

In this study, nine peculiar areas in the Sesia Val Grande Geopark were identified on the base of the prevailing landscape and its propensity to develop a sustainable relationship between man and the hydro-geosystemic services (Figure 10):

- Area 1: Vigezzo Valley: Landscape of hydrogeological instability
- Area 2: Valley and Piana del Toce: Landscape of alluvial dynamics
- Area 3: San Bernardino and San Giovanni Intra: Landscape of torrential dynamics
- Area 4: Monte Rosa: Landscape of Alpine glacialism
- Area 5: Alpe di Mera: Landscape of deep gravitational instability
- Area 6: Val Mastallone, Upper Val Strona: Landscape of deep valley incisions
- Area 7: Trivero-Val Ponzone: Landscape of the springs
- Area 8: Monte Fenera and Borgosesia: Karst Landscape
- Area 9: High Po Plain: Landscape of the deep aquifers of the Upper Po Plain

Once identified, hydro-geodiversity areas were compared with the geosite location in order to validate the correspondence between them and the areas of hydro-geodiversity, as well as to verify their representativeness and to test the functioning of the qualitative-quantitative procedure previously applied.

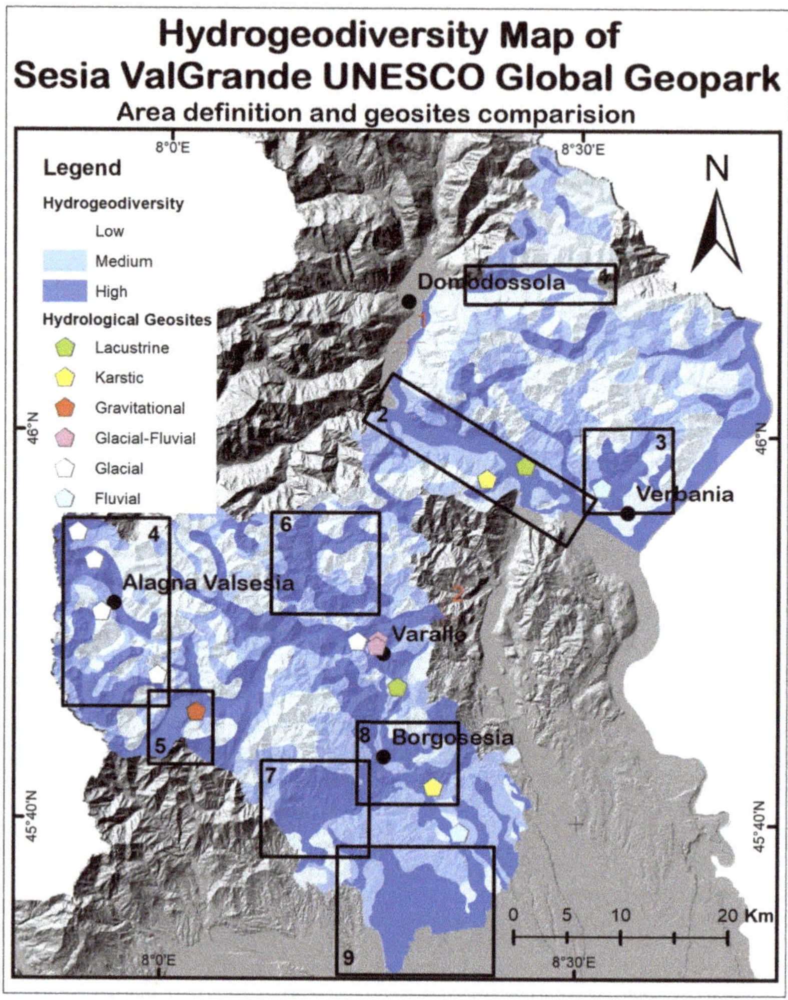

Figure 10. The final solution of the hydro-geodiversity assessment reclassified into three classes with natural breaks in the area definition and hydrogeosites [30].

Geosites from the list of that identified for candidacy to the UNESCO program of the Sesia Val Grande Geopark [31], as well as those extrapolated from the ISPRA national inventory of geosites [32] were selected.

Only geosites with a significance in terms of hydro-geodiversity were chosen. In particular, 25 geosites in this selection were plotted on the final map. All geosites (with the exception of 1) fell into areas with high hydro-geodiversity.

Despite this good correspondence, it should be noted that in some areas there are more than one geosite (areas 4, 8) and in other areas, geosites are classified with a high hydro-geodiversity. Occasionally, we noted the total lack of geosites (areas 1, 2, 6,7). If we were analyze the features of geosites, we would note that the categories of representativeness expressed are fluvial, glacial, gravitational, karst, and lacustrine.

Geosites that represent and test the aquifer dynamics and the relationship between the geological structure and the concentration of springs are missing. However, this aspect is considered fundamental in the hydro-geodiversity classification. This is not surprising and is in line with the tendency to underestimate hidden geosystem features, such as underground processes. This is also demonstrated by recent results of a systematic literature review [33] that show how goods and services derived from the subsurface are underrepresented in the contemporary literature on ecosystem services.

Based on the results and the comparison with the current state of geoconservation of the study area, area 7 (landscape of springs) and area 9 (landscape of the deep aquifers of the Upper Po Plain) are the most important areas in terms of hydro-geosystemic services, as they are directly related to the withdrawal and consumption of water (e.g. drinking water, for agriculture, for breeding). They are also the areas in which human impact is deeper and where there are no instances of hydrogeological protection sufficient for a good preservation. Therefore, more studies and insights about these issues is needed.

Author Contributions: Conceptualization, L.P., M.G., M.L.; methodology, L.P., M.G., M.L.; software, L.P. and G.C.; validation, D.A.D.L.; formal analysis, G.C.; investigation, L.P., M.G., M.L. and G.C.; data curation, G.C.; writing—original draft preparation, L.P. and M.L.; writing—review and editing, L.P. and M.L.; supervision, M.G.; project administration, L.P.; funding acquisition, M.G.

Funding: This study was supported and funded by the Compagnia San Paolo bank foudation in partnership with University of Torino, under the funding program 2016; the project "GeoDIVE—From rocks to stones, from landforms to landscapes" was funded with grant #CSTO169034.

Conflicts of Interest: The authors declare no conflicts of interest.

References

1. Johnson, D.R.; Ruzek, M.; Kalb, M. What is Earth System Science. In Proceedings of the 1997 International Geoscience and Remote Sensing Symposium, Singapore, 4–8 August 1997; pp. 688–691. Available online: https://www.researchgate.net/publication/3700802_What_is_Earth_system_science (accessed on 29 July 2019).

2. Gray, M. *Geodiversity: Valuing and Conserving Abiotic Nature*, 2nd ed.; Wiley Blackwell: Chichester, UK, 2013; pp. 1–3, 75–150.

3. Sharples, C. *Concepts and Principles of Geoconservation (Version 3)*; Tasmanian Parks & Wildlife Service: Hobart, Tasmani, Austrilia, 2002; p. 55.

4. Gray, M. Geodiversity: The backbone of geoheritage and geoconservation. In *Geoheritage: Assessment, Protection and Management*; Reynard, E., Brilha, J., Eds.; Elsevier: Amsterdam, The Netherlands, 2018; pp. 13–24.

5. Simić, S. Hydrological Heritage within the protection of geodiversity in Serbia: Legislation history. *J. Geogr. Inst. Jovan Cvijić SASA* **2011**, *61*, 17–32. [CrossRef]

6. Kebede, S.; Kebede, A. The role of geodiversity on the groundwater resource potential in the upper Blue Nile River Basin, Ethiopia. *App. Geochem.* **2005**, *20*, 1658–1676. [CrossRef]

7. Downs, P.W.; Gregory, K.J. *Evaluation of River Conservation Sites: The Context for a Drainage Basin Approach*; O' Halloran, D., Green, C., Stanley, M., Knill, J., Eds.; The Geological Society: London, UK, 1994; pp. 139–143.

8. Blue, B.; Brierley, G. Geodiversity in riverscapes: Emerging understandings from the Qinghai-Tibetan plateau. *J. Geogr. Sci.* **2012**, *5*, 775–792.

9. Araujoo, A.M.; Pereira, D.Í. A New Methodological Contribution for the Geodiversity Assessment: Applicability to Ceará State (Brazil). *Geoheritage* **2018**, *10*, 591–605. [CrossRef]

10. Zwoliński, Z.; Stachowiak, J. Geodiversity map of the Tatra National Park for geotourism. *Quaest. Geogr.* **2012**, *31*, 99–107. [CrossRef]

11. Serrano, E.; Ruiz, P. Geodiversity: A theoretical and applied concept. *Geogr. Helv.* **2007**, *62*, 140–147. [CrossRef]

12. Sesia Valgrande Geopark UGG. Available online: www.sesiavalgrandegeopark.it/geoparco/carta-d-identita.html (accessed on 5 October 2019).

13. Autorità Bacino del Po. Available online: https://www.provincia.novara.it/ContrattoFiumeAgogna/monografiaPAI.pdf (accessed on 5 October 2019).

14. Carraro, G. Geodiversity Evaluation Related to Water Resources in Sesia Val Grande UNESCO Global Geopark. Master's Thesis, University of Torino, Torino, Italy, 2019.

15. ARPA Piemonte, Alveotipi. Available online: http://webgis.arpa.piemonte.it/ags101free/services/geologia_e_dissesto/BDGeo100_AlveoTipi_Portate/MapServer/WMSServer (accessed on 5 October 2019).

16. Castiglioni, B. *L'Italia nell'Età Quaternaria*; Touring Club Italiano: Milano, Italy, 1940.

17. Comitato Glacioogico Italiano. Available online: http://repo.igg.cnr.it/ghiacciaiCGI/ghiacciai_new.html (accessed on 5 October 2019).

18. Brack, P.; Ulmer, P.; Schmid, S.M. A crustal magmatic system from the Earth mantle to the Permian surface—Field trip to the area of lower Valsesia and Val d'Ossola (Massiccio dei Laghi, Southern Alps, Northern Italy). *Swiss Bull. Angew. Geol.* **2010**, *15*, 3–21.

19. *Dossier Acque Sotterranee*; Ufficio Federale dell'Ambiente, Delle Foreste e del Paesaggio (UFAFP): Bern, Switzerland, 2003; pp. 22–31.

20. Reynard, E.; Brilha, J. *Geoheritage: Assessment, Protection and Management*; Elsevier: Amsterdam, The Netherlands, 2018; pp. 27–52.

21. Piana, F.; Fioraso, G.; Irace, A.; Mosca, P.; d'Atri, A.; Barale, L.; Falletti, P.; Monegato, G.; Morelli, M.; Tallone, S.; et al. Geology of Piemonte Region (NW Italy, Alps–Apennines interference zone). *J. Maps* **2017**, *13*, 395–405. [CrossRef]

22. Regione Piemonte, SIRI, Sistema Informativo Risoprse Idriche. Available online: http://www.regione.piemonte.it/siriw/cartografia/mappa.do (accessed on 5 October 2019).

23. ARPA Piemonte, Piano Paesaggistico Regionale 2017. Available online: http://webgis.arpa.piemonte.it/ppr_storymap_webapp/ (accessed on 5 October 2019).

24. ARPA Piemonte, online data. Available online: http://webgis.arpa.piemonte.it (accessed on 5 October 2019).

25. Autorità di Bacino distrettuale del fiume Po, online data. Available online: https://adbpo.gov.it/open-data/ (accessed on 5 October 2019).

26. Corine Land Cover. Available online: https://www.copernicus.eu/en (accessed on 5 October 2019).

27. Maturana, H.R.; Varela, F.J. *Autopoiesi e Cognizione: La Realizzazione del Vivente*; Marsilio: Padova, Italy, 1985; p. 205.

28. Business Performance Management Singapore, AHP Online System—AHP-OS, Multi-criteria Decision Making Using the Analytic Hierarchy. Available online: https://bpmsg.com/ahp/ahp.php (accessed on 5 October2019).

29. Forte, J.P.; Brilha, J.; Pereira, D.; Nolasco, M. Kernel Density Applied to the Quantitative Assessment of Geodiversity. *Geoheritage* **2018**, *10*, 205–217. [CrossRef]

30. *ArcGIS*, version 10.5; Environmental Systems Research Institute (ESRI): Redlands, CA, USA, 2019.

31. Sesia -Val Grande Geopark Memorandum of Understanding, Annex 5. 2013. Available online: http://www.sesiavalgrandegeopark.it/doc/Annex5_resubmitted.pdf (accessed on 5 October 2019).

32. ISPRA—Istituto Superiore per la Protezione e la Ricerca Ambientale, Censimento nazionale geositi. Available online: www.isprambiente.gov.it/it/progetti/suolo-e-territorio-1/tutela-del-patrimonio-geologico-parchi-geominerari-geoparchi-e-geositi/il-censimento-nazionale-dei-geositi (accessed on 5 October 2019).

33. Van Ree, C.C.D.F.; van Beukering, P.J.H.; Boekestijn, J. Geosystem Services: A hidden link in ecosystem mangement. *Ecosyst. Serv.* **2017**, *26*, 58–69. [CrossRef]

Article

Impact of the October 2018 Storm Vaia on Coastal Boulders in the Northern Adriatic Sea

Sara Biolchi [1], Cléa Denamiel [2], Stefano Devoto [1,*], Tvrtko Korbar [3], Vanja Macovaz [4], Giovanni Scicchitano [5], Ivica Vilibić [2] and Stefano Furlani [1]

[1] Department of Mathematics and Geosciences, University of Trieste, 34128 Trieste, Italy; sbiolchi@gmail.com (S.B.); sfurlani@units.it (S.F.)
[2] Institute of Oceanography and Fisheries, 21000 Split, Croatia; cdenamie@izor.hr (C.D.); vilibic@izor.hr (I.V.)
[3] Croatian Geological Survey, HR-10000 Zagreb, Croatia; tkorbar@hgi-cgs.hr
[4] Macovaz Studio, 34147 Trieste, Italy; macovazvanja@gmail.com
[5] Studio Geologi Associati T.S.T., 95030 Pedara, Italy; scicchitano@studiogeologitst.com
* Correspondence: stefano.devoto2015@gmail.com

Received: 15 August 2019; Accepted: 22 October 2019; Published: 25 October 2019

Abstract: Boulder detachment from the seafloor and subsequent transport and accumulation along rocky coasts is a complex geomorphological process that requires a deep understanding of submarine and onshore environments. This process is especially interesting in semi-enclosed shallow basins characterized by extreme storms, but without a significant tsunami record. Moreover, the response of boulder deposits located close to the coast to severe storms remains, in terms of accurate displacement measurement, limited due to the need to acquire long-term data such as ongoing monitoring datasets and repeated field surveys. We present a multidisciplinary study that includes inland and submarine surveys carried out to monitor and accurately quantify the recent displacement of coastal boulders accumulated on the southernmost coast of the Premantura (Kamenjak) Promontory (Croatia, northern Adriatic Sea). We identified recent boulder movements using unmanned aerial vehicle digital photogrammetry (UAV-DP). Fourteen boulders were moved by the waves generated by a severe storm, named Vaia, which occurred on 29 October 2018. This storm struck Northeast Italy and the Istrian coasts with its full force. We have reproduced the storm-generated waves using unstructured wave model Simulating WAves Nearshore (SWAN), with a significant wave height of 6.2 m in front of the boulder deposit area. These simulated waves are considered to have a return period of 20 to 30 years. In addition to the aerial survey, an underwater photogrammetric survey was carried out in order to create a three-dimensional (3D) model of the seabed and identify the submarine landforms associated with boulder detachment. The survey highlighted that most of the holes can be considered potholes, while only one detachment shape was identified. The latter is not related to storm Vaia, but to a previous storm. Two boulders are lying on the seabed and the underwater surveys highlighted that these boulders may be beached during future storms. Thus, this is an interesting example of active erosion of the rocky coast in a geologically, geomorphologically, and oceanologically predisposed locality.

Keywords: rocky coast; extreme waves; active erosion; geohazard; Croatia

1. Introduction

This study investigates the movement of the boulders located in the southern sector of the Premantura (Kamenjak) Promontory (Croatia, southern Istria Peninsula) after an extremely severe storm named "Vaia" which struck during 29 October 2018 [1]. Terrestrial, aerial and underwater surveys were performed to analyze the possible geomorphological effects of the storm on the boulder field in terms of movements of boulders, as well as their disappearance and the appearance of new ones.

In 2018, between Saturday, October 27 and the early hours of Tuesday, October 30, large sectors of the Italian and the northern Adriatic coasts were hit by one of the most intense, complex, and damaging storms in recent decades as the result of the passage of an exceptionally strong cyclone named Vaia. Monday, October 29, was characterized by violent gusts of Sirocco, (a southerly wind blowing across the Adriatic Sea), storm surges, and extraordinarily high tides in the northernmost Adriatic Sea combined with severe rainfall, particularly in the northeastern Alps. The storm caused 16 casualties in an area from Trentino (northern Italy) to Campania (southern Italy) and severe damage totaling more than two billion euros. Tens of thousands of customers were still without electricity two days after the event, especially in the Trentino, Veneto, and Friuli areas (Northeast Italy).

We investigated the relationship between this extreme Sirocco storm and an extensive boulder accumulation located at the southernmost tip of the Istrian Peninsula, in Premantura, Croatia [2]. This study provides the results of two years of monitoring using unmanned aerial vehicle (UAV) based multitemporal photogrammetric image comparison and the outputs of underwater environmental monitoring using a three-dimensional (3D) model obtained through photogrammetric analysis of underwater pictures. Field activities were carried out, assisted by a UAV survey, approximately two weeks after storm Vaia. These surveys revealed that the boulder positions had changed with respect to the previous field investigations carried out in November 2017. For example, the isolated boulder, K8, widely described by Biolchi et al. [2] had been moved towards the northwest and rotated.

The transport of rocky boulders in the Mediterranean and more generally along oceanic coasts has been widely studied. It has been documented that waves associated with stormy events and major tsunami events can generate a force strong enough to detach boulders from the ground and transport them along the shore. Despite boulder motion being mostly attributed to tsunamogenic activities in the literature, several authors have recently proven that the impact of storm waves is, in fact, the primary mechanism for boulder detachment and transport [3–9]. Moreover, at higher latitudes, Orviku et al. [10] explained how the decomposition of sea ice and its drifting onshore can transport and accumulate boulders over 1 m in diameter, especially during the spring after ice melting.

Marriner et al. [9] analyzed tsunami and storm data, contained in the EM-DAT (Emergency Events Database) for the 1900 to 2015 period, and found that up to 90% of tsunami attributions of high-energy events in the Mediterranean coastal records should be reconsidered. Cox et al. [11,12] and Cox [13] reported on strong evidence that even some mega boulders in the open ocean setting (Northern Atlantic) were displaced along with the cliff platforms by storm waves.

This study presents a multidisciplinary approach based on the hypothesis that it was storm Vaia that impacted the coastal boulders. This approach, which integrates inland and submarine surveys, as well as numerical modeling aims to:

1. Detect possible transports triggered by storm Vaia and monitor the boulder movements using unmanned aerial vehicle digital photogrammetry (UAV-DP);
2. Examine the submarine environment by testing a new methodology that can be useful for future comparisons after future storms;
3. Produce accurate wave results for the period during the peak of the storm in order to confirm that the boulder motion could have been locally triggered by breaking waves.

The focus of this work is to study the relationship between a previously documented storm and coastal boulder movements, highlighting that a severe storm is sufficient to move or initiate boulder transport both from the sea bottom and in a subaerial environment. The work can be included in the scientific debate concerning the explanation of coastal boulder detachment and movement in terms of "storm vs. tsunami". A clear relationship between a severe storm that occurred during 2014 and the emplacement of a large boulder in the southern Istria peninsula in Premantura was provided by [2]. This study, performed in the same area after a different storm, confirms that sea wave energies are capable of moving or detaching new boulders from the submarine floor.

1.1. Study Area

The studied boulder accumulation is located on the rocky coast of the Premantura Promontory (Kamenjak Nature Park) in the southernmost sector of the Istrian peninsula (Croatia, northern Adriatic Sea). The promontory is formed by a succession of stratified Late Cretaceous carbonate rocks, dipping gently towards the east [2] (Figure 1).

Its southeastern-most tip is named "Jugo Promontory", after the powerful southeasterly wind (jugo in Croatian) and consists of a stratified, shallow-marine Cretaceous limestone succession. The succession is characterized by alternations of thin-bedded (10–30 cm), fine-grained peloidal packstones, and thick-bedded (50-150 cm) mudstones and wackestones containing algal oncoids and rare rudist shells, typical of the lower part of the Gornji Humac Formation, which dates back to the Turonian [14]. The Kamenjak Nature Park is characterized by elevations that are up to 50 m above sea level, rounded bays, pocket beaches, and small islands. The south and east coasts of the park are characterized by gentle slopes, which generally follow the dip angles of the bedrock strata.

At the studied site (southern part of Jugo Promontory), the washed ("white") coastal zone is up to 70 m wide. The zone is the narrowest (40 m) at the highest elevation region, in the central part of the study site. The bed dip direction and the dipping angle on the southwestern part is 88/12, and boulders are grouped along the washed zone, situated 50 to 70 m from the sea, along the border with the vegetated zone, figured in Biolchi et al. [2]. There are a few solitary boulders that are considered the youngest. More than 950 mostly meter-sized boulders (volumes from 0.385 to 11.440 m^3) were recognized on the promontory during a former survey [2], with an estimated total volume of ~2000 m^3.

The submerged part of the coast is of similar morphology, and gently dipping limestone ramps continue under the sea. Cliffs with heights of a few meters developed locally (out of the studied localities) and are related to massive rock masses within the succession of the limestone rocks and to small local faults. Thus, the maximal runup during extreme storms is defined by the height of the breaking waves and the gentle dip angle of the coastal ramp and is marked by the boundary of the vegetated zone along the gentle coast that is around 70 m away from the sea [2].

The Adriatic Sea is a semi-enclosed basin in the northern Mediterranean, with the following dominant winds: Sirocco (jugo in Croatian) blowing from the southeast, Bora (bura in Croatian) blowing from the northeast, and Libeccio blowing from the southwest (Figure 1c) [15]. The Sirocco has the longest fetch, and therefore generates the highest waves in the northern Adriatic Sea [16]. Significant and maximum wave heights of 5.3 m and 10.8 m, respectively, were measured over a roughly 10-year interval (1978–1986) about 50 km southwest of the Premantura Promontory. The other two winds generate lower waves in the area under investigation. The Sirocco may also generate extreme storm surges and sea levels in the northern Adriatic Sea [17], particularly between November and February, which can lead to the flooding of coastal regions. Together with the Adriatic seiche [18] and tides, extreme sea levels can reach up to 1.5 m above mean sea level [19,20]. On top of the tides and the storm surges, tsunami-like waves of meteorological origin, meteotsunamis, may further raise the sea level along the open coastline of Istria by a few tens of centimeters [21,22].

Although they have been found to impact coastal boulders in various parts of the world's oceans, no significant seismic tsunamis have been reported in the northern Adriatic Sea over the last two hundred years [23,24]. The worst-case hazard scenarios provide for a maximum tsunami height no higher than 20 cm [25].

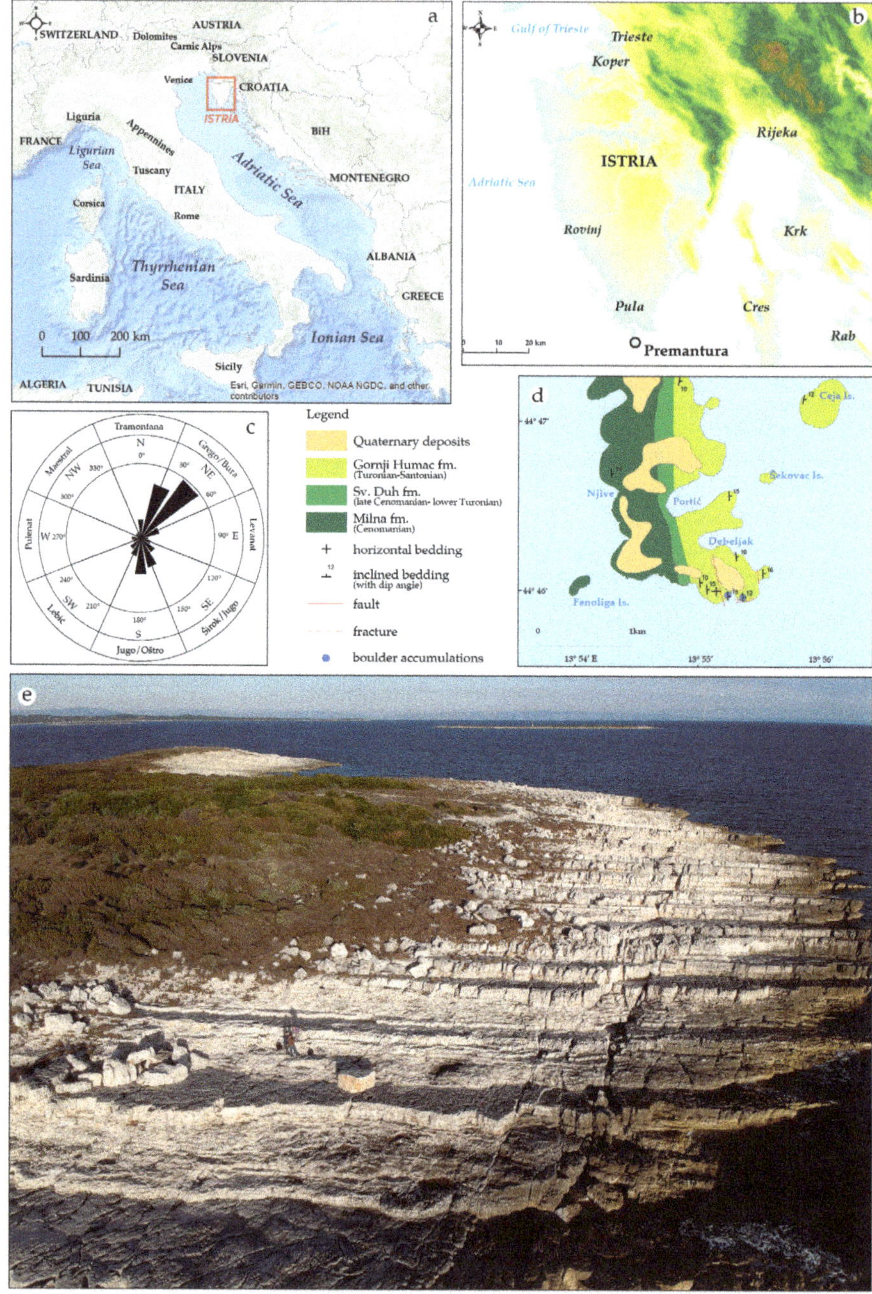

Figure 1. (**a**) Location of the Istrian peninsula, (**b**) location of the Premantura study area, (**c**) Windrose, (**d**) geological map of the South part of the Premantura (Kamenjak) Promontory, and (**e**) oblique view of the study area (Jugo promontory and Cape Kamenjak) where the K8 boulder and clusters of boulders spread along rocky coast are clearly visible.

1.2. Storm Vaia

At the time of this study, storm Vaia had not yet been described in the literature and the only available information came from reports provided by the Italian Meteorological Society [1] and the numerical results (Figure 2) obtained during this study with the Adriatic Sea and Coast (AdriSC) modeling suite presented in more details in Section 3.3.

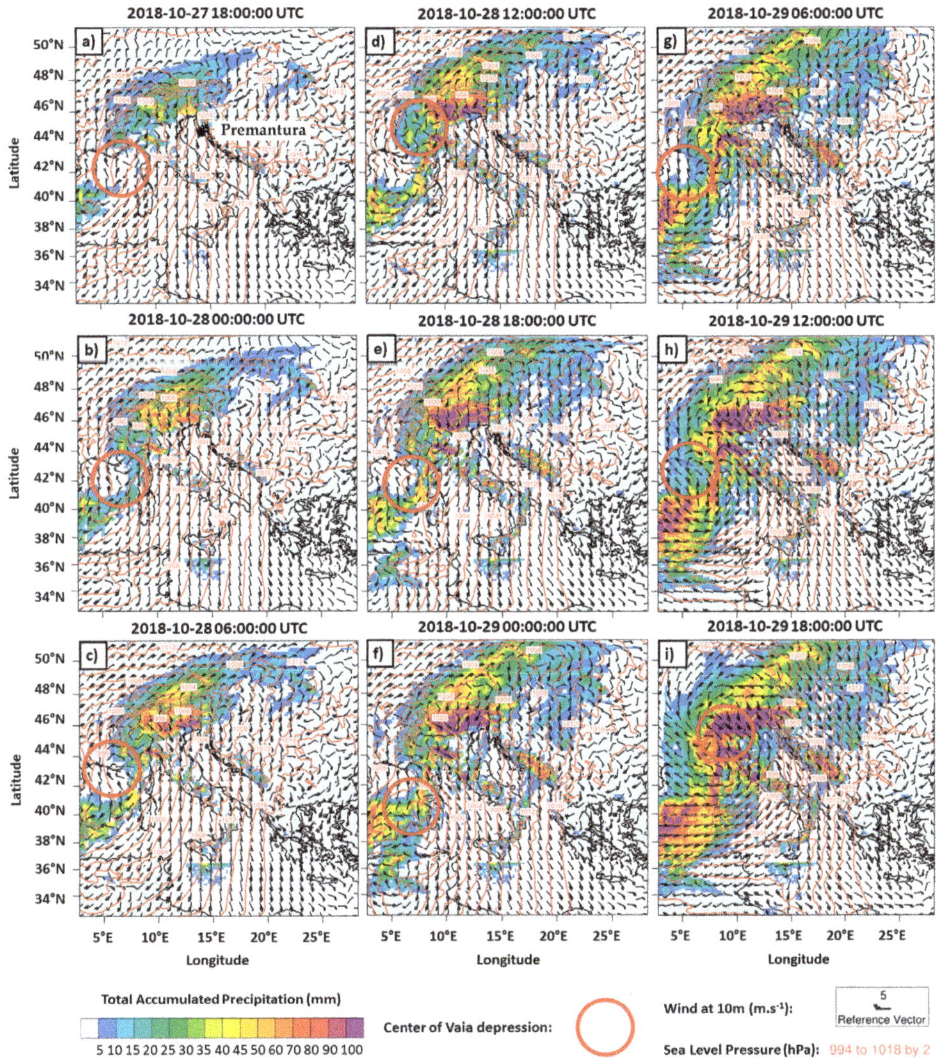

Figure 2. Evolution of the accumulated rain, the sea level pressur, and the winds (in chronological order from panel (**a–i**)) during storm Vaia using the numerical results obtained with the WRF-15km model from the AdriSC modeling suite. The center of the depression is represented with a red circle which highlights the track of the storm.

Atmospherically speaking, the Vaia depression developed on Saturday, 27 October 2018, within an extensive low-pressure trough stretching from the Baltic to the western sector of the Mediterranean Sea.

This depression held its position at sea between the Gulf of Lion, the Balearics, and Sardinia until the morning of Monday, October 29. This first phase of the storm was marked by humid Libeccio air currents coming from between the south and southwest, accompanied by intense rainfall events in the northern Apennines and mountainous areas from northwestern Italy to the Carnic Alps (northeastern Italy). By midnight on October 28, more than 300 mm of rain had already fallen in many areas from the Prealps above the city of Brescia to the mountains of Friuli, with a noticeable increase in river flows. After a pause of a few hours, the storm's second phase, fueled by the arrival of the first major cold front of the 2018 season in the Alps, developed on the morning of Monday, October 29. During the course of the day storm Vaia underwent rapid deepening (about 17 hPa in 18 h), almost classifiable as "explosive cyclogenesis" (the threshold for which is considered a pressure decrease of 24 hPa in 24 h), with the center of the depression moving from the west of Corsica to the northeast during the afternoon and then to northwestern Italy in the evening (red circles in Figure 2). This deepening of the storm was accompanied by the following: (1) a brutal reinforcement of the winds across Italy, (2) a major strengthening of the Sirocco, and (3) the development of violent self-regenerating storm cells in the area of Sardinia, and the Tyrrhenian and Ligurian Seas. In the afternoon and evening of the October 29th a southeasterly windstorm reached the Adriatic basin while a heavy rain began to fall again on the already saturated soils of the Alps and Prealps but also in Northwest Italy due to its more southeasterly airflow.

The Vaia depression will be remembered not for the rainfall, but for the violence of the Sirocco that blew between morning and afternoon of Monday, October 29, when it swung round to the Libeccio in the evening, starting with Italy's western sea areas (i.e., measured wind speeds reached 102 km/h at Rome Ciampino airport, 119 km/h at Genoa airport; 128 km/h in Lugano in southern Switzerland, 128 km/h on the Valles Pass in the Dolomites, 148 km/h at Capo Carbonara in southeastern Sardinia, 155 km/h at the Colle di Cadibona near Savona in NW Italy, 171 km/h at both La Spezia and Follonica in Tuscany, and 200 km/h on Monte Rest in the Carnic Prealps in northeastern Italy). The violent southerly gusts of wind, which persisted for many hours along considerable lengths of coastline, raised devastating storm surges, particularly in Liguria, with severe damage to the coastal roads and railways, buildings, and tourist facilities with dozens of boats destroyed in ports. On the evening of Monday, October 29, a buoy belonging to Liguria's Environmental Protection Agency lying offshore from Capo Mele recorded a maximum wave height of 10.3 m with a very long period of 11 s, an indicator of highly destructive wave power along the coast. Also noteworthy was the episode of high water in Venice and along the northern Adriatic coast, including the Istrian peninsula. In particular, the tide gauge at Punta della Salute (at one end of Venice's Grand Canal) measured a maximum of 156 cm at 14:10 on 29 October, a value exceeded by only three other events in the historical series since 1872: 1 February 1986 (158 cm), 22 December 1979 (166 cm), and the infamous 4 November 1966 (194 cm).

2. Materials and Methods

The investigations on the boulder accumulation located along the southern coast of the Premantura (Kamenjak) Promontory have been developed through a multidisciplinary approach, which envisaged inland surveys, submarine investigations, and wave modeling.

2.1. Inland Surveys

Field activities have been performed since 2017 and include traditional geomorphological and geological investigations and UAV surveys. Geomorphological and geological activities have included extensive fieldwork and are reported in [2]. The above-cited paper includes a list of boulders and their axis lengths.

UAVs are widely used in various fields of geoscience [26–28] and offer major benefits through their ability to provide high-resolution photographic images from reduced flight times [29]. The digital photogrammetry (DP) technique enabled the reconstruction of high-resolution orthophotographs and a digital elevation model (DEM) of a large sector of the Premantura area. This reconstruction was done through the processing of 135 images collected by a DJI Phantom drone [TM] (DJI, Nanshan District, Schenzen, China) in 2017 (flight altitudes between 20 and 60 m).

These images were processed using Agisoft Photoscan Professional software version 1.4.0 (Agisoft, St. Petersburg, Russia) and the results were processed in 2018 before storm Vaia in a GIS environment using QGIS version 2.18 Las Palmas (QGIS Development Team, QGIS Geographic Information System. Open Source Geospatial Foundation Project).

One of the main outputs of the photogrammetric processing was a detailed map of the position of 950 boulders spread along the Jugo Promontory, as shown in [2].

In order to carry out a multitemporal photogrammetric survey devoted to recognizing possible movements of blocks triggered by storm Vaia and other future storm events, we performed three drone campaigns in November 2018 (two weeks after storm Vaia) and 2019, using a quadcopter DJI Spark drone[TM] (DJI, Nanshan District, Schenzen, China), equipped with a 12MP camera. DJIFlightplanner software[TM] (AeroScientific, Blackwood, Australia) and Litchi software[TM] (VC Technology, London, UK) assisted in the choice of 2019 flight plans and permitted the operator to easily set the altitude, radius, number of waypoints, speed, and directions.

We performed two flights on 15 November 2018, and six flights at different altitudes on 30 April 2019, and 14 June 2019. The DJI Spark drone[TM] executed automatically and autonomously the 2019 flight plans using the Litchi software[TM] installed on the remote controller device.

Table 1 lists which sector of the study area was investigated in each flight and the main characteristics of the UAV surveys.

Table 1. Main characteristics of unmanned aerial vehicle (UAV) flights carried out in 2018 and 2019.

Survey	Date	Altitude (m)	Number of Pictures	Sector Investigated	GSD (cm/pixel)
1	15/11/2018	30	165	W	1.039
2	15/11/2018	30	266	W	1.039
1	14/06/2019	28	294	W	0.935
2	30/04/2019	30	235	W and central	1.039
3	30/04/2019	20	229	W and central	0.693
4	30/04/2019	30	314	E	1.039
5	30/04/2019	61	253	Entire	2.078
6	30/04/2019	35	338	Central and E	1.212

In order to define the ground control points (GCPs) for the production of georeferenced aerial images, the flights carried out in 2019 were accompanied by a differential global navigation satellite system (DGNSS) survey [30]. The GCPs were mainly located on the west side of the boulder accumulation where we had noticed the recent movement of the isolated boulder named K8 in [2], and, we assumed, further possible movements of a few solitary boulders because their distribution was closer to the coastline. The master GCP was located in an area that may not be reached by any boulders, whereas 6 DGNSS bolts accompanied by black-yellow photogrammetric targets were spread in positions clearly identifiable from aerial photos taken in previous UAV campaigns. These DGNNS points were crucial for the quantification of possible boulder motion, with a maximum error of 15 cm calculated on the maximum discrepancy between the reconstructed coordinates and the one measured at the edge of the area.

The images were processed using 3DF Zephyr Aerial software[TM] version 4.007 (3Dflow, Verona, Italy) and detailed orthophotographs of 2018 and 2019 were computed. The best quality photos were taken on the later flights of 2018 and 2019 and for this reason we devoted our attention to these orthophotographs to analyze the western and central sector of the study area. These photos were imported into the QGIS software, which permitted the evaluation of possible boulder rotation and translation transport.

The distance between two known and measurable points on the ground, such as the pothole described in [2] and the old position of boulder K8 (both clearly visible on aerial photos of 2018 and 2019), was measured using a QGIS tool. This measurement was compared with the real distance measured directly on the ground on 30 April 2018. The difference of approximately 10 cm validates the 15 cm accuracy of the model computed by the 3DF Zephyr Aerial software[TM].

2.2. Submarine Investigations

In order to reconstruct the 3D model of the seabed, underwater images were collected via a snorkel survey in June 2019. Photogrammetric procedures are a common and rapid technique to acquire metric measurements in a submarine environment using the combination of field operations and post-processing [31,32]. Submerged photogrammetry requires a precise preparation and setting of the images to reduce water refraction and distortion. Underwater photography is vulnerable to the different refraction coefficient of the marine water, which produces a reduction of the field of view (FOV) and a complete change in the optical parameter of the lens [33]. To reduce the changes in focal length, we used a dome glass. Conversely, this increases the distortion of the system significantly. In this work, a "quick and dirty" procedure was applied to evaluate the limitations of a low-cost approach using GoPro action cameras together with the capability of acquiring good quality data snorkeling from the surface. Action cameras are usually equipped with wide-angle lenses and the option of selecting a narrow FOV via software adjustment, losing a large percentage of the sensor resolution and resulting in a low-quality set of photos. We operated in full resolution to manage the wide FOV resulting in strong distortion and a huge color cast at the picture boundaries.

The 3D model was assembled in Agisoft Metashape [TM] (Version 1.5.0) starting from a selection of 916 of the 1200 images produced by a GoPro Hero 6 Black[TM] (GoPro, San Mateo, California, USA) set in a six inch waterproof plastic case.

In order to remove the color cast and the large amount of suspension present at the picture corners, a mask was applied to each photo. This strategy removed almost one-third of the pixels belonging to the area with the strongest distortion and permitted a precise alignment.

To better cover the submarine area, an S-shape route was followed with a time-lapse camera set to 0.5 s. Therefore, the overlap of the underwater images was higher than in the aerial photogrammetry. This procedure avoided inaccurate orientation of the cameras at different depths of the seabed [34].

2.3. Wave Modeling

The numerical modeling was carried out using the Adriatic Sea and Coast (AdriSC) modeling suite, that was jointly developed within the ADIOS [35] and MESSI [36] projects with the aim of accurately representing the processes driving the Adriatic's atmospheric and oceanic circulation at different temporal and spatial scales. This modeling suite consists of: (1) a basic module that deals with the coupled ocean and atmospheric general circulation and (2) a nearshore module that provides high-resolution fields during extreme events.

The basic module is based on the Coupled Ocean–Atmosphere–Wave–Sediment Transport (COAWST) modeling system [37]. The module is built around the Model Coupling Toolkit (MCT), which exchanges data fields and dynamically couples the Weather Research and Forecasting (WRF) atmospheric model, the Regional Ocean Modeling System (ROMS), and the Simulating WAves Nearshore (SWAN) model. The basic module was set up with (1) two different nested grids of 15 km and 3 km resolution used in the WRF model and covering the central Mediterranean area and the Adriatic-Ionian region, respectively, and (2) two different nested grids of 3 km and 1 km resolution used in the ROMS and SWAN models and covering the Adriatic-Ionian region (the same grid as the WRF model) and the Adriatic Sea, respectively.

The nearshore module additionally accounts for the nearshore bathymetry changes and combines the WRF model with the fully coupled ADCIRC-SWAN unstructured off-line model [38]. In this module, the hourly results from the WRF 3 km grid obtained with the basic module were downscaled to a 1.5 km grid covering the Adriatic Sea, while the hourly results from the 1 km ROMS grid and the 10 min results from the 1 km SWAN grid are used to force the unstructured mesh of the ADCIRC-SWAN model.

The AdriSC modeling suite was installed and fully tested on the European Centre for Medium-Range Weather Forecast (ECMWF) high-performance computing facilities. More details on the AdriSC modeling suite setup can be found in [39].

In this study, in order to reproduce the storm which took place in the Adriatic Sea on 29 October 2018, the SWAN model was set up in both modules to be coupled with the oceanic and atmospheric models (i.e., WRF, ROMS, and ADCIRC). In addition, the computation of the bottom stress of the ocean models (respectively, ROMS and ADCIRC) was updated in order to consider the spatial distribution of the sediment grain size at the bottom of the Adriatic Sea, extracted from the Adriatic Seabed database [40], and the wave effects. To reproduce the storm as accurately as possible, the basic module was set up to run for three days between 27 October and 30 October 2018, with initial conditions and boundary forcing provided by (1) the 6 hourly ERA-Interim reanalysis fields [41], (2) the daily analysis MEDSEA-Ocean fields [42], and (3) the hourly MEDSEA-Wave fields [43]. The nearshore module, forced by the results of the basic module, was set up to run between midday on 28 and 30 October 2018.

In this study, only the wave results from the unstructured SWAN model of the nearshore module (hereafter referred to as AdriSC unSWAN) were analyzed.

Due to the lack of precise bathymetry data and resolution in the nearshore area where the waves are breaking (i.e., surf zone) and the known limitations of the physics used in the SWAN model (e.g., the parametrization of the wave breaking), the wave heights modeled by the AdriSC unSWAN module during the event were extracted off the surf zone (ideally in deep water) and the Sunamura and Horikawa equation [44] was applied to evaluate the wave height at the breaking point, Equation (1):

$$\frac{H_b}{H_0} = (tan\beta)^{0.2} * \left(\frac{H_0}{L_0}\right)^{-0.25} \tag{1}$$

where H_b is the breaking wave height; H_0 and L_0 are the wave height and the wave length off the surf zone (ideally in deep water), respectively; and β is the slope of the seabed near the coast (i.e., in the area where the waves are expecting to break). Finally, the breaking wave height obtained was compared with the hydrodynamic model proposed by Nandasena et al. [45] which is extensively used in the literature to define the storm wave heights capable of detaching and transporting coastal boulders.

3. Results

3.1. Onshore Surveys

The area where the limestone boulders are distributed is located along the southern coastal section of the Jugo Promontory, mostly between the sea and the vegetated zone. The southwestern coast of Cape Kamenjak is directly exposed to the Sirocco-induced waves. The dip direction and dip angles of the limestone beds are 88° and 12°, respectively. An indistinct joint system has developed along the bed strike (i.e., generally running North-South), whereas a distinct open fractures system generally strikes east and west (with a mean dip direction and dipping angle measure of 350°/85°) as per meter-scale distances. Thus, quadrangular limestone fragments are formed by the fracture network together with layer planes [2]. The seven tonne boulder (K8), here renamed boulder #1, is characterized by its unusual orange surface coloring due to karst weathering. Its elevation is 2 m above sea level. Before the storm, boulder #1 was located at 27 m distance from the coastline and was oriented with the longer axis facing the main wave direction. Given its isolated position and intense orange color, it is clearly distinguishable in aerial and terrestrial images. The occurrence of very fresh subrecent biogenic carbonate encrustations on its southern and upper surfaces, mainly produced by coralline algae and serpulids, as well as by more fragile barnacle shells, attests to its marine origin. Its deposition has been reconstructed by [2] and is the result of a severe Sirocco storm between January and February 2014.

During the field activities carried out on 15 November 2018, just two weeks after storm Vaia storm, the shift of boulder #1 was the most evident and visible impact of the storm (Figure 3).

Figure 3. (**a**) Oblique view of boulder #1 and (**b**) the red rectangle indicates its past position on the limestone pavement.

As a result of the print that was left by boulder #1 after four years lying on the limestone pavement and the multitemporal comparison of UAV-derived orthophotos, its displacement was measured as being approximately 3 m towards the northwest and with a counterclockwise rotation of 18°.

The UAV surveys in 2018 and 2019 were carried out to identify and quantify any other boulder movements. The comparison between the 2017 and 2018 UAV orthophotos showed that 14 of them had changed their position or had been rotated and revealed the appearance of a new small boulder. Figure 4 highlights the boulders that had moved, plotted on the 2018 UAV orthomosaic, and compared to their 2017 positions.

Figure 4. Spatial distribution of 14 boulders affected by the 29 October storm Vaia. The base image was taken during the November 2018 UAV flight. The star indicates the presence onshore of a new limestone boulder detached and moved from the seafloor during the storm. No movements were recorded after 15 November 2018, and 30 April 2019.

All the boulder that had moved were measured using field measurements (sizes, position, and distance from the coast), UAV-obtained images, and 3D models. These 3D models are crucial for a desk-based geomorphological analysis, such as boulder recognition, and size measurement. Figure 5 displays a view from Flight #2, performed on 30 April 2019.

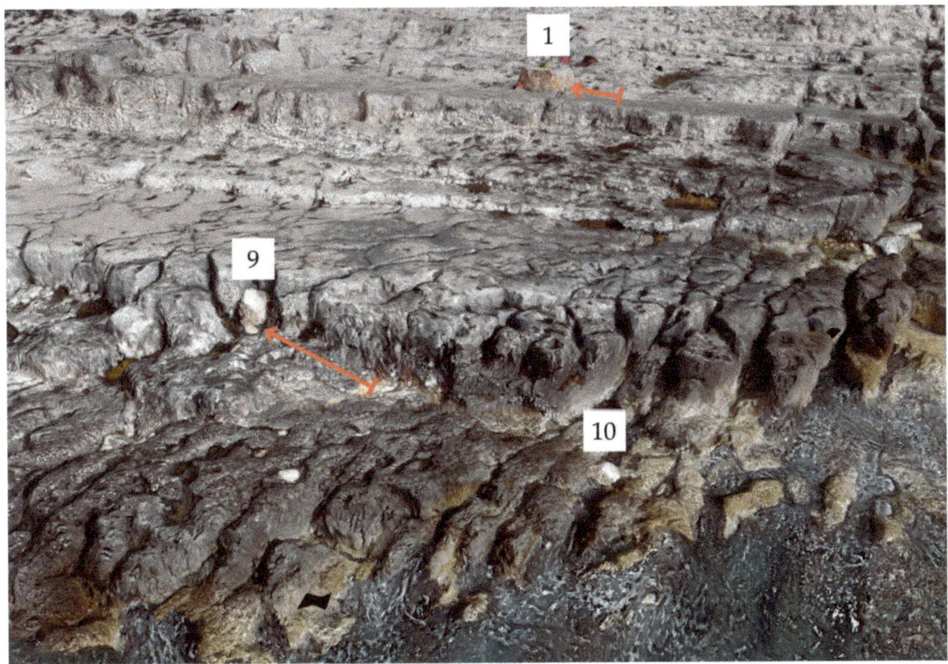

Figure 5. A view of the three-dimensional (3D) model obtained during Flight #2 performed on 30 April 2019. The red arrows show the shift of boulders #1 and #9 and the position of the new boulder #10.

Table 2 lists the boulders that were affected by the storm including: size, quantification of translational movement (with an accuracy of 1 m), and rotation characterization. Moreover, an estimation of the wave height required to move a boulder in a subaerial scenario according to the Nandasena model is reported.

Boulders #2, #3, #4, #7, and #8 are trapped in two different clusters populated by tens of blocks and have mainly been rotated. Boulders #5 and #6 are close to the coastline and their motions were interrupted by a persistent east–west oriented joint that crosses the limestone promontory. The discontinuity is clearly visible in Figure 4. This movement caused a rotation of Boulder #5 and its fragmentation. Boulder #11 was deeply fragmented by severe waves and probably by collisions with other fragments, whereas boulders #12, #13, and #14 have mainly been rotated and fragmented. This behavior is related to the presence of other boulders and the abovementioned persistent discontinuity that inhibited their movements. The same scenario took place with boulders #5 and #6. The western part of Figure 4 shows well the severe impact of the last storm. The yellow star indicates the presence of a new limestone boulder (boulder #10), detached from the seafloor and transported by the waves. North of this new boulder (Figure 5), Boulder #9 was moved about 5 m and is now embedded within a niche.

Table 2. List of boulders moved or rotated by the storm.

Boulder (#)	Axis a; b; c (m)	Displacements (m)	Direction of Movement/Rotation	Rotation	Nandasena Wave Height (m)	Notes
1	2.25; 1.65; 0.95	3	NW	Affirmative	14,22	Isolated
2	0.9; 0.9; 0.7	No transport	-	Toppled	10,48	Cluster
3	1.2; 0.6; 0.6	1	E	Affirmative	8,98	Cluster
4	1.4; 0.8; 0.7	No transport	-	Affirmative	10,48	Cluster
5	1.7; 0.9; 0.7	2	W	Affirmative	10,48	Trapped
6	2.1; 1.9; 0.7	No transport	-	Affirmative	10,48	Trapped
7	2.0; 1.4; 1.3	No transport	E	Affirmative	19,46	Cluster
8	2.0; 1.1; 0.9	No transport	-	Affirmative	13,47	Cluster
9	1.6; 0.7; 0.5	5	N	Negative	7,48	Trapped
10	1.3; 0.6; 0.6	>5	Probably N	Affirmative	8,98	New; isolated block
11	2.4; 0.9; 0.8	No transport	-	Affirmative	11,98	Isolated
12	0.8; 0.5; 0.4	0.5	SW	Affirmative	5,99	Trapped
13	0.6; 0.4; 0.3	No transport	-	Affirmative	4,49	Trapped
14	0.6; 0.4; 0.3	No transport	-	Affirmative	4,49	Trapped

Conversely, no movement or rotation of the boulders was detected between mid-November 2018 and 30 April 2019, either in the field or after the comparison of the orthophotographs. During this period, no severe storm affected the Istrian coasts.

3.2. Submerged Landforms from Photogrammetry

The surface of the seabed exhibits a tabular shape, due to the limestone bedding (Figure 6a), and is pitted by several holes, ranging from a few centimeters to about 2 m in size. Most of them are potholes (Figure 6c), occasionally filled by rounded cobbles.

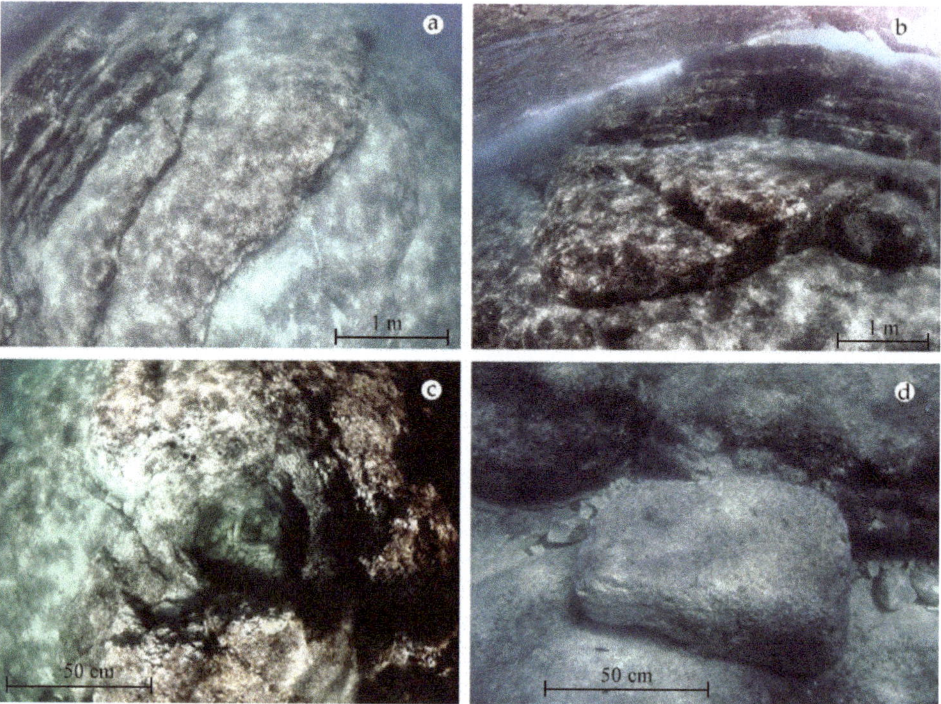

Figure 6. (**a**) Limestone beds below the mean sea level, (**b**) a detachment scar, (**c**) the largest pothole in the submerged area, and (**d**) isolated boulder.

An oblong-shaped hole, about 2 m in length and 1 m in width, is less colonized by marine life than the surrounding seabed and lies offshore from the limestone bed hosting boulder #1.

The 3D model shows a sloping seabed, reaching its maximum depth at about 50 m from the coastline. Landward, the seabed is cut by a persistent joint parallel to the coastline (the same as described in the previous paragraph), while a limestone bed borders the eastern side of the study area (Figure 7).

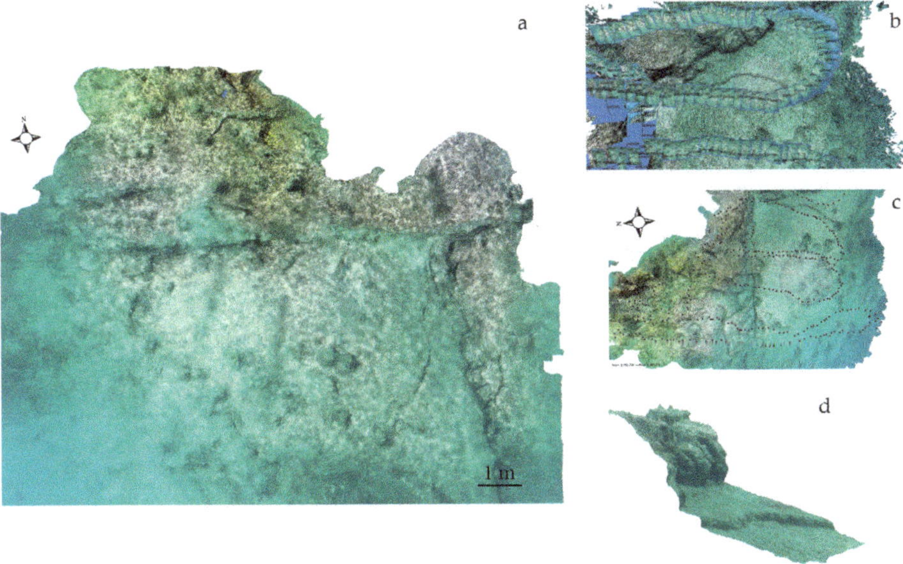

Figure 7. (**a**) A 3D model of the seafloor at the Premantura site, (**b**) the roughly S-shape route followed during the snorkel survey, (**c**) a 3D sketch of a sector of the submerged area, and (**d**) a submerged profile. The submerged landforms on the seafloor are clearly visible in the model, particularly the limestone beds, potholes, and other rounded abrasional landforms.

3.3. Wave Modeling of the 29 October 2018, Adriatic Sea Storm

In order to evaluate the skills of the AdriSC modeling suite to reproduce storm Vaia, the unSWAN wave model results were, first, compared with both state-of-the-art wave model running on the entire Mediterranean Sea and wave measurements from buoys located along the Croatian coastline.

The qualitative comparison across the entire Adriatic Sea of the AdriSC unSWAN significant wave height and peak period with the operational MEDSEA-Wave results (Figure 8) during the peak of the storm (i.e., 29 October 2018, at around 20:00) shows that the AdriSC unSWAN model is capable of capturing the wave conditions during the storm. However, it is fundamental to notice that the peak period in the northern Adriatic Sea is higher for the AdriSC unSWAN model than for the MEDSEA-Wave results. This might be due to the difference in wind force, the physics of the models or the difference in resolution, which leads to a better representation of the nearshore geomorphology of the Adriatic Sea by the AdriSC unSWAN model than the 4 km MEDSEA-Wave model.

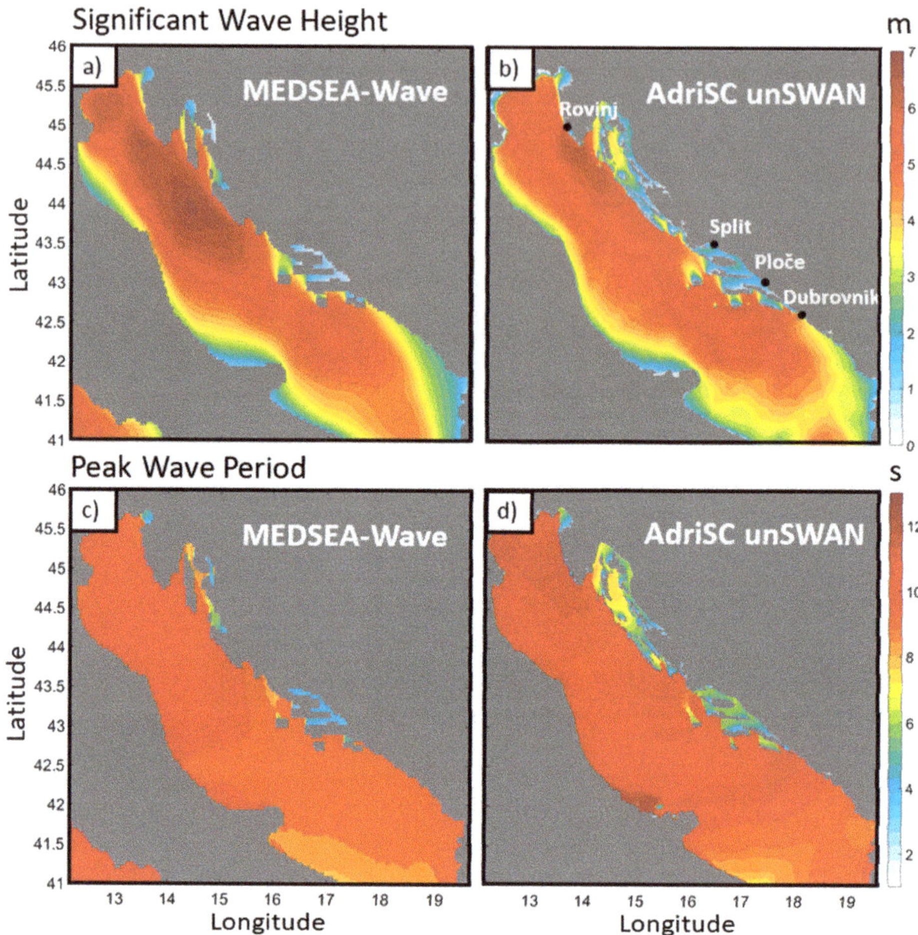

Figure 8. (**a**) Significant wave height using MEDSEA-Wave, (**b**) significant wave height using AdriSC unSWAN, (**c**) peak period using MEDSEA-Wave, and (**d**) peak period using AdriSC unSWAN. All four models are related to 29 October 2018, at 20:00.

In Table 3, the AdriSC unSWAN maximum significant wave height and its associated peak period during the 29 October 2018 storm are compared at four different nearshore locations (Rovinj, Split, Dubrovnik, and Ploče in Figure 8) with the measurements obtained by the Croatian Hydrographic Institute (Hrvatski hidrografski institut, HHI) and reported in [46].

Table 3. Comparison of the wave parameters (significant wave, height, and peak period) measured by the Croatian Hydrographic Institute at four different locations (Rovinj, Split, Dubrovnik and Ploče) together with the AdriSC unSWAN model results during the peak of the storm on 29 October 2018.

	Significant Wave Height [m]		Peak Period [s]	
Location	**Measured**	**unSWAN**	**Measured**	**unSWAN**
Rovinj	4.57	4.51	9.1	10.9
Split	1.80	1.87	5.7	5.8
Dubrovnik	4.55	4.16	9.5	9.7
Ploče	1.13	1.10	4.7	4.7

In general, the AdriSC unSWAN model shows some skill in reproducing the measurements at the four nearshore locations along the Croatian coastline, however, the unSWAN model overestimates the peak period in Rovinj by nearly 2 s and underestimates the significant wave height in Dubrovnik by about 0.5 m (Table 3). The overestimation of the peak period of the AdriSC unSWAN model in Rovinj is consistent with the results obtained in Figure 8, and thus the model is likely to have generally overestimated the peak period in the northern Adriatic Sea during storm Vaia on 29 October 2018.

Given the qualitative and quantitative comparisons performed, the AdriSC unSWAN model is thought to have reasonably reproduced the storm on 29 October 2018, and can be used to assess the impact of the waves on the boulders of the Premantura area. Near Premantura (Figure 9a), the AdriSC unSWAN model has a resolution ranging from 100 m at the coastline to 1 km further offshore, and a bathymetry that captures the main geomorphological features of the seabed, however, the islands of Fenera, Ceja, and Bodulas are too small to be included in the mesh and are each represented with one point with a depth of 0 (zero) m (Figure 9b).

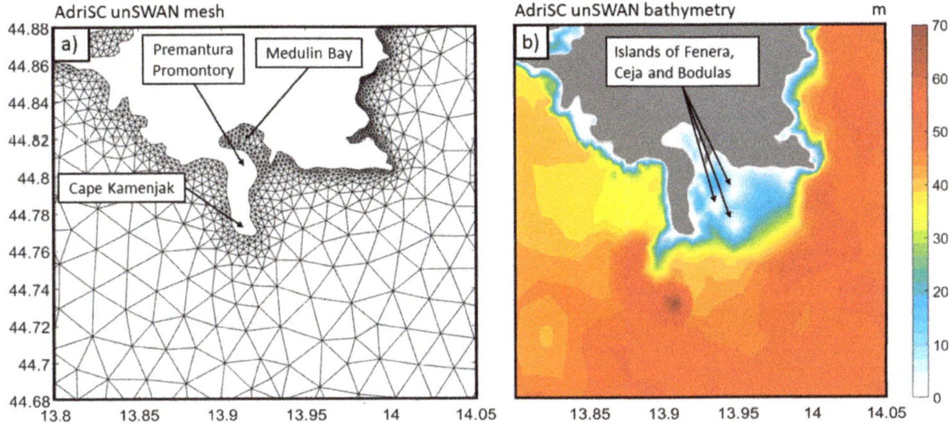

Figure 9. AdriSC unSWAN mesh structure (panel **a**) and bathymetry (panel **b**) in the vicinity of the Premantura area.

Results from the AdriSC ADCIRC and unSWAN models during the storm on 29 October 2018, for the study area, are presented in Figure 10. The modeled storm produces sea surface elevations up to 0.5 m in most of the area (Figure 10a), except in Medulin Bay, where it reaches 1 m due to seiche activity. Currents (Figure 10b) reach their maximum strength in coastal regions. They are particularly strong at the tip of the Premantura Promontory, where they were modeled with velocities up to 1.7 m/s. Such strong currents are presumably induced by strong waves inclined to the coastline. Significant wave heights reached 7 m (Figure 10c), with a peak period of 12 s (Figure 10d) and a wavelength of 105 m (Figure 10e) in the wider Premantura area. These wave parameters, particularly the significant wave height, are greater than any recorded measurement in the northern Adriatic Sea and, following an analysis performed 50 km west from the Premantura Promontory [47], may be considered as occurring once every 20 to 30 years.

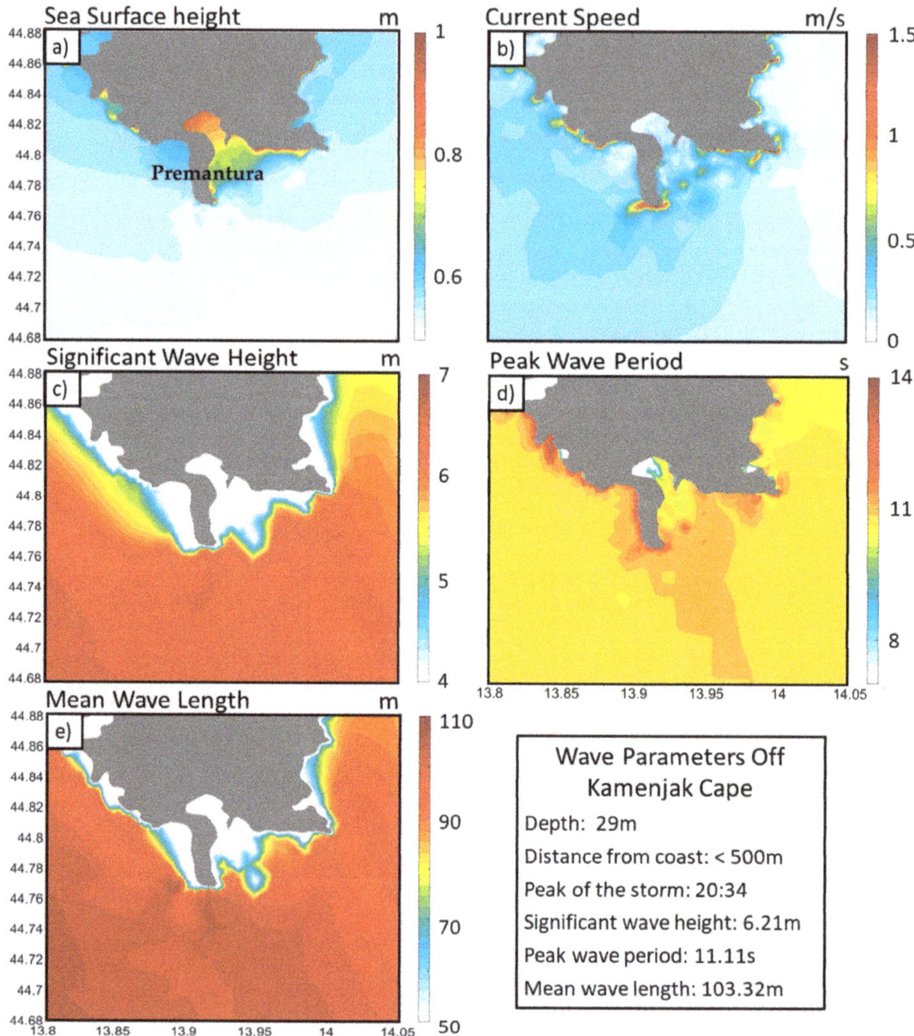

Figure 10. Spatial variability of (1) ADCIRC maximum sea surface height (panel **a**) and maximum current speed (panel **b**) and (2) unSWAN maximum significant wave height (panel **c**), maximum peak period (panel **d**), and maximum mean wave length (panel **e**) obtained in the vicinity of the Premantura Promontory during the storm of 29 October 2018.

To investigate the boulder motions on the Premantura Promontory (as the nearshore bathymetry of the model may not be accurate due to a lack of both resolution and accuracy of the bathymetry data), the wave parameters were extracted using a point off Cape Kamenjak, at a depth of about 29 m (Figure 10) and the wave height at the breaking point was calculated using the Sunamura and Horikawa equation [44]. On 29 October at 20:34 the maximum significant wave height reached 6.2 m with an associated maximum wave height of 10.8 m, peak period of 11.1 s, and mean wave length of 103.3 m. Assuming the same bottom slope as in [2], the wave height at the breaking point for the storm on 29 October 2018, is 13.6 m, which is higher than the theoretical wave heights needed to move or transport the boulders presented in Table 2, except for Boulder #7 that requires wave heights of more than 19 m according to Nandasena [45].

4. Discussion

The comparison of orthomosaics obtained from UAV images, repeated annually, can be a very useful tool for monitoring the variation of the boulder positions. This methodology has been successfully applied in both North Atlantic [11–13,48] and Mediterranean boulder sites [49]. UAV-derived orthophotographs and digital surface models (DSMs) can provide excellent data and information on coastal boulder patterns. Orthophotographs allow for the mapping of a axes and b axes, including their orientation, whereas precise values for c axes and boulder volume can be taken from the DSM [50]. In recent years, after an initial decade when the analyses were mostly dedicated to explaining boulder detachment and transport mechanisms and distinguishing between those of tsunamigenic and those of storm origin, boulder studies are now more oriented towards the observation of boulder movements and reassemblage after exceptional storm events [13,50]. In addition, Hastewell et al. [51] have proposed an innovative technique for boulder movement reconstruction, using radio frequency Identification (RFID) tagging and DGNNS technology. A single RFID tag is inserted inside a boulder and can be activated with an electromagnetic signal emitted by a pole antenna moved by an operator also equipped with a backpack reader and a handheld computer.

On the Premantura Promontory, where the origin of the boulders has been ascribed to storm events, totally excluding those of tsunami origin [2], and where boulders analysis began only recently (end of 2016), a proper monitoring network including orthophotograph comparison, both aerial and submarine, geomorphological observation, instrumental monitoring techniques, and numerical modeling, is most definitely needed as severe storms are forecast to increase in the near future [52].

Following the long-term monitoring experience of coastal landslides affecting the northwestern part of Malta [53–55], a network of GNSS benchmarks has also been installed in Premantura. At the same time, the submarine 3D model, presented here, could be the starting point for monitoring the effects of future storms in the underwater environment, that represents the source area for the quarrying and detachment of future boulders. Finally, every day since early 2019, the AdriSC operational component has been providing a 48 h forecast of high-resolution wave parameters, which can be used to detect in advance whether or not a potential storm can move boulders.

By comparing UAV-derived orthophotographs, we detected the movement of 13 boulders and the emplacement of a new one after a severe storm. The western part of the boulder field suffered the intensity of the last storm most, whereas the central and eastern part did not show any evidence of boulder movements. This lack of movement is related to the structural setting of the promontory, where, in the central and eastern sectors the boulders (probably older) are located at higher elevations. Thus, active erosion occurs in the western part of the Jugo Promontory (Cape Kamenjak).

By observing boulder movements (Figure 4), the maximum inundation flow of the storm waves capable of transporting boulders was estimated to about 50 m of distance from the coast (that has been measured considering wave directions and orientation of the limestone ramps). The displacement followed the azimuth of the limestone beds strikes that, in turn, have conditioned the detachment of rocky material forming scarps throughout the past. These scarps, together with flat, gently inclined limestone bedding planes (pavements), have acted as ramps for boulder transport. Isolated boulders, such as boulder #1 (K8 in [2]), are those that have been subject to the largest movements, while the boulders gathered in clusters were only rotated or toppled, but remained in position, trapped by other boulders or the abovementioned scarps, fractures, and faults. The latter, in particular, have been enlarged over time by coastal karst and marine weathering, as well as by the removal of small portions of rocky material. Some boulders exhibited signs of fragmentation, due to collisions with other boulders or against the rocky scarps.

The submerged scenario, accurately obtained using digital photogrammetric reconstructions, is a high energy environment, with fresh detachment scarps and rounded or sub-rounded potholes, where decimetric cobbles and boulders lie and move, causing the abrasion of the rocky seafloor. The topographical setting of the seafloor reflects the coastal geomorphology and topography, where limestone beds alternate with scarps and faults. With respect to the previous underwater geomorphological surveying, any recent detachment scar has been noted, while a clear detachment niche has been reconstructed in detail. The 3D submarine model obtained, reconstructed using digital photogrammetric analysis of underwater pictures, will be the starting point for future monitoring of the submerged environment, both before and after exceptional storms, that may affect the Adriatic Sea even more given the future climate predictions [56].

After three years of studies and surveys, we can state that storm wave erosion at the Premantura site is active. The appearance of new boulders (especially the seven tonne boulder #1 during the 2014 storm) attests to a dynamic process of sea erosion that is continuously changing the coastal setting. As already suggested by [2], the erosion process was more intense, in the past, when the coast was initially brought into contact with the sea, and the removal of rocky material due to the repetitive action of waves began. Radiocarbon dates provided by [2] suggests that this process has been active over several centuries.

Regarding the numerical modeling, the AdriSC modeling suite presented some skills to reproduce storm Vaia and the unSWAN wave model provided high-resolution wave parameters in the vicinity of Premantura with an unstructured mesh capturing the global geomorphology of the area. The models also allowed for the quantification of the coastal dynamics, particularly the total elevation, including the wave set up and the wave-induced currents, which may significantly impact the wave field when reaching values as high as those modeled off the Premantura Promontory. However, the wave results indicate that the storm produced an estimated maximum wave height of 13.6 m at the breaking point near the area of interest, which was not enough to move Boulder #7. The hydrodynamic equations were used to have an idea about the wave height. The model of Nandasena is mostly used and generally presents a good fit with the measured waves, such as in our study. Of course, the model has its limits and does not offer a precise wave height. The parameters of the equations, such as the lift coefficient, which was calculated using laboratory experiments by [57] for certain particle size and density conditions, are probably not applicable to the local geological, topographical, and climatic conditions where boulders move.

This study shows the following: (1) the need to accurately survey the nearshore bathymetry in the vicinity of the Premantura Promontory (e.g., via shipborne single-beam, multi-beam, side-scan sonar sensors or airborne laser scanning bathymetric surveys) and (2) the limitation of using the SWAN model which, contrarily to the XBeach model [58] — which is developed for wave propagation, sediment transport and morphological changes of the nearshore area, beaches, dunes and backbarriers during storms, cannot correctly reproduce the dynamics in the swash zone (i.e., the land-ocean boundary) where some of the boulders are located. This study provides the first attempt to model the wave conditions responsible for the boulder motion during storm events in the Premantura Promontory, but more modeling efforts and better bathymetry data will be required in the future to truly quantify the wave effects on the boulder dynamics, particularly in the swash zone as described in [59,60]. In addition, the return period of the wave height modeled during storm Vaia, in the northern Adriatic Sea, is about 20 to 30 years. Given that the most extreme storms could, although not significantly, tend to increase slightly in the northern Adriatic Sea in future climate scenarios [61], this might, together with the increase in mean sea level, increase the erosion of rocky coasts. Additionally, other areas, such as the lowlands along the coasts of the northern Adriatic Sea (some of which are subsiding) [56], or the coastal cities with substantial cultural heritage [62], may even be more endangered by the combination of mean sea level and wave activity. Therefore, both these factors should be included in any assessment of the vulnerability of the northern Adriatic Sea to climate change.

Finally, taking into account all the geomorphological observations, the aerial orthophotographic comparison, the wave data in a time range of 1 year (from November 2017 to November 2018), and the wave results from the AdriSC model during the storm, the recent boulder movements and rearrangements can be definitively linked to the 29 October, storm Vaia.

5. Conclusions

The Premantura Promontory represents a unique example of an extensive coastal boulder accumulation in the northern Adriatic Sea triggered and rearranged by storm waves, as reported by [2] and confirmed in the present study.

The occurrence of this kind of boulder deposit depends on various factors (i.e., the discontinuity network, marine wave direction, and coastal exposure) and for these reasons the integration of underwater and onshore surveys is crucial in the understanding of the processes involved in boulder transport.

The integration of geomorphological surveys and multitemporal UAV-DP permitted the identification of 13 boulders that were moved or rotated and the emplacement of a new boulder during the severe storm that hit the northern Adriatic coasts on 29 October 2018, i.e., storm Vaia.

The western part of the promontory suffered the greatest impacts of the waves as a result of the structural settings of the limestone layers that acted as favorable pavements for the movements of the boulders.

The wave model confirmed that the storm waves had the potential to move these boulders during the storm Vaia event.

Boulder movement on the Premantura Promontory is, thus, linked to the frequency of exceptional storms capable of generating extreme wave heights. Following previous studies carried out in the Adriatic Sea, the return period of the waves modeled during storm Vaia is roughly 20 to 30 years, however, in the last four years, at least two major extreme storms have affected the northern Adriatic Sea (2014 and 2018), causing breaking waves exceeding 10 m height and capable of causing boulder movements.

We demonstrated that the tsunamogenic origin for coastal boulders movements is not required, especially where and when severe waves repeatedly hit the coast. In fact, in the study area the topographical and the geomorphological setting, together with the exposition of southern Istria towards the most severe winds and waves of the Adriatic Sea, favoured boulder detachment and accumulation.

Ongoing research should mainly focus on the improvements of the DP procedure in submarine environments and on the analysis of outputs of DGNNS surveys. Survey zero was carried out on 30 April 2019, when six GNSS benchmarks were installed on six boulders already displaced by storm Vaia. Moreover, the underwater 3D model could be the basis for future comparisons of submerged environments after future storms.

Author Contributions: S.D. and G.S. operated the UAV surveys; S.B., S.F., and V.M. carried out diving operations; the geomorphological and geological surveys were carried out by S.B., S.D., S.F., and T.K.; C.D. and I.V. performed the wave modeling. Original idea, S.F., S.D. and S.D.; Conceptualization, all authors; Writing—original draft, all authors; Supervision, S.F.; Writing-review & editing, S.D. and S.B.

Funding: This work was carried out within the framework of the Geoswim Project (Resp. Stefano Furlani, University of Trieste), the MOPP–Medflood INQUA-CMP 1603 Project, and the Croatian Science Foundation projects GEOSEKVA (Grant IP-2016-06-1854) and ADIOS (Grant IP-2016-06-1955).

Acknowledgments: The authors are grateful to Falck family for their financial support and thank Matteo Mantovani, Linley Hastewell, Andrea Morosetti, and Valeria Vaccher for field surveys and monitoring support. Acknowledgement is also made for the support of the ECMWF staff, as well as for the ECMWF's computing and archive facilities used in this research. The ECMWF ERA Interim dataset is available at [63].

References

1. Nimbus Web. Available online: http://www.nimbus.it/eventi/2018/181031TempestaVaia.htm (accessed on 22 March 2019).
2. Biolchi, S.; Furlani, S.; Devoto, S.; Scicchitano, G.; Korbar, T.; Vilibić, I.; Šepić, J. The origin and dynamics of coastal boulders in a semi-enclosed shallow basin: A northern Adriatic case study. *Mar. Geol.* **2019**, *411*, 62–77. [CrossRef]
3. Raji, O.; Dezileau, L.; Grafenstein, U.V.; Niazi, S.; Snoussi, M.; Martinez, P. Extreme sea events during the last millennium in the northeast of Morocco. *Nat. Hazards Earth Syst. Sci.* **2015**, *15*, 203–211. [CrossRef]
4. Pepe, F.; Corradino, M.; Parrino, N.; Besio, G.; Lo Presti, V.; Renda, P.; Calcagnile, L.; Quarta, G.; Sulli, A.; Antonioli, F. Boulder coastal deposits at Favignana Island rocky coast (Sicily, Italy): Litho-structural and hydrodynamic control. *Geomorphology* **2018**, *303*, 191–209. [CrossRef]
5. Biolchi, S.; Furlani, S.; Antonioli, F.; Baldassini, N.; Deguara, J.C.; Devoto, S.; Di Stefano, A.; Evans, J.; Gambin, T.; Gauci, R.; et al. Boulder accumulations related to extreme wave events on the eastern coast of Malta. *Nat. Hazards Earth Syst. Sci.* **2016**, *16*, 719–756. [CrossRef]
6. Scicchitano, G.; Pignatelli, C.; Spampinato, C.R.; Piscitelli, A.; Milella, M.; Monaco, C.; Mastronuzzi, G. Terrestrial Laser Scanner techniques in the assessment of tsunami impact on the Maddalena peninsula (south-eastern Sicily, Italy). *Earth Planets Space* **2012**, *64*, 889–903. [CrossRef]
7. Deguara, J.C.; Gauci, R. Evidence of extreme wave events from boulder deposits on the south-east coast of Malta (Central Mediterranean). *Nat. Hazards* **2017**, *86*, 543–568. [CrossRef]
8. Piscitelli, A.; Milella, M.; Hippolyte, J.-C.; Shah-hosseini, M.; Morhange, C.; Mastronuzzi, G. Numerical approach to the study of coastal boulders: The case of Martigues, Marseille, France. *Quat. Int.* **2017**, *439*, 52–64. [CrossRef]
9. Marriner, N.; Kaniewski, D.; Morhange, C.; Flaux, C.; Giaime, M.; Vacchi, M.; Goff, J. Tsunamis in the geological record: Making waves with a cautionary tale from the Mediterranean. *Sci. Adv.* **2017**, *3*, e1700485. [CrossRef]
10. Orviku, K.; Jaagus, J.; Tõnisson, H. Sea ice shaping the shores. *J. Coast. Res.* **2011**, 681–685.
11. Cox, R.; Zentner, D.B.; Kirchner, B.J.; Cook, M.S. Block ridges on the Aran Islands (Ireland): Recent movements caused by storm waves, not tsunamis. *J. Geol.* **2012**, *120*, 249–272. [CrossRef]
12. Cox, R.; Jahn, K.L.; Watkins, O.G.; Cox, P. Extraordinary boulder transport by storm waves, and criteria for analysing coastal boulder deposits. *Earth-Sci. Rev.* **2018**, *177*, 623–636. [CrossRef]
13. Cox, R. Very large boulders were moved by storm waves on the west coast of Ireland in winter 2013–2014. *Mar. Geol.* **2019**, *412*, 217–219. [CrossRef]
14. Gušić, I.; Jelaska, V. *Stratigrafija Gornjokrednih Naslaga Otoka Brača u Okviru Geodinamske Evolucije Jadranske Karbonatne Platforme = Upper Cretaceous Stratigraphy of the Island of Brač within the Geodynamic Evolution of the Adriatic Carbonate Platform*; Institute of Geology: Zagreb, Croatia, 1990.
15. Grisogono, B.; Belušić, D. A review of recent advances in understanding the mesoand microscale properties of the severe Bora wind. *Tellus A* **2009**, *61*, 1–16. [CrossRef]
16. Smirčić, A.; Gačić, M.; Dadić, V. Ecological study of gas fields in the northern Adriatic: Surface waves. *Acta Adriat.* **2002**, *37*, 17–34.
17. Vilibić, I. The role of the fundamental seiche in the Adriatic coastal floods. *Cont. Shelf Res.* **2006**, *26*, 206–216. [CrossRef]
18. Cerovečki, I.; Orlić, M.; Hendershott, M.C. Adriatic seiche decay and energy loss to the Mediterranean. *Deep Sea Res. Part I: Oceanogr. Res. Pap.* **1997**, *44*, 2007–2029. [CrossRef]
19. Međugorac, I.; Pasarić, M.; Orlić, M. Severe flooding along the eastern Adriatic coast: The case of 1 December 2008. *Ocean Dyn.* **2015**, *65*, 817–830. [CrossRef]
20. Raicich, F. Long-term variability of storm surge frequency in the Venice Lagoon: An update thanks to 18th century sea level observations. *Nat. Hazards Earth Syst. Sci.* **2015**, *15*, 527–535. [CrossRef]
21. Šepić, J.; Vilibić, I.; Fine, I. Northern Adriatic meteorological tsunamis: Assessment of their potential through ocean modeling experiments. *J. Geophys. Res. Oceans* **2015**, *120*, 2993–3010. [CrossRef]
22. Vilibić, I.; Šepić, J. Destructive meteotsunamis along the eastern Adriatic coast: Overview. *Phys. Chem. Earth Parts A/B/C* **2009**, *34*, 904–917. [CrossRef]

23. Tinti, S.; Maramai, A.; Graziani, L. The New Catalogue of Italian Tsunamis. *Nat. Hazards* **2004**, *33*, 439–465. [CrossRef]

24. Fago, P.; Pignatelli, C.; Piscitelli, A.; Milella, M.; Venerito, M.; Sansò, P.; Mastronuzzi, G. WebGIS for Italian tsunami: A useful tool for coastal planners. *Mar. Geol.* **2014**, *355*, 369–376. [CrossRef]

25. Tiberti, M.M.; Lorito, S.; Basili, R.; Kastelic, V.; Piatanesi, A.; Valensise, G. Scenarios of Earthquake-Generated Tsunamis for the Italian Coast of the Adriatic Sea. *Pure Appl. Geophys.* **2008**, *165*, 2117–2142. [CrossRef]

26. Casella, E.; Rovere, A.; Pedroncini, A.; Stark, C.P.; Casella, M.; Ferrari, M.; Firpo, M. Drones as tools for monitoring beach topography changes in the Ligurian Sea (NW Mediterranean). *Geo-Mar. Lett.* **2016**, *36*, 151–163. [CrossRef]

27. Casella, E.; Collin, A.; Harris, D.; Ferse, S.; Bejarano, S.; Parravicini, V.; Hench, J.L.; Rovere, A. Mapping coral reefs using consumer-grade drones and structure from motion photogrammetry techniques. *Coral Reefs* **2017**, *36*, 269–275. [CrossRef]

28. Anzidei, M.; Scicchitano, G.; Tarascio, S.; De Guidi, G.; Monaco, C.; Barreca, G.; Mazza, G.; Serpelloni, E.; Vecchio, A. Coastal retreat and marine flooding scenario for 2100: A case study along the coast of Maddalena Peninsula (southeastern Sicily). *Geogr. Fis. Din. Quat.* **2018**, *41*, 5–16.

29. Francioni, M.; Salvini, R.; Stead, D.; Coggan, J. Improvements in the integration of remote sensing and rock slope modelling. *Nat. Hazards* **2018**, *90*, 975–1004. [CrossRef]

30. Gili, J.; Corominas, J.; Rius, J.M. Using Global Positioning System techniques in landslide monitoring. *Eng. Geol.* **2000**, *55*, 167–192. [CrossRef]

31. Benjamin, J.; McCarthy, J.; Wiseman, C.; Bevin, S.; Kowlessar, J.; Moe Astrup, P.; Naumann, J.; Hacker, J. Integrating Aerial and Underwater Data for Archaeology: Digital Maritime Landscapes in 3D. In *3D Recording and Interpretation for Maritime Archeology*; Springer: Cham, Switzerland, 2019; pp. 211–232. ISBN 978-91-27-35725-9.

32. Gaglianone, G.; Crognale, J.; Esposito, C. Investigating submerged morphologies by means of the low-budget "GeoDive" method (high resolution for detailed 3D reconstruction and related measurements). *Acta Imeko* **2018**, *7*, 50–59. [CrossRef]

33. Drap, P. Underwater Photogrammetry for Archaeology. In *Special Applications of Photogrammetry*; da Silva, D.C., Ed.; InTech: Rijeka, Croatia, 2012; pp. 111–136.

34. Nocerino, E.; Neyer, F.; Grun, A.; Troyer, M.; Menna, F.; Brooks, A.J.; Capra, A.; Castagnetti, C.; Rossi, P. Comparison of Diver-Operated Underwater Photogrammetric Systems For Coral Reef Monitoring. *ISPRS* **2019**, *42*, 143–150. [CrossRef]

35. ADIOS. Available online: www.izor.hr/ADIOS (accessed on 21 July 2019).

36. MESSI. Available online: www.izor.hr/messi (accessed on 21 July 2019).

37. Warner, J.C.; Armstrong, B.; He, R.; Zambon, J.B. Development of a Coupled Ocean-Atmosphere-Wave-Sediment Transport (COASWST) Modeling System. *Ocean Model.* **2010**, *35*, 230–244. [CrossRef]

38. Dietrich, J.C.; Tanaka, S.; Westerink, J.J.; Dawson, C.N.; Luettich, R.A.; Zijlema, M.; Holthuijsen, L.H.; Smith, J.M.; Westerink, L.G.; Westerink, H.J. Performance of the Unstructured-Mesh, SWAN+ADCIRC Model in Computing Hurricane Waves and Surge. *J. Sci. Comput.* **2012**, *52*, 468–497. [CrossRef]

39. Denamiel, C.; Šepić, J.; Ivanković, D.; Vilibić, I. The Adriatic Sea and Coast modelling suite: Evaluation of the meteotsunami forecast component. *Ocean Model.* **2019**, *135*, 71–93. [CrossRef]

40. The Adriatic Seabed. Available online: http://instaar.colorado.edu/~{}jenkinsc/dbseabed/coverage/adriaticsea/adriatico.htm (accessed on 21 July 2019).

41. Balsamo, G.; Albergel, C.; Beljaars, A.; Boussetta, S.; Brun, E.; Cloke, H.; Dee, D.; Dutra, E.; Muñoz-Sabater, J.; Pappenberger, F.; et al. ERA-Interim/Land: A global land surface reanalysis data set. *Hydrol. Earth Syst. Sci.* **2015**, *19*, 389–407. [CrossRef]

42. Pinardi, N.; Allen, I.; Demirov, E.; De Mey, P.; Korres, G.; Lascaratos, A.; Le Traon, P.Y.; Maillard, C.; Manzella, G.; Tziavos, C. The Mediterranean Ocean Forecasting System: First phase of implementation (1998–2001). *Ann. Geophys. (0992-7689) (Eur. Geophys. Soc.)* **2003**, *21*, 3–20. [CrossRef]

43. Ravdas, M.; Zacharioudaki, A.; Korres, G. Implementation and validation of a new operational wave forecasting system of the Mediterranean Monitoring and Forecasting Centre in the framework of the Copernicus Marine Environment Monitoring Service. *Nat. Hazards Earth Syst. Sci.* **2018**, *18*, 2675–2695. [CrossRef]

44. Sunamura, T.; Horikawa, K. Two-Dimensional Beach Transformation Due to Waves. *Coast. Eng. Proc.* **1974**, *1974*, 920–938.

45. Nandasena, N.A.K.; Paris, R.; Tanaka, N. Reassessment of hydrodynamic equations: Minimum flow velocity to initiate boulder transport by high energy events (storms, tsunamis). *Mar. Geol.* **2011**, *281*, 70–84. [CrossRef]

46. Hrvatski Hidrografski Institut. Available online: http://www.hhi.hr/pr/onenews/250 (accessed on 21 July 2019).

47. Leder, N.; Smirčić, A.; Vilibić, I. Extreme values of surface wave heights in the northern Adriatic. *Geofizika* **1999**, *15*, 1–13.

48. Pérez-Alberti, A.; Trenhaile, A.S. An initial evaluation of drone-based monitoring of boulder beaches in Galicia, north-western Spain. *Earth Surf. Process. Landf.* **2015**, *40*, 105–111. [CrossRef]

49. Turner, D.; Lucieer, A.; De Jong, S.M. Time Series Analysis of Landslide Dynamics Using an Unmanned Aerial Vehicle (UAV). *Remote Sens.* **2015**, *7*, 1736–1757. [CrossRef]

50. Boesl, F.; Engel, M.; Eco, R.C.; Galang, J.B.; Gonzalo, L.A.; Llanes, F.; Quix, E.; Brückner, H. Digital mapping of coastal boulders—High-resolution data acquisition to infer past and recent transport dynamics. *Sedimentology* **2019**. [CrossRef]

51. Hastewell, L.J.; Schaefer, M.; Bray, M.; Inkpen, R. Intertidal boulder transport: A proposed methodology adopting Radio Frequency Identification (RFID) technology to quantify storm induced boulder mobility. *Earth Surf. Process. Landf.* **2019**, *44*, 681–698. [CrossRef]

52. Gonzáles-Alemán, J.J.; Pascale, S.; Gutierrez-Fernandez, J.; Murakami, H.; Gaertner, M.A.; Vecchi, G.A. Potential Increase in Hazard from Mediterranean Hurricane Activity with Global Warming. *Geophys. Res. Lett.* **2019**, *46*, 1754–1764. [CrossRef]

53. Devoto, S.; Forte, E.; Mantovani, M.; Pasuto, A.; Piacentini, D.; Soldati, M. Integrated monitoring of lateral spreading phenomena along the north-west coast of the Island of Malta. In *Landslide Science and Practice—Volume 2: Early Warning, Instrumentation and Monitoring*; Margottini, C., Canuti, P., Sassa, K., Eds.; Springer-Verlag: Berlin, Germany, 2013; Volume 2, pp. 235–242.

54. Mantovani, M.; Devoto, S.; Forte, E.; Mocnik, A.; Pasuto, A.; Piacentini, D.; Soldati, M. A multidisciplinary approach for rock spreading and block sliding investigation in the north-western coast of Malta. *Landslides* **2013**, *10*, 611–622. [CrossRef]

55. Soldati, M.; Devoto, S.; Prampolini, M.; Pasuto, A. The Spectacular Landslide-Controlled Landscape of the Northwestern Coast of Malta. In *Landscapes and Landforms of the Maltese Islands, World Geomorphological Landscapes*; Gauci, R., Schembri, J.A., Eds.; Springer Nature Switzerland: Basel, Switzerland, 2019; pp. 167–178. [CrossRef]

56. Lionello, P.; Galati, M.B.; Elvini, E. Extreme storm surge and wind wave climate scenario simulations at the Venetian littoral. *Phys. Chem. Earth* **2012**, *40*, 86–92. [CrossRef]

57. Eistein, H.A.; El-Samni, E.A. Hydrodynamic Fources on a Rough Wall. *Rev. Modern Phys.* **1979**, *21*, 520–524. [CrossRef]

58. XBeach. Available online: https://xbeach.readthedocs.io/en/latest/index.html (accessed on 5 March 2019).

59. Kennedy, A.B.; Mori, N.; Zhang, Y.; Yasuda, T.; Chen, S.-E.; Tajima, Y.; Pecor, W.; Toride, K. Observations and Modeling of Coastal Boulder Transport and Loading During Super Typhoon Haiyan. *Coast. Eng. J.* **2016**, *58*. [CrossRef]

60. Lorang, M.S. Predicting Threshold Entrainment Mass for a Boulder Beach. *J. Coast. Res.* **2000**, *16*, 432–445.

61. Bock, Y.; Wdowinski, S.; Ferretti, A.; Novali, F.; Fumagalli, A. Recent subsidence of the Venice Lagoon from continuous GPS and interferometric synthetic aperture radar. *Geochem. Geophys. Geosyst.* **2012**, *13*. [CrossRef]

62. Hinkel, J.; Jaeger, C.; Nicholls, R.J.; Lowe, J.; Renn, O.; Peijun, S. Sea-level rise scenarios and coastal risk management. *Nat. Clim. Chang.* **2015**, *5*, 188–190. [CrossRef]
63. ECMWF. Available online: https://www.ecmwf.int/en/forecasts/datasets/archive-datasets/reanalysis-datasets/era-interim (accessed on 5 March 2019).

Article

Coastal Vulnerability Assessment along the North-Eastern Sector of Gozo Island (Malta, Mediterranean Sea)

Angela Rizzo [1], Vittoria Vandelli [2,*], George Buhagiar [3], Anton S. Micallef [4,5] and Mauro Soldati [2]

[1]   REgional Models and geo-Hydrological Impacts (REMHI Division), Centro Euro-Mediterraneo sui Cambiamenti Climatici (CMCC), 73100 Lecce, Italy; angela.rizzo@cmcc.it

[2]   Department of Chemical and Geological Sciences, University of Modena and Reggio Emilia, 41125 Modena, Italy; mauro.soldati@unimore.it

[3]   Research and Planning Section, Marine and Storm Water Unit, Public Works Department, FRN 1700 Floriana, Malta; george.buhagiar@gov.mt

[4]   Euro-Mediterranean Centre on Insular Coastal Dynamics (ICoD), University of Malta, MSD 2080 Msida, Malta; anton.micallef@um.edu.mt

[5]   Institute of Earth Systems, University of Malta, MSD 2080 Msida, Malta

*   Correspondence: vittoria.vandelli@unimore.it

Received: 2 April 2020; Accepted: 12 May 2020; Published: 15 May 2020

Abstract: The coastal landscape of the Maltese Islands is the result of long-term evolution, influenced by tectonics, geomorphological processes, and sea level oscillations. Due to their geological setting, the islands are particularly prone to marine-related and gravity-induced processes, exacerbated by climate change. This study aligns different concepts into a relatively concise and expedient methodology for overall coastal vulnerability assessment, taking the NE sector of Gozo Island as a test case. Geomorphological investigation, integrated with analysis of marine geophysical data, enabled characterization of coastal dynamics, identifying this stretch of coast as being potentially hazardous. The study area features a high economic value derived from tourist and mining activities and natural protected areas, that altogether not only make coastal vulnerability a major concern but also the task of assessing it complex. Before introducing the methodology proposed for overall vulnerability assessment, an in-depth revision of the vulnerability concept is provided. The evaluation was carried out by using a set of key indicators related to local land use, anthropic and natural assets, economic activities, and social issues. Results show that the most critical areas are located east of Marsalforn including Ramla Bay, an important tourist attraction hosting the largest sandy beach in Gozo. The method combines physical exposure and social vulnerability into an overall index. It proves to be cost effective in data management and processing and is suitable for the identification and assessment of overall vulnerability of coastal areas to consequences of climate- and marine-related processes, such as coastal erosion, landslides and sea level rise.

Keywords: coastal morphodynamics; climate change; vulnerability index; Gozo; Malta

---

1. Introduction

Coastal zones are host to very dynamic and complex environmental systems, subject to the direct and indirect influence of a number of factors that have contributed to their evolution over time. The present-day landscape of coastal areas is the combined result of interactions between natural agents that include physical factors inherent in the system and external climatic and marine forces [1,2]. Human activity also plays an important role in shaping coastal dynamics, often exerting additional pressures that may dominate over natural processes [3]. During the past decades, the rate of human occupation along littoral areas has risen significantly [4] and currently, in Europe, about 86 million people are estimated to live within 10 km from the coastline [5]. Furthermore, approximately one-third of the Mediterranean population is concentrated along its coastal regions and around 120 million inhabitants are concentrated in coastal hydrological basins located in the southern region of the Mediterranean Sea [6]. Coastal areas are the transitional zone between the aquatic and the terrestrial ecosystems and they have an environmental intrinsic value on account of their high level of biological diversity, which supports the provision of several ecosystem services essential for human well-being [7,8]. In view of these considerations, the sustainable conservation of coastal areas is a worldwide issue and coastal vulnerability evaluation and risk assessment are of paramount importance for integrated coastal management [9].

Coastal hazards, including marine-related and gravity-induced processes such as landslides, coastal erosion, storm water runoff and coastal flooding, have different impacts on coastal values, due to differing local geomorphological and anthropogenic settings. The impact of these processes is expected to increase with climate change. The assessment reports of the Intergovernmental Panel on Climate Change [10–13] emphatically forecast increase in both the frequency and the severity of extreme weather/climate events, combined with sea level rise, which altogether will undoubtedly impact coastal areas more adversely.

The Sendai Framework for Disaster Risk Reduction 2015–2030 [14], whose implementation is overseen by the United Nations Office for Disaster Risk Reduction (UNDRR), has recognized risk assessment as an important preliminary action on the basis of its relevance internationally. This points towards the prioritisation of the role of "understanding disaster risk" and "enhancing disaster preparedness" and promotes the use of "foresight" and "scenarios" for improving the level of preparedness for existing, emerging and new types of risk. At the same time, the identification of suitable climate adaptation strategies, which also account for future scenarios, is essential for increasing the resilience of coastal areas and ensuring the long-term conservation of coastal natural services and anthropic activities [15].

At the European level, risk assessment and mapping have been included in a number of legislative instruments, such as the Floods Directive [16], which requires member states to provide maps and indications for risk management, prescribing specific requirements for climate change impact evaluation.

As highlighted in the Special Report on "managing the risks of extreme events and disasters to advance climate change adaptation" published by the Intergovernmental Panel on Climate Change (IPCC) [17], vulnerability has been a pivotal concept for disaster risk reduction since the 1970s. Initial studies in this field, carried out by Baird et al. [18], O'Keefe et al. [19], Lewis [20], and Hewitt [21], introduced the concept of disaster risk as the combination of the probability of occurrence of a hazardous event with its negative consequences. Over time, the concept of vulnerability has been interpreted in different ways [22–25] reflecting how the approaches and conceptual models for its assessment differ among the scientific research communities [23,25].

This paper proposes and applies a methodological approach that integrates physical exposure and social vulnerability into an "overall vulnerability index", to identify areas that can be negatively affected by climate- and marine-related processes, such as coastal erosion, landslides and sea level rise.

2. Conceptual Framework

Given the spectrum of possible conceptual interpretations of "vulnerability" and other related terms such as "exposure", it is deemed essential that, as of primacy, and prior to proposing the methodology adopted in this research for the overall assessment, the range of most-commonly used concepts of vulnerability and exposure coined in the contexts of both climate change and disaster risk management are first introduced here, before clarifying which concept and definition are adopted by this study. The disaster risk community identifies the notion of vulnerability as one of the elements required for risk assessment [26]. In this scientific context, "risk" is defined by exposure to a hazard, which is a potentially damaging physical event or phenomenon, and vulnerability, which denotes the relationship between the severity of the hazard and the degree of damage caused to the exposed element [27]. Specifically, according to the United Nations International Strategy for Disaster Reduction [26], vulnerability is defined in a qualitative way as "the characteristics and circumstances of a community, system or asset that makes it susceptible to the damaging effects of a hazard". In the same document, exposure is also defined qualitatively as "people, property, systems, or other elements present in hazard zones that are thereby subject to potential losses".

In earlier definitions, vulnerability was considered as the degree of loss of an element at risk after the occurrence of a natural process [28]. On one hand, the above-mentioned and more recent (but not the latest) definitions consider the notion of vulnerability as denoting a pre-existing condition related to the characteristics of the elements at risk and gives less emphasis to the process [29]. On the other hand, in the Climate Change Adaptation (CCA) context, more focus is placed on the concept of vulnerability defined as a function of exposure, sensitivity and adaptive capacity [10,30]. Therefore, in the risk context, the definition of vulnerability is quite similar to that describing the sensitivity of the system's components in the climate approach [10], while in the climate change community it is defined in a similar way to the concept of risk used in the Disaster Risk Reduction (DRR) context [31].

Only in more recent years have the United Nation Institutions (UNDRR, IPCC) had a key role in converging on a common and shared definition of the vulnerability concept in CCA and DRR fields. In fact, the IPCC integrates the different conceptualizations of vulnerability and provides its upgraded and clearer definition in the glossary of the Fifth Assessment Report (AR-5, [11]), in which vulnerability expresses "the propensity or predisposition to be adversely affected. Such predisposition constitutes an internal characteristic of the affected elements (or societies) and includes the concepts to cope with, resist, and recover and the lack of capacity to cope and adapt to the adverse effects of a physical event". Indeed, in this latter AR, the definition of risk (as a whole) is expressed as being the result of the interaction between vulnerability, exposure and hazard, whereby the distinction between the contribution of vulnerability and exposure to risk has been made clearer, which is a view also shared with the DRR community. This paper adopts a method of combining "physical exposure" information and "social vulnerability" data, which is in line with this evolution and convergence of concepts of vulnerability and exposure, across the above discourses.

In parallel to the vulnerability concept development, several actions taken in the last couple of decades have also focused on the development of methods and tools for supporting decision-makers in the reduction of coastal hazards' impacts. Particularly noteworthy is the index-based approach, which was introduced by Gornitz at the beginning of the 1990s [32,33]. This is mainly based on indirect analysis supported by photointerpretation and topographic maps and is still the most commonly used method for coastal vulnerability assessment at regional levels [34–40].

At the European level, a number of transnational research projects have been funded with the aim of improving scientific knowledge and increasing the awareness of decision-makers on coastal vulnerability, the potential damages of marine processes and the effectiveness of coastal adaptation strategies (such as EUROSION, MICORE, RISC-KIT, ANYWHERE, OPERANDUM). In particular, the main outputs of the RISC-KIT Project are represented by a set of tools that allow identification of the most vulnerable areas (hot-spots) by means of a coastal index [41,42] and then selection of suitable measures for increasing coastal adaptation and therefore favouring risk reduction [43].

The evolution and emergence of concepts of vulnerability and exposure as being different features is warranted in certain situations, to provide a better understanding of "why" and "how" societal assets and values may be under threat or at risk under different scenarios of exposure. This can provide guidance to planners and policy makers in identifying suitable adaptation measures and actions, but it also makes risk assessment far more complex.

Some factors that contribute to the vulnerability of society can be expressed in terms of the dependence of societal wellbeing on certain exposed physical assets or other values that are at risk. Examples of this are infrastructure and utilities, which underpin societal wellbeing to different degrees, depending on their nature. This component of vulnerability is distinct from vulnerability associated with the inherent characteristics of society. A community relying on physical elements for its wellbeing can be deemed as being vulnerable to the extent (or level) of its dependence on those specific physical elements located in a hazard-prone area. This creates different levels of "physical exposure", in direct relation to their level of importance and to the degree to which societal wellbeing is dependent on and would be affected by losses to them. Thus, "physical vulnerability" is considered as one part of "overall vulnerability", while "social vulnerability" (attributed to inherent societal factors) is considered as another component of "overall vulnerability".

In this context, the research presented here aims to evaluate the overall vulnerability, and to refine the analysis and understanding of two distinct components, physical vulnerability (representing exposure from dependence on physical elements) and social vulnerability to a given set of external climate- and marine-related processes. Mirroring physical exposure as physical vulnerability and combining it with social vulnerability into an overall vulnerability index, an expedient and cost-effective method for the identification (and assessment) of the overall vulnerability of coastal areas at risk is provided.

The selected coastal study area, located along the NE sector of the Island of Gozo (Maltese archipelago), is one for which considerable applicative research has already been undertaken in order to showcase the high geological and geomorphological significance of its coastal landscapes [44]. Although the area is known to be particularly susceptible to several coastal hazards (erosion, storm surges, floods, sea level rise, landslides), it still lacks a detailed and refined appraisal of its natural and anthropic coastal elements, based on an analysis of assets and values that could potentially be at risk. Approaching the assessment of vulnerability of this coastal area from the perspective of the two components of social vulnerability and physical vulnerability provides a sharper analytical insight into overall vulnerability. Furthermore, the analytical methods and tools used in this study provide a methodological template for overall vulnerability assessment in the form of an Overall Vulnerability Index (OVI) that can be computed and used, with relative ease at different scales.

The scientific literature review carried out as part of this study reveals a gap in the area of assessment of vulnerability. Studies carried out to date have focused on the overall evaluation of hazard and susceptibility along the Maltese coastal sectors [45], rather than on the methodological assessment of vulnerability per se. Integrated approaches have been applied along the NW coast of Malta for landslide hazard assessment [46–49]. Landslide susceptibility assessment assisted by Persistent Scatterers Interferometry (PSI) was carried along the same stretch of the NW coast of Malta [49,50]. Based on the assumption that the identification and analytical mapping of the exposed elements represents a key step for the evaluation of the coastal risk, this study tries to bridge this gap by proposing a relatively simple yet reliable, cost-effective and easily replicable procedure for the assessment of coastal vulnerability. Nonetheless, while not being overly complex, the proposed approach provides an analytical discernment of different components of overall vulnerability that, when mapped geographically, provide sufficient guidance as to which areas are most vulnerable and on what account.

Following on from the methodological approaches proposed in previous studies [42,51–54], this study formulates and applies a research method for overall vulnerability assessment, which is based on the following steps: (i) identification of the main exposed elements (natural and anthropic) located in the investigated area; (ii) definition of their relative exposure level, in economic and ecological terms; (iii) assessment of the social vulnerability of the population living in the investigated area; (iv) calculation of the overall vulnerability by means of a combined index. As highlighted in [55] the use of an index as an evaluation tool requires the definition, weighting and aggregation of a number of indicators, which are defined as variables that are "an operational representation of a characteristic or quality of a system" [56,57].

In detail, the method applied here foresees a set of indicators for the evaluation of the exposure level of local land uses (considering both the presence of economic activities as well as natural protected areas) and anthropic assets (such as transport networks and main utilities). Furthermore, the social context that characterizes the investigated area is also taken into account, in order to include social vulnerability in terms of population capacity to respond to and cope with a hazardous event in the overall evaluation of vulnerability. The combination of physical and social vulnerability provides the overall index, which expresses the level of overall vulnerability.

The use of distinct physical and social vulnerability indicators, and the definition of an Overall Vulnerability Index (OVI), has the main advantage of summarizing complex issues making them more easily interpretable, facilitating decision making and communication among stakeholders [58]. Meanwhile, the distinction between the two components at analysis stage provides a more in-depth understanding of specific anthropic elements that contribute to different risk levels.

3. Study Area

3.1. Geological and Geomorphological Setting

The Maltese archipelago is located in the central Mediterranean Sea and comprises the main islands of Malta, Gozo and Comino (Figure 1). The coastal landscape (Figure 2) is the result of long-term evolution under the influence of tectonic activity, geomorphological processes and sea level oscillations. Due to their geological and geomorphological setting, these islands are particularly prone to different marine-related and gravity-induced processes such as landslides, coastal erosion, storm water runoff and sea level rise, enhanced by ongoing climate change. Multidisciplinary research and integrated investigations have been carried out to better understand the evolution of the geomorphological processes within the archipelago in the wider dynamic scenario of ongoing climate change as a way towards the reduction of risks associated with these processes [47,59–61].

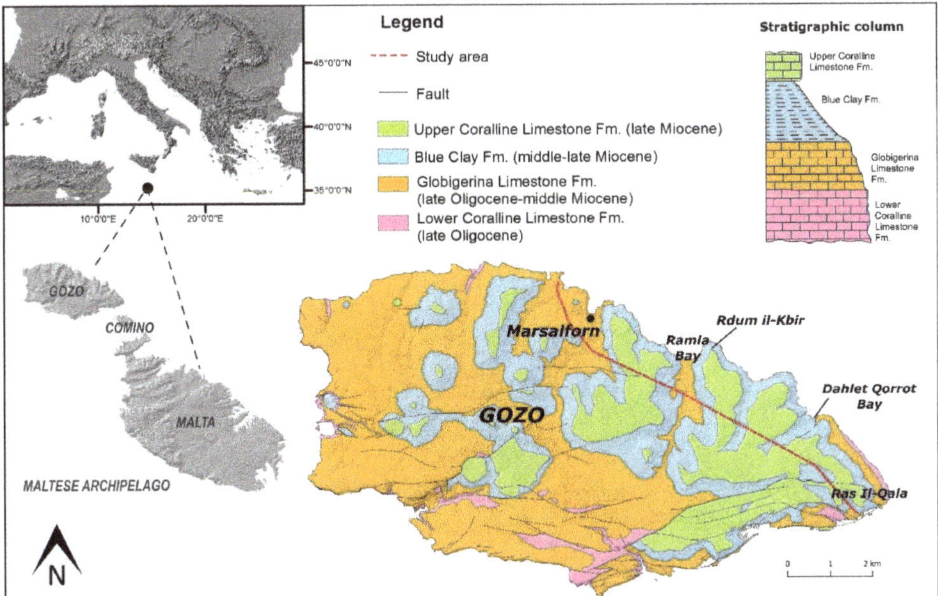

Figure 1. Geological setting of the Maltese archipelago (cf. [62]) and location of the study area, bounded by the NE coastline of Gozo and the red dashed line parallel to it on the inland side (after [44], modified).

Figure 2. Main geomorphic features of the study area: (**a**) block slides (west of Daħlet Qorrot Bay); (**b**) rock fall at the bottom of a limestone plateau and earth flow/slide affecting the underlying clayey terrain (between San Blas Bay and Daħlet Qorrot Bay); (**c**) plunging cliff between Daħlet Qorrot Bay and Ras il-Qala; (**d**) sloping coast between Daħlet Qorrot Bay and Ras il-Qala; (**e**) shore platform east of Marsalforn Bay; (**f**) Ramla Bay pocket beach; (**g**) cliff shaped in Blue Clay east of Marsalforn Bay; (**h**) built-up coast of Marsalforn Bay. After [44], modified.

Previous studies of the Island of Gozo have focused on general aspects of coastal features [60,63–66]. Only a few papers deal with the specific geomorphological aspects of the island [67–69], some of them referring to its rich geoheritage [70–73]. A detailed geomorphological map of the investigated stretch of coast (NE Gozo), based on geomorphological and geological field surveys integrated with the analysis of marine geophysical data, has been published recently [44]. The geomorphological map is accompanied by two other maps that show land use and the distribution of coastal geomorphotypes (Figure 3). Such documents constituted the basis for carrying out the vulnerability analysis presented below.

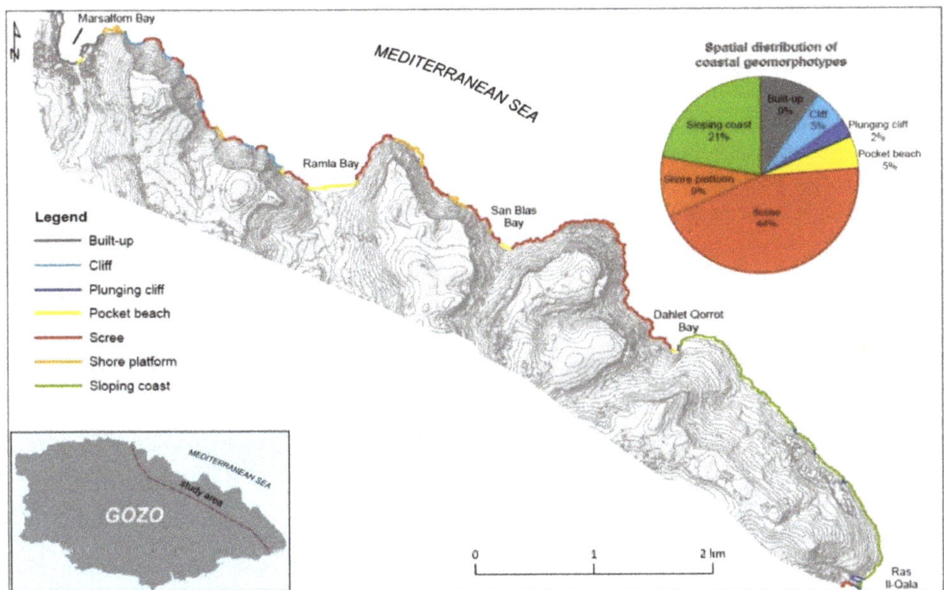

Figure 3. Distribution of coastal morphotypes in the study area (after [44], modified).

From a geomorphological perspective, the investigated coastal stretch is characterized by limestone plateaus bounded by steep structural scarps which are progressively reshaped by gravitational and/or degradation processes. Clayey slopes, located at the foot of the limestone plateau, accommodate terraced fields of actively used or abandoned agricultural land. Numerous blocks of rock are strewn over the clayey terrain (a unique landscape known as *rdum* in Maltese) that slopes more gently away from the plateau edge.

The investigated coastline is characterized by the alternation of inlets and promontories. The accumulation of sand and mixed grainsize deposits results in the formation of pocket beaches where this corresponds with bays and coves. The large sandy beach of Ramla il-Ħamra Bay ('red sandy beach') is a particularly noteworthy example. It is also partly bounded by dunes on the landward side and is protected as a Special Area of Conservation (SAC) within the Natura 2000 network. The main inhabited centre of the investigated stretch of coast is Marsalforn, with its homonym bay, the latter being intensely urbanised and anthropized, comprising a predominantly built-up coastline.

The landscape of the investigated stretch of coast, and in general of the entire Maltese archipelago, is largely controlled by the different erodibility of the exposed lithostratigraphic units constituted by marine limestones and marls of the late Oligocene-Miocene [74,75]. The outcropping geological formations comprise (from the oldest to the youngest): Lower Coralline Limestone Formation (late Oligocene, Chattian), Globigerina Limestone Formation (late Oligocene—middle Miocene, late Chattian—Langhian), Blue Clay Formation (middle-late Miocene, Serravallian—Tortonian) and Upper Coralline Limestone Formation (late Miocene, Tortonian—early Messinian).

The Lower Coralline Limestone Formation consists of pale grey, hard, shallow marine biomicrites and biosparites [76], and outcrops in a restricted coastal stretch forming subvertical cliffs. Few Lower Coralline plunging cliffs (*sensu* [77]) are found in the investigated stretch of coast. They host roof notches with an asymmetric shape, recognized by Furlani et al. [78]. Lower Coralline Limestone is more commonly found in sloping coast formations, as typified in the stretch between Dahlet Qorrot Bay and Ras il-Qala.

The Lower Coralline Limestone Formation underlies the Globigerina Limestone Formation, the latter being younger and more erodible with respect to the former. The Globigerina Limestone

Formation consists of flattened areas along the coast, and includes yellowish, fine-grained, planktonic foraminiferal limestones. This formation features prominently as shore platforms along the investigated stretch of coast, where it outcrops above sea level and where the overlying softer blue clay layer has been eroded away, with examples at the Rdum il-Kbir promontory and on the eastern side of Marsalforn Bay. The Blue Clay Formation, overlying the Globigerina Limestone Formation, consists of grey, soft marls, clays and silty sands, forming gentle slopes. Sea cliffs shaped in Blue Clay can be also found within the study area. Finally, the youngest unit outcropping in the study area is the Upper Coralline Limestone Formation. This layer forms the plateaus at the top and is frequently weathered into steep cliffs and well-developed karst topography. It is also the source of the blocks of rock strewn onto the clayey slopes that slope away from the plateau edge more gently. A schematic stratigraphic coloumn is shown in Figure 1.

The NE coast of Gozo is particularly prone to gravity-induced processes such as rock spreads, block slides, rock falls and earth flows/slides. This is mainly due to the tectonic and geological features of the area which is characterized by a dense network of joints and fractures [79] and by the superposition of terrains with different geomechanical behaviour (cf. [80]). In fact, the brittle limestone plateaus overlying clayey terrains enhance the fracturing of the plateaus and the development of lateral spreading locally evolving into block sliding [44,46,81–85].

Moreover, clayey slopes are more prone to shallow earth flows and earth slides [86], while limestone cliffs are affected by rock falls, which has caused scarp retrogression over time.

Geomorphological investigations, relevant to the submerged area along the investigated stretch of coast [44], have revealed that block slides and earth flow/slide runout continue locally below the sea level, reaching ca. −20 m of depth. This is also in agreement with evidence from other submerged areas along the Maltese Islands [44,87]. Evidence from landslide dating in the NW coast of Malta infer that these deposits were emplaced in a subaerial environment during sea level lowstands and subsequently became submerged during the post-glacial marine transgression [61].

3.2. Social, Economic and Tourist Setting

Studies that include geoheritage assessment and geosite inventory highlight that the Maltese archipelago is considered as an attractive geotourist destination due to the strong interaction between natural and cultural aspects (cf. [71,88]). Data from the World Travel and Tourism Council (WTTC) for the year 2017, retrieved from Selmi et al. [88], show that 27.1% of Malta's GDP and 28.3% of total Maltese employment (corresponding to 55,000 jobs), were accounted for by activities directly related to and induced by travel and tourism [89,90]. Taking the type of tourism into consideration, data referring to 2017 show that the majority of tourists (almost 85%) visited Malta for holidays, while a very low percentage visited the country for business and other purposes (8% and 7%, respectively). More recent data show that the number of tourists is constantly increasing (standing at 2.6 million in 2018). The WTTC forecast that the travel and tourism contribution to national GDP will rise to 34.6% by 2027.

The tourism on the Island of Gozo is highly dependent on tourism activity of the main island, Malta [91], given that all of Gozo's tourist traffic necessarily passes through mainland Malta. Tourism is a significant source of income and employment and it is one of the primary contributors to the Gozo economy. Gozo's tourism also relies heavily on domestic tourists (with 400,000 domestic tourists or visitors per year coming from Malta) and on one-day trips of international tourists [92]. In 2018, almost 100,000 guests including resident and non-resident spent an average of three to four nights in one of the accommodation facilities of Gozo [93].

The Gozo Island attracts many tourists, especially during the summer period, for its environmental, cultural and geological heritage. Moreover, senior and middle-aged foreign residents more commonly chose the Island of Gozo as a place for their retirement, mainly for its relative peacefulness and quiet, combined with its array of scenic features (Gozo is colloquially known as "the place where time stood still").

The coastal sector investigated by this study includes four administrative districts: Żebbuġ (7.6 km^2), Xagħra (7.6 km^2), Nadur (7.2 km^2) and Qala (5.9 km^2), which are characterized by extensive urban development. The study area is host to two renowned tourist destinations, Marsalforn Bay and Ramla Bay, both of high importance to Gozo tourism. Marsalforn Bay is one of the main inhabited centres located on the Gozo coast. The availability of a significant number of accommodation facilities, shops, restaurants and several diving centres contribute to a dense tourist population and to lucrative businesses in Marsalforn, especially in summer.

Total guests in the Gozo and Comino region increased by 12.6% to 97,781 in 2017, while total nights spent went up by 11.1% to 347,943 when compared to the previous year [94]. Within a continuous upward trend in the last five years or so, the Gozo and Comino region recorded a strong growth in terms of inbound tourist arrivals in 2017 [94]. It is worth noting that among the top five localities where inbound tourists to Gozo stayed longest [94], two are within the study area, with important tourist destinations such as Marsalform and Ramla Bay.

Ramla Bay is the largest sandy beach in Gozo. It is characterized by golden-reddish sand, which lends it its name in Maltese (Ramla il-Ħamra) and makes this beach peculiar and unique to the Maltese Islands. The area around the beach is also of archaeological and historical interest, hosting Roman remains lying beneath the sand, a submerged seawall, fugass and battery defensive structures constructed by the Knights of St. John of Jerusalem in the mid-18th century, and the famous Calypso Cave looking over the western side of the beach.

Furthermore, the study area as a whole includes areas of high natural and ecological importance, hosting two wide natural protected zones, Għajn Barrani (located west of Marsalforn, including Ramla Bay) and Il-Qortin tal-Magun u l-Qortin il-Kbir (located close to Dahlet Qorrot Bay), both included in the Natura 2000 network as SAC. Finally, it should be noted that the high economic value of the study area is also derived from specific land uses in certain areas, such as quarrying, which occupy a surface of 0.13 km^2.

4. Materials and Methods

The method proposed and applied in this research for the assessment of coastal vulnerability is in line with the most recent index-based approaches generally used for this kind of analysis. In detail, the method relies on the outcomes of other research in the field [41,95] and is based on a new approach for the evaluation of indicators referring to: (i) the potentially exposed categories of assets (defined as "physical indicators"); and (ii) a number of parameters related to the social context (defined as "social indicators").

The indicators are interpreted here as operational representations (cf. [56,57]) of the physical and social characteristics of the area. Each physical indicator comprises different categories of assets, to which a score representing an increasing level of exposure was attributed. In order to tailor the approach to the local settings of the study area, the exposure level assigned to each category of assets was defined based on expert judgment. It is a widely shared opinion that the scoring and aggregation of indicators into indices may have a large impact on the resulting rankings and, consequently, on decision-making [55].

The spatial overlay of the physical indicators provides an estimate of the physical vulnerability level, which ranges from very low to very high vulnerability. Meanwhile, the various social indicators refer to vulnerability of the socio-economic aspects. These are also classified into five levels ranging from very low to very high, to provide a measure of the social vulnerability level for the investigated area. The overlay of the two sets of aggregated physical and social indicators enables an overall zonation that shows grades of the combined physical-socioeconomic vulnerability, defined as "overall vulnerability", thus identifying which are the most vulnerable stretches of the investigated coastal area.

Specifically, the method applied here comprises the following four main steps:

1. Definition of the landward limit of the coastal area to be investigated;
2. Classification of physical and social indicators into levels;
3. Data overlay and computation of an Overall Vulnerability Index;
4. Overall coastal vulnerability zonation and representation on a map.

The analyses are supported by GIS tools, which allow identification of the exposed coastal assets (i.e., natural and semi-natural environments, buildings, infrastructure, and agriculture), their combination with social data, and the calculation and mapping out of the overall vulnerability levels.

4.1. Definition of the Landward Limit of the Coastal Area to be Investigated

The area investigated was delimited according to the definition of a RICE area (Radius of Influence of Coastal Erosion and Flooding) proposed in the framework of the EUROSION project [95] for the identification of the coastal areas potentially impacted by marine-related process. The limit was set at a maximum distance of 500 m inland, or reaching a maximum of 5 m a.s.l. from the coastline. Furthermore, a buffer area of 100 m inland from the edge of the scarps formed in the Upper Coralline Limestone—which are extensively affected by landslide processes—was also added to the area defined in the above manner, in order to specifically account for landslide hazard. Information about landslide distribution provided by the geomorphological map of the study area (cf. [44]) was used specifically for this purpose, so as to identify the areas prone to slope instability.

The resulting landward limit was simplified and shown as a dashed red line in Figure 3, in order to provide a more linear indication of the investigated coastal sector.

4.2. Classification of Physical and Social Indicators

This step comprises the identification of data required for the evaluation and classification into five different levels, from very low to very high, for each of the two sets of indicators (physical and social indicators).

The evaluation of the anthropic and natural assets potentially exposed to coastal hazards is based on the elaboration of the following indicators: land use, transport network, and utilities. A GIS layer has to be created for each indicator in order to identify polygons representing the spatial unit to which the related exposure level is assigned, indicating its specific numerical value, ranging from 1 (very low exposure) to 5 (very high exposure). These layers are first converted to a raster format (5 × 5 m) and then overlaid to estimate the physical vulnerability level, by considering the maximum exposure value assigned to each cell.

At the detailed level, the land use information was collected from the land use map available in Prampolini et al. [44], in which land use categories were classified as: (i) natural and semi-natural areas, (ii) agricultural areas, and (iii) artificial surfaces. This land use map was supplemented with additional detailed information in order to include strategic elements, such as civil protection posts, police stations, emergency posts, fire corps posts, port authorities' posts, hospitals, and schools. The identification of these elements was supported by photointerpretation of the most accurate maps available for the study area and verified in the field during the investigation.

An exposure level (ranging from 1, very low exposure, to 5, very high exposure) was assigned to each land use category based on expert judgement. The highest exposure level was attributed to the areas occupied by settlements, constructions, and human activities while the lowest level was assigned to abandoned agricultural zones. Details concerning the land use indicator and the exposure levels related to each category are shown in Table 1.

Table 1. Physical indicators and expert-based exposure level (ranging from value 1 to value 5) assigned to each category of assets.

Physical Indicators (Value)	Very Low/Null Exposure (1)	Low Exposure (2)	Medium Exposure (3)	High Exposure (4)	Very High Exposure (5)
Land Use	Abandoned agricultural area. Bare rock	Agricultural area; Green urban areas; Natural and vegetated area, Terraced agricultural field; Land principally occupied by agriculture with significant areas of natural vegetation	Residential area; Dump; Quarry; Cemetery	Beaches; Dunes; Sand	Historical and archaeological site; Natural protected area (SCI); Strategic elements; Entertainment (commerce, finance, business, recreational, leisure, and sport)
Transport	Absence of transport network or highly degraded transport network	Footway; Path; Track; Steps	Tertiary road; Living street; Residential; Services	Secondary road	Primary road
Utilities	Absence of utilities	Mainly local and small utilities	Street lighting	11kV overhead line	Substations; Feeder pillar

The transport network information was retrieved from Open Street Map downloaded from [96] as a shapefile. In this case, the roads were classified as follows: primary road, secondary road, tertiary road, inhabited street, residential, services, footway, path, track, and steps; the greater exposure level was attributed to roads of national or international importance. For each polyline element, a buffer distance was built (up to 20 m of total width for the primary road) in order to take into account a proper area of pertinence. Details concerning the exposure levels defined for the transport indicator are shown in Table 1.

The electricity network was taken into account for the utilities. The spatial distribution of these elements is available on Malta Inspire Geoportal [97]. In this case, a buffer zone up to 10 m width (10 m for substations, feeder pillars, and overhead lines, and 5 m for street lighting) was considered for each element. Details concerning the exposure levels associated with each electricity element are indicated in Table 1.

The social indicators allow characterization of the districts located in the investigated coastal area by evaluating, directly and indirectly, the social characteristics of the population living in the zone and therefore prone to be affected by coastal hazards. Generally, the social vulnerability information is obtained from census data. In this study, social data was downloaded from the Inspire Geoportal [97] and integrated with information available in the special report recently published by the Malta National Statistics Office (NSO) [98]. The social indicators are gauged from the economic budget allocated by the Government for supporting the population. The social vulnerability analysis carried out here is based on the assumption that higher allocation of social schemes corresponds to higher vulnerability of the population living in the investigated districts.

Specifically, the following indicators were taken into account: health care, disability, old age, children, and unemployment. These aspects represent some of the higher-risk groups of society, which social services and budget allocations aim to protect. The NSO provides a classification of these indicators in five classes that were here converted in a numerical value ranging from 1 to 5. The available data is provided by the NSO at district level and thus a numerical value was assigned to each district as a whole. The social vulnerability was then calculated as geometric mean of the values attributed to each indicator and then classified into five levels (from 1, for very low vulnerability, to 5, for very high vulnerability) by means of an equal interval classification method. A detailed description of the social indicators used in this study is reported in Appendix A.

The physical and social indicators identified are weighted equally, as in most of the indicator-based studies [99], meaning that each individual indicator has the same influence on the final calculation of the overall vulnerability.

4.3. Data Overlay and Computation of an Overall Vulnerability Index

Once all the required data related to the physical and social vulnerability are collected and expressed in five levels, they are overlaid by means of a specific GIS tool. The overall vulnerability calculation is therefore the result of the aggregation of the physical vulnerability and social vulnerability data, the combination of which provides the definition of the Overall Vulnerability Index (OVI) as follows:

$$\text{Overall Vulnerability Index} = (\text{Physical vulnerability} \times \text{Social vulnerability})\,\hat{}\,0.5 \qquad (1)$$

Finally, the Overall Vulnerability Index is classified into five levels of increasing vulnerability (from very low to very high vulnerability) by means of the equal interval classification method.

5. Results

The results of this study are represented by: (i) GIS-based maps that show the spatial distribution of the natural and anthropic exposed elements and their classification in terms of physical vulnerability; (ii) social vulnerability level estimated for each investigated coastal district; (iii) overall vulnerability map of the NE part of the Gozo Island; (iv) areal extent and relative percentage of each vulnerability level. The land use classification (Figure 4) enables the evaluation of the surface occupied by different categories of natural and anthropic elements (cf. Table 1 in Section 4.2).

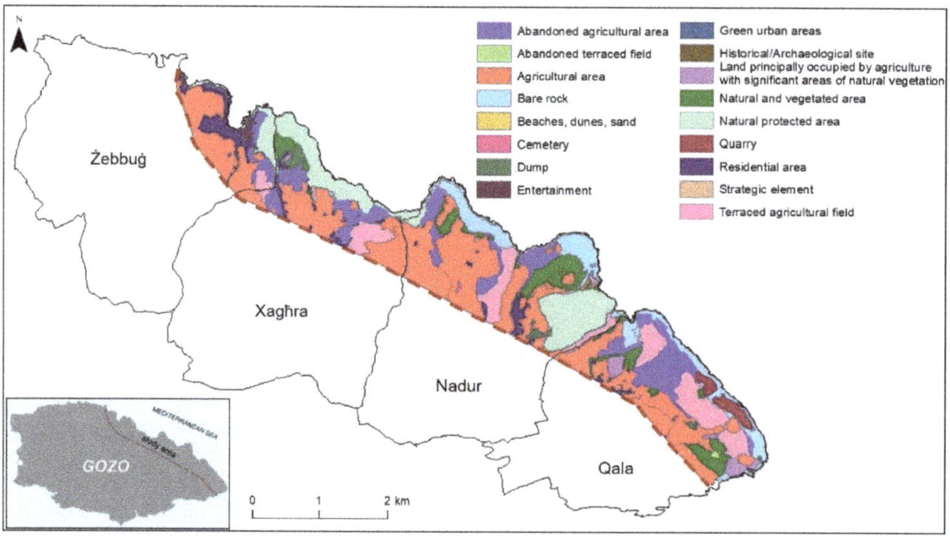

Figure 4. Detailed land use map. The red dashed line identifies the inland boundary of the study area. The districts' boundaries are also indicated.

In detail, 3.6 km^2 are occupied by agricultural areas, 1.2 km^2 by abandoned agricultural areas, 1 km^2 by terraced agricultural field, 1.2 km^2 by natural protected areas, 0.8 km^2 by bare rock, 0.7 km^2 by natural and vegetated areas, and, 0.5 km^2 by residential areas. Finally, quarrying and land occupied mainly by agriculture with significant areas of natural vegetation occupy 0.1 km^2. The remaining surface area (corresponding to 602,268.2 m^2) is occupied by the following categories: strategic elements (including a Police station), historical/archaeological sites (The Tower of Ta Sopu, west of Daħlet Qorrot Bay, and Saint Anthony's Battery at Ras il-Qala), green urban areas, beaches, dunes, sand, abandoned terraced fields, entertainment zones, and a cemetery. These values are indicated as percentages in Figure 5.

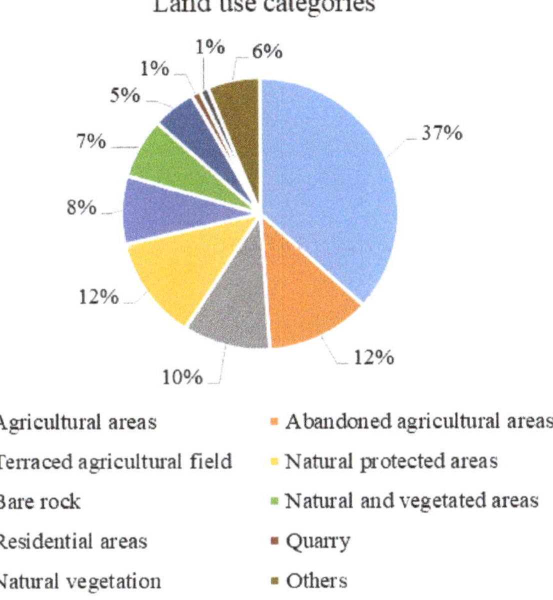

Figure 5. Percentage of the area occupied by each land use category according to the classification shown in Figure 4.

The interpretation and spatial representation of the overlaid physical indicators (cf. Section 4.2) (cf. Table 1) is representative of the different levels of physical vulnerability, given that the physical exposure levels are correlated directly to the level of value of the assets at risk.

This enables the zonation and calculation of what are, in effect, the extent of the areas with different levels of physical vulnerability, expressed in square kilometres and in percentages of the total surface investigated. The spatial interpretation of physical vulnerability is mapped in Figure 6 while the corresponding numerical values are reported in Table 2.

The numerical values assigned to each of the social vulnerability indicators for the investigated coastal districts are summarised in Table 3.

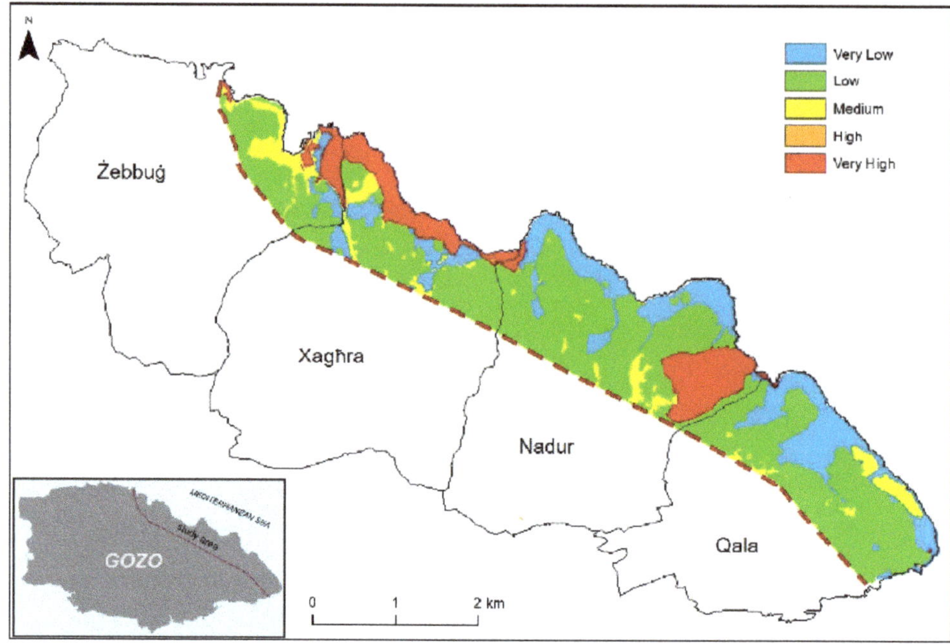

Figure 6. Physical vulnerability map resulting from data overlay of the physical indicators (land use, transport and utilities). The red dashed line identifies the inland boundary of the study area.

Table 2. Areal distribution of each physical vulnerability level resulting from the overlay and aggregation of the physical indicators (land use, transport network and utilities).

Physical Vulnerability	Surface (km^2)	Surface (%)
Very low	2.0	21.7
Low	5.4	57.8
Medium	0.7	7.4
High	0.005	0.1
Very high	1.2	13.0

Table 3. Social indicators evaluated for the four coastal districts of the study area. The method for attributing the numerical value assigned to each indicator is indicated in the methodological paragraph (cf. Section 4.3).

Social Indicators	Żebbuġ District	Xagħra District	Nadur District	Qala District
Health care indicator	3	2	3	3
Disability indicator	3	5	4	5
Old age indicator	2	3	3	3
Family/Children indicator	3	3	3	3
Unemployment indicator	5	2	2	2
Population (number of inhabitants, 2016)	2043	4029	4001	1885

The aggregation of the social vulnerability indicators (cf. Table 3) enables the evaluation of the social vulnerability level for each of the investigated districts (Figure 7). Specifically, two districts (Nadur and Xagħra) are characterized by medium vulnerability and two (Qala and Żebbuġ) by high vulnerability.

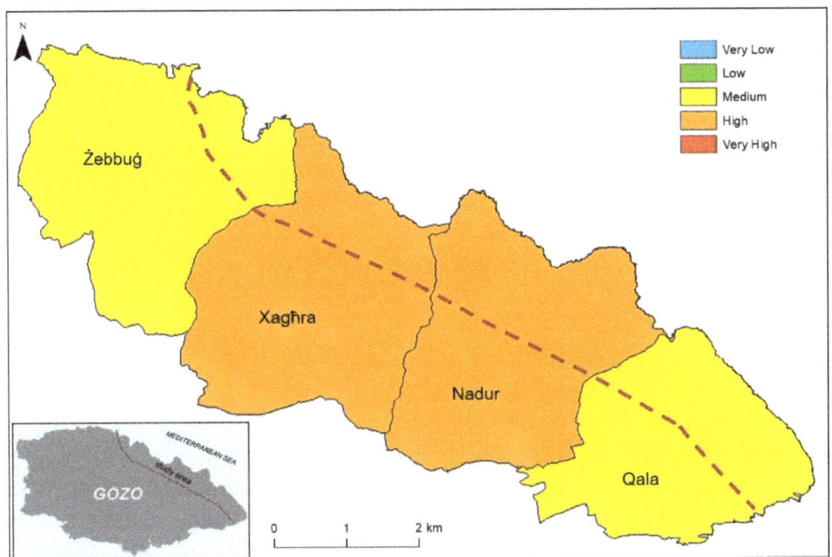

Figure 7. Social vulnerability map of the four administrative districts located in NE Gozo. The red dashed line indicates the inland boundary of the study area.

Finally, the overlay of the physical vulnerability levels with the social vulnerability levels enables a computation (and mapping) of the Overall Vulnerability Index for different polygons in the study area (cf. Section 4.3). It provides the basis for evaluating the variation in the overall vulnerability levels across the investigated coastal sector and enables this to be represented spatially (Figure 8).

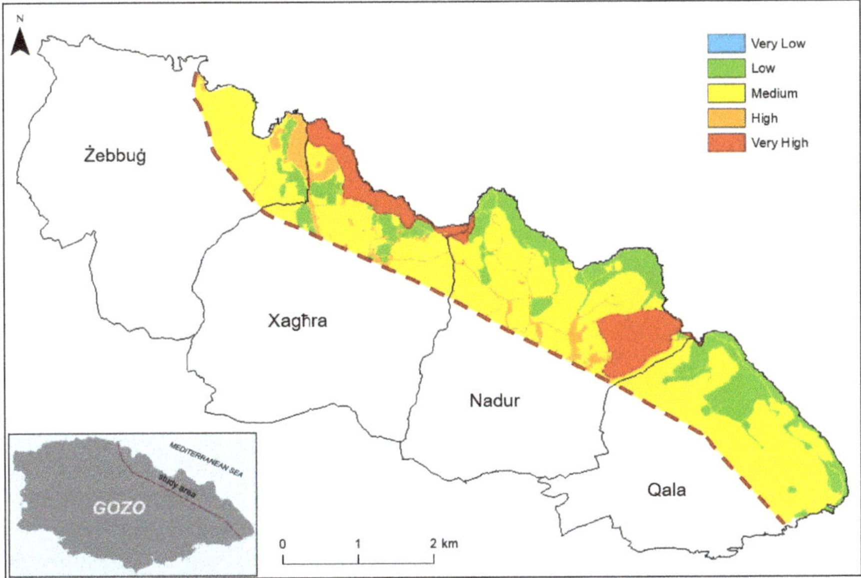

Figure 8. Overall vulnerability map resulting from the spatial aggregation of the physical vulnerability levels and the social vulnerability levels over the area. The red dashed line indicates the inland boundary of the study area.

In sum, 1.9 km^2 are occupied by areas showing a low overall vulnerability level, 5.6 km^2 are occupied by areas showing a medium overall vulnerability level, 0.7 km^2 are occupied by areas with a high overall vulnerability level and, finally, areas characterized by a very high overall vulnerability level cover 1 km^2 (Table 4).

Table 4. Areal distribution of overall vulnerability levels that are the result of the combination and overlay of spatial distributions of the physical and social vulnerability factors.

Overall Vulnerability	Surface (km^2)	Surface (%)
Very low	-	-
Low	1.9	20.4
Medium	5.6	61.3
High	0.7	7.3
Very high	1	11

6. Discussion

The proposed index-based method allows zoning of the investigated coastal stretch into different levels of vulnerability to climate- and marine-related processes. Results show that the study area is divided in four zones only (from low to very high vulnerability), since there are no areas with a very low level of vulnerability.

The coastal sectors located east of Marsalforn Bay, which include Ramla Bay and the area on the western side of Daħlet Qorrot Bay, show the highest overall vulnerability level. This is due to the combination of very high physical vulnerability levels pertaining to the presence of two Natura 2000 sites (cf. Sections 3.1 and 5) and high social vulnerability levels for the Xagħra and Nadur districts. The high social vulnerability of Xagħra and Nadur is explained by the fact that each of the latter districts accounts for more than 4000 inhabitants, which is twice the number of inhabitants of the two other districts considered, Żebbuġ and Qala, for which a medium social vulnerability is assigned.

The coastal sector surrounding Marsalforn Bay is mainly characterized by a medium level of overall vulnerability as a result of the combination of a medium physical vulnerability level, explainable by the presence of residential areas, and, a medium social vulnerability level obtained for the Żebbuġ district. Furthermore, the innermost sector of the study area shows a prevailing medium overall vulnerability level, resulting from the combination of a low to very low physical exposure level, owing to the presence of cultivated and abandoned agricultural fields respectively, and a high social vulnerability level resulting for the districts of Xagħra and Nadur, as discussed above. The eastern part of the investigated sector, with bare outcrops of rock, hosts areas characterized by a low overall vulnerability.

The index-based method here proposed can be considered a suitable approach for the identification of coastal areas that are most vulnerable to consequences of different climate- and marine-related processes. Since the method adopted here combines vulnerability and exposure, the results represent an overall vulnerability assessment. It relies on the evaluation of the two main components that are used to define the overall vulnerability: physical and the social vulnerability, thus accounting for both the physical assets potentially exposed to climate and marine processes and the social aspects of the local population. This approach is in line with the convergence of different terms by the IPCC in its Fifth Assessment Report (AR-5, [11]).

The application of the proposed OVI method has allowed identification of a number of operational advantages. First of all, as the analysis is based on a sequence of steps, the method is relatively simple and easy to apply. It has been found to be cost-effective on account of the possibility of managing and processing the acquired data in a GIS environment with relative ease. Furthermore, it does not require intense field work, since it relies on indirect analysis, supported by existing geological-geomorphological information that is either readily available in the literature or easily collected. It can therefore be applied to wider coastal stretches, as in the case of other index-based approaches [40,100–103]. It should be underlined that the applied method can take advantage of data sets and information concerning the natural and anthropic assets which are generally freely accessible and downloadable, e.g., from national geoportals available in most of the countries (at least across Europe), and from open databases, such as OpenStreetMap. Therefore, the method may have a wide range of applicability at different scales. Large scale analyses are based on detailed information about exposed/vulnerable elements and population (such as the land use and vulnerability maps proposed in this study). Meanwhile, small scale vulnerability maps concern wide areas, taking into account only the most prominent and spatially persistent exposed/vulnerable elements and regional data about the exposed/vulnerable population (as in the case of EUROSION Project), which accounted for all the European countries and provided a comprehensive European-level data repository at scale 1:100,000 [95].

The OVI method provides a sound approach suitable for the identification (and assessment) of vulnerable areas and sectors, even as an expedient and cost-effective scoping stage to identify which area may be analysed more specifically and at greater expense. It provides a useful tool for an overall time-efficient and cost-effective approach to focus limited research resources onto where they are needed most.

The presented research is a pilot-study and a first-ever combined overall vulnerability assessment carried out for any area in the Maltese archipelago, and it is promises to provide new contributions at different levels and in different ways. At the localised level, the proposed method, and the results obtained from it, are promising as they reveal the potential applicability of this method for other (and potentially wider) coastal areas of the Maltese archipelago. Moreover, even at the local level, the overall vulnerability assessment could be easily updated by using any newly available or more detailed data regarding the social vulnerability of the investigated districts, as and when this becomes available. The relatively simple combined approach makes it more possible to keep the OVI up to date, even at the localised level. By way of example, it is relevant that during 2016, Malta's social protection outlay rose by €70.3 million in comparison to 2015 and that Old Age and Sickness/Health Care witnessed the biggest increases in social outlay [98]. The NSO [94] published statistical data on the basis of six different regions, from which localities can be studied specifically as socio-economic parameters change. This means that the social vulnerability level of the investigated district could be expected to rise in the next decades, influencing the overall vulnerability assessment. At the national level, the method could be applied to provide a first-round assessment of overall coastal vulnerability, to identify the most vulnerable stretches of coastal areas and values, and then to follow this up with a more detailed qualitative and quantitative analysis and comparison of risk assessments for specific areas and sectors, for different hazard types.

Running parallel to other research at the European level, further studies could include the spatial susceptibility analysis, including each of the coastal hazardous processes already identified here as affecting the investigated stretch of coast (e.g., erosion, sea level rise, landslides). As already done for other coastal zones in Europe [39,40,54], the investigated area should be classified into zones with different susceptibility, accounting for their proneness to be potentially affected by the specific impacts from extreme events and related processes pertaining to the particular hazard types (e.g., erosion, sea level rise, landslides). The susceptibility to different climate- and marine-related hazards could then be coupled with the vulnerability data derived by this study to perform a complete risk analysis (as done for example in [25,53]).

In this context, it is worth noting that the Maltese archipelago is situated centrally in the Mediterranean Sea, which has been classified as one of the regions most sensitive to climate change and, therefore, it is considered as a hot-spot area [104]. Climate change projections for the Mediterranean region [105] show that the area is experiencing an increasing temperature with consequent change in spatial and temporal distribution of weather/climate extremes [106]. More intense events, with alternating and more severe drought and precipitation, are expected to trigger and exacerbate erosive and mass movement processes [11,13], affecting the spatial distribution of the susceptibility to these types of events and hazards.

A direct consequence of global temperature increase is the rise in sea level, the direct impacts of which on coastal systems become manifest with larger rates of erosion, increase in flooding events, wetland loss, and saltwater intrusion [107]. Studies regarding the reconstruction of sea level changes in the Mediterranean Sea during the last 2000 years have shown that, in tectonically stable Mediterranean areas, the sea rose about 1.1 m [108,109] while for the last two decade the estimated rise was of about 3 cm/decade [110]. However, differential vertical land movements, including uplift and subsidence, characterize the Mediterranean coasts and, for this reason, the trends of sea level rise in the Mediterranean Sea have a large spatial variability [111]. Taking into account the role of terrestrial ice melt, steric effects and glacial isostatic adjustment, the future total Mediterranean averaged sea level rise has been estimated to be between 9.8 and 25.6 cm by 2040–2050, depending on the Representative Concentration Pathways scenario [112]. Sea level projections, obtained by coupling modelled eustatic trends with local ground movements, are in general used for supporting the assessment of the future coastal risk to sea level rise, which is aimed at reducing its impacts on natural coastal environments and human economic activities [3].

Based on the assumption that the investigated coastal area could potentially be affected by all the above mentioned climate-related hazards, further research activities should focus on risk analysis and mapping, as already done with reference to hazards related to sea level rise in several parts of Europe [113,114] and worldwide ([115–119] and reference therein). In this context, the results obtained in this study can prepare the ground for a comprehensive risk assessment, combining the identification of the exposed assets and the evaluation of vulnerability levels with a quantitative assessment of the hazardous processes affecting the area.

Finally, it is worth noting that the study is in line with the methodological approach proposed by the European Environment Agency (EEA) for the identification and implementation of climate change adaptation strategies (European Climate Adaptation Platform Climate-ADAPT). Specifically, in the EEA approach, six steps of analysis are indicated: (i) preparing the ground for adaptation; (ii) assessing risk and vulnerability; (iii) identifying adaptation options; (iv) assessing adaptation options; (v) implementation; (vi) monitoring and evaluation. The vulnerability analysis shown here represents a useful tool for addressing Step 1 and Step 2, while further activities should be planned with the aim of developing and tailoring the most suitable adaptation actions for the protection of the natural ecosystem and the maintenance of the anthropic activities.

7. Conclusions

This research represents a first attempt at the evaluation of coastal overall vulnerability in the Island of Gozo (Maltese archipelago) to the main climate- and marine-related processes, such as coastal erosion, landslides and sea level rise, to which the island is particularly prone. The analysis method developed and tested in this study is based on a conceptual interpretation of overall vulnerability, defined as the combination of two distinct components, namely, physical vulnerability and social vulnerability.

The study departs from a discussion of different notions of risk, including commonly used notions of exposure and vulnerability, which are defined qualitatively in the most authoritative sources (cf. Section 2). Given the identified, often conflicting understanding of the concept of vulnerability by various sectors of researchers and policy makers, it was determined that an in-depth consideration of the vulnerability concept was required prior to determining the vulnerability assessment methodology.

The study proposes a method for the assimilation, analysis and computation of overall vulnerability, and for its graphical spatial representation. The method manages to converge different qualitative notions of physical exposure and social vulnerability, and also makes it possible to derive a spatial quantitative distribution of the Overall Vulnerability Index. The latter can be represented in a map showing different coastal vulnerability levels that are easily readable spatially, making it simple to communicate comprehensibly to decision makers.

The Overall Vulnerability Index is calculated by means of an index-based method that is proposed along the lines of approaches developed in the framework of previous research projects and adapted specifically for the context of the study area and to the available information.

The study area was identified as possessing features of high economic value derived from tourist and mining activities and natural protected areas, altogether making coastal vulnerability a major concern. The final results of the analysis reveal that most of the investigated area (61.3%) is characterized by a medium level of overall vulnerability. A very high overall vulnerability level (11%) was obtained for the areas located east of Marsalforn Bay and close to Daħlet Qorrot Bay, including Ramla Bay, mainly owing to the presence of two sites protected as Special Areas of Conservation within the Natura 2000 network (Għajn Barrani and Il-Qortin tal-Magun u l-Qortin il-Kbir sites). A high vulnerability (7.3%) resulted for the main roads, while 20.4% of the total surface is characterized by low vulnerability areas mainly corresponding to abandoned agricultural fields and bare rocks outcrops.

The type of analysis presented here can be easily replicated for other coastal regions because the data sets required for the proposed method are almost always freely available or relatively easy to compile. Furthermore, the wider application of this method to the entire coastal area of the Gozo Island as well as to the coastal zones of the Maltese Islands has the potential to contribute to an overall risk characterization for the entire Maltese archipelago and it provides a useful tool for the identification of the most exposed and vulnerable zones (hotspot areas) that require action for their protection as a matter of priority. Starting from readily available data, which can be processed and mapped in a GIS environment with relative ease, the method proposed and tested here can provide policy makers, as well as coastal management agencies, with a graphical representation of easily comprehensible and useful data to support decision-making processes at both operational (short-term) and strategic (medium-long-term) level, enabling them to devise adaptation and structural protection measures in a more specific and effective manner.

Moreover, the method is cost-effective and time-efficient on account of ease of data processing, and it can also be a precursor for more detailed qualitative and quantitative analysis of risk (from different hazard types) at different scales for the areas or sectors considered to be at greater risk.

Finally, beyond the site-specific results, the method represents an important contribution toward more comprehensive risk assessment, also in terms of its potential transferability and replicability to different hazard types, including the local effects of climate change from extreme weather/climate events and sea level rise that need to be taken into consideration in a timely and effective manner.

Author Contributions: Conceptualization: A.R., V.V., G.B., A.S.M., M.S.; methodology: A.R., V.V., G.B., A.S.M., M.S.; formal analysis: V.V. and A.R.; investigation: M.S., V.V., A.S.M.; data curation: A.R. and V.V.; writing original draft preparation: A.R. and V.V.; writing review and editing: A.R., V.V., G.B., A.S.M., M.S.; visualization: V.V. and A.R.; supervision: M.S.; project administration: M.S. and A.S.M.; funding acquisition: M.S. and A.M.S. All authors have read and agreed to the published version of the manuscript.

Funding: The study has been carried out in the frame of the Project "Coastal risk assessment and mapping" funded by the EUR-OPA Major Hazards Agreement of the Council of Europe (2020–2021). Grant Number: GA/2020/06 n° 654503. Scientific responsible: Anton Micallef (ICoD) and Mauro Soldati (Unimore).

Acknowledgments: The authors are grateful to Chris Gauci (Research and Planning Section, Marine and Storm Water Unit, Public Works Department, Floriana, Malta) for information provided about the social vulnerability aspects. The authors would also like to thank Chiara Bordin for helping with data processing in GIS environment.

Conflicts of Interest: The authors declare no conflict of interest.

Appendix A

The evaluation of the social vulnerability for the coastal districts located in the investigated area was based on the calculation of a synthetic index that takes into account a number of indicators that represent social protection schemes that provide, where possible, protection against a single risk or need. For a comprehensive description of the social protection gross expenditure, readers can refer to the social protection report for the period 2012–2016, published in 2019 and directly downloadable at: https://nso.gov.mt/en/publicatons/Publications_by_Unit/Documents/A2_Public_Finance/Social%20Protection%202016.pdf

The social indicators taken into account in this study are:

- The Health Care function, which consists of all those benefits paid to persons during temporary periods of unemployment due to sickness or injury, and health care provided in the framework of social protection;
- The Disability function, which mainly covers cash benefits paid to persons who are below the retirement age and unable to work because of a mental or physical disability;
- The Old Age function, which covers all interventions against the risks linked to retirement and ageing. These include pensions given to a person once they retire from the labour market, lodging in specialized retirement homes and any services provided to persons unable to independently care for themselves.
- The Family/Children function, which includes cash benefits provided to households with children, various childcare services available to families and other social services provided with the specific intention to assist families with children.
- The Unemployment function, which represents benefits paid to either compensate an individual for the loss of his/her gainful employment or to cover the income of persons who retire from employment prior to the statutory age.

References

1. Davidson–Arnott, R. *An Introduction to Coastal Processes and Geomorphology*; Cambridge University Press: Cambridge, UK, 2010; p. 458.
2. Masselink, G.; Gehrels, R. *Coastal Environments and Global Change*; John Wiley & Sons: West Sussex, UK, 2014; p. 448.
3. Nicholls, R.J.; Wong, P.P.; Burkett, V.R.; Codignotto, J.O.; Hay, J.E.; McLean, R.F.; Ragoonaden, S.; Woodroffe, C.D. Coastal systems and lowlying areas. Climate Change 2007. Impacts, Adaptation and Vulnerability. In *Contribution of Working Group II to the Fourth Assessment Report of the Intergovernmental Panel on Climate Change*; Parry, M.L., Canziani, O.F., Palutikof, J.P., van der Linden, P.J., Hanson, C.E., Eds.; Cambridge University Press: Cambridge, UK, 2007; pp. 315–356.
4. World Resources Institute. *Decision Making in a Changing Climate*; United Nations Development Programme World Bank; World Resources Institute: Washington, DC, USA, 2010.
5. European Environment Agency. *The Changing Faces of Europe's Coastal Areas*; EEA Report; European Environment Agency: Copenhagen, Denmark, 2006.
6. European Environment Agency. *Mediterranean Sea Region Briefing—The European Environment—State and Outlook*; EEA Report; European Environment Agency: Copenhagen, Denmark, 2015.
7. Reid, W.V.; Mooney, H.A.; Cropper, A.; Capistrano, D.; Carpenter, S.R.; Chopra, K.; Dasgupta, P.; Dietz, T.; Duraiappah, A.K.; Hassan, R.; et al. *Ecosystems and Human Well–Being: Synthesis*; Millennium Ecosystem Assessment: Washington, DC, USA, 2005.
8. Maes, J.; Teller, A.; Erhard, M.; Liquete, C.; Braat, L.; Berry, P.; Egoh, B.; Puydarrieux, P.; Fiorina, C.; Santos, F.; et al. *Mapping and Assessment of Ecosystems and Their Services. An Analytical Framework for Ecosystem Assessments under Action 5 of the EU Biodiversity Strategy to 2020*; Publications Office of the European Union: Luxembourg, Luxembourg, 2013; pp. 1–58.
9. Gallina, V.; Torresan, S.; Critto, A.; Sperotto, A.; Glade, T.; Marcomini, A. A review of multi–risk methodologies for natural hazards: Consequences and challenges for a climate change impact assessment. *J. Environ.* **2016**, *168*, 123–132. [CrossRef] [PubMed]
10. Intergovernmental Panel on Climate Change. *Climate Change 2007: Mitigation. Contribution of Working Group III to the Fourth Assessment Report of the Intergovernmental Panel on Climate Change*; Metz, B., Davidson, O.R., Bosch, P.R., Dave, R., Meyer, L.A., Eds.; Cambridge University Press: Cambridge, UK, 2007.

11. Intergovernmental Panel on Climate Change. *Climate Change 2014: Impacts, Adaptation and Vulnerability. Contribution of Working Group II to the Fifth Assessment Report of the Intergovernmental Panel on Climate Change. Part A: Global Aspects*; Field, C.B., Barros, V.R., Dokken, D.J., Mach, K.J., Mastrandrea, M.D., Bilir, T.E., Chatterjee, M., Ebi, K.L., Estrada, Y.O., Genova, R.C., et al., Eds.; Cambridge University Press: Cambridge, UK, 2014.

12. Hoegh–Guldberg, O.; Jacob, D.; Taylor, M.; Bindi, M.; Brown, S.; Camilloni, I.; Diedhiou, A.; Djalante, R.; Ebi, K.L.; Engelbrecht, F.; et al. 2018: Impacts of 1.5 °C Global Warming on Natural and Human Systems. In *Global Warming of 1.5 °C*; Masson–Delmotte, V., Zhai, P., Pörtner, H.-O., Roberts, D., Skea, J., Shukla, P.R., Pirani, A., Moufouma–Okia, W., Péan, C., Pidcock, R., et al., Eds.; An IPCC Special Report on the Impacts of Global Warming of 1.5 °C above Pre–Industrial Levels and Related Global Greenhouse Gas Emission Pathways, in the Context of Strengthening the Global Response to the Threat of Climate Change, Sustainable Development, and Efforts to Eradicate Poverty; Intergovernmental Panel on Climate Change: Geneva, Switzerland, 2018.

13. Intergovernmental Panel on Climate Change. Summary for Policymakers. In *Climate Change and Land: An IPCC Special Report on Climate Change, Desertification, Land Degradation, Sustainable Land Management, Food Security, and Greenhouse Gas Fluxes in Terrestrial Ecosystems*; Shukla, P.R., Skea, J., Buendia, E.C., Masson-Delmotte, V., Pörtner, H.-O., Roberts, D.C., Zhai, P., Slade, R., Connors, S., Diemen, R., et al., Eds.; Cambridge University Press: Cambridge, UK, 2019.

14. UNISDR (United Nations International Strategy for Disaster Reduction). *Sendai Framework for Disaster Risk Reduction 2015–2030*; The United Nations Office for Disaster Risk Reduction: Geneva, Switzerland, 2015; Available online: https://www.preventionweb.net/files/43291_sendaiframeworkfordrren.pdf (accessed on 26 April 2020).

15. European Commission. *An EU Strategy on Adaptation to Climate Change*; The European Commission: Brussels, Belgium, 2013; Available online: https://eur-lex.europa.eu/LexUriServ/LexUriServ.do?uri=COM:2013:0216:FIN:EN:PDF (accessed on 26 April 2020).

16. Directive 2007/60/EC of the European Parliament and of the Council of 23 October 2007 on the Assessment and Management of Flood Risks. Available online: https://eur--lex.europa.eu/legal--content/EN/TXT/PDF/?uri=CELEX:32007L0060&from=EN (accessed on 26 April 2020).

17. Field, C.B.; Barros, V.; Stocker, T.F.; Qin, D.; Dokken, D.J.; Ebi, K.L.; Mastrandrea, M.D.; Mach, K.J.; Plattner, G.-K.; Allen, S.K.; et al. *Managing the Risks of Extreme Events and Disasters to Advance Climate Change Adaptation*; A Special Report of Working Groups I and II of the Intergovernmental Panel on Climate Change (IPCC); Cambridge University Press: Cambridge, UK, 2012.

18. Baird, A.; O'Keefe, P.; Westgate, K.; Wisner, B. *Towards an Explanation of and Reduction of Disaster Proneness*; Occasional Paper Number 11; Disaster Research Unit, University of Bradford: Bradford, UK, 1975.

19. O'Keefe, P.; Westgate, K.; Wisner, B. Taking the naturalness out of natural disasters. *Nature* **1976**, *260*, 566–577. [CrossRef]

20. Lewis, J. The vulnerable state: An alternative view. In *Disaster Assistance: Appraisal, Reform and New Approaches*; Stephens, L., Green, S.J., Eds.; New York University Press: New York, NY, USA, 1976; pp. 104–129.

21. Hewitt, K. *Interpretations of Calamity*; Allen & Unwin: London, UK, 1983.

22. O'Brien, K.; Eriksen, S.; Schjolen, A.; Nygaard, L. *What's in a Word? Conflicting Interpretations of Vulnerability in Climate Change Research*; CICERO Working Paper 2004:04; CICERO, Oslo University: Oslo, Norway, 2004.

23. Romieu, E.; Welle, T.; Schneiderbauer, S.; Pelling, M.; Vinchon, C. Vulnerability assessment within climate change and natural hazard contexts: Revealing gaps and synergies through coastal applications. *Sustain. Sci.* **2010**, *5*, 159–170. [CrossRef]

24. Jurgilevich, A.; Räsänen, A.; Groundstroem, F.; Juhola, S. A systematic review of dynamics in climate risk and vulnerability assessments. *Environ. Res. Lett.* **2017**, *12*, 013002. [CrossRef]

25. Mysiak, J.; Torresan, S.; Bosello, F.; Mistry, M.; Amadio, M.; Marzi, S.; Furlan, E.; Sperotto, A. Climate risk index for Italy. *Philos. Trans. R. Soc.* **2018**, *376*, 20170305. [CrossRef] [PubMed]

26. UNISDR. UNISDR Terminology on Disaster Risk Reduction. 2009. Available online: https://www.unisdr.org/files/7817_UNISDRTerminologyEnglish.pdf (accessed on 26 April 2020).

27. Cavallin, A.; Marchetti, M.; Panizza, M.; Soldati, M. The role of geomorphology in the environmental impact assessment. *Geomorphology* **1994**, *9*, 143–153. [CrossRef]

28. UNDRO. *Disaster Prevention and Mitigation—Compendium of Current Knowledge*; United Nations: New York, NY, USA, 1984.

29. Papathoma–Köhle, M.; Gems, B.; Sturm, M.; Fuchs, S. Matrices, curves and indicators: A review of approaches to assess physical vulnerability to debris flows. *Earth Sci. Rev.* **2017**, *171*, 272–288. [CrossRef]

30. Intergovernmental Panel on Climate Change. *Climate Change 1995. Impacts, Adaptations and Mitigation of Climate Change: Scientific-Technical Analyses*; Watson, R.T., Zinyowera, M.C., Moss, H.J.D., Eds.; Second Assessment Report of the Intergovernmental Panel on Climate Change; Cambridge University: Cambridge, UK, 1995.

31. Costa, L.; Kropp, J.P. Linking components of vulnerability in theoretic frameworks and case studies. *Sustain. Sci.* **2013**, *8*, 1–9. [CrossRef]

32. Gornitz, V. Vulnerability of the East Coast, USA to future sea level rise. *J. Coast. Res.* **1990**, *9*, 201–237.

33. Gornitz, V.M.; White, T.W.; Cushman, R.M. Vulnerability of the US to future sea level rise, Coastal Zone 91. In Proceedings of the 7th Symposium on Coastal and Ocean Management, Long Beach, CA, USA, 8–12 July 1991; American Society of Civil Engineers: New York, NY, USA, 1991; pp. 1345–1359.

34. Ojeda–Zújar, J.; Álvarez–Francosi, J.I.; Martín–Cajaraville, D.; Fraile–Jurado, P. El uso de las TIG para el cálculo del índice de Vulnerabilidad costera (CVI) ante una potencial subida del nivel del mar en la costa andaluza (España). *GeoFocus* **2009**, *9*, 83–100.

35. Özyurt, G.; Ergin, A. Application of sea level rise vulnerability assessment model to selected coastal areas of Turkey. *J. Coast. Res.* **2009**, *56*, 248–251.

36. Özyurt, G.; Ergin, A. Improving coastal vulnerability assessments to sea–level rise: A new indicator–based methodology for decision makers. *J. Coast. Res.* **2010**, *26*, 265–273. [CrossRef]

37. McLaughlin, S.; Cooper, J.A.G. A multi-scale coastal vulnerability index: A tool for coastal managers? *Environ. Hazard* **2010**, *9*, 233–248. [CrossRef]

38. Di Paola, G.; Iglesias, J.; Rodríguez, G.; Benassai, G.; Aucelli, P.P.C.; Pappone, G. Estimating coastal vulnerability in a meso-tidal beach by means of quantitative and semi–quantitative methodologies. *J. Coast. Res.* **2011**, *61*, 303–308. [CrossRef]

39. Santos, M.; Del Río, L.; Benavente, J. GIS–based approach to the assessment of coastal vulnerability to storms. Case study in the Bay of Cádiz (Andalusia, Spain). *J. Coast. Res.* **2013**, *65*, 826–831. [CrossRef]

40. Armaroli, C.; Duo, E. Validation of the Coastal storm Risk Assessment Framework along the Emilia–Romagna coast. *Coast. Eng.* **2018**, *134*, 159–167. [CrossRef]

41. Van Dongeren, A.; Ciavola, P.; Martinez, G.; Viavattene, C.; Bogaard, T.; Ferreira, O.; McCall, R. Introduction to RISC–KIT: Resilience–increasing strategies for coasts. *Coast. Eng.* **2018**, *134*, 2–9. [CrossRef]

42. Viavattene, C.; Jiménez, J.A.; Ferreira, O.; Priest, S.; Owen, D.; McCall, R. Selecting coastal hotspots to storm impacts at the regional scale: A Coastal Risk Assessment Framework. *Coast. Eng.* **2018**, *134*, 33–47. [CrossRef]

43. Stelljes, N.; Martinez, G.; McGlade, K. Introduction to the RISC–KIT web based management guide for DRR in European coastal zones. *Coast. Eng.* **2018**, *134*, 73–80. [CrossRef]

44. Prampolini, M.; Gauci, C.; Micallef, A.S.; Selmi, L.; Vandelli, V.; Soldati, M. Geomorphology of the north–eastern coast of Gozo (Malta, Mediterranean Sea). *J. Maps* **2018**, *14*, 402–410. [CrossRef]

45. Micallef, S.; Micallef, A.; Galdies, C. Application of the Coastal Hazard Wheel to assess erosion on the Maltese coast. *Ocean Coast. Manag.* **2018**, *156*, 209–222. [CrossRef]

46. Mantovani, M.; Devoto, S.; Forte, E.; Mocnik, A.; Pasuto, A.; Piacentini, D.; Soldati, M. A multidisciplinary approach for rock spreading and block sliding investigation in the north–western coast of Malta. *Landslides* **2013**, *10*, 611–622. [CrossRef]

47. Mantovani, M.; Devoto, S.; Piacentini, D.; Prampolini, M.; Soldati, M.; Pasuto, A. Advanced SAR interferometric analysis to support geomorphological interpretation of slow–moving coastal landslides (Malta Mediterranean Sea). *Remote Sens.* **2016**, *8*, 443. [CrossRef]

48. Soldati, M.; Devoto, S.; Foglini, F.; Forte, E.; Mantovani, M.; Pasuto, A.; Piacentini, D.; Prampolini, M. An integrated approach for landslide hazard assessment on the NW coast of Malta. In Proceedings of the International Conference: Georisks in the Mediterranean and Their Mitigation, Valletta, Malta, 20–21 July 2015; Galea, P., Borg, R.P., Farrugia, D., Agius, M.R., D'Amico, S., Torpiano, A., Bonello, M., Eds.; Gutemberg Press Ltd.: Tarxien, Malta, 2015; pp. 160–167.

49. Piacentini, D.; Devoto, S.; Mantovani, M.; Pasuto, A.; Prampolini, M.; Soldati, M. Landslide susceptibility modeling assisted by Persistent Scatterers Interferometry (PSI): An example from the northwestern coast of Malta. *Nat. Hazards* **2015**, *78*, 681–697. [CrossRef]

50. Mantovani, M.; Piacentini, D.; Devoto, S.; Prampolini, M.; Pasuto, A.; Soldati, M. Landslide susceptibility analysis exploiting Persistent Scatterers data in the northern coast of Malta. In Proceedings of the International Conference Analysis and Management of Changing Risks for Natural Hazards, Padua, Italy, 18–19 November 2014; pp. 1–7.

51. Viavattene, C.; Jimenez, J.A.; Owen, D.; Priest, S.; Parker, D.; Micou, A.P.; Ly, S. Coastal Risk Assessment Framework Guidance Document. Deliverable No: D.2.3—Coastal Risk Assessment Framework Tool, Risc-Kit Project (G.A. No. 603458). 2015. Available online: http://www.risckit.eu/np4/file/23/RISC_KIT_D2.3_CRAF_Guidance.pdf (accessed on 15 October 2018).

52. Ferreira, O.; Viavattene, C.; Jiménez, J.; Bole, A.; Plomaritis, T.; Costas, S.; Smets, S. CRAF Phase 1, A framework to identify coastal hotspots to storm impacts. Risk Evaluation & Assessment. In Proceedings of the FLOODrisk 2016—3rd European Conference on Flood Risk Management, Lyon, France, 17–21 October 2016. [CrossRef]

53. Aucelli, P.P.; Di Paola, G.; Rizzo, A.; Rosskopf, C.M. Present day and future scenarios of coastal erosion and flooding processes along the Italian Adriatic coast: The case of Molise region. *Envrion. Earth Sci.* **2018**, *77*, 371. [CrossRef]

54. Ballesteros, C.; Jiménez, J.A.; Viavattene, C. A multi–component flood risk assessment in the Maresme coast (NW Mediterranean). *Natl. Hazards* **2018**, *90*, 265–292. [CrossRef]

55. Papathoma–Köhle, M.; Cristofari, G.; Wenk, M.; Fuchs, S. The importance of indicator weights for vulnerability indices and implications for decision making in disaster management. *Int. J. Disaster Risk. Reduct.* **2019**, *36*, 101103. [CrossRef]

56. Birkmann, J. Indicators and criteria for measuring vulnerability: Theoretical bases and requirements. In *Measuring Vulnerability to Natural Hazards: Towards Disaster Resilient Societies*; Birkmann, J., Ed.; UNU Press: Tokyo, Japan, 2006.

57. Fuchs, S.; Frazier, T.; Siebeneck, L. *Vulnerability and Resilience to Natural Hazards*; Fuchs, S., Thaler, T., Eds.; Cambridge University Press: Cambridge, UK, 2018.

58. Nardo, M.; Saisana, M.; Saltelli, A.; Tarantola, S. *Tools for Composite Indicators Building*; European Commission: Brussels, Belgium, 2005.

59. Soldati, M.; Maquaire, O.; Zezere, J.L.; Piacentini, D.; Lissak, C. Coastline at risk: Methods for multi–hazard assessment. *J. Coast. Res.* **2011**, *61*, 335–339. [CrossRef]

60. Foglini, F.; Prampolini, M.; Micallef, A.; Angeletti, L.; Vandelli, V.; Deidun, A.; Soldati, M.; Taviani, M. Late Quaternary Coastal Landscape Morphology and Evolution of the Maltese Islands (Mediterranean Sea) Reconstructed from High–Resolution Seafloor Data. In *Geology and Archaeology: Submerged Landscapes of the Continental Shelf*; Har, J., Bailey, G., Lüth, L., Eds.; Geological Society, Special Publication: London, UK, 2016; Volume 411, pp. 77–95.

61. Soldati, M.; Barrows, T.T.; Prampolini, M.; Fifield, K.L. Cosmogenic exposure dating constraints for coastal landslide evolution on the Island of Malta (Mediterranean Sea). *J. Coast. Conserv.* **2018**, *22*, 831–844. [CrossRef]

62. Oil Exploration Directorate. *Geological Map of the Maltese Islands*; Office of the Prime Minister: Valletta, Malta, 1993.

63. Magri, O. A geological and geomorphological review of the Maltese Islands with special reference to the coastal zone. *Territoris* **2006**, *6*, 7–26.

64. Micallef, A.; Foglini, F.; Le Bas, T.; Angeletti, L.; Maselli, V.; Pasuto, A.; Taviani, M. The submerged palaeolandscape of the Maltese Islands: Morphology evolution and relation to Quaternary environmental change. *Mar. Geol.* **2013**, *335*, 129–147. [CrossRef]

65. Paskoff, R.; Sanlaville, P. Observations geomorphologiques sur le cotes de l'Archipel Maltaise. *Z. Geomorphol.* **1978**, *22*, 310–328.

66. Said, G.; Schembri, J. Malta. In *Encyclopaedia of the World's Coastal Landforms*; Bird, E., Ed.; Springer: Dordrecht, The Netherlands, 2010; pp. 751–759.

67. Galve, J.P.; Tonelli, C.; Gutiérrez, F.; Lugli, S.; Vescogni, A.; Soldati, M. New insights into the genesis of the Miocene collapse structures of the island of Gozo (Malta, central Mediterranean Sea). *J. Geol. Soc.* **2015**, *172*, 336–348. [CrossRef]

68. Mottershead, D.; Bray, M.; Soar, P.; Farres, P.J. Extreme wave events in the central Mediterranean: Geomorphic evidence of tsunami on the Maltese Islands. *Z. Geomorphol.* **2014**, *58*, 385–411. [CrossRef]

69. Soldati, M.; Tonelli, C.; Galve, J.P. Geomorphological evolution of palaeosinkhole features in the Maltese archipelago (Mediterranean Sea). *Geogr. Fis. Din. Quat.* **2013**, *36*, 189–198. [CrossRef]

70. Coratza, P.; Galve, J.P.; Soldati, M.; Tonelli, C. Recognition and assessment of sinkholes as geosites: Lessons from the Island of Gozo (Malta). *Quaest. Geogr.* **2012**, *31*, 22–35. [CrossRef]

71. Coratza, P.; Gauci, R.; Schembri, J.A.; Soldati, M.; Tonelli, C. Bridging Natural and Cultural Values of Sites with Outstanding Scenery: Evidence from Gozo, Maltese Islands. *Geoheritage* **2016**, *8*, 91–103. [CrossRef]

72. Cappadonia, C.; Coratza, P.; Agnesi, V.; Soldati, M. Malta and Sicily Joined by Geoheritage Enhancement and Geotourism within the Framework of Land Management and Development. *Geosci. J.* **2018**, *8*, 253. [CrossRef]

73. Satariano, B.; Gauci, R. Landform loss and its effect on health and well–being: The collapse of the Azure Window (Gozo) and the resultant reactions of the media and the Maltese community. In *Landscapes and Landforms of the Maltese Islands*; Gauci, R., Schembri, J.A., Eds.; Springer: Cham, Switzerland, 2019; Volume 14, pp. 289–303.

74. Baldassini, N.; Di Stefano, A. Stratigraphic features of the Maltese Archipelago: A synthesis. *Nat. Hazards* **2017**, *86*, 203–231. [CrossRef]

75. Gauci, R.; Schembri, J.A. (Eds.) *Landscapes and Landforms of the Maltese Islands*; Springer: Cham, Switzerland, 2019.

76. Pedley, M.; Clarke, M.H. *Limestone Isles in a Crystal Sea: The Geology of the Maltese Islands*; Publishers Enterprises Group: San Gwann, Malta, 2002.

77. Biolchi, S.; Furlani, S.; Devoto, S.; Gauci, R.; Castaldini, D.; Soldati, M. Geomorphological identification, classification and spatial distribution of coastal landforms of Malta (Mediterranean Sea). *J. Maps* **2016**, *12*, 87–99. [CrossRef]

78. Furlani, S.; Antonioli, F.; Gambin, T.; Gauci, R.; Ninfo, A.; Zavagno, E.; Micallef, A.; Cucchi, F. Marine notches in the Maltese islands (central Mediterranean Sea). *Quat. Int.* **2017**, *439*, 158–168. [CrossRef]

79. Alexander, D. A review of the physical geography of Malta and its significance for tectonic geomorphology. *Quat. Sci. Rev.* **1988**, *7*, 41–53. [CrossRef]

80. Soldati, M.; Devoto, S.; Prampolini, M.; Pasuto, A. The spectacular landslide–controlled landscape of the northwestern coast of Malta. In *Landscapes and Landforms of the Maltese Islands*; Gauci, R., Schembri, J.A., Eds.; Springer: Cham, Switzerland, 2019; Volume 14, pp. 167–178.

81. Devoto, S.; Biolchi, S.; Bruschi, V.M.; Furlani, S.; Mantovani, M.; Piacentini, D.; Soldati, M. Geomorphological map of the NW coast of the Island of Malta (Mediterranean Sea). *J. Maps* **2012**, *8*, 33–40. [CrossRef]

82. Devoto, S.; Biolchi, S.; Bruschi, V.M.; González, D.A.; Mantovani, M.; Pasuto, A.; Soldati, M. Landslides Along the North–West Coast of the Island of Malta. In *Landslide Science and Practice: Landslide Inventory and Susceptibility and Hazard Zoning*; Margottini, C., Canuti, P., Sassa, K., Eds.; Springer: Berlin, Germany, 2013; Volume 1, pp. 57–63.

83. Devoto, S.; Forte, E.; Mantovani, M.; Mocnik, A.; Pasuto, A.; Piacentini, D.; Soldati, M. Integrated Monitoring of Lateral Spreading Phenomena Along the North–West Coast of the Island of Malta. In *Landslide Science and Practice: Early Warning, Instrumentation and Monitoring*; Margottini, C., Canuti, P., Sassa, K., Eds.; Springer: Berlin, Germany, 2013; Volume 2, pp. 235–241.

84. Magri, O.; Mantovani, M.; Pasuto, A.; Soldati, M. Geomorphological investigation and monitoring of lateral spreading along the north–west coast of Malta. *Geogr. Fis. Din. Quat.* **2008**, *31*, 171–180.

85. Pasuto, A.; Soldati, M. Lateral Spreading. In *Treatise on Geomorphology*; Shroder, J.F., Ed.; Academic Press: San Diego, CA, USA, 2013; Volume 7, pp. 239–248.

86. Dykes, A.P. Mass movements and conservation management in Malta. *J. Environ. Manag.* **2002**, *66*, 77–89. [CrossRef]

87. Prampolini, M.; Foglini, F.; Biolchi, S.; Devoto, S.; Angelini, S.; Soldati, M. Geomorphological mapping of terrestrial and marine areas, northern Malta and comino (Central Mediterranean sea). *J. Maps* **2017**, *13*, 457–469. [CrossRef]

88. Selmi, L.; Coratza, P.; Gauci, R.; Soldati, M. Geoheritage as a Tool for Environmental Management: A Case Study in Northern Malta (Central Mediterranean Sea). *Resources* **2019**, *8*, 168. [CrossRef]

89. Central Bank of Malta. The Evolution of Malta's Tourism Product over Recent Years. Available online: https://www.centralbankmalta.org/file.aspx?f=72256 (accessed on 17 October 2019).

90. World Travel and Tourism Council. *Travel & Tourism Economic Impact 2018 Malta*; World Travel and Tourism Council: London, UK, 2018.

91. Chaperon, S.; Bramwell, B. Dependency and agency in peripheral tourism development. *Ann. Tour. Res.* **2013**, *40*, 132–154. [CrossRef]

92. Ebejer, J.; Mangion, M.J.; Bingül, M.B.; Kwiatkowska, D. Rural Landscape and Tourism: A Proposed Policy for Sustainable Tourism in Gozo. In Proceedings of the 7th LE: NOTRE Landscape Forum 2018, Gozo, Malta, 20–24 March 2018; Available online: https://www.um.edu.mt/library/oar/bitstream/123456789/40147/1/Article%20on%20Gozo%20tourism%20for%20LeNOTRE%20Landscape%20FINAL%20DRAFT%20for%20OAR.pdf (accessed on 22 November 2019).

93. Malta Tourism Authority. Tourism in Malta—Facts and Figures. 2017. Available online: https://www.mta.com.mt/en/file.aspx?f=32328 (accessed on 16 October 2019).

94. National Statistics Office. *Regional Statistics—Malta*, 2019 ed.; National Statistics Office: Valletta, Malta, 2019.

95. Salman, A.; Lombardo, S.; Doody, P. *Living with Coastal Erosion in Europe: Sediment and Space for Sustainability*; Technical Report; EUCC: Warnemünde, Germany, 2004.

96. Geofabrik. Available online: http://download.geofabrik.de/europe/malta.htm (accessed on 3 July 2019).

97. Malta Inspire Geoportal. Available online: https://msdi.data.gov.mt/geoportal.html (accessed on 3 July 2019).

98. National Statistics Office. Social Protection—Reference Years 2012–2016. Malta, 2019. Available online: https://nso.gov.mt/en/publicatons/Publications_by_Unit/Documents/A2_Public_Finance/Social%20Protection%202016.pdf (accessed on 3 July 2019).

99. Beccari, B. A comparative analysis of disaster risk, vulnerability and resilience composite indicators. *PLoS Curr.* **2016**, *8*. [CrossRef]

100. Del Río, L.; Gracia, F.J. Erosion risk assessment of active coastal cliffs in temperate environments. *Geomorphology* **2009**, *112*, 82–95. [CrossRef]

101. Di Paola, G.; Aucelli, P.P.C.; Benassai, G.; Iglesias, J.; Rodríguez, G.; Rosskopf, C.M. The assessment of the coastal vulnerability and exposure degree of Gran Canaria Island (Spain) with a focus on the coastal risk of Las Canteras Beach in Las Palmas de Gran Canaria. *J. Coast. Conserv.* **2018**, *22*, 1001–1014. [CrossRef]

102. Mattei, G.; Rizzo, A.; Anfuso, G.; Aucelli, P.P.C.; Gracia, F.J. A tool for evaluating the archaeological heritage vulnerability to coastal processes: The case study of Naples Gulf (southern Italy). *Ocean Coast. Manag.* **2019**, *179*, 104876. [CrossRef]

103. Rizzo, A.; Aucelli, P.P.C.; Gracia, F.J.; Anfuso, G. A novelty coastal susceptibility assessment method: Application to Valdelagrana area (SW Spain). *J. Coast. Conserv.* **2018**, *22*, 973–987. [CrossRef]

104. Giorgi, F. Climate change hot-spots. *Geophys. Res. Lett.* **2006**, *33*. [CrossRef]

105. Giorgi, F.; Lionello, P. Climate change projections for the Mediterranean region. *Glob. Planet. Chang.* **2008**, *63*, 90–104. [CrossRef]

106. Hov, Ø.; Cubasch, U.; Fischer, E.; Höppe, P.; Iversen, T.; Gunnar Kvamstø, N.; Zbigniew, W.K.; Rezacova, D.; Rios, D.; Santos, F.D.; et al. Extreme Weather Events in Europe: Preparing for Climate Change Adaptation. Norwegian Meteorological Institute: Oslo, Norway, 2013.

107. Nicholls, R.J. *Impacts of and Responses to Sea-Level Rise. Understanding Sea-Level Rise and Variability*; Church, J.A., Woodworth, P.L., Aarup, T., Wilson, W.W., Eds.; Wiley–Blackwell: Hoboken, NJ, USA, 2010; pp. 17–51.

108. Lambeck, K.; Antonioli, F.; Anzidei, M.; Ferranti, L.; Leoni, G.; Scicchitano, G.; Silenzi, S. Sea level change along the Italian coast during the Holocene and projections for the future. *Quat. Int.* **2011**, *232*, 250–257. [CrossRef]

109. Aucelli, P.P.C.; Cinque, A.; Mattei, G.; Pappone, G. Historical sea level changes and effects on the coasts of Sorrento Peninsula (Gulf of Naples): New constrains from recent geoarchaeological investigations. *Palaeogeogr. Palaeoclimatol. Palaeoecol.* **2016**, *463*, 112–125. [CrossRef]

110. Tsimplis, M.N.; Calafat, F.M.; Marcos, M.; Jordá, G.; Gomis, D.; Fenoglio-Marc, L.; Struglia, S.; Josey, S.; Chambers, D.P. The effect of the NAO on sea level and on mass changes in the Mediterranean Sea. *J. Geophys. Res. Oceans* **2013**, *118*, 944–952. [CrossRef]

111. Anzidei, M.; Lambeck, K.; Antonioli, F.; Furlani, S.; Mastronuzzi, G.; Serpelloni, E.; Vannucci, G. Coastal structure, sea–level changes and vertical motion of the land in the Mediterranean. *Geol. Soc.* **2014**, *388*, 453–479. [CrossRef]

112. Galassi, G.; Spada, G. Sea-level rise in the Mediterranean Sea by 2050: Roles of terrestrial ice melt, steric effects and glacial isostatic adjustment. *Glob. Planet. Chang.* **2014**, *123*, 55–66. [CrossRef]

113. Antonioli, F.; Anzidei, M.; Amorosi, A.; Presti, V.L.; Mastronuzzi, G.; Deiana, G.; Marsico, A. Sea–level rise and potential drowning of the Italian coastal plains: Flooding risk scenarios for 2100. *Quat. Sci. Rev.* **2017**, *158*, 29–43. [CrossRef]

114. Antonioli, F.; Defalco, G.; Moretti, L.; Anzidei, M.; Bonaldo, D.; Carniel, S.; Leoni, G.; Furlani, S.; Presti, V.I.; Mastronuzzi, G.; et al. Relative sea level rise and potential flooding risk for 2100 on 15 coastal plains of the Mediterranean Sea. *Geophys. Res. Abstr.* **2019**, *21*, 5274.

115. Aucelli, P.P.C.; Di Paola, G.; Incontri, P.; Rizzo, A.; Vilardo, G.; Benassai, G.; Buonocuore, B.; Pappone, G. Coastal inundation risk assessment due to subsidence and sea level rise in a Mediterranean alluvial plain (Volturno coastal plain–southern Italy). *Estuar. Coast. Shelf Sci.* **2017**, *198*, 597–609. [CrossRef]

116. Di Paola, G.; Alberico, I.; Aucelli, P.P.C.; Matano, F.; Rizzo, A.; Vilardo, G. Coastal subsidence detected by Synthetic Aperture Radar interferometry and its effects coupled with future sea-level rise: The case of the Sele Plain (Southern Italy). *J. Flood Risk Manag.* **2018**, *11*, 191–206. [CrossRef]

117. Nicholls, R.J.; Cazenave, A. Sea–level rise and its impact on coastal zones. *Science* **2010**, *328*, 1517–1520. [CrossRef] [PubMed]

118. Revell, D.L.; Battalio, R.; Spear, B.; Ruggiero, P.; Vandever, J. A methodology for predicting future coastal hazards due to SLR on the California Coast. *Clim. Chang.* **2011**, *109*, 251–276. [CrossRef]

119. Kulp, S.A.; Strauss, B.H. New elevation data triple estimates of global vulnerability to sea–level rise and coastal flooding. *Nat. Commun.* **2019**, *10*, 1–12.

Article

Barrier Islands Resilience to Extreme Events: Do Earthquake and Tsunami Play a Role?

Ella Meilianda [1,2,*], Franck Lavigne [3,4], Biswajeet Pradhan [5,6,7], Patrick Wassmer [4], Darusman Darusman [8] and Marjolein Dohmen-Janssen [9]

[1] Tsunami and Disaster Mitigation Research Center (TDMRC), Universitas Syiah Kuala, Banda Aceh 23233, Indonesia

[2] Civil Engineering Department, Engineering Faculty, Universitas Syiah Kuala, Banda Aceh 23111, Indonesia

[3] Geography Department, University of Paris 1 Panthéon-Sorbonne, 75005 Paris, France; franck.lavigne@univ-paris1.fr

[4] Laboratoire de Geographie Physique, UMR 8591 CNRS, 1 Place A. Briand, 92190 Meudon, France; patrick.wassmer@lgp.cnrs.fr

[5] Centre for Advanced Modelling and Geospatial Information Systems (CAMGIS), University of Technology Sydney, Ultimo, NSW 2007, Australia; biswajeet.pradhan@uts.edu.au

[6] Center of Excellence for Climate Change Research, King Abdulaziz University, P.O. Box 80234, Jeddah 21589, Saudi Arabia

[7] Earth Observation Center, Institute of Climate Change, Universiti Kebangsaan Malaysia, Bangi 43600 UKM, Selangor, Malaysia

[8] Post-Graduate School, Universitas Syiah Kuala, Banda Aceh 23111, Indonesia; darusman@unsyiah.ac.id

[9] Water Engineering & Management (WEM), Faculty of Engineering & Technology, University of Twente, 7522 NB Enschede, The Netherlands; c.m.dohmen-janssen@utwente.nl

[*] Correspondence: ella_meilianda@unsyiah.ac.id

Citation: Meilianda, E.; Lavigne, F.; Pradhan, B.; Wassmer, P.; Darusman, D.; Dohmen-Janssen, M. Barrier Islands Resilience to Extreme Events: Do Earthquake and Tsunami Play a Role? *Water* **2021**, *13*, 178. https://doi.org/10.3390/w13020178

Received: 30 October 2020
Accepted: 8 January 2021
Published: 13 January 2021

Publisher's Note: MDPI stays neutral with regard to jurisdictional claims in published maps and institutional affiliations.

Abstract: Barrier islands are indicators of coastal resilience. Previous studies have proven that barrier islands are surprisingly resilient to extreme storm events. At present, little is known about barrier systems' resilience to seismic events triggering tsunamis, co-seismic subsidence, and liquefaction. The objective of this study is, therefore, to investigate the morphological resilience of the barrier islands in responding to those secondary effects of seismic activity of the Sumatra–Andaman subduction zone and the Great Sumatran Fault system. Spatial analysis in Geographical Information Systems (GIS) was utilized to detect shoreline changes from the multi-source datasets of centennial time scale, including old topographic maps and satellite images from 1898 until 2017. Additionally, the earthquake and tsunami records and established conceptual models of storm effects to barrier systems, are corroborated to support possible forcing factors analysis. Two selected coastal sections possess different geomorphic settings are investigated: (1) Lambadeuk, the coast overlying the Sumatran Fault system, (2) Kuala Gigieng, located in between two segments of the Sumatran Fault System. Seven consecutive pairs of comparable old topographic maps and satellite images reveal remarkable morphological changes in the form of breaching, landward migrating, sinking, and complete disappearing in different periods of observation. While semi-protected embayed Lambadeuk is not resilient to repeated co-seismic land subsidence, the wave-dominated Kuala Gigieng coast is not resilient to the combination of tsunami and liquefaction events. The mega-tsunami triggered by the 2004 earthquake led to irreversible changes in the barrier islands on both coasts.

Keywords: Sumatra; barrier island; earthquake; tsunami; land subsidence; liquefaction; morphology resilience; GIS

1. Introduction

Major earthquakes that have occurred in history have caused fatalities and losses beyond the direct earthquake shaking. Looking at historical losses from 1900 onward,

183

around 1% of damaging earthquakes in history have caused about 93% of deaths worldwide. Within those events, secondary effects, such as a tsunami, liquefaction, fire, and the impact of nuclear power plants, have caused 40% of economic losses and deaths [1]. Coastal areas situated at the peripheral of tectonic subduction zones, such as the Sunda Trench, Peru-Chile Trench, and Japan Trench, are subject to multiple threats of co-seismic and post-seismic secondary effects. The triggered massive landslides, liquefaction, and tsunami have caused severe casualties and damage to the affected areas. The Central Sulawesi tectonic earthquake in September 2018, for instance, has triggered multiple secondary effects, including localized tsunami, landslides, and liquefaction [2]. The Great Eastern Japan Earthquake in April 2011 triggered the tsunami, liquefaction, fire, and the impact of a nuclear power plant [1]. The Great Sumatran Earthquake in December 2004 triggered both mega-tsunamis and land subsidence, all of which have caused significant fatalities and losses and required long-term recovery.

Eighty percent of the world's coasts are rocky [3], characterized by a sedimentary-deficit, and a complex morphology [4]. On the basis of satellite inventory, Stutz and Pilkey [5] reveal that 20,783 km of shoreline are occupied by 2149 barrier islands worldwide, which is 37% longer than formerly thought. Indonesia is an archipelagic country with an estimated 91,363.65 km of coastline [6], making it the second longest coastline in the world after Canada [4]. The Indonesian archipelago has 44 barrier islands, which equals 2% of the global figure. Their distribution is strongly related to sea-level history in addition to the influence of tectonic settings. Administratively, 80% of districts and municipalities in Indonesia are situated at the coastal areas, making the coastline the multi-purpose zone in the daily life and socio-economic development of the archipelagic nation.

The shoreline along the Sumatra island of Indonesia mostly consists of shore-parallel lagoonal ecosystems separated by chenier-type barrier islands, with crest height on average no more than 2 m [7]. In the temperate zone, such as in the United States, barrier islands are under tremendous pressure of rising sea level and increasing storminess. On the other hand, the barrier islands and beach ridges located in a tectonically active coastal region, such as the entire length of the west and north Sumatra of Indonesia, are equally under tremendous pressure of the potential tsunami, co-seismic and post-seismic land level changing, and liquefaction as the secondary effects of the tectonic fault and subduction earthquakes.

Studies about the influence of the secondary effects of tectonic activities on the barrier island and spit morphological development are currently relatively limited. This study aims to investigate the morphological resilience of the barrier islands in responding to tsunamis, co-seismic subsidence, and liquefaction. Spatial analysis was utilized to detect shoreline changes from the multi-source datasets of the multi-decadal time scale, including old topographic maps and satellite images from 1898 until 2017. As the next step, we identified the history of major or moderate earthquakes that occurred in each consecutive period. Subsequently, we corroborated those earthquake events with the reported impact of the earthquakes on the coastal areas. We also utilized the established conceptual models of storm effects to barrier systems to interpret possible controlling factors causing the remarkable morphological changes observed at two selected coastal sections. The results will be substantial in building a conceptual foundation for interplaying seismic-related forcing factors with the littoral transport regime that determines the morphological resilience of a coastal area.

1.1. Impact of Earthquakes on Coastal Systems

Investigation of a massive earthquake event has always been appealing for scientists. It is the opportunity to improve the understanding of various factors resulting in casualties and damage. Short-term responses of coastal morphology to tsunamis have been widely discussed in the previous studies through various approaches. Some works have combined models of tsunami wave propagation and inundation, and spatial analysis of the damage in the tsunami-affected area [8–10]. Others focused on the geomorphological [11–14], or ecological [10,15] impacts of the tsunami. Additionally, a few tens of meters of land subsidence were also observed as the immediate response of coastal areas to seismic activities was observed early after the occurrence of the megathrust earthquake of December 2004 along the north and west coast of Sumatra Island [16–19]. An extended period investigation to the post-tsunami coastal recovery in Aceh has also been conducted, including further monitoring on changes of shoreline position and land-level changes [20–22], of land use [23], and the vulnerability and living condition of the coastal inhabitants [8,24].

Liquefaction has yet to be discussed thoroughly, concerning the impact of the megathrust 2004 Sumatra earthquake in Banda Aceh. Nevertheless, some indication of liquefaction indeed observed in Port Blair, the capital city of the Andaman and Nicobar Island of India, where various types of buildings, particularly of the typical reinforced concrete ones, have experienced settlement [25]. Such building failures by settlement were also observed and reported in Banda Aceh [26]. Recently, the soil mechanism of the liquefaction potential in Banda Aceh city after the earthquake was evaluated by [27] using the semi-empirical Idriss Method to quantify the liquefaction potential. The result suggests that most of the highly built subdistricts in Banda Aceh are prone to liquefaction. It is noteworthy that the liquefaction impact is not only triggered by the shaking from the megathrust Sumatra–Andaman earthquakes but also potentially by the activation of the Great Sumatran Fault System [28].

Following the megathrust Sumatran earthquake and tsunami of 26 December 2004, a series of geodetic monitoring campaigns were conducted between 2005 and 2015, to monitor the development of the land-level changes of the west coast of Aceh since the 2004 tsunami [21]. The results showed that after experiencing abrupt co-seismic subsidence, the beach experienced uplift with a rate of 27 mm/year since late 2005. This number is an order of magnitude higher than the rate of eustatic sea-level rise, which is around 4–12 mm/year at the Indonesian waters [29]. The results demonstrate a possible relationship between the development of the new frontier beach ridge, the immediate co-seismic land subsidence, and the probable post-seismic rebound (uplift) associated with the viscoelastic mantle relaxation a few years following the mega earthquake.

Overall, the previous studies suggest that the secondary effects of seismic activity, either associated with the off-shore megathrust subduction or mainland tectonic faults activity, play a crucial role in altering the morphological development of the tectonically active coastal area.

1.2. Barrier Islands and Spits Morphological Resilience

Several studies of large barrier systems, including barrier island or beach ridges and sand spits, have been well-studied, primarily when associated with extreme storm events [30–35]. They have shown to constitute a valuable archive of coastal evolution and their morphology, and internal structures contain information on both past relative sea levels and past storminess activity. Barrier islands and sand spits are considered exemplars of coastal resilience [30,31] and critically essential ecosystems to protect the low-lying area behind them [36]. Under the influence of storms, large barrier systems, such as barrier island and beach ridges, are inherently resilient landforms as long as they can internally recycle sediment to maintain overall landform integrity [37], the rate of sea-level rise is not excessive, and there is no sediment deficit [38]. Nott et al. [33] suggest that the building of beach ridges is being slowed down by a decrease in sediment supply to the coast due to less frequent river floods during periods of reduced storminess. While during a storm event barrier islands may experience breaching due to the intensive funneling of the overwash flows into specific throats [39], a calmer wave regime promotes littoral drift, allowing sand spits to grow extensively [30].

Arguably, a tsunami event may contribute to the geomorphological imprint in the evolution of a tectonically active coast in a way similar to a major storm event. Despite different sources and mechanisms, the effect of the destruction caused by a tsunami may have been analog to that of a storm surge, in particular along the coastline, where barrier islands are the first line of natural coastal protection. Studies of short-term shoreline changes of barrier islands at Banda Aceh coast of Sumatra Island, Indonesia after the 2004 tsunami [13], and those at the Dauphin Island, US after the Hurricane Katrina in 2005 [40] are two comparable studies which demonstrated similar variability of shoreline changes caused by extreme events.

Despite the great diversity of storm mechanisms that strike coastal areas, we consider that the impact of the surging waves to barrier islands is equivalent to that of tsunamis. Sallenger [35] identifies four types of storm regimes and how barrier islands respond to those different forcing magnitudes, which later were re-described in detail by [34]. The four regimes are briefly described in [32] which consist of the "Swash regime", "Collision regime", "Overwash regime", and "Inundation regime," which involve the morphological effects of migration, escarpment, breaching, and submergence of the barrier islands, respectively. The case of sinking effects due to co-seismic subsidence or liquefaction is relatively easy to observe by identifying the narrowing subaerial parts by comparing a pair of maps or satellite images of consecutive years. However, it is almost impossible to determine the cause of the sinking by solely relying on the spatial analysis from the multi-temporal maps and satellite images. Therefore, we also corroborate the results from previous studies and the other sources of information and accountable reports to support the analysis.

Successful coastal management that includes mitigation of possible negative impacts must be based on an understanding of these patterns of change as natural responses to high-intensity events [41]. It is also essential to understand the coastal geomorphic setting and the geological boundaries before attempting to model the large-scale behavior of these types of coastal systems [42]. Herein, historical data inevitably play a major role in identifying any remarkable, even more so, the irreversible changes in the long-term past.

2. Study Area

We investigate morphological changes of the seaward-most barrier islands and spits along the coastline of the north tip of Sumatra Island, which is situated between 05°16′15″ N and 05°36′16″ N, and between 95°16′15″ E and 95°22′35″ E (Figure 1a,b). The low-lying coastal area behind them is where the capital city of Aceh Province, Banda Aceh situated in the central part, and surrounded by the Aceh Besar district occupying ca. 125 km^2 northern valley of the Barisan mountain range at the north tip of Sumatra Island, Indonesia. The Barisan mountain range is formed as the backbone of the leading-edge Sumatra Island, parallel to the Great Sumatran Fault running parallel with the Sumatra-Andaman subduction zone, or also known as Sunda Trench (Figure 1a). The shoreline stretches ca. 25 km connecting Ujong Pancu headland in the southwest and the Ujong Batee headland in the northeast (Figure 1c). The brackish back-barrier wetland ecosystem, aquacultures, and lowland coastal villages occupy the area of ca. 4 km width from the coastline with elevations are varying of −0.5 m to +2.0 m from the mean sea level. The entire coastal area was severely devastated by the megathrust earthquake of M 9.0, followed by the gigantic tsunami event on 26 December 2004.

The tsunami disaster caused severe casualties and damaged properties, as well as erosion at the coastline, altering the functionality of the ecosystems and the livelihoods of coastal communities [44]. In Banda Aceh, the shoreline retreated as far as ca. 200 m inland, and erosion rates of 30 m^3/m on average (up to 80 m^3/m locally) were calculated [11]. Six months after the tsunami, the shoreline experienced about 15% further retreats from the initial erosion by the tsunami [13].

Major infrastructure, e.g., port basin, coastal revetments, and coastal roads, collapsed or were destroyed, whereas the coastal environment was also profoundly altered [8]. A comparison of a coastal revetment of before and after the 2004's tsunami at Lambadeuk is depicted in Figure 2a,b, respectively. The remnants of the same revetment were stranded offshore after the disaster event (Figure 2b), suggesting the occurrence of local land subsidence. Several preliminary studies suggest that the local land subsidence at Lambadeuk and Ulee Lheue (Figure 1c) was less than 50 cm [16,18]. On the other hand, the barrier islands which used to protect the built areas behind the Kuala Gigieng coast were breached at their weakest sections by the tsunami waves. Figure 2c,d exemplify one of the breaching sections of the barrier island at Kuala Gigieng on the fourth day and six months after the tsunami, respectively.

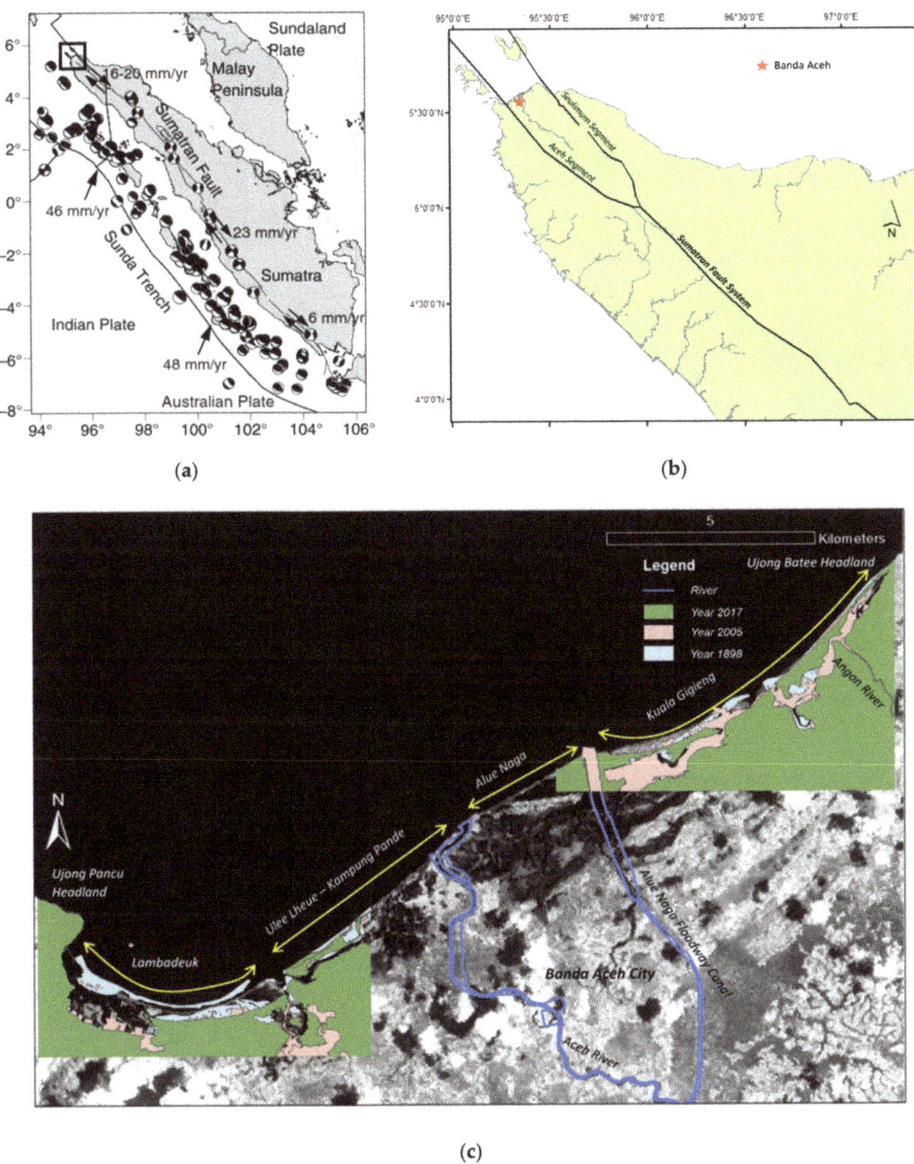

(a)

(b)

(c)

Figure 1. Banda Aceh coast, at the northern tip of Sumatra Island, Indonesia: (**a**) Tectonic settings of Sumatra, compiled from various sources by Hurukawa et al. [43], showing the configuration of the Sumatran fault, slip rates, the plot of Global Centroid Moment tensor (CMT) solutions of shallow earthquakes ($M_w \geq 6.0$ and depth ≤ 60 km) between 1976 and 2012; (**b**) Segments of the Great Sumatran Fault System at the northern Sumatra consists of two bifurcated fault segments, i.e., Aceh Segment and Seulimum Segment. The rectangle shows the investigated coastal area in front of Banda Aceh, the capital of Aceh Province, which is situated between the two fault segments; (**c**) The two investigated coastal sections in this study are Lambadeuk at the southwest and Kuala Gigieng at the northeast since they are the most dynamic and there is less interference by hard-structure coastal protection.

Figure 2. Indication of land subsidence at Lambadeuk, southwest of Banda Aceh coast: (**a**) The revetment located at the upper shoreface to protect the coast from further erosion nine months before the tsunami hit the coast on 26 December 2004; (**b**) The remnant of a seawall post-tsunami. The yellow arrows in both pictures show the same revetment; (**c**) The breaching of barrier island at Kuala Gigieng four days after the tsunami; (**d**) The development of sand spit growth six months after the tsunami, to reconnect the breached barrier islands at Kuala Gigieng.

At the northern Sumatra, the fault system splits into two active segments, i.e., Aceh Segment and Seulimum Segment (Figures 1b and 3a). Natawidjaja and Triyoso [28] found that currently, those segments possess a seismic gap of 325 km and 70 km long in the last 100 years, respectively, and considered them as an alarming hazard potential in the future. The highly populated Banda Aceh city is situated between these two segments (Figure 3a). The coast is facing the Andaman Sea and is semi-embayed by the entraining forearc small islands at the north off-shore from the rough, energetic waves of the Indian Ocean to the west. Lambadeuk exemplifies a relatively broad lowland coastal system overlying a major segment of the active tectonic fault, i.e., Aceh Segment. Kuala Gigieng displays a marine-dominated coastal system. The mainland consists of old parallel coastal ridges and swales. In the last few centuries and beyond, both coasts were naturally protected by the Late Holocene barrier islands as the coastal system's seaward-most land boundary. The alongshore multitemporal bathymetric profiles in Figure 3b illustrate the averagely shallow bathymetry in front of Kuala Gigieng, and the tilting southwest towards Lambadeuk, where considerably deep trenches are observed in two consecutive years in 1893 and 1924. In contrast, the bathymetry around the same location appears to have been remarkably shallow in 2006.

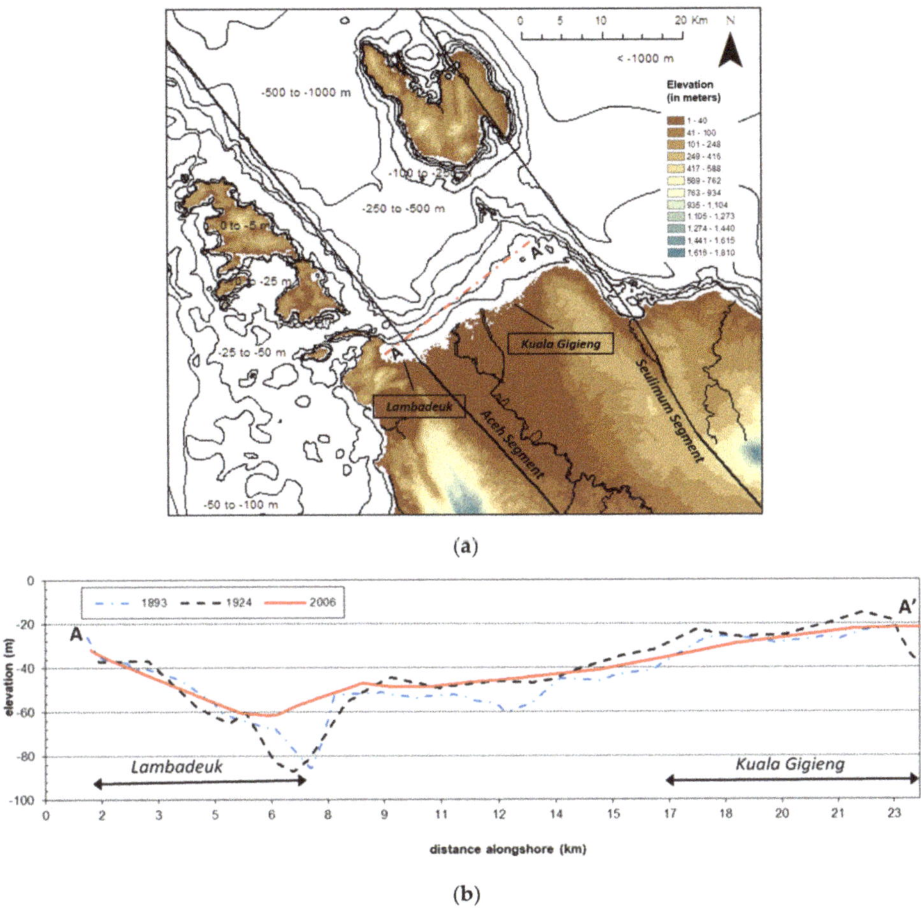

Figure 3. Banda Aceh is situated in the wedge between two tectonic fault segments of the Great Sumatran Fault System, i.e., the Aceh Segment on the southwest and Seulimum Segment on the northeast: (**a**) The map shows the topographic and bathymetric contours of the northern Sumatra region, and the two fault segments; (**b**) The multitemporal alongshore bathymetric profiles in front of Banda Aceh coast show higher elevation at Kuala Gigieng and lower down towards Lambadeuk. The bathymetric data of the year 1893 and 1924 was digitized from the Dutch Colonial Nautical Chart obtained from KITLV, and the 2006 bathymetry was obtained from the bathymetric survey conducted by UP-PSDA of Syiah Kuala University.

The tides along the coast are categorized into a micro-tidal regime which is on average less than one meter high. The typical equatorial monsoonal climate brings about seasonal prevailing wind-induced wave heights and periods variations, i.e., southwesterly during April to September and northeasterly during October to March [22]. The Aceh River is the primary natural river crossing the low-lying coastal city of Banda Aceh (Figure 1c). The river course has been artificially bifurcated at 10 km upstream to a 300 m wide artificial Alue Naga Floodway Canal since the early 1990s, another major outflow dissecting the coastline at present. Both major outlets are the primary sources of sediment supply to the coastal system of the investigated area. Currently, both are regulated by training jetties to maintain navigational depth for fishing boats. Another smaller river that flows at the northeast flank of Banda Aceh plain is the Angon River (Figure 1c), which has been nonmigratory throughout the last century, debouching into the shore-parallel lagoon behind the Kuala Gigieng inlet.

The coastal area was relatively densely populated prior to the 2004 tsunami event, occupied by approximately 250,000 people. Around half of the total population has been reported dead or gone missing [45] after the 2004 mega tsunami. More than a decade since the tsunami event, the coastal city has been rehabilitated and reconstructed. Despite the devastation due to the earthquake and tsunami, people are likely to return to their original living and business locations close to the coast. By 2009, the post-tsunami rehabilitation and reconstruction program in Banda Aceh had established resettlement at the coastal area up to 91% [46], along with all the necessary infrastructure such as a ferry port, religious and administrative buildings, schools, housing areas, and a sanitary landfill [23]. Apart from having been recovered from the tsunami event, the coastal area has been subjected to frequent flooding during high spring tides [47,48]. The coastal area is also vulnerable to hydrometeorological hazards, exceptionally high risk to the future sea-level rise [23]. We recently investigated the scenarios of coastal inundation due to the slow-onset projected sea-level rise in the next couple of centuries. The results show that the increasing number of built areas closer to the coastline, despite past tsunami experience, are potentially subject to tremendous loss due to seawater inundation in the next couple of centuries [23]. Such conditions provoke socio-economic and environmental vulnerability for the entire coastal area [49].

3. Materials and Methods

Historical records of particular events leading to morphological alteration are indispensable resources for better understanding the chronological implication to the state of coastal morphology. Previous studies have taken a similar approach by using the history of a few large-scale tectonic events to investigate the meso-term coastal changes (e.g., [50–52]). Herein, we consider that a multi-decadal timescale is appropriate to capture any remarkable changes in the coastal morphology induced by those extreme events associated with seismic activity, and also suitable for coastal management planning [53].

3.1. Spatial Data

We digitized shorelines from various data sources, i.e., from Colonial topographic maps of the 19th century and more recent satellite images of different spatial resolutions, the details of which are listed in Table 1.

Table 1. Spatial data sources for shoreline change detection.

Data Type	Date/Time	Spatial Scale/Resolution	Source
Topographic maps	1898 1924	1:20,000 1:50,000	KITLV
Satellite images	5 June 1967 23 March 1989 8 March 2000 30 June 2005	2.0 m pixel resolution 30 m pixel resolution 30 m pixel resolution 2.5 m pixel resolution	KeyHole-7/USGS Landsat TM/LAPAN Landsat ETM/LAPAN NORAD survey/SIM Centre–BRR NAD
	10 January 2017	1.5 m pixel resolution	IKONOS/LAPAN
Photogrammetric topographic map	June 2005	0.5 m contour interval	NORAD survey/JICA/SIM Centre–BRR NAD

All data sources were geo-referenced to a master map (i.e., ortho-rectified aerial photo acquired in June 2005 from the NORAD survey) processed in ArcGIS 10.6.1 to have a common horizontal datum, projection, and coordinate system. Figure 4 shows the samples of the multisource and multitemporal data used in this study.

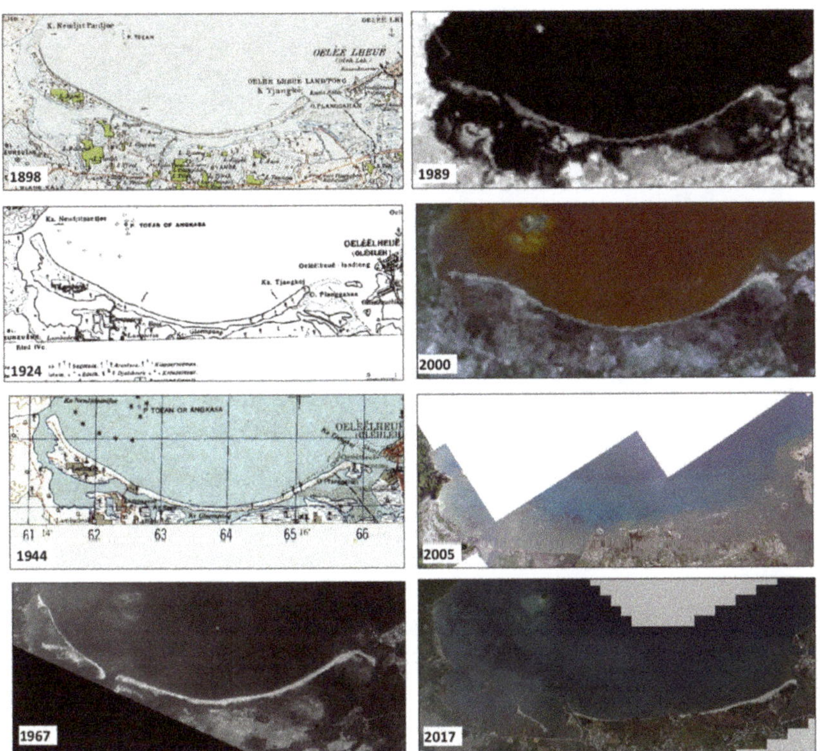

Figure 4. Samples of the multisource topographic maps and satellite images of the Lambadeuk coast, displaying a highly varying data type and quality. Information about the data details can be found in Table 1.

3.2. Historical Records of Earthquake and Tsunami Events

To support our spatial analysis using the historical maps and satellite images, we corroborate the records and results of studies of the earthquake events in the Indian Ocean/Andaman Sea region and along the northern part of the Great Sumatran Fault System. The underlying active fault system along the Sumatra Island is one of the most significant active faults in the world, with slip rates ranges from 10 to 27 mm/year [28]. Natawidjaja and Triyoso [28] identified that the Great Sumatran Fault has at least 19 segments along the fault zone. More than a dozen large earthquakes have occurred historically in the past two centuries associated with the fault zone. Similarly, Hurukawa et al. [43] concluded in their study that almost all the earthquakes occurred since 1892 at Sumatra Island were located at Sumatran Fault, with high seismic hazard mainly occurred over the entire northern part of Sumatra Island during the period of 1942–2003.

Here, we limit the area of earthquake influence by setting up a regional boundary within which we identify any major (M $\geq$ 7.0) and moderate (6.5 $\leq$ M < 7.0) earthquake events records, as depicted in Figure 5. The historical earthquake records are obtained primarily from the NGDC/WDS Global Historical Tsunami Database of the National Centers for Environmental Information (NCEI) database [54]. Additionally, records were also obtained from the Catalogue of Significant and Destructive Earthquakes 1821–2017 for the earthquakes associated with the Great Sumatran Fault [55]. From the database we found five remarkable earthquake events which occurred within our investigation period, which are enlisted in Table 2. The locations of the earthquake epicenters are depicted in Figure 5.

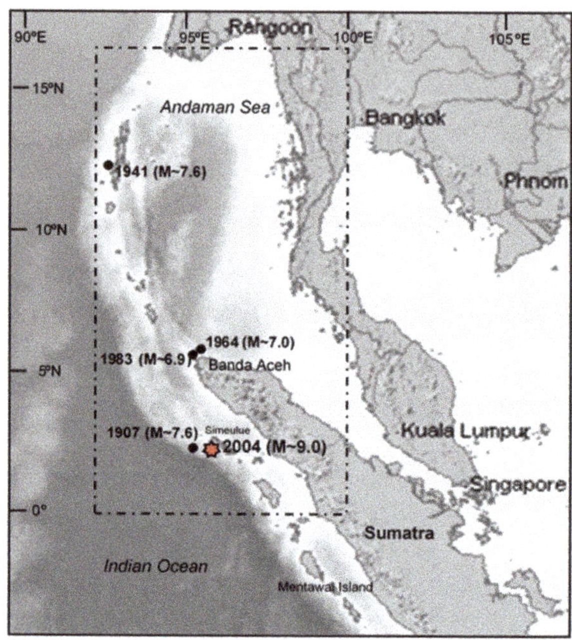

Figure 5. Regional earthquake occurrences at the northern Sumatra during the investigation period in the present study (1898–2017). The dashed-line rectangle shows the regional boundary of influence of the major (M ≥ 7.0) and moderate (6.5 ≤ M < 7.0) earthquake records. Records of earthquakes M ≥ 7.0 accompanied with the validity of tsunami occurrences are retrieved from the earthquake catalog by [54], and the tectonic earthquakes associated with the Great Sumatran Fault obtained from [55].

Table 2. Earthquakes leading tsunami events affecting Banda Aceh in the period of 1804 to 2004. Geographical boundaries displayed in Figure 4.

Date	Latitude (° North)	Longitude (° East)	Focal Depth (km)	Location	EQ Magnitude	Tsunami Validity *
4 January 1907	2	94.5	50	NW Sumatra (Simeulue Island)	7.6	4
26 June 1941	12.5	92.5	20	Andaman Sea	7.6	4
02 April 1964	5.6	95.4	60	NW Sumatra	7.0	3
04 April 1983 **	5.7	94.7	78	NW Sumatra	6.9	No evidence
26 December 2004	3.3	96	30	Off West Sumatra	9.0	4

Note: * Data source from [54]; 4 = definite tsunami; 3 = probable tsunami; 2 = questionable tsunami; 1 = very doubtful tsunami; 0 = event like a seiche. ** Data source from [55].

The NCEI database is accompanied by the reports of the past earthquakes and their associated resulting hazards (e.g., tsunamis, earthquake damage, liquefaction, etc.) that were mostly archived by Soloviev and Go [56] and also by several other scientific documents and papers. Here, the tsunami event records were gathered from scientific and scholarly sources, regional and worldwide catalogs, tide gauge reports, individual event reports, and unpublished works. Descriptions of tsunami events with various occurrence validity levels associated with some earthquake records were used in this study to identify the number of possible tsunami events in the last couple of centuries [57,58]. The level of validity ranges from 1 (less probable occurrence) to 4 (most probable occurrence). It also specifies the tsunami events that occurred at specific locations from where any tsunami waves would have propagated towards their surrounding coastlines. A complete discussion of possible errors can be found in [57].

3.3. Storminess

There is merely a little overview of the climatic condition in the last century in the investigated area. Verstappen [7] reported that, in general, the climatic condition at the northern tip of Sumatra island was relatively dry in the last century. The equatorial position of Indonesia shelters it from tropical cyclones that often devastate the Philippines, Sri Lanka, Bangladesh, and the coastal zone of western Australia. The recent studies [59–61] of the historical cyclones over the Bay of Bengal and the Andaman Sea reveals that the pathways of the cyclones revolve around the latitudes of higher than 8° N. Accordingly, the coastal areas at the northern tip of Sumatra (5° N) are assumed to experience much weaker tropical storm events. Thus, remarkable breaching events of barrier islands, such as those identified in the present study, are unlikely as the results of tropical storms, which is particularly important to keep in mind when analyzing the possible forcing factors responsible for observable morphological changes in this study.

3.4. Regular Wave Climate and Littoral Transport Rate

We obtained the wave climate of the Banda Aceh coast by converting the eleven-year daily wind data records from 1995 to 2005 from the National Meteorology, Climatology and Geophysics Agency (BMKG), and subsequently translated into statistical wave heights and periods as functions of wind velocity, duration, and fetch. Surface and tidal currents are not well recorded in this coastal region. The tides moderately range of 1.00 m from MHWL to MLWL. From the data, we found that the prevailing monsoonal wave directions were from the northwest and the northeast. Based on the resulting wave data, we then roughly estimate the littoral transport rate by using the simple CERC formula [62] at both Lambadeuk and Kuala Gigieng coasts. Estimates of net littoral transport along both coasts will be discussed in Section 4.

3.5. Analysis of Morphological Changes

Morphological changes of the seaward most barrier islands and spits in seven consecutive periods were analyzed in this study at the Lambadeuk and Kuala Gigieng coasts. Each period consists of shorelines of the seaward most barrier islands and spits of two consecutive years. Any changes found to be significant were measured as indicative figures to estimate the amount of displacement or growth of the morphological features.

4. Results

There is a substantial difference in geomorphic settings between Lambadeuk and Kuala Gigieng coasts. Lambadeuk, at the southwest flank of Banda Aceh city, is a transgressive coast built up during the Late Holocene. Typically, a transgressive Holocene coastal barrier underwent continuous erosion and roll-over so that the oldest washover stratigraphies are likely to have been erased [32]. On the other hand, the Kuala Gigieng coast at the northeastern flank is a regressive coast, which typically consists of coast-parallel ridges and swales environment. One can also observe consecutive ridge-swale morphology preservation at the highly energetic and marine-predominant coastal areas along the western coasts of Aceh, Sumatra, e.g., in [21,22].

Figure 6 displays the consecutive observation periods in this study, along with the identified earthquake events (or the absence of those) that occurred in each period. Each of the barrier islands and spits at the Lambadeuk and Kuala Gigieng have uniquely evolved and altered its shape, position, and vulnerability to seismic-related hazards impacts. The most significant changes are evident from a sequential comparison of the island and spit geometries and rates of areal changes. Figure 7a–j depicts the overall results of the delineation of both coasts' morphological features, covering the shorelines of barrier islands, sand spits, and the back-barrier intertidal areas. The sub-aerial parts of the morphological features are color-coded, and the sub-aqueous parts appear in white. Here, we display the shoreline delineations of 1898 and 2005 in every figure panels to provide references of the morphological states at the initial date and the date a few months after the 2004 tsunami event. The latter is represented by the shorelines delineated from the high-resolution aerial photograph acquired in June 2005 (Table 1). In between those years, we then observe and analyze the successive pairs of barrier shoreline changes.

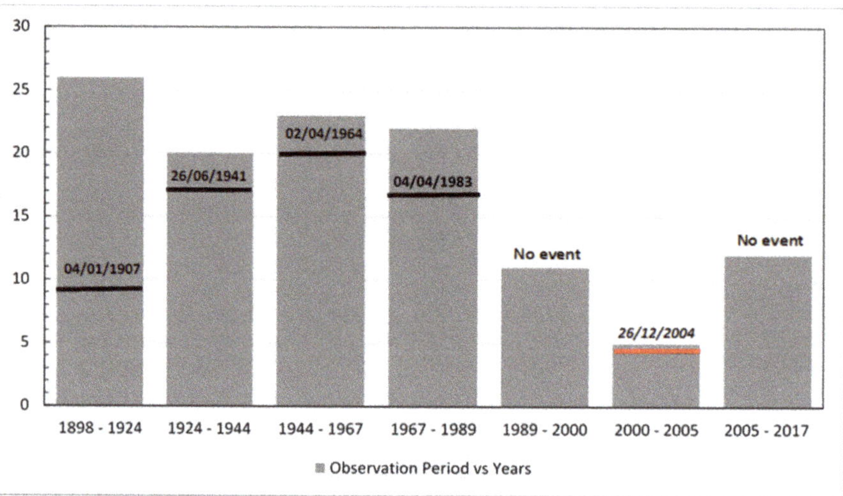

Figure 6. The observation of morphological changes of the northern tip coast of Sumatra Island, consists of seven periods with various intervals depending on the data sets available for comparison. The horizontal axis shows the consecutive periods of observation from 1898 until 2017, and the vertical axis is the length of each successive period of observation. Periods with remarkable earthquake events are marked by the bold black line indicating the year the event occurred, which shows how far away from the pair of observation years the event is. The horizontal red line in 2000–2005 indicates 26 December 2004, marking the earthquake that triggered the mega-tsunami.

We quantify the rate of change of the barrier islands' morphology by simply substituting the polygon areas of a pair of barriers' perimeters from two consecutive observation years, and dividing the result by the length of the period in between. To avoid misleading quantification of the rate of change, we exclude the quantification of the areal changes of the back-barrier lagoon and intertidal morphology. The misled quantification may come from sediment deposition to the back-barrier area from the barrier breaching and from the water inundation extend over the intertidal areas at the time the satellite image was acquired. Nevertheless, for the observation purposes, we still display the delineation of the morphological features at the back-barrier during each observation years, as was delineated from the datasets.

4.1. Rate of Change of Barrier Islands Morphology

From the wave data analysis, we obtained the seasonal prevailing monsoonal waves annually. The southwest monsoon occurs between April and September and is characterized by relatively rough waves coming from the northwest at the Banda Aceh coast. About 53% of the waves approach the coast with a significant wave height of 1.0 m with a period of 3 s. During the northeast monsoon between October and March, the climate tends to be milder, with 30% of the waves approaching from the northeast with a significant wave height of less than 1 m and a period of 4.5 s. Based on the resulting wave data, we then roughly estimate the littoral transport rate by using the simple CERC (CERC stands for Coastal Engineering Research Center, U.S. Army Engineer Waterways Experiment Station, Vicksburg, MS) formula [62], which results in the estimate of net littoral transports at a rate of +0.30 hectares/year and +1.08 hectares/year at Lambadeuk and Kuala Gigieng, respectively. The prevailing direction of net littoral sediment transport at both coasts is directed southwest, with positive rates suggesting the coasts are accretional.

Tables 3 and 4 display the rates of land gain or loss at Lambadeuk and Kuala Gigieng as the results of the barrier islands' morphological responses during the consecutive periods of observations. The largest amount of land loss is observed for all the barrier systems in 2000–2005 due to the impact of the 2004 earthquake and tsunami event, i.e., −73.54 hectare and −128.73 hectare at Lambadeuk and Kuala Gigieng, respectively. These are equivalent to a 100% and 59% loss of land with rates of −14.71 hectares/year and −25.75 hectares/year, respectively. Both rates are an order of magnitude higher than the estimate of the littoral transport rate under regular wave climate.

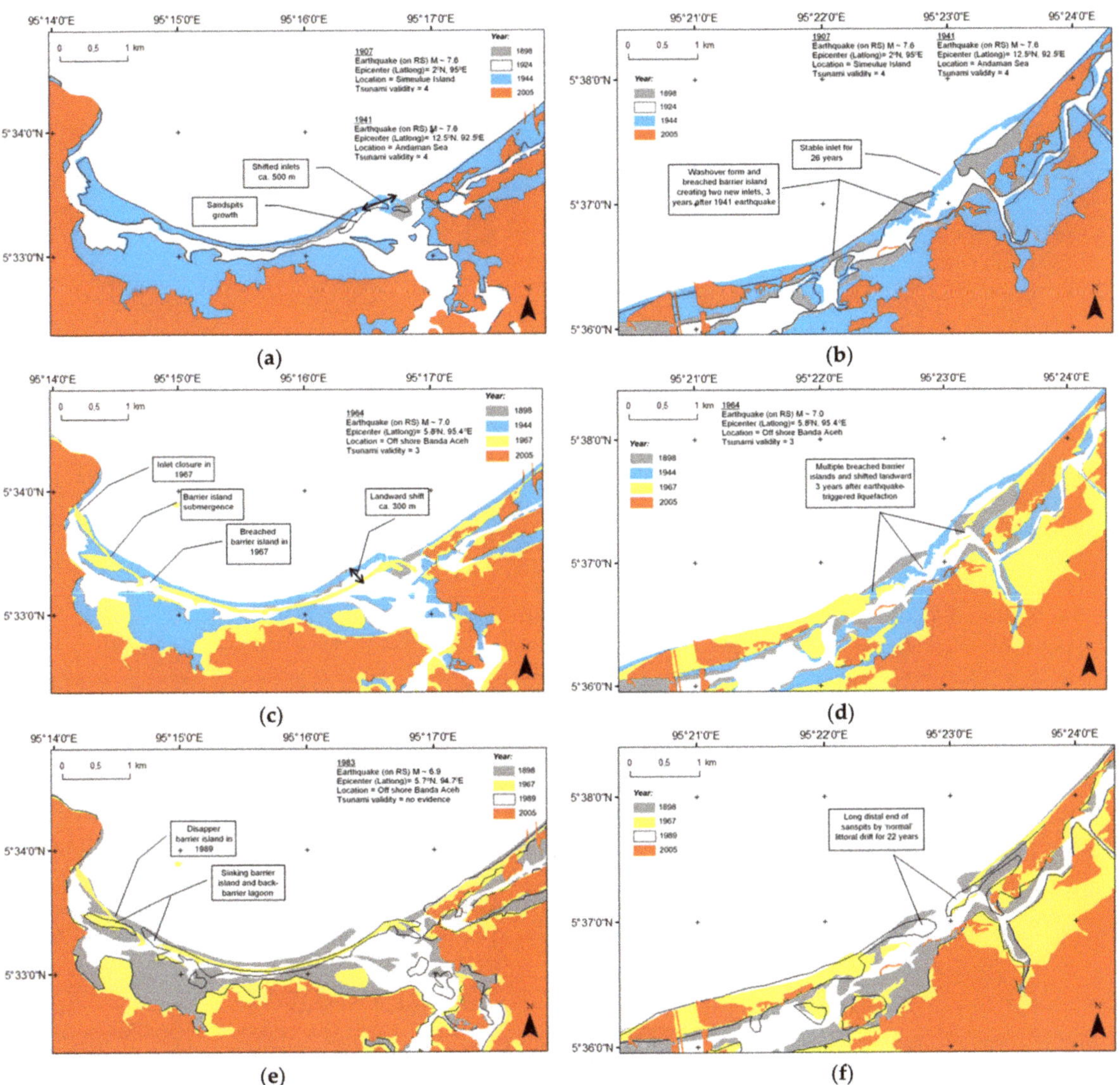

Figure 7. *Cont.*

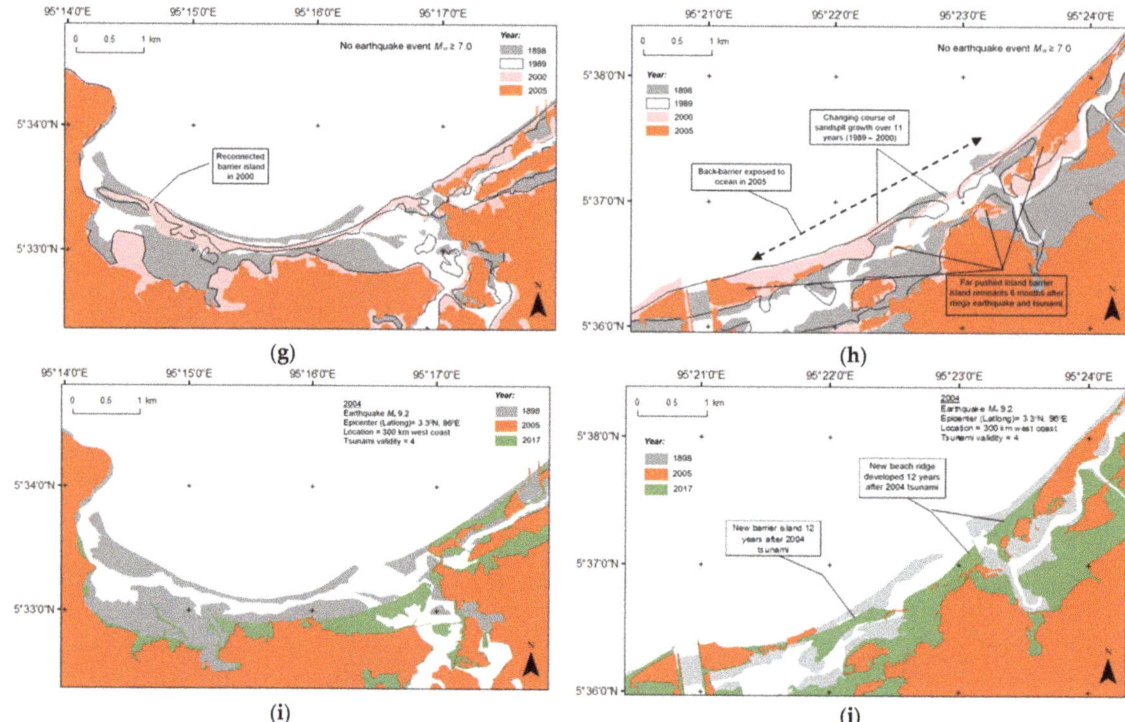

Figure 7. Observation of barrier island morphological changes from the consecutive pairs of shoreline delineation of Lambadeuk and Kuala Gigieng coastal section, at the northern tip of Sumatra Island: (**a**,**b**) Changes at Lambadeuk and Kuala Gigieng, respectively, in two periods 1898–1924 (26 years), and 1924–1944 (20 years). Major earthquakes occurred in both periods; i.e., on 4 January 1907 of *M* 7.6 which is most likely to have triggered a tsunami event (tsunami validity = 4), and on 26 June 1941, which triggered a far-field tsunami event; (**c**,**d**) Changes at Lambadeuk and Kuala Gigieng, respectively, in 1944–1967 (23 years), during which a major earthquake occurred on 2 April 1964, of which the epicenter was relatively close to Kuala Gigieng, and reportedly triggered tsunami and liquefaction; (**e**,**f**) Changes at Lambadeuk and Kuala Gigieng in 1967–1989 (22 years), during which a moderate earthquake occurred on 4 April 1983 of *M* 6.9, and the epicentre was just off-shore from the Lambadeuk coast. No reports on any tsunami event. However, Lambadeuk reportedly experienced tectonic land subsidence; (**g**,**h**) Between 1989 and 2000 (11 years), no significant earthquake occurred at both Lambadeuk and Kuala Gigieng; (**i**,**j**). Dramatic changes occurred in the period of 2000–2005. During this short period, the mega-tsunami triggered by one of the largest magnitude earthquakes to have occurred in the modern history occurred on 26 December 2004, of M~9.0.

Table 3. Average long-term historical rates of land area change for both coasts for selected periods at Lambadeuk. Rates are in hectares/year. Positive numbers indicate land gain, and negative numbers indicate a land loss.

Period	Length of Period (Years)	Areal Changes (Hectares)	Rates of Changes (Hectares/Year)	Percentage of Land Loss/Gain
1989–1924	26	+8.59	0.33	13%
1924–1944	20	+0.93	+0.05	1%
1944–1967	23	−31.79	−1.38	−41%
1967–1989	22	+9.69	+0.44	21%
1989–2000	11	+17.61	+1.60	31%
2000–2005	5	−73.54	−14.71	−100%
2005–2017	12	0.00	0.00	0%

Table 4. Average long-term historical rates of land area change for both coasts for selected periods at Kuala Gigieng coast. Rates are in hectares/year. Positive numbers indicate land gain and negative numbers indicate land loss.

Period	Length of Period (Years)	Areal Changes (Hectares)	Rates of Change (Hectares/Year)	Percentage of Land Loss/Gain
1989–1924	26	0.00	0.00	0%
1924–1944	20	−49.06	−2.45	−20%
1944–1967	23	−7.94	−0.35	−4%
1967–1989	22	+14.07	+0.64	8%
1989–2000	11	+21.30	+1.94	11%
2000–2005	5	−128.73	−25.75	−59%
2005–2017	12	+43.49	+3.62	49%

In the century before the mega-tsunami, Lambadeuk experienced significant land loss during 1944–1967 by −41%; nevertheless, it was compensated by 21% and 31% of land gain in the subsequent two consecutive periods (Table 3). Overall, Lambadeuk on average gained 5% extra land during the last century, only to face a complete loss of barrier island due to the 2004 tsunami until the present. In contrast, Lambadeuk experienced land loss of merely −1% averagely during the last century prior to the 2004 tsunami. The loss of barrier island area by −59% during the mega-tsunami has been compensated by 49% in 2017 since the tsunami.

4.2. Multitemporal Morphological Changes

The following is our analysis of the morphological changes of barrier islands and spits observable in Figure 7, and estimates of the total land loss or gain as the results of those changes based on the quantitative analysis described in Section 4.1. Following this analysis, the interpretation of possible controlling factors for those changes will be discussed in Section 5.

4.2.1. Period 1898–1924–1944

In these consecutive periods, the main barrier island at Lambadeuk is in a relatively stable position in 1898–1924 and 1924–1944 (Figure 7a). Despite a major earthquake occurring in 1907, the barrier islands at Lambadeuk and Kuala Gigieng show a relatively stable state of morphology, which is observable by comparing the shorelines of 1898 and 1924 (Figure 7a,b). The growth of barrier spits extended from the eastern end of the barrier island at Lambadeuk suggests no extreme forcing, which could have caused remarkable changes in the barrier's morphology, such as landward migrating or breaching. A relatively remarkable barrier breaching occurred, however, at the adjacent coastal section at Ulee Lheue (the eastern barrier island next to Lambadeuk in Figure 7a). The possible cause is inconclusive in this study, as there was no report of any tsunami or land subsidence occurred as a result of the major earthquake, at the investigated area; despite a well-known tsunami event which was reported at Simeulue Island at the southwestern offshore of Sumatra.

In 1924–1944, a barrier spit was growing further east as far as ca. 500 m, and almost connected to the remnant of a major breach of the Ulee Lheue barrier island (Figure 7a). In 1944, the spit was modified, along with slightly narrowing and counterclockwise re-orientation of the barrier island's eastern part. Meanwhile, at Kuala Gigieng, the barrier islands (appear in blue in Figure 7b) had undergone significant changes. Two new inlets were created while the entire sub-aerial part of the barrier island has shifted landward at a maximum distance of ca. 300 m at the central part, and the section at the northeast at the same time moved seaward at more or less the same distance. Furthermore, this new barrier island formation seems to be narrowing entirely, resulting in a total land loss of −49.06 hectares with an average rate of about −2.45 hectares/year (Table 4).

4.2.2. Period 1944–1967

In 1944–1967, the barrier islands and spits of both investigated coasts were significantly changing compared to the previous period. In 1967, both coasts experienced a relatively far landward shift, i.e., around 300 m distance, while the narrowing subaerial parts of the barrier features are remarkable (Figure 7c,d). Additionally, the barrier spits were breached at multiple locations at Kuala Gigieng (Figure 7d). A tectonic earthquake associated with the Seulimum segment activation occurred in 1964 with a magnitude M 6.7 [28,43], of which the epicenter was merely 73 km northeastern off-shore of Kuala Giging coast (Figure 5). The remarkable features of narrowing and breached barrier islands and spits remain observable in 1967, i.e., in merely three years after the earthquake event.

4.2.3. Period 1967–1989

In 1967–1989, most of the barrier islands and spits on both coasts became relatively stable, i.e., no barrier island migration was found. Nevertheless, the westernmost part of the barrier at Lambadeuk that appeared in 1944 disappeared in 1967 (Figure 7e), suggesting submergence of the barrier island. In contrast, at Kuala Gigieng, the barrier islands were in a stable position throughout the whole period (Figure 7f). It is noteworthy that in this period, there was a moderate earthquake of magnitude M 6.9 which occurred in 1983, of which the epicenter was located just off-shore from Lambadeuk (Figure 5 and Table 2), suggesting that the earthquake was associated with the Aceh Segment which was underlying the coast.

4.2.4. Period 1989–2000

There were no major or moderate earthquakes recorded in 1989–2000 (Table 2 and Figure 5). At Lambadeuk, the barrier island maintained its position during the 11 years, showing a stable and mature barrier spits at the eastern end. The previously submerged and breached barrier island in 1989 at the western end has been reconnected since (Figure 7g). At Kuala Gigieng, elongated barrier spits were developed, interestingly, in the opposite direction from the development that occurred during the last period. Although the barrier islands and spits position show no significant migration, a narrowing land area is apparent. The further growth of the right-hand barrier spit to the southeast appeared in 2000 (appeared in pink in Figure 7h) extending further southeast as far as 700 m from the tip of barrier spit in 1989.

4.2.5. Period 2000–2005–2017

The years between 2000 and 2005 were the remarkable period where the great earthquake triggering the mega-tsunami occurred and severely destroyed the northern tip coast of Sumatra Island on 26 December 2004. The change of morphological features along the coast was expectedly enormous. Figure 7i,j display the massive overwash of the entire investigated coast leading to several breaches, and the remnants of the barrier islands were shifted landwards. The stabilized barrier spit which appeared in 2000 had disappeared entirely (Figure 7i,j).

Twelve years after the 2004 tsunami, the shoreline in 2017 reveals a striking difference in both of the investigated coasts in responding to the tsunami waves. At Lambadeuk, after the disappearance of the entire barrier island (Figure 7i), the shoreline started to develop from the former lagoon's inner shore, which has since been exposed to the sea. Kuala Gigieng now has two relatively permanently open inlets because of the barrier island and barrier spit breaching during the 2004 tsunami (Figure 7j).

5. Discussion

Our investigation on the morphological changes of barrier islands and spits at two coastal settings along the northern tip of Sumatra Island revealed several new findings on the possible controlling factors associated with the seismic activity that change barrier island morphology. The intervals between periods of observation of morphological changes are of less than 30 years, which fits the return period of major earthquakes in the region of Sumatra. The impact on the coastal morphology can be in the form of barriers breaching, sinking, landward migrating, and landward bending of spits development. The possible interplaying forcing factors to those morphological changes are discussed in the following subsections.

5.1. Tsunami Overwash

In 1944, the barrier island at Lambadeuk remained stable compared to its morphological state in 1924, with the additional growth of barrier spit (Figure 7a). In contrast, the barrier islands at Kuala Gigieng experienced a landward migration, breaching, and the growth of barrier spit bending landwards (Figure 7b). Such morphological pattern is typical for barrier islands washed over by storm events or hurricanes, e.g., Hurricane Sandy and Hurricane Katrina [40]. Energy dissipation is then mostly achieved through the overwash bore running over the barrier, flattening the barrier crest profile, and depositing off-shore originating material over the back-barrier zone [32]. Provided that the height of the barrier island was typically less than 2 m high [7], the forcing factor which possibly controls such morphological impact is a wave coming from the sea. Waves equal to or slightly higher than the height of a barrier island, with a fairly long period, may have overtopped some of the lowest points of the barrier island. Analogously to a storm event, among the four regime types of impacts caused by storm waves on a barrier island [32], the observable morphological changes at Kuala Gigieng in 1944 may fall into the "Overwash regime".

In 1924–1944, there were no earthquake events associated with either the Sunda Trench and the Great Sumatran Fault system that occurred near the northern tip of Sumatra Island in this period. However, a major earthquake of M 7.6 occurred in 1941 at the western coast of Car Nicobar Island, with its epicenter at 12.1° N and 92.5° E, or around 834 km northwest from Banda Aceh (see location in Figure 5). NCEI [54] recorded a high score validity tsunami event associated with this earthquake (Table 2). From the historical reports [63,64], the earthquake generated a tsunami throughout the Andaman Sea and the Indian Ocean. Despite the lack of reliable records, for instance from the tidal gauge, which was in operation at the time, some local newspapers in India reported that a tsunami was witnessed along the eastern coast of India. At the Nicobar and Andaman Islands, the wave heights were reported as high as 0.75–1.75 m. It was estimated that 3000 to 5000 people were killed in Sri Lanka and on the east coast of India [65]. Local newspapers believed to have mistaken the reported deaths and damage to a storm surge.

In support of these reports, a numerical model developed by [66] was to simulate the Andaman–Sumatra tsunami propagation of tsunami triggered by the earthquake that occurred in 1941. The result reveals an agreement with the documented observation reports that the earthquake felt over a wide area covering the eastern and southern Andaman Sea region (i.e., the northern coast of Sumatra). The model results show that the tsunami has reached the coast of Nagapittinam on the west coast of India after 165 min, with run-up heights in the range of 0.95–1.25 m, and the north tip of Sumatra at 120 with tsunami wave height was less than 1.00 m [66].

It is noteworthy to mention that following the earthquake, two events with magnitude 6.0 struck within 24 h after the main shock of 27 June 1941, and there were 14 aftershocks of magnitude up to 6.0 until January 1942 [67]. At Lambadeuk coast, McKinnon [68] reported that a local informant indicated that the location of the village of Lambaro, located at the western end of the barrier island of Lambadeuk (Figure 7a), had been moved three times within living memory, the last time being early in the Japanese occupation (1941–1942) when the inhabitants were forcibly removed from the shoreline and resettled at the foot of the surrounding hills. Thus, the far-field tsunami event in 1941 was most likely responsible for changes of the barrier islands and spits at Kuala Gigieng in 1924–1944. The nearshore bathymetry may have been controlling the tsunami arrival at the shoreline of Lambadeuk and Kuala Gigieng. The deepening bathymetry (Figure 3b) may help to dampen the incoming tsunami wave energy at Lambadeuk, while the shallow bathymetry at Kuala Gigieng may have amplified the tsunami wave force.

5.2. Combined Liquefaction and Tsunami Overwash

The observation period of 1944–1967 exemplifies the period where the impact of both liquefaction and tsunami occurred, and the effects are observable on both coasts. The barrier islands and spits delineated from the satellite image in 1967 in Figure 7c,d show the state of the coast three years after a major earthquake event on 2 April 1964 of magnitude M 7.0, which epicenter located at 5.6° N and 95.4° E (see Table 2), or ca. 12 km northeastern off-shore of Banda Aceh City. The barrier islands and spits at both coasts experienced hundreds of meters of landward migration (and slightly reoriented clockwise at Lambadeuk), multiple-barrier breaching, especially at Kuala Gigieng coast, as well as a narrowing of subaerial parts of barriers, most probably as a result of land subsidence induced by liquefaction (Figure 7b,c). Surprisingly, the multiple inlets as the results of breaching events remained open even after three years of development, suggesting that littoral transport and sediment supply under the regular wave climate have not made up for the loss of land caused by the tsunami and liquefaction.

Soloviev [56] reported that an earthquake that occurred on 2 April 1964 had caused considerable damage to adobe buildings, the ground cracked open, the ground subsided, mud and sand gryphon appeared. Such impacts of ground shaking are comparable to our observation on the 2016 Pidie Jaya tectonic fault earthquake of M 6.5, in Pidie Jaya district on the northern coast of Aceh Province [69]. In Pidie Jaya earthquake, black sandy mud emerged through small cracks opening in the internal floors of houses. Both coasts are comparable for their similar type of wave-dominated sandy coastal area with typical soil type of alluvium containing gravels, sands, and muds [70]. Such soil structure provides some degree of tremor amplification [69], which creates saturation on sandy layers and increases pore water pressure, leading to liquefaction [71].

NCEI [54] recorded a high-validity event of the tsunami (validity 3), mostly based on a report by [56], in which a combination of a tidal (tsunami) wave and the locals observed land subsidence of half a meter at Ulee Lheue, east of Lambadeuk barrier island (Figure 7c). The multiple breachings that appeared at Kuala Gigieng in 1967 (Figure 7d) may have been further submerged by the already breached barrier islands in the earlier period (1924–1944), suggesting an effect of liquefaction combined with a tsunami overwash.

5.3. Co-Seismic Tectonic Subsidence

The observation period of 1967–1989 reveals a unique contrasting development of barrier system morphology between the two observed coasts. Contrary to the long-distance growth of sand spits to their distal length observed at Kuala Gigieng (Figure 7f), the western side of barrier island at Lambadeuk (Figure 7e) appeared to have been heavily subsided, that the slightly more than 1 km barrier island connected to the Ujong Pancu headland which appeared in 1967 and had disappeared entirely in the satellite image of 1989. This left a piece of subaerial part detached from the main barrier island. Such contrasting changes most likely have something to do with tectonic activities within the period.

A moderate tectonic earthquake occurred on 4 April 1983 with magnitude M 6.9 [72], and the epicenter was located at 70 km northwest off-shore of Banda Aceh (Figure 5, Table 2). The earthquake caused significant damage to buildings in Banda Aceh city, with the Mercalli scale recorded as category VI [55]. There was no report from the locals of any tsunami occurrence or liquefactions such as those which occurred in 1964. It is also noteworthy that the submerged area is close to the underlying Aceh Segment, suggesting tectonic subsidence responsible for the submergence. McKinnon [73] investigated the archaeological artifacts at this location, which suggests localized tectonic subsidence in the vicinity of the Sumatran Fault System, which appeared to be dramatically evident around Lambadeuk. Strong evidence of submergence came from the rectangular structure discernible during the low tides, which happened to be a former mosque foundation that remained visible underneath the water in an aerial photograph from 1978. An interview with a local informant suggested that in the last 80 years towards the first Japanese occupation in 1942, a sunk off the coast at least two to three meters had been occurring [73]. The shoreline had retreated about 150–200 m at a village called Lambaro, which was originally located at the Lambadeuk barrier island. This village, which had already been moved three times, was eventually rebuilt about 12 km inland at present. Bearing in mind that this area has evidently been submerged in two consecutive periods, i.e., 1944–1967 and 1967–1989, this suggests that Lambadeuk is not resilient to multiple tectonic subsidence.

5.4. Dormant Period of Major and Moderate Earthquake

Among the entire periods of our investigation, there was also a period where major or moderate earthquakes were absent, which was during 1989–2000 (Figure 7g,h). Based on the work of Nott et al. [33], one might have expected that less intense storm periods would have induced a slowing down of the development of beach ridges (in this case, the barrier islands). Instead, we note that the reconnection of the previously breached barrier islands occurred during this period (Figure 7g,h). Moreover, the further growth of the barrier spit at Kuala Gigieng towards the opposite direction from the previous period (Figure 7h) suggests that the prolonged net littoral drift prevailing southwestern direction is most likely responsible for the observed changing course spit growth.

Despite the stable development of barrier islands morphology during the non-event periods by the littoral process, the development of barrier islands depends on the continuous sediment supply from the rivers. A consistent prevailing longshore drift can promote such stable elongated barrier spit growth in a considerably long period, without any remarkable disruption such as by tectonic events [74]. A continuous sediment supply from the major outlets may have facilitated the growth of long barrier spits to their distal lengths, whereas the seasonal change of littoral drift between the alternating rough west monsoon and calm northeast monsoon may promote the spits' balanced growth.

5.5. Mega Tsunami Overwash

The great earthquake of 26 December 2004 has generated not only a tsunami but also land subsidence along the northern tip coast of Sumatra, particularly along its western coast. There were only a few meters of land subsidence that occurred in Banda Aceh [16,18]. The locals also observed liquefaction effects of clayed soils in many areas at Banda Aceh city [8]. Despite these, the far-field gigantic tsunami waves of more than 10 meters in height arrived at the northern tip of Sumatra and were most definitely the controlling factors of the remarkable changes of the barrier islands at Lambadeuk and Kuala Gigieng. This includes the extensive disappearance of the barrier islands (Figure 7i,j), which had long been preserved as the natural coastal protection to the back-barrier ecosystem since the Late Holocene.

Even after 14 years of post-tsunami develpment, the coastal area remains exposed to the ocean with the absence of a new barrier island. Clearly, the 2004 tsunami event has been the primary forcing factor responsible for the disappearance of most parts of the barrier islands at Lambadeuk and Kuala Gigieng (Figure 7g,h). For the case of a tsunami, out of the four types of storm regimes and prediction of beach changes proposed by [35], the presumable tsunami overwash that occurred in the 2004 event most probably falls into regime 4: "Inundation regime". Here, the barrier islands were overtopped by the tsunami, of which the height is considerably higher than the crest of the barrier island, overtopping the barrier island and flattening the barrier topography.

5.6. Fluvial and Lagoon Systems

The three major outlets supplying sediments to the coastal area are the Aceh River, Alue Naga Floodway Canal, and Angon River. All of them have been generally stable in their position throughout the entire investigated period, even after the 2004 tsunami. Small river streams flowing into the lagoon system have not been migrating from their original position during the period of investigation, particularly those the locations of which were not directly associated with the tectonic fault (e.g., Sumatran Fault) and the length of the river streams are relatively short, i.e., less than 2 km. Combined with wave actions, in particular at the wave-dominated Kuala Gigieng coastal system, the sediment supply may have been actively contributed to the relatively quick reconnection of the barrier island breaching and the overall recovery of the associated barrier islands.

Verstappen [7] characterized the Sumatran river system, whereby the river discharge was affected by the monsoon. The sand fraction of the material carried off by the rivers and reaching the sea usually is considerably small as a result of the intense chemical weathering inherent in the prevailing humid tropical environment. Moreover, in the locations where neo-tectonism is dominant, the prevailing climatic conditions are less important compared to those at the more seismically stable locations.

5.7. The Morphological Resilience of Tectonically-Active Coasts

The present study reveals that major and moderate earthquakes with a return period of 20 to 30 years at the north tip of Sumatra Island of Indonesia have the potential to trigger tsunamis and co-seismic land subsidence and liquefaction. Figure 8 summarizes the varying land areas of barrier islands from one period to another subject to the presence of major or moderate earthquakes or the absence of them at Lambadeuk and Kuala Gigieng. In the previous discussions, we have been able to identify several different cases of morphological changes of barrier islands, which were driven by various controlling factors associated with major and moderate earthquake events in the last century.

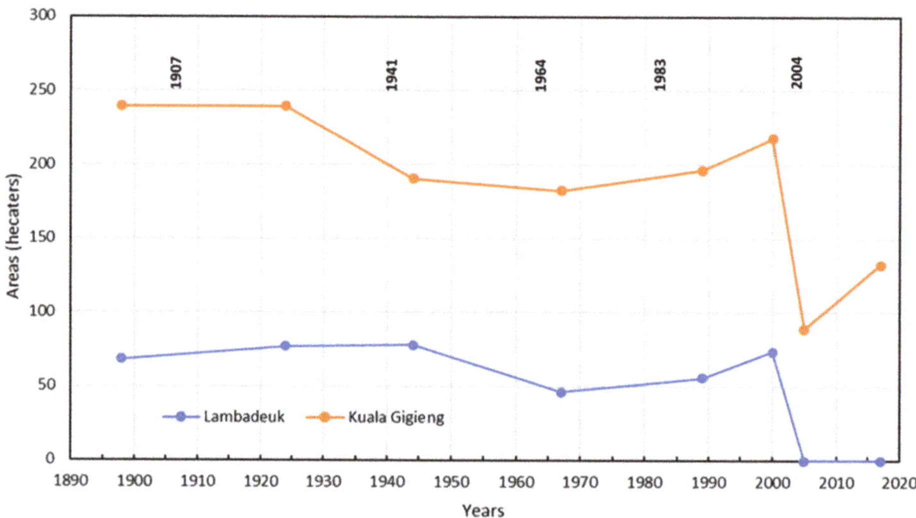

Figure 8. Historical land area trends for the Lambadeuk and Kuala Gigieng barrier islands to the timing of major earthquakes that impacted the islands.

The mega-tsunami of 2004 remarks the fundamental changes in terms of coastal morphological characteristics, i.e., from a low-lying wetland ecosystem naturally protected by barrier islands and spits into a wave-exposed intertidal coastal area. The sediment transport under the present day's wave regime is unlikely capable of restoring the removal of those barrier islands that was built during the Late Holocene, most probably under an extremely different wave regime. Thus, we may conclude that both Lambadeuk and Kuala Gigieng are not resilient to a mega-tsunami.

For the case of barrier islands directly associated with the tectonic fault system, a combination of co-seismic subsidence, liquefaction, and tsunami can dramatically change the barrier island morphology, such as Lambadeuk. As for barrier islands located away from the tectonic fault system, such as Kuala Gigieng, the tsunami and liquefaction are responsible for modifying the barrier morphology, either independently or simultaneously. The breaching, landward migration, and the sinking barrier islands which occurred in the consecutive periods have been proven irreversible, despite a temporary growth of extended barrier spits at a sediment-rich environment. The unrecovered breaching, the submergence, as well as the disappearance of barriers observed even before the event of mega-tsunami of 2004, suggest that the barrier islands are not resilient enough against the recurrence of seismic events.

Evidently, within a century, the recurrence of several major earthquakes, either associated with subduction or tectonic fault zones, may trigger secondary effects, which potentially reduce the functionality of barrier islands and spits as natural coastal protection. The recurrence of small tsunamis, tectonic land subsidence, and liquefaction falls within an engineering timescale, which, therefore, should ultimately be taken into account in managing such a tectonically active coastal settings.

6. Conclusions

The present study investigates the resilience of barrier islands and spits because they protect low-lying coastal areas with high economic and environmental value. This is particularly true in archipelagic countries like Indonesia. Spatial analysis of barrier islands and spits is delineated in GIS, utilizing a multitemporal and multisource data series from the old Colonial topographic maps to the most recent high-resolution satellite images of 1898 to 2017, which encompasses a multidecade timescale observation. The earthquake and tsunami records and established conceptual models of storm effects to barrier systems are corroborated to support possible forcing factor interpretations. We found that tsunami, co-seismic subsidence, and liquefaction are the secondary effects of moderate or major earthquake occurrences that mostly control the state of the modern barrier islands and spits morphology in the investigated coastal area. Those controlling factors intermittently disrupt and interplay with sediment transport otherwise induced by regular wave climate, and eventually alter the coastal morphology development trend in the long term. The records of earthquake occurrences in front of the Banda Aceh coast suggest a return period of 20 to 30 years of major and moderate earthquakes, most of which triggered tsunami events (e.g., earthquakes in 1907, 1941, and 1964), and have far weaker wave energy than the one that occurred in December 2004. Evidently, liquefaction and co-seismic subsidence are the other controlling factors contributing to the remarkable morphological changes, which in some cases may be coupled with tsunami events. The results demonstrate that the semi-protected embayed Lambadeuk coast has been progressively lost its barrier islands due to repeated co-seismic subsidence. The wave-dominated Kuala Gigieng coast is not resilient to the combination of tsunami and liquefaction events. The mega-tsunami triggered by the 2004 earthquake has led to irreversible changes in both coasts' barrier systems. The barrier system is supposed to be a natural protection measure for the ecosystem and the economic value of the protected back-barrier domain. The exposed coastal mainland due to the barrier system disappearance may increase the risk of disaster by tsunamis, co-seismic subsidence, and liquefaction in the future. A further comprehensive evaluation of the resilience of the tectonically active coast, therefore, needs to be done. For such a dynamic tectonically active coastal area, investment in coastal protection to protect the invaluable ecosystem, and economic activities on the back-barrier domain will, therefore, be challenging. The results of the present study imply that in managing the coastal area where seismicity is highly active, despite small magnitudes, tsunamis, liquefaction, and land subsidence should be considered well.

Author Contributions: Conceptualization, E.M., F.L. and B.P.; methodology, E.M., F.L., M.D.-J.; software, E.M., F.L.; validation, E.M., F.L., P.W.; formal analysis, E.M.; investigation, E.M., P.W.; resources, E.M., D.D.; data curation, E.M.; writing—original draft preparation, E.M.; writing—review and editing, B.P., F.L., M.D.-J.; visualization, E.M.; supervision, M.D.-J., F.L.; project administration,

D.D.; funding acquisition, E.M., D.D. All authors have read and agreed to the published version of the manuscript.

Funding: This research was funded by the Indonesian Ministry of Research, Technology and Higher Education (Ristekdikti) through Scheme for Academic Mobility and Exchange (SAME) Program and World Class Professor Program (WCP) Scheme A.

Institutional Review Board Statement: Not applicable.

Informed Consent Statement: Not applicable.

Data Availability Statement: The data presented in this study are available on request from the corresponding author.

Conflicts of Interest: The authors declare no conflict of interest.

References

1. Daniell, J.E.; Schaefer, A.M.; Wenzel, F. Losses Associated with Secondary Effects in Earthquakes. *Front. Built Environ.* **2017**, *3*, 30. [CrossRef]
2. Bradley, K.; Mallick, R.; Andikagumi, H.; Hubbard, J.; Meilianda, E.; Switzer, A.D.; Du, N.; Brocard, G.; Alfian, D.; Benazir, B.; et al. Earthquake-triggered 2018 Palu Valley landslides enabled by wet rice cultivation. *Nat. Geosci.* **2019**, *12*, 935–939. [CrossRef]
3. Sunamura, T. *Geomorphology of Rocky Coasts*; Wiley: Chichester, UK, 1992.
4. Bird, E.C.F. *Coastal Geomorphology: An Introduction*, 2nd ed.; John Wiley & Sons: West Sussex, UK, 2011.
5. Stutz, M.L.; Pilkey, O.H. Open-Ocean Barrier Islands: Global Influence of Climatic, Oceanographic, and Depositional Settings. *J. Coast. Res.* **2011**, *272*, 207–222. [CrossRef]
6. Sui, L.; Wang, J.; Yang, X.; Wang, Z. Spatial-Temporal Characteristics of Coastline Changes in Indonesia from 1990 to 2018. *Sustainability* **2020**, *12*, 3242. [CrossRef]
7. Verstappen, H.T. *Outline of the Geomorphology of Indonesia: A Case Study on Tropical Geomorphology of a Tectogene Region*; ITC Publication: Enschede, The Netherlands, 2000; Volume 79.
8. Lavigne, F.; Paris, R.; Grancher, D.; Wassmer, P.; Brunstein, D.; Vautier, F.; Leone, F.; Flohic, F.; De Coster, B.; Gunawan, T.; et al. Reconstruction of tsunami inland propagation on December 26, 2004 in Banda Aceh, Indonesia, through field investigations. *Pure Appl. Geophys.* **2009**, *166*, 259–281. [CrossRef]
9. Narayana, A.C.; Tatavarti, R.; Shinu, N.; Subeer, A. Tsunami of December 26, 2004 on the southwest coast of India: Post-tsunami geomorphic and sediment characteristics. *Mar. Geol.* **2007**, *242*, 155–168. [CrossRef]
10. Paris, R.; Lavigne, F.; Wassmer, P.; Sartohadi, J. Coastal sedimentation associated with the December 26, 2004 tsunami in Lhok Nga, west Banda Aceh (Sumatra, Indonesia). *Mar. Geol.* **2007**, *238*, 93–106. [CrossRef]
11. Paris, R.; Wassmer, P.; Sartohadi, J.; Lavigne, F.; Barthomeuf, B.; Desgages, E.; Grancher, D.; Baumert, P.; Vautier, F.; Brunstein, D.; et al. Tsunamis as geomorphic crises: Lessons from the December 26, 2004 tsunami in Lhok Nga, West Banda Aceh (Sumatra, Indonesia). *Geomorphology* **2009**, *104*, 59–72. [CrossRef]
12. Umitsu, M.; Tanavud, C.; Patanakanog, B. Effects of landforms on tsunami flow in the plains of Banda Aceh, Indonesia, and Nam Khem, Thailand. *Mar. Geol.* **2007**, *242*, 141–153. [CrossRef]
13. Meilianda, E.; Dohmen-Janssen, C.M.; Maathuis, B.H.P.; Hulscher, S.J.M.H.; Mulder, J. Short-term morphological responses and developments of Banda Aceh coast, Sumatra Island, Indonesia after the tsunami on 26 December 2004. *Mar. Geol.* **2010**, *275*, 96–109. [CrossRef]
14. Yunus, A.P.; Shahabi, H.; Avtar, R.; Narayana, A.C. Shoreline and Coastal Morphological Changes Induced by the 2004 Indian Ocean Tsunami in the Katchal Island, Andaman and Nicobar—A Study Using Archived Satellite Images. In *Coastal World Heritage Sites*; Dou, J., Yunus, A.P., Santiago-Fandino, V., Eds.; Springer: Cham, Switzerland, 2016; Volume 14, pp. 65–77.
15. Wong, P.P. Rethinking post-tsunami integrated coastal management for Asia-Pacific. *Ocean. Coast. Manag.* **2009**, *52*, 405–410. [CrossRef]
16. Gibbons, H.; Gelfenbaum, G. Astonishing wave heights among the findings of an international tsunami survey team on Sumatra. In *Sound Waves*; USGS: Reston, VA, USA, 2005.
17. Borrero, J.C. Field survey of northern Sumatra and Banda Aceh, Indonesia after the tsunami and earthquake of 26th December 2004. *Seismol. Res. Lett.* **2005**, *76*, 312–320. [CrossRef]
18. Meltzner, A.J.; Sieh, K.; Abrams, M.; Agnew, D.C.; Hudnut, K.W.; Avouac, J.-P.; Natawidjaja, D.H. Uplift and subsidence associated with the great Aceh-Andaman earthquake of 2004. *J. Geophys. Res. Space Phys.* **2006**, *111*, 02407. [CrossRef]
19. Subarya, C.; Chlieh, M.; Prawirodirdjo, L.; Avouac, J.-P.; Bock, Y.; Sieh, K.; Meltzner, A.J.; Natawidjaja, D.H.; McCaffrey, R. Plate-boundary deformation associated with the great Sumatra–Andaman earthquake. *Nat. Cell Biol.* **2006**, *440*, 46–51. [CrossRef] [PubMed]
20. Sihombing, Y.I.; Adityawan, M.B.; Chrysanti, A.; Widyaningtias; Farid, M.; Nugroho, J.; Kuntoro, A.A.; Kusuma, M.A. Tsunami Overland Flow Characteristic and Its Effect on Palu Bay Due to the Palu Tsunami 2018. *J. Earthq. Tsunami* **2019**, *14*. [CrossRef]

21. Monecke, K.; Templeton, C.K.; Finger, W.; Houston, B.; Luthi, S.; McAdoo, B.G.; Meilianda, E.; Storms, J.E.; Walstra, D.J.; Amna, R.; et al. Beach ridge patterns in West Aceh, Indonesia, and their response to large earthquakes along the northern Sunda trench. *Quat. Sci. Rev.* **2015**, *113*, 159–170. [CrossRef]

22. Monecke, K.; Meilianda, E.; Walstra, D.-J.; Hill, E.M.; McAdoo, B.G.; Qiu, Q.; Storms, J.E.; Masputri, A.S.; Mayasari, C.D.; Nasir, M.; et al. Postseismic coastal development in Aceh, Indonesia—Field observations and numerical modeling. *Mar. Geol.* **2017**, *392*, 94–104. [CrossRef]

23. Meilianda, E.; Pradhan, B.; Syamsidik; Comfort, L.K.; Alfian, D.; Juanda, R.; Syahreza, S.; Munadi, K. Assessment of post-tsunami disaster land use/land cover change and potential impact of future sea-level rise to low-lying coastal areas: A case study of Banda Aceh coast of Indonesia. *Int. J. Disaster Risk Reduct.* **2019**, *41*, 101292. [CrossRef]

24. Leone, F.; Lavigne, F.; Paris, R.; Denain, J.-C.; Vinet, F. A spatial analysis of the December 26th, 2004 tsunami-induced damages: Lessons learned for a better risk assessment integrating buildings vulnerability. *Appl. Geogr.* **2011**, *31*, 363–375. [CrossRef]

25. Kaushik, H.B.; Jain, S.K. Impact of Great December 26, 2004 Sumatra Earthquake and Tsunami on Structures in Port Blair. *J. Perform. Constr. Facil.* **2007**, *21*, 128–142. [CrossRef]

26. Tobita, T.; Iai, S.; Banta, C.; Wimpie, A. Reconnaissance report of the 2004 great sumatra-andaman, Indonesia, Earthquake: Damage to geotechnical works in Banda Aceh and Meulaboh. *J. Nat. Disaster Sci.* **2006**, *28*, 35–41.

27. Jalil, A.; Fathani, T.F.; Satyarno, I.; Wilopo, W. A study on the liquefaction potential in banda aceh city after the 2004 sumatera earthquake. *Int. J.* **2020**, *18*, 147–155.

28. Natawidjaja, D.H.; Triyoso, W. The sumatran fault zone—From source to hazard. *J. Earthq. Tsunami* **2007**, *1*, 21–47. [CrossRef]

29. BAPPENAS. Scientific Basis: Analysis and Projection Sea Level Rise and Extreme Weather Event Report. In *ICCSR Report*; Ministry for National Development Planning: Lusaka, Zambia, 2010; p. 89.

30. Anh, N.Q.D.; Tanaka, H.; Tinh, N.X.; Viet, N.T. Sand spit morphological evolution at tidal inlets by using satellite images analysis: Two case studies in Vietnam. *J. Sci. Technol. Civ. Eng. (STCE) NUCE* **2020**, *14*, 17–27. [CrossRef]

31. Kombiadou, K.; Costas, S.; Carrasco, A.R.; Plomaritis, T.A.; Ferreira, Ó.; Matias, A. Bridging the gap between resilience and geomorphology of complex coastal systems. *Earth Sci. Rev.* **2019**, *198*, 102934. [CrossRef]

32. Goslin, J.; Clemmensen, L.B. Proxy records of Holocene storm events in coastal barrier systems: Storm-wave induced markers. *Quat. Sci. Rev.* **2017**, *174*, 80–119. [CrossRef]

33. Nott, J.; Forsyth, A.; Rhodes, E.; O'Grady, D. The origin of centennial- to millennial-scale chronological gaps in storm emplaced beach ridge plains. *Mar. Geol.* **2015**, *367*, 83–93. [CrossRef]

34. Masselink, G.; van Heteren, S. Response of wave-dominated and mixed energy barriers to storms. *Mar. Geol.* **2014**, *352*, 321–347. [CrossRef]

35. Sallenger, A.H.S. Storm impact scale for barrier island. *J. Coast. Res.* **2000**, *16*, 890–895.

36. Brantley, S.T.; Bissett, S.N.; Young, D.R.; Wolner, C.W.V.; Moore, L.J. Barrier Island Morphology and Sediment Characteristics Affect the Recovery of Dune Building Grasses following Storm-Induced Overwash. *PLoS ONE* **2014**, *9*, e104747. [CrossRef]

37. Long, A.J.; Waller, M.; Plater, A. Coastal resilience and late Holocene tidal inlet history: The evolution of Dungeness Foreland and the Romney Marsh depositional complex (U.K.). *Geomorphology* **2006**, *82*, 309–330. [CrossRef]

38. Stephan, P.; Suanez, S.; Fichaut, B. Long-, mid- and short-term evolution of coastal gravel spits of Brittany, France. In *Sand and Gravel Spits*; Randazzo, N., Jackson, D., Cooper, A., Eds.; Springer: Basel, Switzerland, 2015; pp. 275–288.

39. Morton, R.A.; Sallenger, A.H.S. Morphological impacts of extreme storms on sandy beaches and barriers. *J. Coast. Res.* **2003**, *19*, 560–573.

40. Sallenger, A.H.S.; Wright, C.W.; Howd, P.; Doran, K.; Guy, K. Extreme Coastal Changes on the Chandeleur Islands, Louisiana, During and After Hurricane Katrina. In *U.S. Geological Survey Scientific Investigations Report*; U.S. Geological Survey: Reston, VA, USA, 2009; pp. 27–36.

41. Woodroffe, C.D. The Natural Resilience of Coastal Systems: Primary Concepts. In *Managing Coastal Vulnerability*; McFadden, L., Penning-Rowsell, E., Nicholls, R.J., Eds.; Elsevier: Amsterdam, The Netherlands, 2007; pp. 45–60.

42. Riggs, S.R.; Cleary, W.J.; Snyder, S.W. Influence of inherited geologic framework on barrier shoreface morphology and dynamics. *Mar. Geol.* **1995**, *126*, 213–234. [CrossRef]

43. Hurukawa, N.; Wulandari, B.R.; Kasahara, M. Earthquake History of the Sumatran Fault, Indonesia, since 1892, Derived from Relocation of Large Earthquakes. *Bull. Seism. Soc. Am.* **2014**, *104*, 1750–1762. [CrossRef]

44. Kanagaratnam, U.; Schwarz, A.M.; Adhuri, D.; Dey, M.M. Mangrobe rehabilitation in West Coast of Aceh—Issues and perspectives. *NAGA WorldFish Cent. Q.* **2006**, *29*, 10–18.

45. Waltham, T. The Asian Tsunami disaster, December 2004. *Geol. Today* **2005**, *21*, 22–26. [CrossRef]

46. Syamsidik, S.; Oktari, R.S.; Munadi, K.; Arief, S.; Fajri, I.Z. Changes in coastal land use and the reasons for selecting places to live in Banda Aceh 10 years after the 2004 Indian Ocean tsunami. *Nat. Hazards* **2017**, *88*, 1503–1521. [CrossRef]

47. Al'ala, M.; Rasyif, T.M.; Fahmi, M. Numerical simulation of ujong seudeun land separation caused by the 2004 indian ocean tsunami, aceh-indonesia. *Sci. Tsunami Hazards* **2015**, *34*, 159–172.

48. Tursina; Syamsidik, S. Numerical simulations of land cover roughness influence on tsunami inundation in Ulee Lheue Bay, Aceh-Indonesia. In *IOP Conference Series: Earth and Environmental Science*; IOP Publishing: Bristol, UK, 2017; Volume 56, p. 12009.

49. Dominey-Howes, D.; Papathoma, M. Validating a Tsunami Vulnerability Assessment Model (the PTVA Model) Using Field Data from the 2004 Indian Ocean Tsunami. *Nat. Hazards* **2007**, *40*, 113–136. [CrossRef]

50. Dougherty, A.J.; Choi, J.-H.; Dosseto, A. Prograded Barriers + GPR + OSL = Insight on Coastal Change over Intermediate Spatial and Temporal Scales. *J. Coast. Res.* **2016**, *75*, 368–372. [CrossRef]

51. Van Maanen, B.; Nicholls, R.J.; French, J.R.; Barkwith, A.; Bonaldo, D.; Burningham, H.; Murray, A.B.; Payo, A.; Sutherland, J.; Thornhill, G.D.; et al. Simulating mesoscale coastal evolution for decadal coastal management: A new framework integrating multiple, complementary modelling approaches. *Geomorphology* **2016**, *256*, 68–80. [CrossRef]

52. Leont'Yev, I.O. Predicting shoreline evolution on a centennial scale using the example of the vistula (Baltic) spit. *Oceanology* **2012**, *52*, 700–709. [CrossRef]

53. Cowell, P.; Roy, P.; Jones, R. Simulation of large-scale coastal change using a morphological behaviour model. *Mar. Geol.* **1995**, *126*, 45–61. [CrossRef]

54. National Geophysical Data Center; NOAA; NCEI. NGDC/WDS Global Historical Tsunami Database. 2020. Available online: https://www.ngdc.noaa.gov/hazard/tsu_db.shtml (accessed on 30 October 2020). [CrossRef]

55. BMKG. *Catalogue of Significant and Destructive Earthquakes 1821–2017*; Pusat Gempa Bumi dan Tsunami Kedeputian Bidang Geofisika, BMKG: Jakarta, Indonesia, 2018.

56. Soloviev, S.L.; Go, C.N. *A Catalogue of Tsunamis on the Western Shore of the Pacific Ocean*; Academy of Sciences of the USSR, Ed.; Nauka Publishing House: Moscow, Russia, 1974; p. 439.

57. Lee, W.H.K.; Meyers, H.; Shimzaki, K. *Historical Seismograms and Earthquakes of the World*; Academic Press, Inc.: San Diego, CA, USA, 1988.

58. Kanamori, H.; Rivera, L.; Lee, W.H.K. Historical seismograms for unravelling a mysterious earthquake: The 1907 Sumatra Earthquake. *Geophys. J. Int.* **2010**, *183*, 358–374. [CrossRef]

59. Pentakota, S.; Seelanki, V.; Kolusu, S. Role of Andaman Sea in the intensification of cyclones over Bay of Bengal. *Nat. Hazards* **2018**, *91*, 1113–1125. [CrossRef]

60. Sahoo, B.; Prasad, K.B. Assessment on historical cyclone tracks in the Bay of Bengal, east coast of India. *Int. J. Climatol.* **2016**, *36*, 95–109. [CrossRef]

61. Siddiki, U.R.; Islam, M.N.; Ansari, M.N.A. Cyclonic track analysis using GIS over the Bay of Bengal. *Int. J. Appl. Sci. Eng. Res.* **2012**, *1*, 689–701.

62. CERC. Volume I: Technical Reference; Volume II: User's Guide. In *Automated Coastal Engineering System*; Department of the Army Waterway Experiment Station, Ed.; Corps of Engineers: Vicksburg, MS, USA, 1992.

63. Murty, T.S.; Rafiq, M. A tentative list of tsunamis in the marginal seas of the North Indian Ocean. *Nat. Hazards* **1991**, *4*, 81–83. [CrossRef]

64. Bilham, R.; Engdahl, R.; Feldl, N.; Satyabala, S.P. Partial and Complete Rupture of the Indo-Andaman Plate Boundary 1847–2004. *Seism. Res. Lett.* **2005**, *76*, 299–311. [CrossRef]

65. Pararas-Carayannis, G. Earthquake and Tsunami of June 26, 1941 in the Andaman Sea and the Bay of Bengal. 2005. Available online: http://www.drgeorgepc.com/Tsunami1941AndamanIslands.html (accessed on 30 October 2020).

66. Srivastava, K.; Kumar, R.K.; Swapna, M.; Rani, V.S.; Dimri, V.P. Inundation studies for Nagapattinam region on the east coast of India due to tsunamigenic earthquakes from the Andaman region. *Nat. Hazards* **2011**, *63*, 211–221. [CrossRef]

67. Murty, T. Storm surges—Meteorological ocean tides. In *Bulletin of the Fisheries Research Board of Canada*; National Research Council of Canada: Ottawa, ON, Canada, 1984.

68. Fearnley, S.; Miner, M.; Kulp, M.; Bohling, C.; Martinez, L.; Penland, S. Hurrican Impact and Recovery Shoreline Change Analysis and Historical Island Configuration. In *Sand Resources, Regional Geology, and Coastal Processes of the Chandeleur Islands Coastal System: An Evaluation of the Breton National Wildlife Refuge*; Lavoie, D., Ed.; U.S. Geological Survey: Reston, VA, USA, 2009; pp. 7–26.

69. Idris, Y.; Cummins, P.; Rusydy, I.; Muksin, U.; Syamsidik; Habibie, M.Y.; Meilianda, E. Post-Earthquake Damage Assessment after the 6.5 Mw Earthquake on December, 7th 2016 in Pidie Jaya, Indonesia. *J. Earthq. Eng.* **2019**, 1–18. [CrossRef]

70. Bennet, J.D.; Cameron, D.R.; Bridge, D.M.; Clarke, M.G.; Djunuddin, A.; Ghazali, S.A.; Thomson, S.J. *Geologic 1:250,000 Map of Banda Aceh Quadragle, Sumatra*; Geological Research and Development Centar (GDRC), Ed.; Direktorat Geologi: Bandung, Indonesia, 1983.

71. Ishihara, K.; Yoshimine, M. Evaluation of Settlements in Sand Deposits Following Liquefaction During Earthquakes. *Soils Found.* **1992**, *32*, 173–188. [CrossRef]

72. Volcano Discovery. Light mag. 4.7 Earthquake—Northern Sumatra, Indonesia on Friday, 8 April 1983. Available online: https://www.volcanodiscovery.com/earthquakes/quake-info/3801900/mag4quake-Apr-8-1983-northern-Sumatra-Indonesia.html (accessed on 30 October 2020).

73. McKinnon, E.E. Beyond Serandib: A note on Lambri at the northern tip of Aceh, Indonesia. In *Southeast Asia Program*; Cornell University: Ithaca, NY, USA, 1988; pp. 103–121.

74. Anthony, E.J. Patterns of Sand Spit Development and Their Management Implications on Deltaic, Drift-Aligned Coasts: The Cases of the Senegal and Volta River Delta Spits, West Africa. In *Coastal Research Library*; Randazzo, G., Jackson, D., Cooper, A., Eds.; Springer: Cham, Switzerland, 2015; p. 344.

Article

Assessing Fine-Scale Distribution and Volume of Mediterranean Algal Reefs through Terrain Analysis of Multibeam Bathymetric Data. A Case Study in the Southern Adriatic Continental Shelf

Fabio Marchese [1,2,*], Valentina Alice Bracchi [1,2], Giulia Lisi [1], Daniela Basso [1,2], Cesare Corselli [1,2] and Alessandra Savini [1,2,3]

[1]   Department of Earth and Environmental Sciences, University of Milano-Bicocca, Piazza della Scienza 4, 20126 Milano, Italy; valentina.bracchi@unimib.it (V.A.B.); g.lisi@campus.unimib.it (G.L.); daniela.basso@unimib.it (D.B.); cesare.corselli@unimib.it (C.C.); alessandra.savini@unimib.it (A.S.)

[2]   Department of Earth and Environmental Sciences, CoNISMa Local Research Unit of Milano-Bicocca, Piazza della Scienza 4, 20126 Milano, Italy

[3]   MaRHE Center (Marine Research and High Education Center), Magoodhoo Island, Faafu Atoll, Maldives

*   Correspondence: fabio.marchese1@unimib.it

Received: 21 November 2019; Accepted: 2 January 2020; Published: 4 January 2020

Abstract: In the Mediterranean Sea, crustose coralline algae form endemic algal reefs known as Coralligenous (C) build-ups. The high degree of complexity that C can reach through time creates notable environmental heterogeneity making C a major hotspot of biodiversity for the Mediterranean basin. C build-up can variably modify the submarine environment by affecting the evolution of submerged landforms, although its role is still far from being systematically defined. Our work proposes a new, ad-hoc semi-automated, GIS-based methodology to map the distribution of C build-ups in shallow coastal waters using high-resolution bathymetric data, collected on a sector of the southern Apulian continental shelf (Southern Adriatic Sea, Italy). Our results quantitatively define the 3D distribution of C in terms of area and volume, estimating more than 103,000 build-ups, covering an area of roughly 305,200 m^2, for a total volume of 315,700 m^3. Our work firstly combines acoustic survey techniques and geomorphometric analysis to develop innovative approaches for eco-geomorphological studies. The obtained results can contribute to a better definition of the ocean carbon budget, and to the monitoring of local anthropogenic impacts (e.g., bottom trawling damage) and global changes, like ocean warming and acidification. These can affect the structural complexity and total volume of carbonate deposits characterizing the Mediterranean benthic environment.

Keywords: geomorphometry; seafloor mapping; spatial analysis; algal reefs; bioconstruction volume; multibeam bathymetry

---

1. Introduction

Calcareous algal reefs are 3D biogenic-carbonate structures mostly built by calcareous red algae (CCA) that typify the seascape, from roughly 10 m of water depth down to those areas where reef-forming algae can thrive in dim light conditions [1–4]. In the warmer oceans, CCA may contribute to the complex of *coral reefs* or, on a small scale, may form independent bioherms. In temperate waters, like the Mediterranean Sea, CCA form endemic algal reefs known as Coralligenous (C) buildups, which represent the second most important hot-spot of biodiversity in the Mediterranean Sea after *Posidonia oceanica* meadows [4]. C is able to produce large deposits of biogenic calcium carbonate [5] according to climate change, oceanographic conditions, accommodation space, terrigenous supply variations and human-caused impact [6–10].

C geomorphological expression plays a crucial key-role in shaping the Mediterranean seascape and affecting its evolution through geological times, as it constitutes the most massive reefs along the Mediterranean continental shelf [11]. Nevertheless, a precise distribution of C is still uncomplete [11–14]. C morphotypes have been categorized mainly in two groups, banks and rims [4,15,16] that is based upon the nature of the substrate. Among banks, several different terms have been used to distinguish diverse local morphotypes (heads, blocks, patches, or banks [17]; vertical pillar, [18]; algal reefs, [19,20]; mound-shaped algal banks, [21]; minute reef aggregation and superficial layer formations, [22]; columns and ridges, [14]). A first attempt to use shape geometry descriptor to extract C from backscatter data was proposed by [23], but only recently a new categorization of C morphotypes, based on such a quantitative technique, has been proposed [11]. Moreover, C is responsible for the production and storage of a large amount of carbonate, but its contribution to the ocean carbon budget is still poorly quantified.

We apply geographic information system (GIS)-based analysis on digital elevation models (DEMs), obtained from high-resolution bathymetry data collected by multibeam echosounder (MBES) to investigate the variety of seaforms created by C build-ups, reaching scales and resolutions that are comparable to subaerial geomorphology studies [24].

We propose a new method for C feature extraction, designed upon the principals of geomorphometry and using established algorithms for surface analysis [25,26], applied to a well-studied C area located offshore Puglia Adriatic coasts (Italy) where a diverse type of C build-ups occur [11,14]. Our work focused on: (1) the development of a technique for automated identification and extraction of C build-ups; (2) the characterization of the spatial and morphological distribution of C; (3) the quantification of the total volume of selected C build-ups in relation to the field's subsurface geology, as determined from sub-bottom profiler (SBP) data.

Exploring and imaging submarine landforms requires the development of innovative approaches in using methods and techniques originally developed for the terrestrial environment (i.e., geomorphometry). In particular, we focus on providing quantitative information on C habitat as carbonate deposits on the Mediterranean shelf, which can be used as a baseline to which other C areas may be compared quantitatively.

2. Data and Methods

2.1. Data Acquisition and Processing

High-resolution acoustics data ground-truthed by video records were obtained by ship-based research surveys performed within the framework of BioMAP project (BIOcostruzioni Marine in Puglia, -P.O. FESR 2007/2013) that promoted actions in order to map and monitor C habitats along the Apulian shelf (southern Adriatic margin and northern Ionian margin, Mediterranean Sea) [27]. The survey has been carried out using a small research boat (Calafuria ISSEL, property of CoNISMa) in March and April 2013, acquiring 16 nautical miles of MBES survey lines, covering the seafloor over an area of 1.7 km^2, at depths ranging from 5 to 40 m and using a Teledyne RESON SeaBat 8125 (455 kHz) MBES (Figure 1). Teledyne PDS2000^{TM} software was used to acquire and process bathymetric data producing very high-resolution DEMs (0.5 m × 0.5 m). The correct position of the acquired data has been assured by a Hemisphere Crescent R-Series dGPS, and an IXSEA OCTANS motion sensor and gyro. Water sound velocity was obtained by a Valeport MiniSVS sound velocity sensor and by a Teledyne Reson SVP15 sound velocity profiler.

A set of high-resolution SBP lines have been also acquired on board the same research boat in an area NW of the Bari port (Figure 1), using a parametric Innomar 2000 (SES 2000), with an operating frequency of 10 kHz and 5 kHz. SBP data were processed using ISE Post-Processing Software, converted into SGY format and analyzed using PETREL (Schlumberger).

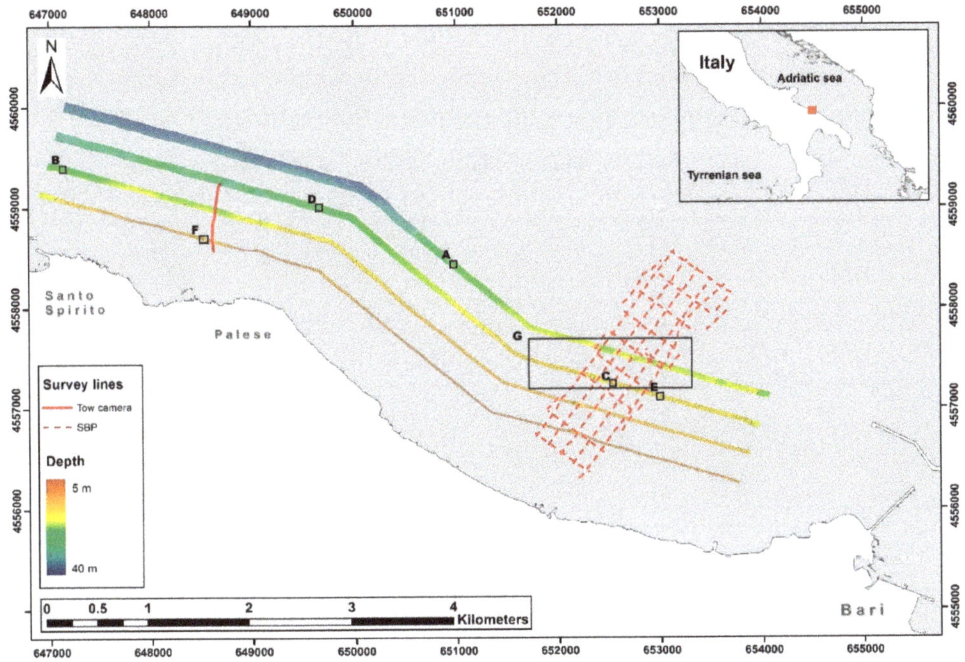

Figure 1. Location of the study area, bathymetric model for inset from [28]. Shaded relief map of the high-resolution multibeam bathymetry data acquired over the study area. Red track lines indicate the SBP survey and the trawled camera route. A, B, C, D, E, F, G areas are magnifications for the other figures in this paper.

The study area has also been surveyed by video inspections to ground-truth remote dataset, using an underwater trawled camera (Quasi-Stellar © Elettronica Enne) (Figure 2).

Figure 2. Examples of underwater images of Coralligenous build-ups made by the trawled camera. Following the classification proposed by [11]: (**A**) Discrete Relief, (**B**) Hybrid Bank and (**C**) Tabular Banks.

2.2. Geomorphometric Analysis and Extraction of C Build-ups

The entire spatial dataset (DEMs, survey routes and video tracks) was integrated into ArcGIS™ and ad-hoc geomorphometric analysis was performed with SAGA (System for Automated Geoscientific Analysis [29]). In particular, basic terrain parameters were extracted from selected areas (Figure 3) and the Topographic Position Index (TPI) proposed by [30] has been performed at the finest possible scale (1 inner radius and 5 outer radii), according to the DEM resolution (i.e., 0.5 × 0.5 m grid cell size) [30–32] TPI is actually only one of a vast array of morphometric parameters based on neighboring areas that can be useful in DEM analysis [32]. In particular, positive TPI values well represent locations that are higher than the average of their surroundings, as defined by the neighborhood (ridges), whereas negative TPI values represent locations that are lower than their surroundings (valleys). TPI values near zero are either flat areas (where the slope is near zero) or areas of constant slope (where the slope of the point is significantly greater than zero). We considered as a minimum TPI value 0.3 and therefore, not all cells under this value were considered as C builds-up. This approach facilitates the extraction of C build-ups from surrounding seafloor and reduces the occurrences of DEM artifacts. TPI scale (1–5) and value (0.3) were selected through a trial-and-error method in order to maintain the high-resolution of the extraction on which it depends the volume computation. Resulting rasters were re-classified and converted to obtain polygonal vector format. The remaining artifacts were manually deleted.

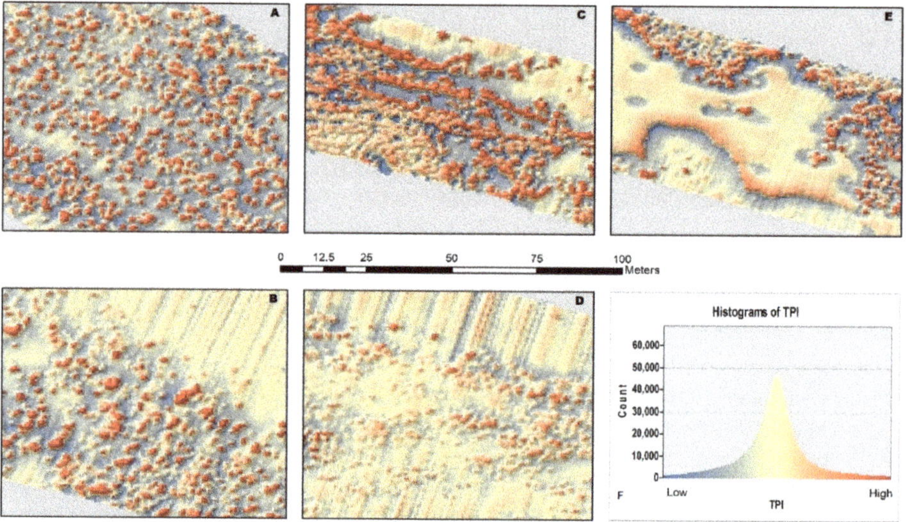

Figure 3. Topographic Position Index (TPI) maps, the color scale applied to the map indicate TPI value (low value in blue and high value in red) for the entire bathymetric dataset: (**A–E**) refer to the boxes in Figure 1, (**F**) histogram of the distribution for TPI value.

For each C polygon the shape index (SI) [33–35] was calculated that measures the complexity of patch shape compared to a circle shape. [11] demonstrates how SI is a useful tool to describe a seafloor landscape characterized by distinct C morphotypes.

The approach here proposed in performing the geomorphometric analysis was designed to specifically detect C build-ups and determine their 3D extension over the surrounding seafloor.

2.3. Acoustics Profiles Analysis

SBP dataset was analyzed using PETREL (Schlumberger) software in order to map the thickness of the C build-ups (Figure 4) along all survey lines and to explore the seismostratigraphic relationships between C build-ups and the substrate over which C has settled down. The level of details offered by our SBP data allowed us to assume that the depth of the substrate over which C growth at the time of its settlement is likely the same as the present-day seafloor that surrounds C build-ups. We, therefore, considered the seabed surrounding C build-ups as the base surface from which we extracted C build-ups as described in detail within the next section (Section 2.4).

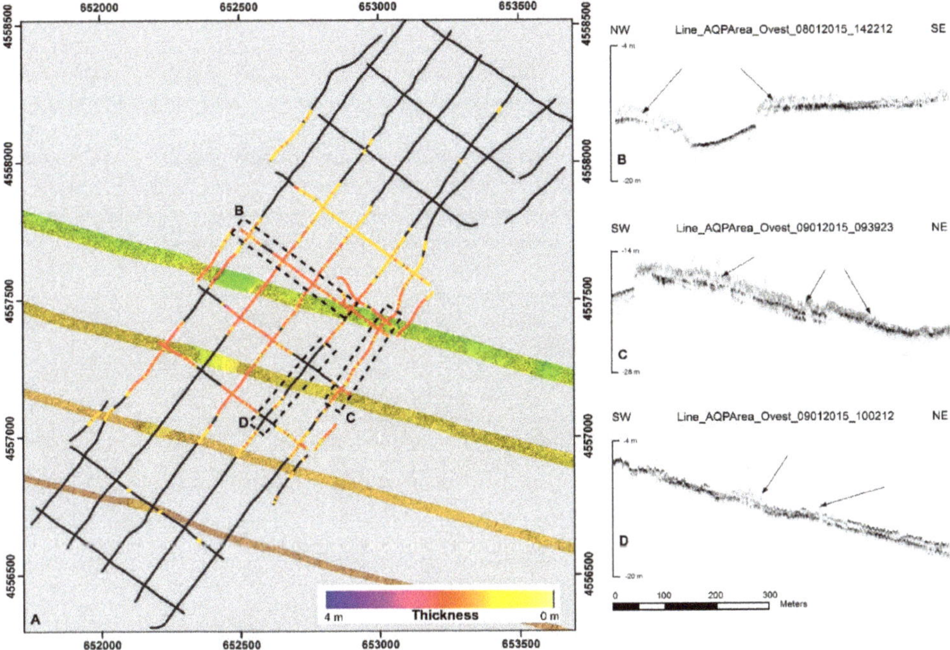

Figure 4. (**A**) Thickness map of the C build-ups calculated using SBP interpretation. (**B–D**) profiles represent examples of C echotype in the area (black arrows).

2.4. Volume Estimate

The volume calculation was developed by the creation of a reference surface without C build-ups that was created for each DEM through ESRI ArcGIS geostatistical analysis toolbox. The interpolation function, used for the creation of the reference surface, was the natural neighbor [36]. A natural neighbor interpolation algorithm is a weighted average interpolation technique not affected by anisotropy or variation in the data density because the selection of the neighbors is based on the configuration of the data [37]. It does not infer trends and will not produce peaks, pits, ridges or valleys that are not already represented by the input samples [37]. The surface passes through the input samples and is smooth everywhere except at the locations of the input samples. It adapts locally to the structure of the input data, requiring no input from the user pertaining to search radius, sample count, or shape. We decided to use the natural neighbors interpolator with the aim of leaving a coarser morphology, thus avoiding smoothing effects given by other methodologies such as Spline or Kriging [38]. The volume computational was performed by the comparison of the analyzed DEMs with the corresponding reference surface in order to calculate the volume of all detected C build-ups. ArcGIS provides the cut/fill tool that summarizes the areas and volumes of change from a cut-and-fill operation, if the input raster surfaces are coincident and with the same cell size. The attribute table of the output raster presents all changes in surface volumes generated by the cut/fill computation. Positive values for the volume difference indicate regions of the original raster surface that have been cut (material removed). Negative values indicate areas that have been filled (material added).

3. Results

3.1. Coralligenous Extraction Accuracy

The model extracted 172.548 polygons but only 103,494 positive morphologies (Figure 5) were finally related to C build-ups by visual inspection of the hillshade bathymetry and validation from video observations collected within the study area. Most of these artifacts have been attributed to: (a) bad roll correction that creates false positive elongated structures (Figure 5b,d), (b) occurrence of *Posidonia oceanica* dead matter on the shallow area that results in reliefs with typical sub-circular shape [39,40] (Figure 5e), and (c) artifacts concentration on DEM discontinues on its boundaries.

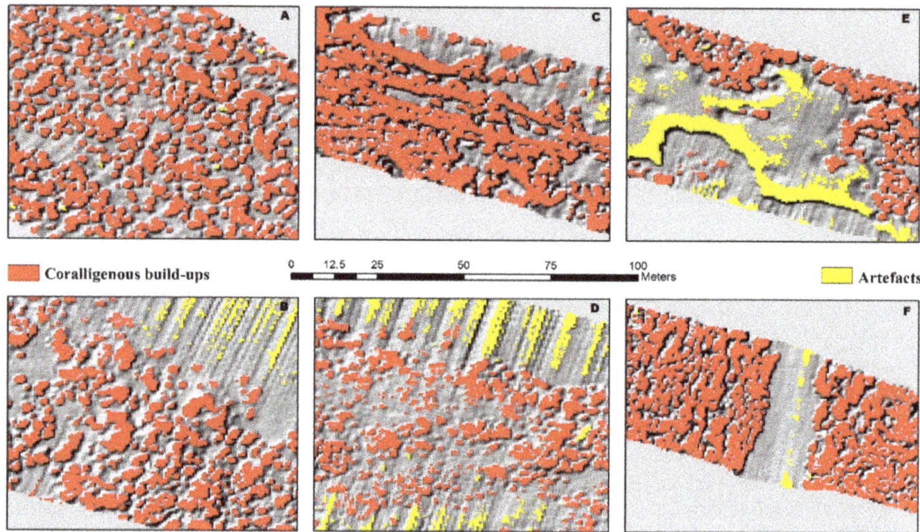

Figure 5. Result from geomorphometric analysis. (**A–F**) refer to the boxes in Figure 1. Red polygons represent coralligenous build-ups extracted, polygons in yellow are examples of artifacts filtered from the study.

3.2. Acoustics Profile Characterization

The 38 SBP profiles analyzed in our study are mostly characterized by an indistinct echotype [41], where few discontinuous reflectors can be detected, in agreement with that previously observed by [11]. When associated with the occurrence of C build-ups, the first return from the seafloor is imaged by an indistinct, low-amplitude, and highly transparent reflector, whereas lateral continuity and overall amplitude increase crossing the surrounding seafloor. A similar seismic structure for algal reefs was described in [22], where the underlying acoustic basement was also visible. Interestingly, in our surveyed area, C build-ups are in contact with an erosional truncation that marks the top of the underneath sequence (Figure 6). The erosional truncation can be associated to aerial erosion of past highstand marine deposits, or to the ravinement surface formed at the onset of the post-glacial sea-level rise, that on the South Adriatic shelf margin enhanced transgressive erosion of older deposits [42]. Both the reflectors (the seafloor and the erosional truncation) were digitalized and exported to calculate the thickness of C frameworks for all survey lines (Figure 6). In the entire area, C shows a width between 0.1 m to 2.6 m with an average of 1.04 m (Figure 6).

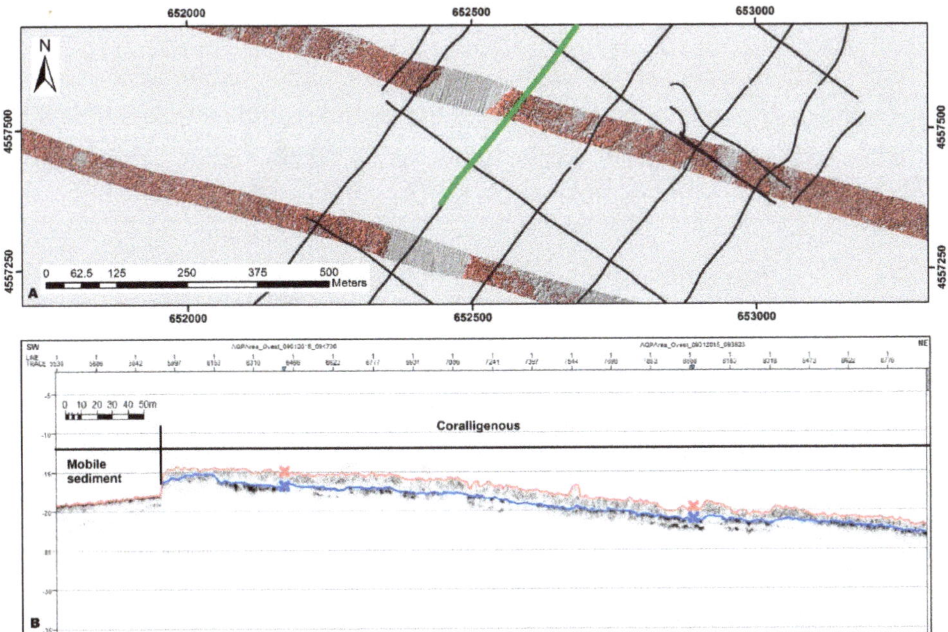

Figure 6. (**A**) SBP dataset with a selected profile (in green) crossing red polygons representing C build-ups obtained from the geomorphometric analysis. (**B**) Selected SBP profile showing in detail the echotype related to C occurrences and the erosional truncation that marks the top of the underneath sequence (blue horizon). Blue and red X represent intersect profiles.

3.3. Area and Volume

TPI parameters extracted the distribution of the C build-ups with high resolution in terms of perimeter boundary. Volume calculation is heavily dependent on the reference surface, consequently dependent on the interpolation algorithm (a large data gap could not be filled in with a good resolution by interpolation). Volume calculation was performed following two different approaches, according to the different C morphotypes described by [11] and mapped in our survey area. In order to not force the interpolation and therefore the volume calculation, especially for tabular bank morphotypes, SI of C morphotypes was used as a threshold in selecting the most appropriate approach. In particular: (Table 1, Figure 7):

- Discrete relief and hybrid bank [11]: SI $\leq$ 2, medium area 1.5 m^2 and maximum area 53 m^2. Volume computation has been performed using the cut and fill tool.
- Tabular bank [11]: SI > 2, medium area of 25 m^2 and maximum area of 1272 m^2. An average thickness of 1.49 m has been considered for volume computation, as calculated from the SBP analysis.

Table 1. Classification of C polygons modified from [11] based on SI and results in terms of area and volume.

Coralligenous Morphotypes	SI	Average Thickness (m)	Area (m^2)	Volume (m^3)
Discrete Relief and Hybrid Banks	$\leq$2	0.33 (TPI)	119,018.54	38,267.70
Tabular Banks	>2	1.49 (SBP)	186,209.99	277,452.88

The *cut and fill* tool was also used to analyze the performance of the segmentation process and the interpolation algorithm. The segmentation process directly affects the extraction of C build-ups (expected to be positive) as it will cut the DEM and consequently how the surrounding cells will influence the interpolation algorithm for the reference surface creation. In this way, with the extraction of positive structures from the DEM, positive balance was expected after the cut and fill tool for a correct volume computation. In our analysis, only in the deepest DEM, 0.3% presents a negative volume balance. This could be attributed to a small group of cells on the border of DEM that has been influenced by the surrounding lack of data.

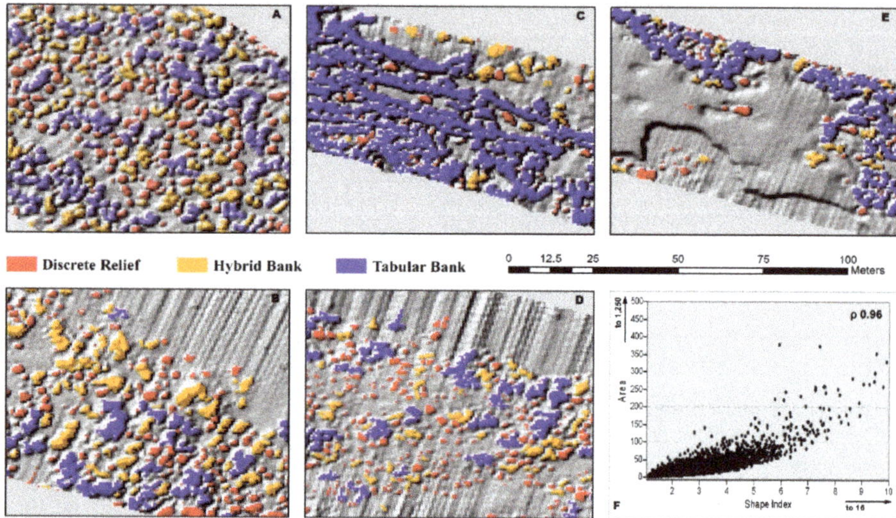

Figure 7. Distribution of C features extracted with TPI colored by morphotypes, following the classification proposed in [11]. (**A**–**E**) refer to the boxes in Figure 1, they are details from the entire dataset and show: Red polygons for discrete relief, yellow polygons for hybrid bank and blue polygons for tabular bank. (**F**) Scatterplot indicates the direct correlation between area and SI, Spearman's ρ = 0.96.

4. Discussion

4.1. TPI-Based Feature Extraction

The results of this work illustrate the importance of combining high-resolution acoustic techniques and geomorphometric analysis in order to have a preliminary quantitative characterization of C habitat distribution. To our knowledge, this is the first attempt to quantify the volume of Mediterranean algal reefs. Conventionally, segmentation of MBES data sets into defined submarine landforms is carried out manually despite the process might be highly subjective, slow and potentially inaccurate [43,44]. On the contrary, geomorphometric techniques (e.g., terrain attributes, feature extraction, automated classification) can objectively characterize seabed terrain from the coastal zone to the deep sea [45–52], although a common and standardized technique for automated seabed classification has never been developed [53]. Only recently other works successfully develop object-oriented methods applying object-based image analysis (OBIA) or considering a complete set of remote data in order to precisely characterize targeted landforms on the seabed that will document the extension of biodiversity hotspots [52,54–56]. In this study, we considered a C build-up as a discrete feature in both two and three-dimensional spaces in order to use a geomorphometric parameter for object-oriented modeling and extraction from the seafloor. The method developed here (Figure 8) intentionally chose a geomorphometric algorithm commonly available in existing terrain analysis software to achieve maximum applicability.

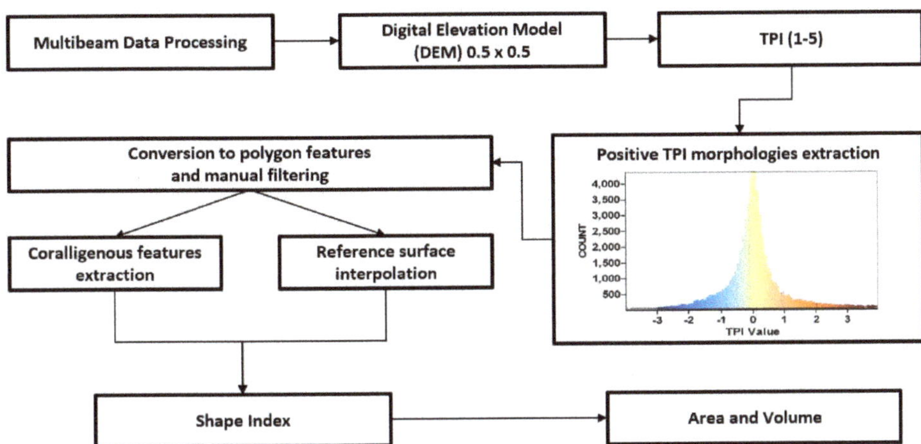

Figure 8. Conceptual model of the workflow for the extraction of volume and area of C features.

Variability of C morphotypes [11] poses several challenges to their automated extraction from DEM. The 3D complexity reached by C tabular banks [11] hampers the automatic detection of their edges, and consequently the creation of the hypothetical base surface for volume calculation. Since C build-ups raise from the surrounding seafloor (up to 4 m in height), their detection could be performed by slope analysis that is in general efficiently used to operate an accurate segmentation of isolated, small-scale features [47,48]. Nevertheless, the slope does not efficiently work for the tabular banks that usually have a lateral extension of more than 1 km^2. In this case, although the slope is efficient in detecting the boundary of C (determined by a high difference in elevation) it is not able to include within the segmentation process the inner part of tabular banks, where the high tridimensional complexity makes difficult the creation of a continuous polygon. On the other hand, geomorphometric parameters that have a higher performance in detecting the 3D complexity, such as the rugosity index (i.e.,: vector ruggedness measure [57,58] or terrain ruggedness index [59]) are able to better define the entire distribution of C tabular banks, although they fail in giving an accurate estimation of the dimension of smaller morphotypes like discrete relief. TPI represents, therefore, a good compromise for C morphotypes detection, since it measures the relative topographic position of the central point as the difference between the elevation at this point and the mean elevation within a predetermined neighborhood.

The time-consuming operation in manual filtering the erroneous polygonal could be definitely avoided using a high quality larger multibeam coverage without artifacts. The accuracy of the detection of C build-ups can be appreciated by the results of the cut and fill tool explained in the previous section.

4.2. Volume Computation

Volume computation is obviously constrained by the substrate depth from which C build-ups started to grow. Our approach considered the seafloor surrounding all detected C build-ups, as the original substrate over which C build-ups started to develop. The analysis of SBP dataset and the associated resolution supports our assumption since seismic profiles clearly show the continuity of the erosional truncation, detected at the base of C build-ups, with the surrounding seafloor bare of build-ups, below less than 2 meters of mobile sediment (visible also on videos, Figure 2). We considered two different approaches in computing the volume of C build-ups, according to the two different groups of C morphotypes (Table 1). Our volume computation strategy took into consideration the influence of the 3D geometry of C build-ups on the performance of the computational process, in order to not force the interpolation algorithms. For smaller data gaps (associated with C build-ups with SI $\leq$ 2) we used the results coming from the cut and fill tool. The volume of C tabular banks (with SI > 2, and therefore polygons with larger areas), that represent big gaps to solve by interpolation, was calculated using the average of C thickness value obtained from SBP data analysis. The analysis of the acoustic profiles was critical in supporting our volume computation strategy for tabular banks. Despite the considerable relief that C can reach in the survey area (up to 4 m if calculated after TPI extraction), SBP data show a maximum thickness of 2.5–3 m (only a few cells up to 4 m, Figure 4) decreasing to 1–1.5 m in the inner part of the C tabular banks.

4.3. Coralligenous Growth Rate and Age

Since present-day C depth distribution roughly matches those areas of the seafloor that were emerged during the last glacial maximum, C build-ups likely started to occur during the last transgression [60]. The C growth rate has been estimated by radiocarbon dating with a mean growth rate of 0.19 mm yr^{-1}, and a range of 0.11 to 0.26 mm yr^{-1} for C build-ups sampled between 10–35 m of water depth [61,62]. The results of this work suggest that a tridimensional quantitative characterization of C habitat could be related to the C growth rate. Using observed thickness values (Table 1) and assuming a prevailing vertical growth direction, we obtained respectively an age of at least 1.7 kyrs BP for discrete relief and hybrid banks, and 7.8 kyrs BP for tabular banks. Both these results are in agreement with the general estimated Mediterranean C build-ups age [4,60–62]. Finally, we calculate the C volume production *per* year that results in 22.04 m^3 yr^{-1} for discrete relief and hybrid banks and 35.38 m^3 yr^{-1} for the tabular banks morphotypes.

5. Conclusions

We demonstrate that our ad-hoc semi-automated, GIS-based methodology is useful to portray the three-dimensional complexity of C build-ups on a sector of the southern Apulian continental shelf and to quantify the extent to which C builders shaped the submarine environment, affecting the evolution of submerged landforms.

Notably, the maps and volume computation obtained from our work represents the first quantitative data supporting an estimation of the contribution of C build-ups to the ocean carbon budget. Our dataset also yields new details about the genesis of C build-ups (i.e., their potential relationships with drowned continental shelf morphologies), by imaging the development of C tabular banks in the study area.

Finally, despite the lack of a comprehensive investigation on the role of quaternary geomorphic processes on the C distribution and growing processes, the obtained results will support any future exploration of the relationship between framework morphologies and changes observed in the associated biological community, which still represent a major gap in knowledge in eco-geomorphological research of Mediterranean algal reefs.

Author Contributions: F.M. collected the entire dataset, performed and designed the methods. F.M. wrote and coordinated the manuscript with the significant contribution of A.S. V.A.B. and D.B. supervised the Coralligenous sections. F.M. and G.L. processed the SBP data. C.C. supervised the BIOMap Project. A.S. supervised the research. All authors contributed substantially to the discussion of the results and revision of the manuscript. All authors have read and agreed to the published version of the manuscript.

Funding: This study was financed by the BIOMAP project (P.O FESR 2007/2013, Puglia Region, Italy).

Acknowledgments: We are grateful to Michele Panza for the assistant during Calafuria Issel surveys (2013). SBP lines are property of Acquedotto Pugliese SPA and have been acquired in the framework of a commercial survey executed by the CoNISMa Local Research Unit of Milano-Bicocca. We thank Acquedotto Pugliese SPA for the scientific use of these data. We also thank three anonymous referees for their valuable comments on sections of this manuscript. FM and VAB are funded through a post-doctoral fellowship in Earth Sciences by the University of Milano-Bicocca. This article is also an outcome of Project MIUR—Dipartimenti di Eccellenza 2018–2022.

Conflicts of Interest: The authors declare no conflict of interest.

References

1. Laborel, J. Le concrétionnement algal "coralligène" et son importance géomorphologique en Méditerranée. *Recueil des Travaux de la Station Marine d'Endoume* **1961**, *23*, 37–60.

2. Bellan-Santini, D.; Lacaze, J.-C.; Poizat, C. Les biocénoses marine et littorales de Méditerranée, synthèse, menaces et perspectives. In *Collection Patrimoines Naturels*; Muséum National d'Histoire Naturelle: Paris, France, 1994; Volume 19, ISBN 2-86515-091-7.

3. Bressan, G.; Babbini, L.; Ghirardelli, L.; Basso, D. Bio-costruzione e bio-distruzione di Corallinales nel mar Mediterraneo. *Biologia Marina Mediterranea* **2001**, *8*, 131–174.

4. Ballesteros, E.; Avançats, E.; Csic, D.B. Mediterranean coralligenous assemblages: A synthesis of present knowledge. *Oceanogr. Mar. Biol. Annu. Rev.* **2006**, *44*, 123–195.

5. Basso, D. Carbonate production by calcareous red algae and global change. *Geodiversitas* **2012**, *34*, 13–33. [CrossRef]

6. Schlager, W. Depositional bias and environmental change-important factors in sequence stratigraphy. *Sediment. Geol.* **1991**, *70*, 109–130. [CrossRef]

7. Schlager, W. Accommodation and supply—A dual control on stratigraphic sequences. *Sediment. Geol.* **1993**, *86*, 111–136. [CrossRef]

8. Betzler, C.; Brachert, T.C.; Braga, J.C.; Martin, J.M. Nearshore, temperate, carbonate depositional systems (lower Tortonian, Agua Amarga Basin, southern Spain): Implications for carbonate sequence stratigraphy. *Sediment. Geol.* **1997**, *113*, 27–53. [CrossRef]

9. Hong, J. Impact of the pollution on the benthic coralligenous community in the Gulf of Fos, northwestern Mediterranean. *Bull. Korean Fish. Soc.* **1983**, *16*, 273–290.

10. Cinelli, F.; Balata, D.; Piazzi, L. Threats to coralligenous assemblages: Sedimentation and biological invasions. In Proceedings of the 3th Mediterranean Symposium on Marine Vegetation, Marseille, France, 27–29 March 2007; pp. 42–47.

11. Bracchi, V.A.; Basso, D.; Marchese, F.; Corselli, C.; Savini, A. Coralligenous morphotypes on subhorizontal substrate: A new categorization. *Cont. Shelf Res.* **2017**, *144*, 10–20. [CrossRef]

12. UNEP RAC/SPA. State of knowledge of the geographical distribution of the coralligenous and other calcareous bio-concretions in the Mediterranean. In Proceedings of the Ninth Meeting of Focal Points for SPAs, Floriana, Malta, 3–6 June 2009.

13. Martin, C.S.; Giannoulaki, M.; De Leo, F.; Scardi, M.; Salomidi, M.; Knitweiss, L.; Pace, M.L.; Garofalo, G.; Gristina, M.; Ballesteros, E.; et al. Coralligenous and maërl habitats: Predictive modelling to identify their spatial distributions across the mediterranean sea. *Sci. Rep.* **2014**, *4*, 5073. [CrossRef]

14. Bracchi, V.; Savini, A.; Marchese, F.; Palamara, S.; Basso, D.; Corselli, C. Coralligenous habitat in the Mediterranean Sea: A geomorphological description from remote data. *Ital. J. Geosci.* **2015**, *134*, 32–40. [CrossRef]

15. Pérès, J.M.; Picard, J. Nouveau manuel de bionomie benthique de la Mer Méditerranée. *Recueil des Travaux de la Station Marine dEndoume Bulletin* **1964**, *31*, 5137.

16. Laborel, J. Marine biogenic constructions in the Mediterranean. A review. *Sci. Rep. Port-Cros Natl. Park* **1987**, *13*, 97–126.

17. Sarà, M. *Research on Benthic Fauna of Southern Adriatic Italian Coast*; Final Scientific Report; Istituto e museo di zoologia e anatomia comparata, University of Bari: Modena, Italy, 1968; pp. 1–55.

18. Di Geronimo, I.; Di Geronimo, R.; Rosso, A.; Sanfilippo, R. Structural and taphonomic analysis of a columnar coralline algal build-up from SE Sicily. *Geobios* **2002**, *35*, 86–95. [CrossRef]

19. Bosence, D.W.J. Coralline algal reef frameworks. *J.—Geol. Soc. Lond.* **1983**, *140*, 365–376. [CrossRef]

20. Bosence, D.W.J. The "Coralligene" of the Mediterranean—A Recent analog for Tertiary coralline algal limestones. In *Paleoalgology*; Springer: Berlin/Heidelberg, Germany, 1985; pp. 216–225.

21. Toscano, F.; Sorgente, B. Rhodalgal-bryomol temperate carbonates from the Apulian shelf (Southeastern Italy), relict and modern deposits on a current dominated shelf. *Facies* **2002**, *46*, 103–118. [CrossRef]

22. Georgiadis, M.; Papatheodorou, G.; Tzanatos, E.; Geraga, M.; Ramfos, A.; Koutsikopoulos, C.; Ferentinos, G. Coralligène formations in the eastern Mediterranean Sea: Morphology, distribution, mapping and relation to fisheries in the southern Aegean Sea (Greece) based on high-resolution acoustics. *J. Exp. Mar. Bio. Ecol.* **2009**, *368*, 44–58. [CrossRef]

23. Fakiris, E.; Papatheodorou, G. Quantification of regions of interest in swath sonar backscatter images using grey-level and shape geometry descriptors: The TargAn software. *Mar. Geophys. Res.* **2012**, *33*, 169–183. [CrossRef]

24. Hughes Clarke, J.E.; Mayer, L.A.; Wells, D.E. Shallow-water imaging multibeam sonars: A new tool for investigating seafloor processes in the coastal zone and on the continental shelf. *Mar. Geophys. Res.* **1996**, *18*, 607–629. [CrossRef]

25. Pike, R.J. Geomorphometry—Process, practice and prospect. *Zeitschrift fur Geomorphologie NF SupplementBand* **1995**, *101*, 221–238.

26. Pike, R.J.; Evans, I.S.; Hengl, T. Geomorphometry: A brief guide. *Geomorphometry Concepts Softw. Appl.* **2009**, *33*, 3–30.

27. Campiani, E.; Foglini, F.; Fraschetti, S.; Savini, A.; Angeletti, L.; Bracchi, V.; Cardone, F.; Chimienti, G.; Marchese, F.; Mastrototaro, F.C.; et al. Conservation and management of coralligenous experience from The BIOMAP Project. In Proceedings of the GEOHAB Meeting, Lorne, Australia, 5–9 May 2014.

28. Becker, J.J.; Sandwell, D.T.; Smith, W.H.F.; Braud, J.; Binder, B.; Depner, J.; Fabre, D.; Factor, J.; Ingalls, S.; Kim, S.-H.; et al. Global Bathymetry and Elevation Data at 30 Arc Seconds Resolution: SRTM30_PLUS. *Mar. Geod.* **2009**, *32*, 355–371. [CrossRef]

29. Conrad, O.; Bechtel, B.; Bock, M.; Dietrich, H.; Fischer, E.; Gerlitz, L.; Wehberg, J.; Wichmann, V.; Böhner, J. System for Automated Geoscientific Analyses (SAGA) v. 2.1.4. *Geosci. Model Dev.* **2015**, *8*, 1991–2007. [CrossRef]

30. Weiss, A. Topographic position and landforms analysis. In Proceedings of the Poster Presentation ESRI User Conference, San Diego, CA, USA, 9–15 July 2001; pp. 227–245.

31. Guisan, A.; Zimmermann, N.E. Predictive habitat distribution models in ecology. *Ecol. Model.* **2000**, *135*, 147–186. [CrossRef]

32. Wilson, J.P.; Gallant, J.C. Digital Terrain Analysis. *Terrain Anal. Princ. Appl.* **2000**, *6*, 1–27.

33. Lang, S.; Blaschke, T. *Landschaftsanalyse Mit GIS Bearbeitet von*; Ulmer-UTB: Stuttgart, Germany, 2007; ISBN 9783825283476.

34. Crowley, J.M. Landscape Ecology, by R.T.T. Forman & M. Godron. John Wiley & Sons, 605 Third Avenue, New York, NY 10158, USA: Xix + 620 pp., figs & tables, 24 × 17 × 3.5 cm, hardbound, US $38.95, 1986. *Environ. Conserv.* **1989**, *16*, 90.

35. McGarigal, K.; Marks, B.J. *FRAGSTATS: Spatial Pattern Analysis Program for Quantifying Landscape Structure*; General Technical Report; US Department of Agriculture, Forest Service: Washington, DC, USA, 1995.

36. Sibson, R. A Brief Description of Natural Neighbour Interpolation. In *Interpreting Multivariate Data; Proceedings of Looking at Multivariate Data, Sheffield, UK, 24–27 March 1980*; University of Sheffield: Sheffield, UK, 1981; p. 374.

37. Watson, D.F. Contouring: A guide to the analysis and display of spatial data. In *Contouring: A Guide to the Analysis and Display of Spatial Data*; Pergamon Press: London, UK, 1992; ISBN 0080402860.

38. Tarolli, P.; Sofia, G.; Dalla Fontana, G. Geomorphic features extraction from high-resolution topography: Landslide crowns and bank erosion. *Nat. Hazards* **2010**, *61*, 65–83. [CrossRef]

39. Tomasello, A.; Luzzu, F.; Maida, G.; Orestano, C.; Pirrotta, M.; Scannavino, A.; Calvo, S. Detection and mapping of Posidonia oceanica dead matte by high-resolution acoustic imaging. *Ital. J. Remote Sens.* **2009**, *41*, 139–146. [CrossRef]

40. Innangi, S.; Barra, M.; Di Martino, G.; Parnum, I.M.; Tonielli, R.; Mazzola, S. Reson SeaBat 8125 backscatter data as a tool for seabed characterization (Central Mediterranean, Southern Italy): Results from different processing approaches. *Appl. Acoust.* **2015**, *87*, 109–122. [CrossRef]

41. Damuth, J.E.; Hayes, D.E. Echo character of the East Brazilian continental margin and its relationship to sedimentary processes. *Mar. Geol.* **1977**, *24*, 73–95. [CrossRef]

42. Ridente, D. Late pleistocene post-glacial sea level rise and differential preservation of transgressive "sand ridge" deposits in the Adriatic sea. *Geosciences* **2018**, *8*, 61. [CrossRef]

43. Cutter, G.R.; Rzhanov, Y.; Mayer, L.A. Automated segmentation of seafloor bathymetry from multibeam echosounder data using local fourier histogram texture features. *J. Exp. Mar. Bio. Ecol.* **2003**, *285*, 355–370. [CrossRef]

44. Bishop, M.P.; James, L.A.; Shroder, J.F.; Walsh, S.J. Geospatial technologies and digital geomorphological mapping: Concepts, issues and research. *Geomorphology* **2012**, *137*, 5–26. [CrossRef]

45. Lecours, V.; Dolan, M.F.J.; Micallef, A.; Lucieer, V.L. A review of marine geomorphometry, the quantitative study of the seafloor. *Hydrol. Earth Syst. Sci.* **2016**, *20*, 3207–3244. [CrossRef]

46. Micallef, A.; Le Bas, T.P.; Huvenne, V.A.I.; Blondel, P.; Hühnerbach, V.; Deidun, A. A multi-method approach for benthic habitat mapping of shallow coastal areas with high-resolution multibeam data. *Cont. Shelf Res.* **2012**, *39*, 14–26. [CrossRef]

47. Savini, A.; Vertino, A.; Marchese, F.; Beuck, L.; Freiwald, A. Mapping cold-water coral habitats at different scales within the Northern Ionian Sea (central Mediterranean): An assessment of coral coverage and associated vulnerability. *PLoS ONE* **2014**, *9*, e87108. [CrossRef]

48. Bargain, A.; Marchese, F.; Savini, A.; Taviani, M.; Fabri, M.C. Santa Maria di Leuca province (Mediterranean Sea): Identification of suitable mounds for cold-water coral settlement using geomorphometric proxies and maxent methods. *Front. Mar. Sci.* **2017**, *4*, 338. [CrossRef]

49. Misiuk, B.; Lecours, V.; Bell, T. A multiscale approach to mapping seabed sediments. *PLoS ONE* **2018**, *13*, e0193647. [CrossRef]

50. Diesing, M.; Green, S.L.; Stephens, D.; Lark, R.M.; Stewart, H.A.; Dove, D. Mapping seabed sediments: Comparison of manual, geostatistical, object-based image analysis and machine learning approaches. *Cont. Shelf Res.* **2014**, *84*, 107–119. [CrossRef]

51. Prampolini, M.; Blondel, P.; Foglini, F.; Madricardo, F. Habitat mapping of the Maltese continental shelf using acoustic textures and bathymetric analyses. *Estuar. Coast. Shelf Sci.* **2018**, *207*, 483–498. [CrossRef]

52. Janowski, L.; Trzcinska, K.; Tegowski, J.; Kruss, A.; Rucinska-Zjadacz, M.; Pocwiardowski, P. Nearshore Benthic Habitat Mapping Based on Multi-Frequency, Multibeam Echosounder Data Using a Combined Object-Based Approach: A Case Study from the Rowy Site in the Southern Baltic Sea. *Remote Sens.* **2018**, *10*, 1983. [CrossRef]

53. Micallef, A.; Berndt, C.; Masson, D.G.; Stow, D.A.V. A technique for the morphological characterization of submarine landscapes as exemplified by debris flows of the Storegga Slide. *J. Geophys. Res.* **2007**, *112*. [CrossRef]

54. Fakiris, E.; Blondel, P.; Papatheodorou, G.; Christodoulou, D.; Dimas, X.; Georgiou, N.; Kordella, S.; Dimitriadis, C.; Rzhanov, Y.; Geraga, M.; et al. Multi-frequency, multi-sonar mapping of shallow habitats-efficacy and management implications in the National Marine Park of Zakynthos, Greece. *Remote Sens.* **2019**, *11*, 461. [CrossRef]

55. Ismail, K.; Huvenne, V.A.I.; Masson, D.G. Objective automated classification technique for marine landscape mapping in submarine canyons. *Mar. Geol.* **2015**, *362*, 17–32. [CrossRef]

56. Lucieer, V.; Lamarche, G. Unsupervised fuzzy classification and object-based image analysis of multibeam data to map deep water substrates, Cook Strait, New Zealand. *Cont. Shelf Res.* **2011**, *31*, 1236–1247. [CrossRef]

57. Hobson, R.D. Surface roughness in topography: Quantitative approach. In *Spatial Analysis in Geomorphology*; Chorley, R.J., Ed.; Harper and Row: New York, NY, USA, 1972; pp. 221–245.

58. Sappington, J.M.; Longshore, K.M.; Thompson, D.B. Quantifying Landscape Ruggedness for Animal Habitat Analysis: A Case Study Using Bighorn Sheep in the Mojave Desert. *J. Wildl. Manag.* **2007**, *71*, 1419–1426. [CrossRef]

59. Riley, S.J.; DeGloria, S.D.; Elliot, R. A Terrain Ruggedness Index that Qauntifies Topographic Heterogeneity. *Intermt. J. Sci.* **1999**, *5*, 23–27.

60. Lo Iacono, C.; Savini, A.; Basso, D. Cold-Water Carbonate Bioconstructions. In *Submarine Geology*; Springer: Cham, Switzerland, 2018; pp. 425–455.

61. Sartoretto, S. Structure Et Dynamique D'Un Nouveau Type De Bioconstruction a Mesophyllum Lichenoides (Ellis) Lemoine (Corallinales, Rhodophyta). *Comptes Rendus de l'Académie des Sciences. Série 3, Sciences de la Vie* **1994**, *317*, 156–160.

62. Sartoretto, S.; Verlaque, M.; Laborel, J. Age of settlement and accumulation rate of submarine "coralligène" (−10 to −60 m) of the northwestern Mediterranean Sea; relation to Holocene rise in sea level. *Mar. Geol.* **1996**, *130*, 317–331. [CrossRef]

 *water*

Article

Sensing the Submerged Landscape of Nisida Roman Harbour in the Gulf of Naples from Integrated Measurements on a USV

Gaia Mattei [1,*], Salvatore Troisi [1], Pietro P. C. Aucelli [1], Gerardo Pappone [1], Francesco Peluso [1] and Michele Stefanile [2]

[1] Department of Science and Technology, Parthenope University Naples, 80143 Naples, Italy; salvatore.troisi@uniparthenope.it (S.T.); pietro.aucelli@uniparthenope.it (P.P.C.A.); gerardo.pappone@uniparthenope.it (G.P.); francesco.peluso@uniparthenope.it (F.P.)
[2] Department of Asia, Africa and Mediterranean, L'Orientale University, 80134 Naples, Italy; michelestefanile@gmail.com
* Correspondence: gaia.mattei@uniparthenope.it; Tel.: +39-081-5476635

Received: 11 October 2018; Accepted: 15 November 2018; Published: 19 November 2018

Abstract: This paper shows an interesting case of coastal landscape reconstruction by using innovative marine robotic instrumentation, applied to an archaeological key-site in the Campi Flegrei (Italy), one of the more inhabited areas in the Mediterranean during the Roman period. This active volcanic area is world famous for the ancient coastal cities of Baiae, Puteoli, and Misenum, places of military and commercial excellence. The multidisciplinary study of the submerged Roman harbour at Nisida Island was aimed at reconstructing the natural and anthropogenic underwater landscape by elaborating a multiscale dataset. The integrated marine surveys were carried out by an Unmanned Surface Vehicle (USV) foreseeing the simultaneous use of geophysical and photogrammetric sensors according to the modern philosophy of multi-modal mapping. All instrumental measurements were validated by on-site measurements performed by specialised scuba divers. The multiscale analysis of the sensing data allowed a precise reconstruction of the coastal morpho-evolutive trend and the relative sea level variation in the last 2000 years by means of a new type of archaeological sea-level marker here proposed for the first time. Furthermore, it provided a detailed multidimensional documentation of the underwater cultural heritage and a useful tool for evaluating the conservation state of archaeological submerged structures.

Keywords: coastal landscape evolution; Roman harbours; archaeological sea level marker; cultural heritage documentation; unmanned surface vessel; remote sensing of acoustic and optical data; 3D photogrammetric point cloud

---

1. Introduction

The understanding of landscape changes plays a fundamental role in the comprehension of the human occupation. In the history of human settlements along the Mediterranean coasts, an important moment is the foundation of harbours and coastal urbanisation, when the coastal landscapes started to be transformed by human activities [1]. As most of these ancient coastal settlements are today submerged due to relative sea level variations occurring over the last millennia, the challenge of coastal geoarchaeological research is to thoroughly study these submerged archaeological structures [2–7]. This cutting-edge information leads to understanding the impacts of past climate changes on modern populations and the effects of Earth processes at the social level [8]. Moreover, by integrating the results of the geological and geomorphological analysis, geophysical and geomatic survey, and archaeological investigations, a twofold target can be efficiently achieved:

- The protection of the underwater archaeological sites exposed to the waves action after its recent submersion.

- The management of the underwater cultural heritage as a witness of the effects of the ongoing climate changes on the ancient settlements as well as on the coastal modifications.

The interdisciplinary nature of underwater geoarchaeological studies leads to operate at multi-scale levels during both the data acquisition and interpretation phases. In the past, the data acquisition phase was mainly carried out by direct surveys of scuba divers, geologists and archaeologists, as the main operating problem during the coastal surveys was the difficulty navigating close to the shoreline, especially in presence of submerged archaeological remains. Thanks to the innovative technology applied to the miniaturisation of geophysical instruments, the use of small crafts carrying out measurements in very shallow waters areas had a remarkable development. These marine systems, including Unmanned Surface Vessels (USV), are revolutionising our ability to map and monitor the marine environment [9–11].

In the last few years, numerous specific crafts have been designed for surveying in shallow waters. As reported in the literature, they have been equipped with several geophysical instruments to reconstruct the seabed and sub-bottom morphology [9,12]. In [13], a prototype USV (MicroVEGA) ia presented, as a direct result of many years of experience in marine geophysical surveys applied to underwater archaeology. MicroVeGA is a drone conceived to operate in very shallow water where traditional boats have several manoeuvring problems. This vessel is equipped with a GPS, a single beam echosounder, an inertial platform and emerged and submerged cameras.

In this paper, we present a second evolutive step of MicroVEGA project including the installation of an underwater photogrammetric system and the development of multidisciplinary procedures optimised for geoarchaeological studies. In fact, the photogrammetric methods applied to underwater archaeology have significantly improved the deductions around ancient landscapes. The three-dimensional reconstruction of an archaeological site provides relevant information to formulate hypotheses about the main morphological characteristics of the ancient anthropic landscape [14–16]. 3D representations of an underwater site provide an important benefit to archaeologists on the study of a three-dimensional overall picture, which is otherwise difficult to obtain in underwater environments [17–19]. In addition, this representation is a modern instrument to appreciate the underwater cultural heritage not accessible to everyone.

The aim of this study was to reconstruct the submerged geoarchaeological landscape in a Roman harbour of Naples Gulf (Nisida, southern Italy) and to evaluate the main morphometric characteristics of the submerged archaeological site, by using a USV equipped with geophysical and photogrammetric sensors. The research targets are as follows:

- assessing the main coastal changes of Nisida Roman harbour that occurred in the last 2000 years mainly due to the relative sea level variation by means of a multiscale elaboration of a transdisciplinary dataset;

- the detection of a new type of archaeological sea level marker in the case of port-like structures made in hydraulic concrete; and

- the detailed documentation of a submerged archaeological site in the Gulf of Naples with a high cultural value and not accessible to everyone.

2. Geoarchaeological Background

The coasts of the Gulf of Naples have been inhabited since the ancient times and shaped by the continued interplay between anthropogenic and volcanic forcing [5–7,20–22]. It is a half-graben characterised by an NE-trending faults system and totally covered by volcanic units related to Campi Flegrei and Vesuvius volcanoes [5,23]. It is structurally controlled by numerous Quaternary fault systems, NE–SW trending SE-dipping and NW–SE trending SW dipping, related with the last stages of the opening of the Tyrrhenian Sea [24–26].

Between the Middle and Upper Pleistocene, the fault systems were responsible for the development of the half-graben of the Gulf of Naples and Sorrento Peninsula fault block ridge [6,23].

During the Late Holocene, this area was modified by volcanic activity and volcano-tectonic vertical ground movements related to Campi Flegrei and Vesuvius [4–7,20–22,27]. However, the most recent extreme event that profoundly modified the Gulf of Naples coasts certainly was the 79 AD eruption of Vesuvius [28] and the subsequent adverse marine weather conditions [7].

Furthermore, despite the subsiding trend that affected the coasts of Naples Gulf after the emplacement of the Campanian Ignimbrite deposits [5,20,21,29–31], the landscape evolution of this coastal sector in the last 2000 years was mainly controlled by the anthropogenic activity and the construction of several maritime infrastructures. During the Greek times, several colonies were founded on its coasts, such as Pithekoussai (on the island of Ischia), Kymae, Dicearchia and Neapolis. Instead, during the Roman period, this territory was densely occupied by towns with a maritime vocation, such as Puteoli and Neapolis, and other coastal cities, such as Aenaria, Cumae, Herculaneum, Pompeii, Stabiae and Surrentum. In fact, the Greek geographer Strabo described the Gulf (mentioned with the name of Crater) as an uninterrupted sequence of luxurious villas and gardens [32].

This coastal sector is known worldwide for the commercial ports that in Roman times represented a hub for trade with the East and Africa and for the arrival and redistribution of wheat and food for the needs of Rome. In this study, we focused our attention on a minor port, little known from historical sources but of considerable cultural interest as it preserves the founding remains of a pier that is the largest in the Gulf of Naples.

The Roman harbour of Nisida was built in the first century BC and mainly protected by two piers, of which nowadays only some totally submerged witnesses remain (Figure 1), though well-preserved and not buried by recent sediments.

The ancient piers and coastal defence structures during the Roman period—since the first century BC—were mainly composed of alignments of *pilae*—large or tall cubes. The *pilae* are a typical maritime construction evolved along with the development of hydraulic concrete in the Gulf of Naples [33–36]. The use of hydraulic concrete revolutionised the design of harbour and other maritime structures during this period and was well described by Strabo in *Geographia*, Vitruvius in *De Architectura*, Seneca in *De Rerum Natura*, and Plinius in *Naturalis Historia*.

Roman hydraulic concrete consists of large irregular stone or tuff aggregate set into a mortar of lime and sand-like volcanic ash rich in chemically reactive aluminosilicates [37–39]. It is mainly made in pulvis puteolanus (powder from Puteoli), so-called by these historians to identify the area around Puteoli, in the Gulf of Naples, as the source of this volcanic ash. Pozzolana is the modern term to define this volcanic ash. This material, which could be cast and set underwater, began to be used in harbour structures in the second century BC [40]. Since its discovery, this precious powder was exported throughout the Roman Empire because of its owner characteristic to be cast and set underwater and—as Vitruvius said—to be hardened by the sea to a strength "which neither the waves nor the force of the water can dissolve". In fact, when seawater infiltrates in the concrete, it dissolves some of the ash and rather than undermining the structure, the alkali fluid that is left allows minerals to strengthen it [37].

The *pilae* made in Roman concrete were grouped together in a single line (sometimes connected by arches, as at the breakwater at Nisida and Puteoli) or in two overlapping rows to form discontinuous breakwaters or sea defences for a shoreline or at the entrance to a harbour [39]. These cubic structures were built on the seabed with the cofferdam technique. Vitruvius precisely describes this technique in *De Architectura* (2.6.1, 5.12.2–3), but there are several useful shorter comments in Strabo, Pliny, and Dionysios of Halikarnassos [37,41]. The cofferdam sides were formed of oaken stakes (*catenae* and *destinae*, as shown in Figure 2), and filled up with hydraulic concrete.

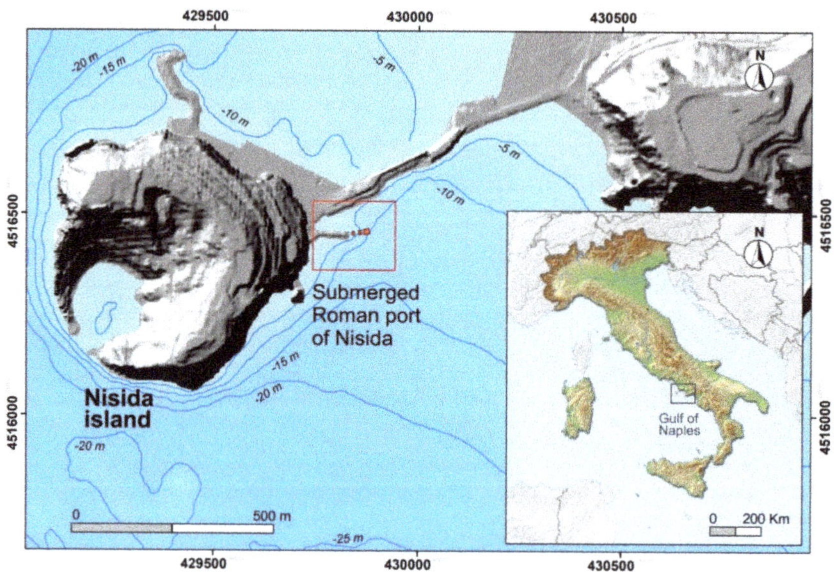

Figure 1. Location map of Nisida Roman harbour with the submerged archaeological structures in red. The shaded relief (altitude 45°, azimuth 315°) in figure derives from a DTM with a spatial grid 2 m × 2 m. The Ministry of the Environment kindly provided the Lidar data used to calculate the DTM.

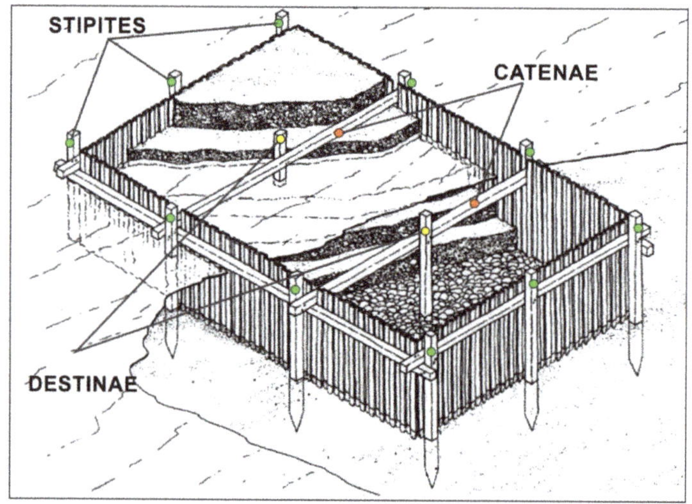

Figure 2. Vitruvius' cofferdam for hydraulic concrete. Reconstruction (after [39]).

This manner of construction can be seen in the ancient representations of the harbours of Puteoli and Baiae on several wall paintings and on a series of engraved glass souvenir vessels found at various sites around the empire [42–44]. As built directly at the sea, these port facilities are also studied as markers of ancient sea levels [45–48].

In this study, the better-preserved *pila* of the Nisida Roman harbour was surveyed with high detail, identifying the traces of the ancient cofferdam, used here to propose a new type of measuring point—valid in the case of port-like structures built in hydraulic concrete cast and set underwater.

Surveyed Area

The Nisida Roman harbour was built in the first century BC—probably around 37 BC—when numerous maritime infrastructures were realised in the Gulf of Pozzuoli with the same building techniques [49]. It was located at the footslope of Nisida Island (Naples City), a tuff cone (3.9 ky BP [21]) partially dismantled by the waves, leaving a passage inside the crater and drawing a small bay (so-called Porto Paone) inhabited since the Roman period. This island, according to the tradition [50], was the *otium* residence of the Roman politician Marcus Iunius Brutus (85 BC–42 BC) and was connected with the land by a tunnel about 4.5 m high (Figure 3a,b), now partially submerged (Figure 3b). The Roman harbour of Nisida was characterised by an *opus pilarum* almost totally destroyed, except for three *pilae* making up the ancient pier nowadays submerged [51] (Figure 3a). Considering the remarkable size of the *pilae* (with edges up to 15 m) and the bathymetry in the area of about 10 m, the harbour probably represented a commercial hub for the suburban sector of Posillipo [49,52].

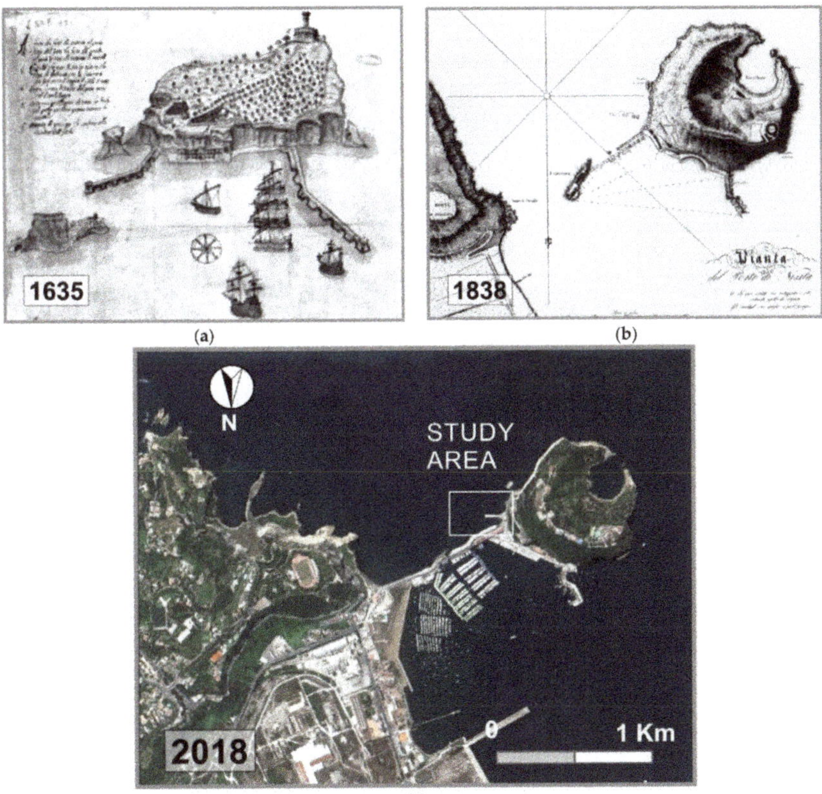

Figure 3. (**a**) Pictorial view of Nisida harbour in 1635 by B. Picchiatti where the Roman *opus pilarum* is clearly visible; (**b**) historical map "PIANTA DEL PORTO DI NISITA" 1838, where the *pilae* are still intact and emerged; and (**c**) present-day aerial photo.

In the last centuries, the main coastal modifications have been caused by the construction of the bridge connecting the Nisida Island to Posillipo hill, as well as the piers in the southern sector of the island (Figure 3c). Two of these piers were built on the remains of Roman port structures, which remained mostly intact up to 1635, as shown in Figure 3a, where the Roman *opus pilarum* is clearly visible. In Figure 3b, instead, it can be observed that the pillars (*pilae*) composing the piers were still intact and emerged, while the connecting arcs had already collapsed. The modern bridge in Figure 3c was built above the semi-submerged Roman tunnel that connected the islet with the mainland.

The investigated area is located in the island SE sector, close to the bridge and the northernmost pier. In detail, it includes the remains of three submerged *pilae* composing the ancient pier; the other *pilae* have been recently covered by the modern cliff.

3. Materials and Methods

3.1. Survey

The study area extends for about 200 m parallel to the coast, and for about 150 m perpendicular to the coast. The bathymetry in this area ranges between the coastline and the isobath of -10 m.

The area is exposed to the southern winds, which create a strong wave motion. In addition, the summer period is often characterised by adverse marine weather conditions (such as algal blooms) that cause high water turbidity. For this reason, the marine surveys were carried out in the winter months, when northern winds create calm sea conditions and good visibility.

The major operational difficulties were caused by the proximity of a marina and the presence of a passage for small boats used to avoid the circumnavigation of the island of Nisida, located right near the investigated area. This condition has made indispensable the support of the Italian Coast Guard to regulate tourist maritime traffic. The aims of the survey were:

- precise reconstruction of the Roman structures still visible;
- evaluation of the conservation state of the more intact *pila*;
- 3D reconstruction of the underwater landscape in the study area; and
- detection of the measuring point for the evaluation of the relative sea level variations in the last 2000 years.

The coastal sector was mainly surveyed by remote sensing methods and the measurements were validated by direct surveys. A small USV was used during the marine surveys (Figure 4a,b,d,e), for both operational and technological reasons. The operative reason derives from the USV manoeuvrability, a characteristic necessary to carry out the planned survey with short and close navigation lines (lines spacing < 2 m) in a small area (15 m × 15 m). The technical reason derives from the equipment of the USV with photogrammetric and geophysical sensors.

The integrated marine survey carried out in the study area included:

- a bathymetric survey;
- a Side scan sonar morphological survey;
- a photogrammetric survey;
- a topographic survey of two submerged Ground Control Points (GCPs, Figure 4c); and
- a direct underwater survey (Figure 4f).

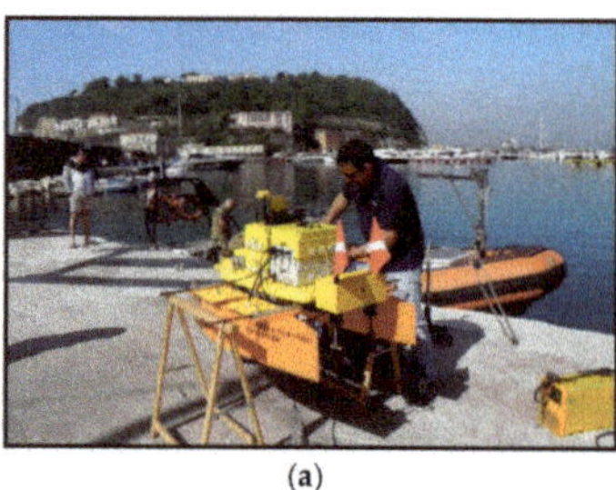
(a)

(b)

Figure 4. *Cont.*

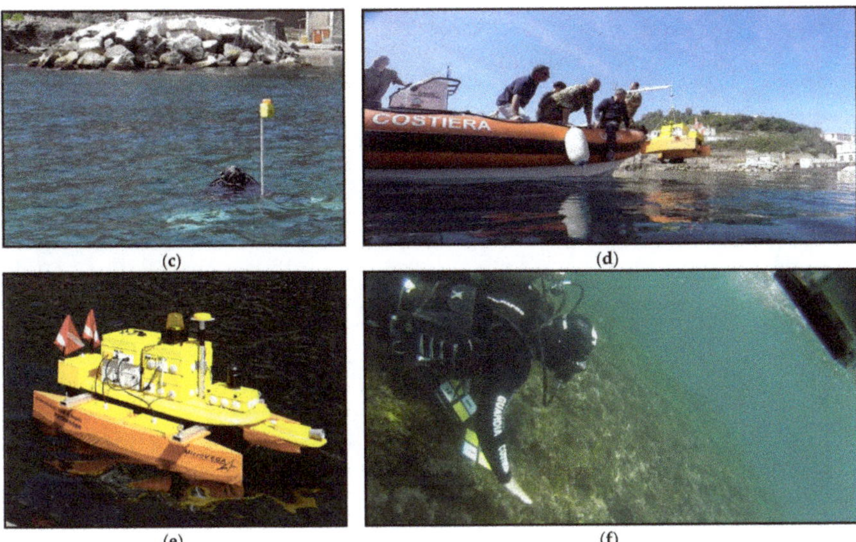

Figure 4. Direct and indirect marine surveys: (**a**) sensors installation on MicroVeGA; (**b**) Semi-submerged Roman gallery; (**c**) topographic survey of a GCP; (**d**) initial operations of the marine survey; (**e**) MicroVeGA in action; and (**f**) initial operations of the topographic survey.

The morpho-bathymetric survey—including a single-beam echo sounder (SBES) and a side scan sonar (SSS) survey—was carried to reconstruct the seabed morphology [53] as well as to precisely map the underwater archaeological structures. To obtain this twofold result, the navigation was planned in two phases (Figure 5):

- A large grid composed of 10 navigation lines perpendicular to the coast and 5 lines parallel, with a linear extension of 2500 m was created (Figure 5a).
- A small grid composed of 12 × 12 navigation lines 24 m long and 2 m spaced, with a linear extension of 600 m was also created (Figure 5a).

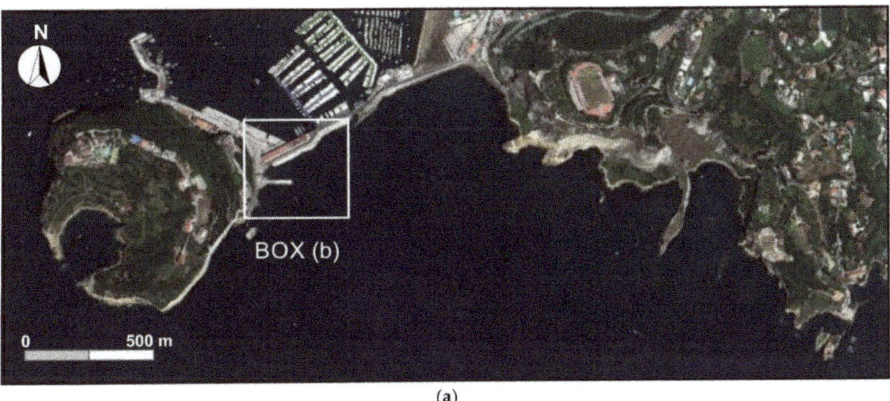

(**a**)

Figure 5. *Cont.*

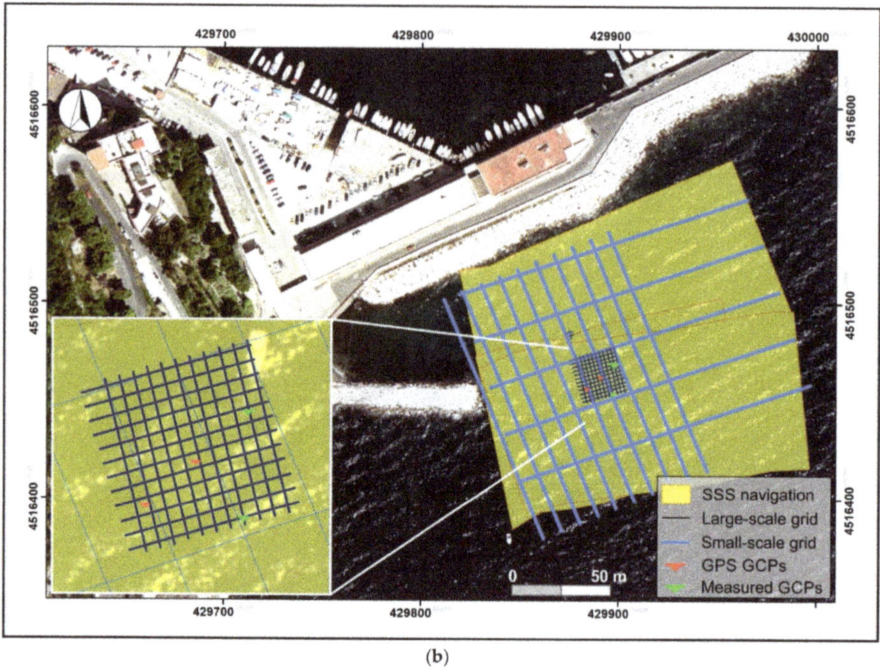

(**b**)

Figure 5. (**a**) Nisida aerial photo with a location map of the study area; and (**b**) navigation grids of morpho-bathymetric and photogrammetric surveys.

During the navigation with a large grid, the side scan sonar and bathymetric data were recorded to characterise and map the seabed morphology by analysing the acoustic signals. During this phase, the underwater cameras recorded georeferenced videos used in post-processing to validate the interpretations of the morpho-acoustic data.

During the navigation with the small grid, the bathymetric and photogrammetric data were recorded. The integration between the bathymetric and photogrammetric measurements provided a 3D reconstruction of the underwater archaeological structure.

As MicroVeGA USV is inspired to the multi-modal mapping operating philosophy, it acquired both acoustic and optical data during the same survey.

In detail, during the integrated marine survey, MicroVeGA USV collected:

- 3100 m of bathymetric data;
- 30,000 m^2 of SSS sonographs; and
- 62 min of high definition videos.

The topographic survey was carried out by a scuba diver of the 2nd Divers Team—Italian Coast Guard of Naples. A GPS fast-static technique was applied in each ground control point to obtain a precise positioning of two markers placed on the submerged *pila* and the mutual distance with two other markers (Figures 4c and 6).

A team of scuba divers, archaeologist and geologists carried out the direct survey to validate the indirect measurements and to video-survey areas of archaeological interest to characterise the submerged structure. The GPS receiver was installed on a graduated range pole positioned on each GCP for a period of 10 min (Figure 4c) to collect single-frequency phase measurements.

The two GPS baselines between a reference ground station located at a distance of about 5 km from the study area and the two GCPs stations had a fixed solution that guaranteed an uncertainty of less than 2 cm in both planimetry and altimetry.

Moreover, spatial distances between the two GCPs positioned during the fast-static GPS survey and other two markers located in the archaeological structure's corners were measured during the survey, and used as constraints in the photogrammetric bundle adjustment procedure.

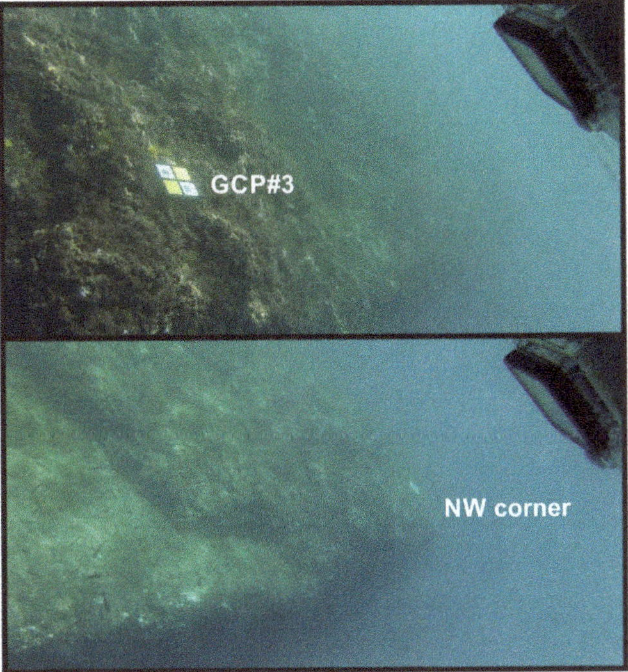

Figure 6. Photos of the underwater archaeological structure by MicroVeGA.

3.2. USV

The study area was surveyed with both direct and indirect methods. The USV (MicroVeGA) used during the indirect surveys—designed exclusively for the geoarchaeological task—is the result of many years of experience in marine geophysical surveys applied to underwater archaeology.

This operational experience has led to the creation of a vehicle with a simple and robust structure but able to effectively carry out the necessary long working sessions at low speed typical of this kind of survey.

MicroVeGA applies the multimodal mapping technique that involves the use of multiple on-board sensors for mapping, localisation, and data collection. All data are broadcast in real time both to a base station and to all operators involved in the research (geophysicist, archaeologist, geomorphologist, etc.).

The main task is the acquisition of data related to the morphology of the seabed to realise three-dimensional landscape models, using geophysical (Single Beam Echo Sounder and Side Scan Sonar) and optical (underwater cameras) instruments.

MicroVeGA platform is a catamaran-type vessel with an overall length of 135 cm, a width of 86 cm and a height of 80 cm (Figure 7, Table 1).

The two hulls are joined by two aluminium crossbars that support a rectangular base of synthetic material (70 cm × 80 cm) on which the payload is positioned.

The payload is divided into four IP-68 waterproof containers, each containing a different type of instrumentation to guarantee the modularity of the survey planning. The operating weight is about 30 kg, although it may vary depending on the operational configuration and the mission profile. The propulsion is entrusted to two brushless electric motors; the vehicle has a maximum speed greater than 2 Knots, while the speed during the data acquisition phase is less than 1 Knots. Operating autonomy with a battery pack is 2 h.

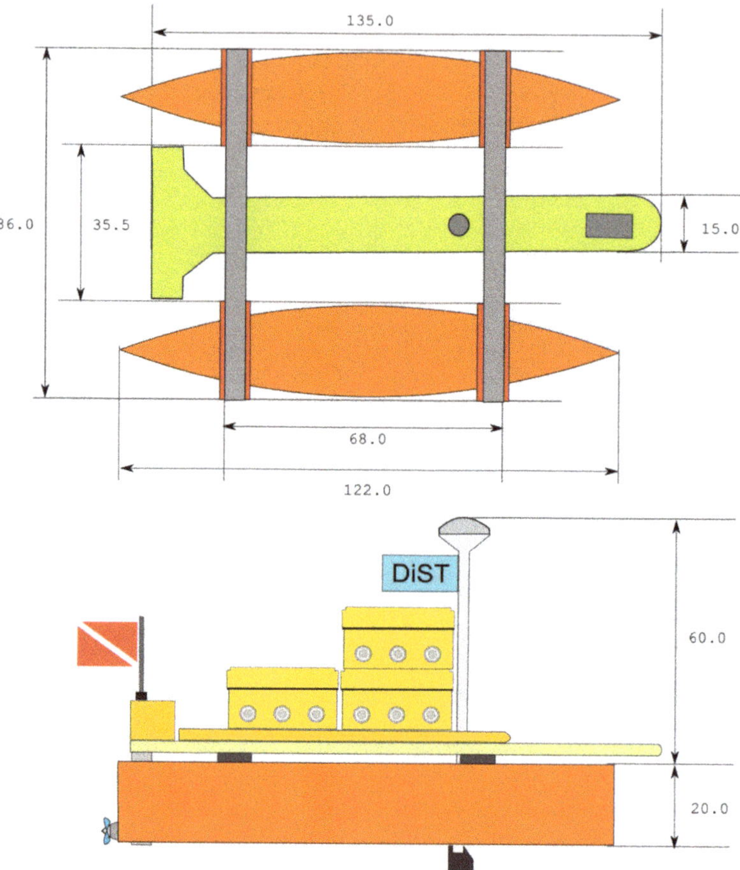

Figure 7. Dimensions (cm) of MicroVeGA catamaran.

Table 1. General characteristics of MicroVeGA drone.

MicroVEGA	General Characteristics
Length	1.35 m
Beam	0.86 m
Height	0.60 m
Empty weight	16 Kg
Max weight	32 Kg
Speed Max/Cruise	1/0.5 m/s
Endurance	2 h at Speed Max
Standard Sensors	GPS, IMU, Compass
Payload	SBES, SSS, CAM 3D
Propulsion	2 × brushless thruster 100 W
Communications	WiFi 5 GHz
Software	Tritech StarFish and TrackStar
Operative System	Windows, Linux and C++
Power	12 V × 35 Ah

The catamaran design was chosen for good stability and optimal loading space for sensors on-board. The low power suggests its use in the very shallow waters and with good weather.

3.2.1. Acquisition Module

MicroVEGA has a payload of sensors able to perform both large- and small-scale geomorphological surveys of submerged coastal environments, in addition to the positioning sensors (GPS, IMU, RC).

The data acquisition payload consists of:

- A 200 KHz digital echosounder;
- A 450 KHz digital side scan sonar; and
- A high definition 3 cam photogrammetric system.

The Single-Beam Echo Sounders (SBES) is an Ohmex with 200 KHz acquisition frequency and 60 m as maximum measured depth, therefore optimised for coastal bathymetric measurements.

The Sonar Tritech Side-Scan StarFish 450C is a small instrument (0.378 m long), optimised for coastal waters (450 KHz CHIRP transmission). The slant range used during the survey is 50 m. The instrument is embedded in the USV and has an offset of a few centimetres from the GPS. Under optimal conditions, the instrument is able to discriminate an object of 0.0254 m (1 inch). The side scan sonar is used to acquire the morphology of both the target and the seabed.

The photogrammetric system installed on-board of MicroVeGA consists of two Xiaomi YI Action cameras and a GoPro Hero 3. Two of them are placed parallel with the vertical axis, with a variable stereoscopic base (b) chosen in relation to the bathymetry of the study area. With our setting for the cameras, ensuring a minimum overlap of 80% during the survey. The third camera (GoPro Hero3) is placed with its axis that forms an angle of about 30° with the seabed. It acquires data from non-covered sectors (Figures 8 and 9).

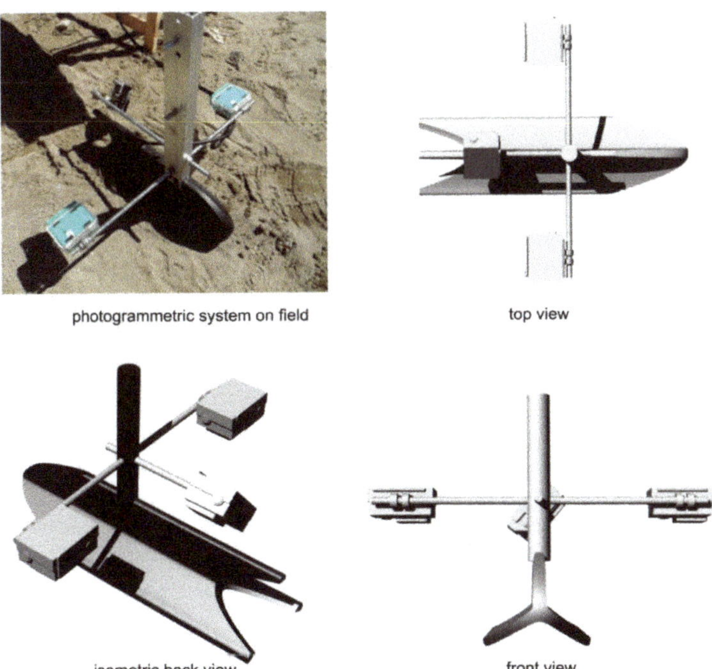

Figure 8. The photogrammetric system installed on board of MicroVeGA.

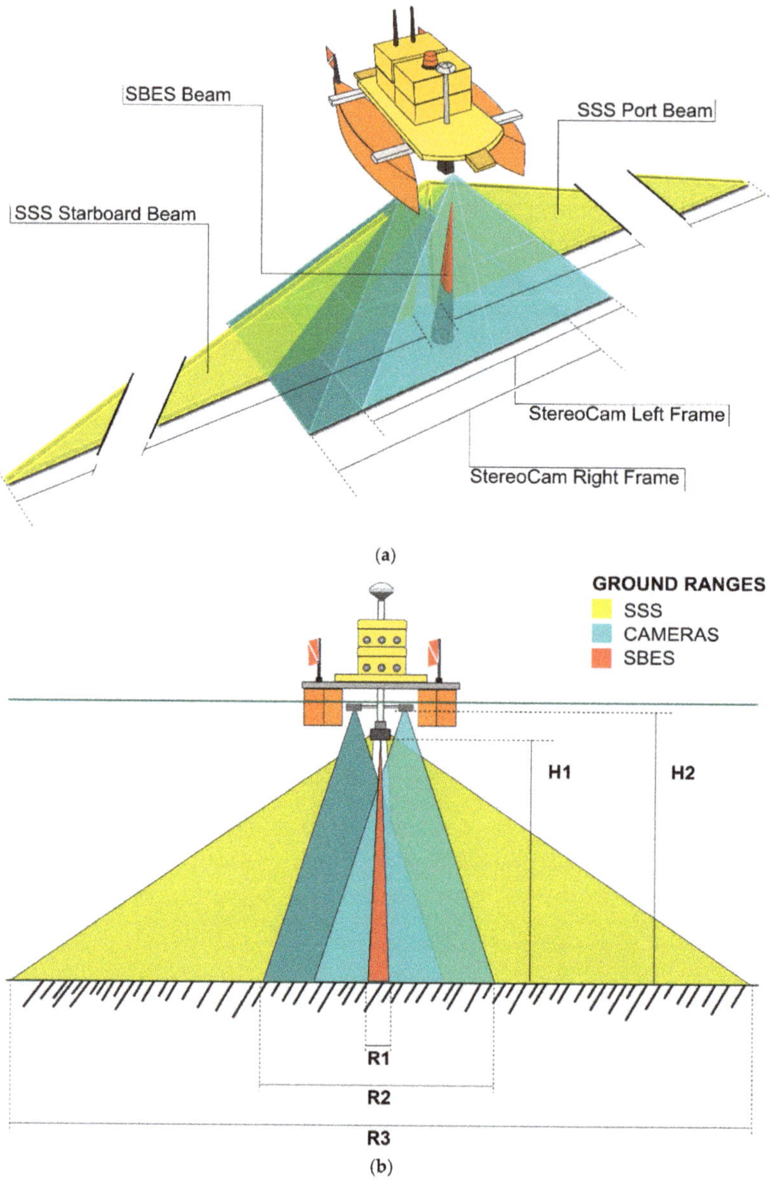

Figure 9. Sketches of ground ranges of acoustical and photogrammetric sensors (figure not in scale): (**a**) 3D reconstruction of the beam opening of the three sensors; and (**b**) 2D reconstruction of the beam opening of the three sensors. R1 is 0.139H1 (with an opening angle of the acoustic beam of 8°); R2 is 1.87H2 + b (where b is the stereoscopic base); and R3 can be set with the SSS software (Starfish Scanline V2.1) and varies between 4 and 300 m.

Videos are synchronised using an acoustic device integrated into the on-board system. It emits sound pulses (beeps) at regular intervals, which are memorised by the cameras' microphone. These acoustic events are stored in the datafile as flags and associated with the position, attitude and depth. During the post-processing phase, the videos captured by the three on-board cameras are synchronised using video editing software. This setup optimised the survey duration, allowing a multi-modal and multidisciplinary interpretation of the target—also thanks to the real-time sharing of all information.

3.2.2. On-Board Computer System and Communications

The core of the MicroVeGA USV is the platform management system (PMS, Figure 10), a customised framework of software applications (Microsoft VB, Linux, and C++) aimed at the full control of data and information flow. The PMS is based on a Mini IT low power single board computer VB7008 x86 with VIA C7-D 1.6 GHz processor, two Arduino MEGA 10-bit microcontrollers and two RaspBerry Pi2 micro-boards, connected via communication ports with the on-board systems (GPS, IMU, Gyro, side scan sonar, echosounder, obstacle detection sensors, temperature control sensors, engine and rudder management system, and survey cameras).

The decision to use a customised system led to full control of the data flow. This choice also facilitated the implementation of several self-built sensors. The PMS has therefore enabled the implementation of new operating methods for the on-site experimentation of low-cost components in the geo-technological field.

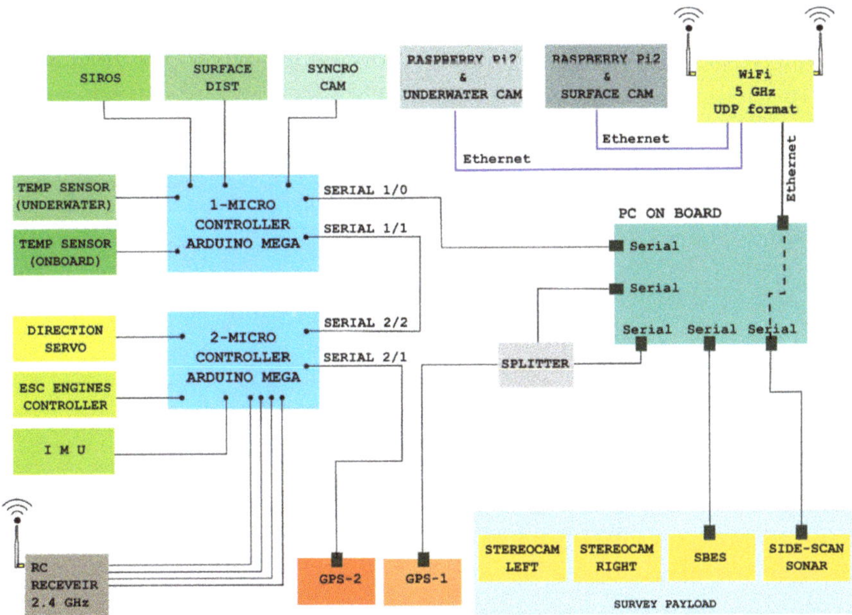

Figure 10. Block diagram of the platform management system (PMS).

The data transmission module is based on a local 5 GHz Wi-Fi network that allows broadcasting in real time all data on multiple devices (PC, tablet, and smartphone). This feature is crucial to carry out a multidisciplinary integrated survey. The Wi-Fi network has an operating range of about 2 km even if during the survey the distances were always within 500 m.

The datafile is recorded on the base station laptops, although a backup of both RAW and pre-formatted data is saved on a mass memory on-board of the USV, normally a 1 TB hard disk. During the surveys, this technological solution proved to be efficient without ever compromising the navigation, data acquisition and video streams of the cameras.

3.2.3. Correction and Quality-Control Module

The quality control module of the acquired data uses a series of components made ad hoc using off-the-shelf components, such as Arduino and Raspberry, and open-source libraries integrated with the appropriate C++ software code.

This module has been designed to evaluate the data precision in order to eliminate in post-processing all data affected by perturbations. Specific threshold values have been set to alerts in real time when the acquired data have a poor quality.

The module mainly evaluates the USV attitude and is based on two components: an acoustic system for the draught measuring and an IMU platform.

The acoustic system for the draught measuring calculates in real time the sinking of the hull and the echosounder's transducer. This value may vary during the survey mainly for two reasons:

- payload changes and therefore variations in the vessel draught; and
- changes in USV attitude during the survey due to meteo-marine forcings.

The limited displacement of the hull makes the USV sinking sensitive even to small variations of loaded cargo. In particular, the MicroVeGA sinking increases by 1 cm for every 2 kg of loaded cargo. The payload variations, frequent in a modular system such as MicroVEGA, can produce noticeable draught differences.

Further variations in the draught can be due to changes in the speed and to weather and sea conditions.

The acoustic system for the draught measuring, therefore, contributes to increasing the bathymetric measurement as well as to improve the navigation of the USV.

This system uses an ultrasonic sensor with a frequency of 1 Hz to measure in real time the distance between the sensor itself and the sea surface, i.e., the sinking of the drone (D1 in Figure 11); it is mounted on a structure parallel to the sea surface.

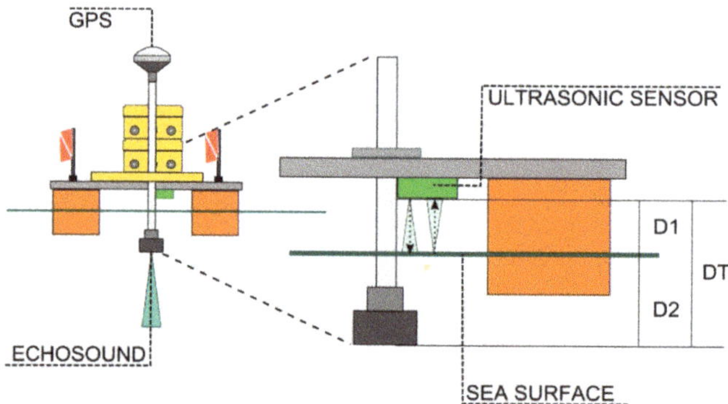

Figure 11. Acoustic system for the draught measuring.

An Arduino microprocessor connected to the ultrasonic sensor sends by a specific routine the draught measurements to the mission software. Draught data are recorded in the integrated datafile, and then broadcasted to the base station in real time

The Nisida survey was carried out in optimal weather conditions (calm sea and no wind). Nevertheless, during a marine survey, occasional events—such as the passage of a vessel—can influence the acquired measurements. Such events, generating waves, can alter the acquired bathymetric measurement.

To evaluate such events, a control system has been developed. This system includes an Arduino MeGA microprocessor, a mass memory on microSD and a 9-axis inertial platform, integrated to the platform management system (PMS).

This system acquires the attitude information of the USV (pitch and roll), allowing discarding bathymetric measurements affected by errors introduced by attitude anomalies.

The device is based on The Pololu MinIMU-9 v3, an inertial measurement unit (IMU) that packs an L3GD20H 3-axis gyro and an LSM303D 3-axis accelerometer and three-axis magnetometer onto a tiny 2.03 × 1.27 cm board, an Arduino MeGA microprocessor and related software.

The embedded software developed in C++ allows the calibration, the setting of the sampling frequency, the data acquisition and the sharing.

The acquired data are stored both on the mass memory of the device (a 16 GB microSD), and transmitted in real time to the base station (datafile).

4. Post-Processing and Data Elaboration

By using MicroVeGA drone, a multimodal mapping of a multidisciplinary dataset has been obtained in the surrounding area of Nisida Roman harbour.

4.1. Bathymetric Data Post-Processing

The bathymetric system was used to measure the depth of archaeological remains and to reconstruct the detailed seabed bathymetry [53]. The depths (D) were referred to the vertical datum of mean sea level (MSL), correcting each measurement (M) with respect to tidal height (h_t) and barometric value (Δhp) obtained from the Naples tide gauge of the National Tide Gauge System because of its close proximity of the survey area:

$$D = M + h_t + \Delta hp$$

where h_t is the tidal height value at the time of measuring, and the barometric correction Δhp is:

$$\Delta hp = (1013 - P) \times 1.023557761$$

where P is the barometric value at the time of measuring.

Finally, the depth measurement was corrected with respect to the transducer submersion, measured by the acoustic system for the draught measuring (see Section 3.2.3).

The bathymetric data were elaborated in a GIS environment to reconstruct the seabed morphology, as described in Section 5.1.

4.2. SSS Data Post-Processing

The side scan sonar system, performing the acoustic mapping of the seabed in shallow waters sectors, is optimised to detect the archaeological remains lying on the seabed and to define the seabed landforms [53], by discriminating the sandy bottom from the rocky one.

In the first instance, all sonographs were processed by using Chesapeake Sonar Web Pro 3.16 software to create a GeoTIFF mosaic and obtain the sonar coverage of the whole area.

This mosaic was elaborated in ArcGIS ArcScene obtaining a 3D view of the submerged acoustical landscape (See Section 5.1).

The analysis of the backscattering signal carried out in this research allows the evaluation of the characteristics of the acoustic reflectors identified: archaeological remains, rocky bottom, and sandy bottom. This analysis was made possible thanks to the automatic use on the SSS of a Time-Varying Gain (TVG) filter, which emphasises the gain for acoustic signals reflected by the structures positioned on the borders of the sonograph.

The trend of the backscattering along a horizontal line (sweep) was analysed since the radiometric values contained in a specific sweep correspond to the intensity of the backscattering recorded by the instrument. To graph the trend of the most significant sweeps, each sonograph was exported in a grey scale image, thus obtaining that the amplitude of the backscattering signal varies between 0 and 255 (See Section 5.1).

4.3. Photogrammetric Data Post-Processing

The photogrammetric survey was carried out simultaneously with the acoustic one, by using the photogrammetric system installed on-board of MicroVeGA.

The photogrammetric 3D model of the upper part of the *pila* was obtained in three steps:

1. The videos at 30 fps recorded by the two Xiaomi cameras (previously calibrated in an underwater environment close to the study area to achieve the inner orientation parameters [54]) were synchronised using the trigger system and the images were extracted using every sixth frame. More than eight-thousand 1920 × 1080 images were thus obtained.
2. The alignment procedure of the images was performed by Agisoft Photoscan software, subdividing them into several strips to reduce calculation times. For each strip, a dense point cloud was extracted and georeferenced by using the coordinates of the markers positioned on the *pila* that were determined by GPS Fast static procedure.
3. The different point clouds were subsequently assembled in a single cloud using the open-source software CloudCompare through the classic ICP procedure [18].
4. The whole point cloud, of almost 10 million points, can be decimated and exported in different formats compatible with GIS applications; the subsampling process was necessary to avoid visualisation problems with poor performance computers.

4.4. Morphometric Analysis of Three-Dimensional Data

The 3D elaboration was applied both to the bathymetric and photogrammetric data in order to obtain a multi-scale reconstruction of the underwater archaeological landscape.

The small-scale data processing was applied to the bathymetric measurements by obtaining a high-resolution sea-bottom digital terrain model (seaDTM), with a spatial grid of 0.1 m × 0.1 m.

The seaDTM was calculated using an ordinary Kriging interpolator of ArcGIS 3D Analyst tool, with a variable search radius.

The large-scale 3D data processing was applied to the photogrammetric point cloud obtaining the precise morphological reconstruction of the upper face of the submerged *pila* (pilaDTM), with a spatial grid 0.01 m × 0.01 m. The pilaDTM (See Section 5.2) was calculated using an inverse distance weighted (IDW) interpolator of ArcGIS 3D Analyst tool, with a variable search radius of the 12 nearest points.

Finally, a slope analysis was applied to the pilaDTM to make a first evaluation of the erosion degree. This analysis followed four steps:

1. Calculate slop in per cent of the pilaDTM.
2. Reclassify slope into four classes (gentle slope, moderate slope, steep slope and very steep slope).
3. Reconstruct a flat surface (topDTM) representing the top of the *pilae* before the action of the erosion processes.
4. Calculate eroded volume between topDTM and pilaDTM.

5. Results

5.1. Small-Scale Data Elaboration

The small-scale data elaboration was applied both to bathymetric and SSS data to obtain a landscape reconstruction of the coastal sector (Figure 12). The seaDTM can be divided into a gentle slope sector between 0 and −10 m and a steeper slope sector with a bathymetry higher than −10 m. On the sub-horizontal seabed, the remains of two *pilae* were precisely georeferenced. However, while the inner *pila* resulted almost completely destroyed, the outer one preserves its original shape. As can be seen in Figure 12, the other five *pilae* represented in the historical maps (Figure 3b) were buried by the modern breakwater.

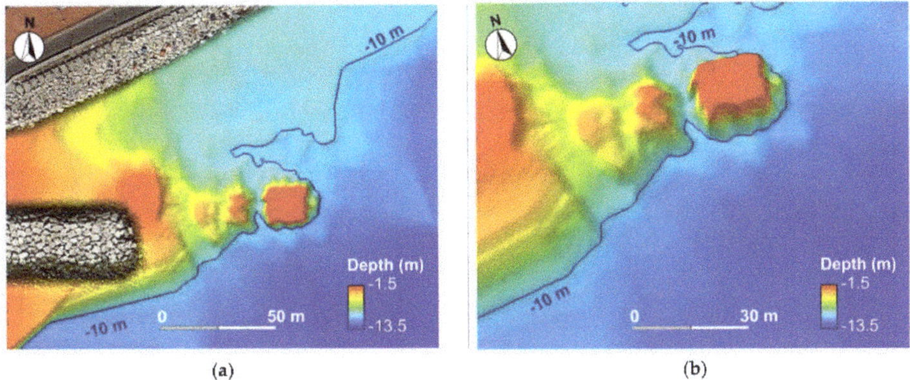

(a)   (b)

Figure 12. (**a**) 2D view of DTM of Nisida coastal archaeological site; and (**b**) 3D view of DTM of Nisida coastal archaeological site.

The SSS mosaic is a realistic picture of the submerged landscape, with several qualitative information on the seabed typology and the conservation state of the archeo-structures (Figures 13 and 14).

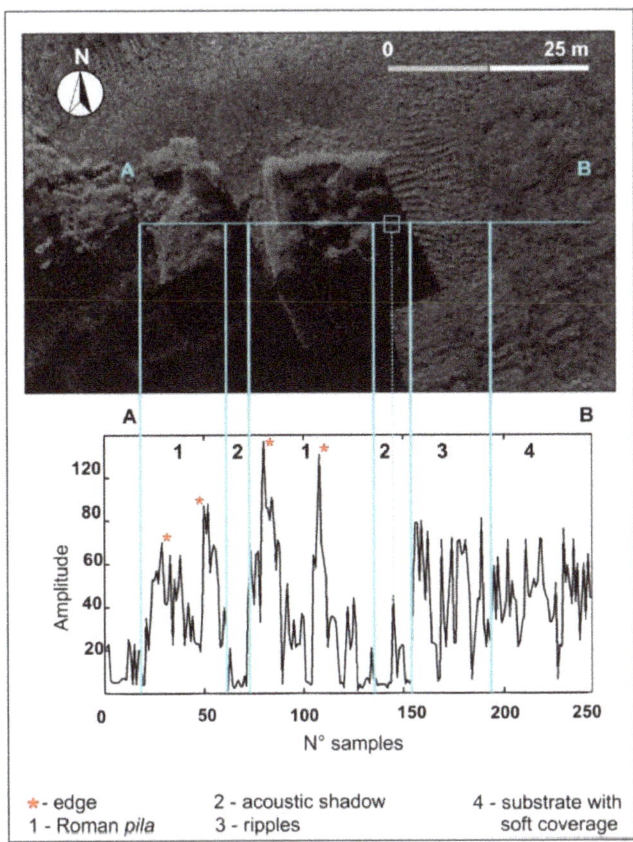

Figure 13. Backscattering signal chart of a sweep passing through the archaeological structures with the amplitude of the acoustic signal expressed in grey scale.

By a numerical point of view, the backscattering analysis of each sonograph allowed an acoustical characterisation of the archaeo-targets and the seabed typology.

As shown in Figure 13, the highest backscattering values are associated with the edges of the archaeological structures (value 140 in the grey scale), which in this case have reflected about 60% of the acoustic signal emitted by the SSS transducer.

The highly reflective acoustic response (max value 80 in the grey scale) of the sub-horizontal surface on which the submerged *pilae* are laid (Sector 4 in Figure 13) can be interpreted as a rocky bottom, covered by a thin sediments layer. The sediment coverage was testified also by the presence of well-aligned ripples (Sector 3 in Figure 13) at the foot of the better-preserved *pila*.

The backscattering signal of the archeo-structures ranges between 10 and 140 amplitude values. This variability is clear evidence of the roughness of the *pilae* upper face due to the erosive effects of the wave's action (Sector 1 in Figure 13).

Finally, the area in Figure 13 between Samples 135 and 155 (on x-axis) is characterised by very low values of signal amplitude, as it is a sector of acoustic shadow, but presents a structure (amplitude value 45 in the grey scale) that emerges from the bottom almost at the same height of the *pila*. This structure can be interpreted as part of the nowadays-collapsed *opus pilarum*.

The extensive analysis of backscattering along with the three-dimensional reconstruction of the seabed morphology allowed a precise mapping of the sub-horizontal surface on which the pier was built during the Roman period with a high reflective acoustic response that can be interpreted as the tufa abrasion platform formed during the Holocene high stand (Figure 14, [21]).

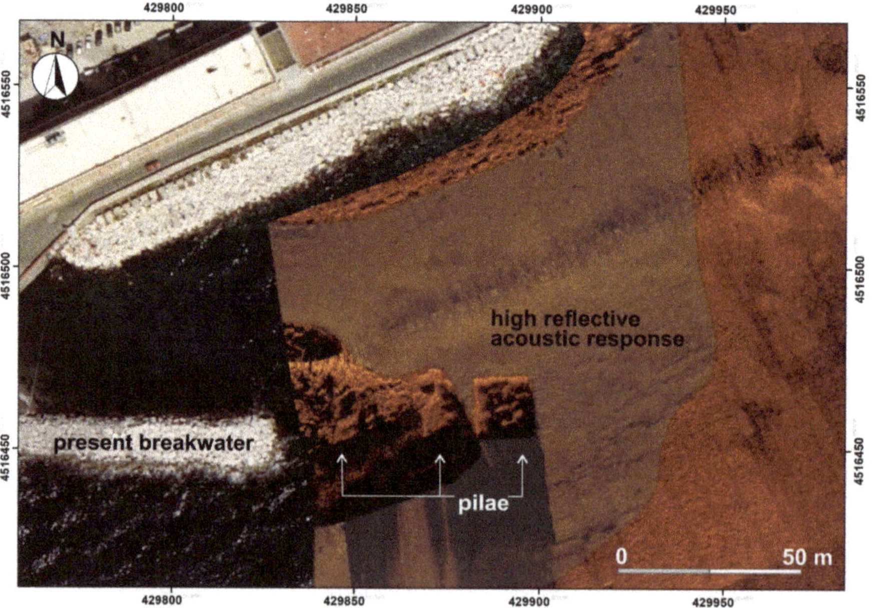

Figure 14. SSS map of the study area, with a seabed characterisation and the location of the archaeological structures.

5.2. Large-Scale Data Elaboration

The large-scale data elaboration—by interpreting bathymetric and photogrammetric data—had a twofold target:

- the reconstruction of the archaeological structures to evaluate the conservation state of the better-preserved *pila* composing the ancient pier as well as to discriminate the erosive effects due to the waves action.
- the detection of a new type of sea level marker measuring point, useful in the case of port-like structures built in the hydraulic concrete cast and set underwater.

The three-dimensional reconstruction of the upper surface of the *pila* (pilaDTM)—deriving from the photogrammetric data interpolation—has clearly demonstrated the more eroded condition of the sectors between N and SE (Figure 15).

The deeper area reaches a depth of −3.0 m in the NE edge of the archeo-structure, as can be observed in the profile BB′ in Figure 15. The areas with a depth range between −3.0 and −4.0 are related to the vertical borders measured during the photogrammetric survey.

By analysing pilaDTM (Figure 15), the higher areas between −1.00 m and −1.5 m of depth represent remnants of former upper face of the *pila*, while the deeper areas on the same face—between −2.5 and −3.5 m—appear rather flat along the N, E and SE borders and can be explained as the sectors more eroded by the sea action.

The slope analysis—here proposed as a quantitative method for the evaluation of the conservation state (Figure 16)—highlights the massive erosive effect due to the wave's action on the top of the *pila*. In fact, the highest sectors—reaching −1 m of depth—are less eroded but are fitted to the less elevated areas by steep forms belonging to the third and fourth classes (steep slope and very steep slope), and occupy 84.7% (Table 2) of the total surface (222.4 m^2). Instead, the more eroded sub-horizontal sectors (7.4 m^2) are 3.3% of the total surface and, for the most part, rests at a depth greater than −2.6 m.

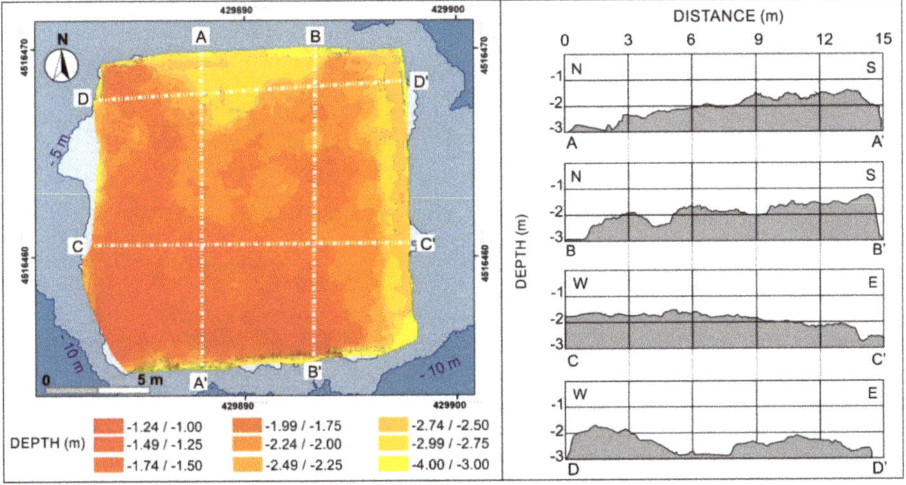

Figure 15. Three-dimensional reconstruction of the archaeological remains obtained by interpolating photogrammetric data.

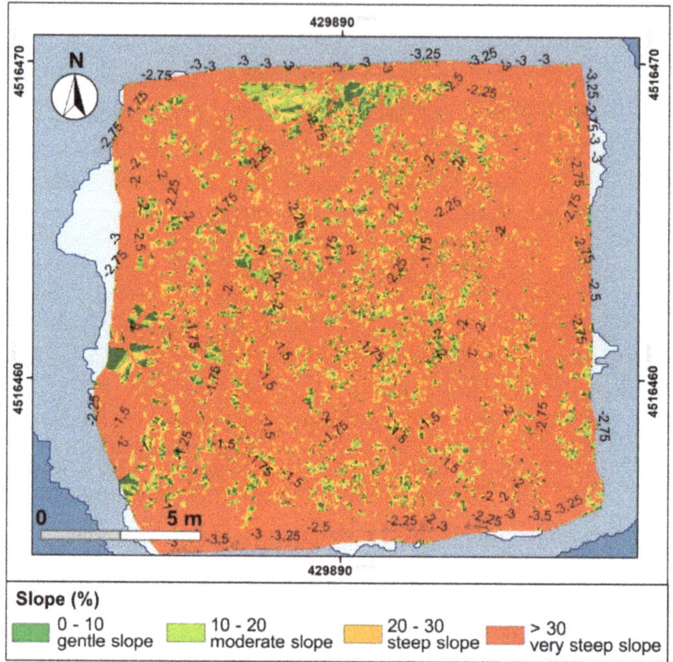

Figure 16. Slope analysis of pilaDTM to evaluate the more eroded areas.

Table 2. Table of slope classes used to classify the morphology of the *pila* upper face and to detect the eroded sectors.

Classes	Slope %	Type	Area (m^2)	Area (%)
1	0–10	Gentle slope	10.3	4.6
2	10–20	Moderate slope	23.6	10.6
3	20–30	Steep slope	26.9	12.1
4	>30	Very steep slope	161.5	72.6

We can affirm that a large part of the upper face of the *pila* presents a rugged surface due to the high erosion degree of the former surface (Figure 17), as also deduced by the backscattering signal analysis of the SSS sonograph.

The volume of the materials eroded by the sea after the modern submersion of the *pila* (Figure 17)—after 1838 AD—was evaluated in at least 240 m^3. It was calculated by using the Cut-Fill tool of ArcGIS, between the pilaDTM and a horizontal surface (topDTM) at -1 m of depth and with the same planar extension of the pilaDTM. The obtained horizontal surface is the surface passing through the less eroded and more elevated point (P). Since in P the cement is covered by a thin vegetation stratum of about 0.14 m (sV), the real submersion of the surveyed *pila* in P (sP) was calculated by a correction:

$$sP = \text{Depth} - sV$$

However, we have also detected several sectors laying at a depth of -2.6 m and presenting a flat morphology. These small flat elements mark the maximum deepening of the erosive processes. Furthermore, they point out the interface between the more erodible concrete laid in subaerial environment (submerged in the last centuries due to the relative sea level rise [20]) and the more cemented and less erodible hydraulic concrete cast and set underwater during the Roman period (Figures 17 and 18a,b).

According to Vitruvius and Lechtman et al. [55], the main characteristic of the pozzolanic hydraulic concrete is to harden in contact with sea water, due to its highly reactive aluminosilicate component (pumice and volcanic ash) that when mixed with lime generates reaction products (gels, rods, fibres and plates) that give strength and bind all the materials together [55].

The subaerial concrete layer here identified overlays the part of the *pila* made in hydraulic concrete—cast and set underwater—that is visible only in the deeper sectors where the subaerial concrete was totally eroded. These low flat areas of the pilaDTM can be interpreted as the top face of the hydraulic concrete hardly erodible by the wave's action and then almost intact.

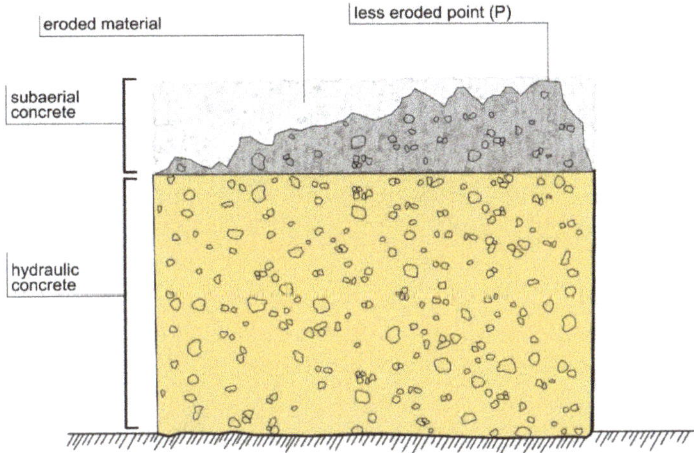

Figure 17. Sketch of the *pila* with the concrete layer before its erosion and the limit between subaerial and hydraulic concrete.

This interpretation was endorsed by the 3D visualisation of the photogrammetric point cloud in RGB colour (Figure 18a). The three areas in the N, NE and SE sectors, previously classified as sub-horizontal surfaces resting at a depth greater than −2.6 m, clearly appeared flat and lighter in colour with respect to the adjacent steep sectors.

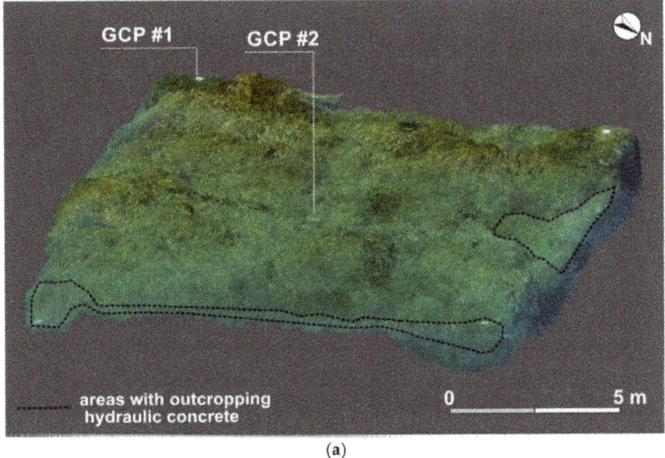

(a)

Figure 18. *Cont.*

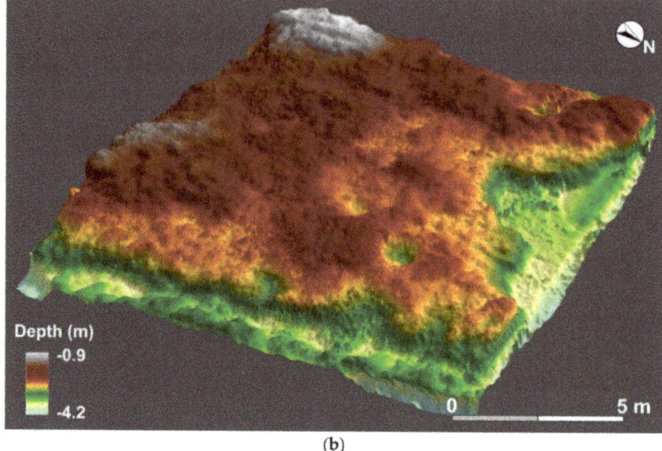

(b)

Figure 18. 3D point cloud with the perimeter of the outcropping areas in hydraulic concrete (light-dark dashed line). (**a**) Cloud point visualisation by using Cloud Compare software; and (**b**) pilaDTM visualisation by using the ArcScene software.

The detection of the upper surface of the hydraulic concrete structure—well preserved at a depth of 2.6 m b.s.l. was the second important result of the large-scale analysis.

In fact, we propose a new type of measuring point (Figure 19), useful in the case of port-like structures built with the cofferdam technique: the limit between the areas in concrete cast and set underwater (hydraulic concrete) and the area in concrete totally laid in subaerial environment (subaerial concrete).

As described by Vitruvius in the famous *De Architectura* treaty in 15 BC, the cofferdams—filled with hydraulic concrete—emerged from the sea level by an amount equal to a wooden board (about 0.5 m). This amount can be used as indicative meaning—the elevation where the RSL indicator was formed or was built with respect to the palaeo-sea level including its uncertainty [45]. Ultimately, by correcting the submersion measurement of the upper limit of the hydraulic concrete with the indicative meaning value, the Roman sea level can be evaluated with a high precision (Figure 19).

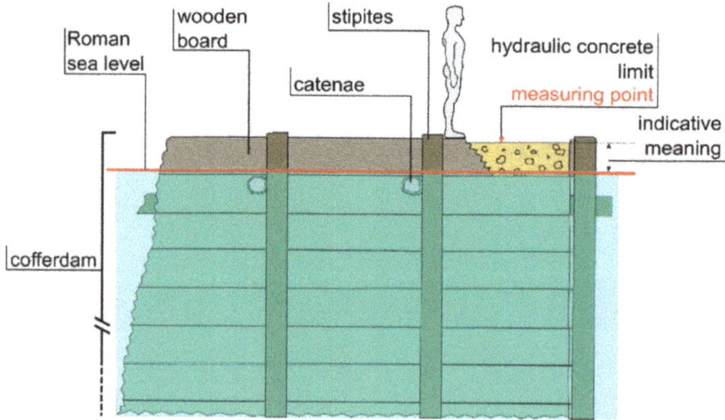

Figure 19. Sketch of the new type of measuring point useful in the case of port-like structures built in hydraulic concrete.

In the case of Nisida Roman harbour, the submersion of hydraulic concrete limit at 2.6 m b.s.l. was corrected with respect to the aforesaid indicative meaning (0.5 m) obtaining a Roman sea level at −3.1 m (Figure 20). The uncertainty can be estimated in 0.2 m that is the mean tidal range for this coastal sector. Instead, the uncertainty due to the construction features of the port facilities, its conservation state and its emersion during the Roman period have been almost zeroed.

We can affirm that this new type of measuring point for ports-like structures built in hydraulic concrete allows for increasing the reliability of this kind of sea level marker.

The 3D point cloud interpolation (pilaDTM) provided another very interesting result, as the detection of several areas of interest by an archaeological point of view. In particular, five traces of the oaken stakes (*catenae* and *destinae*) composing the cofferdam were precisely mapped (Figure 21), four of which are vertical and one is horizontal.

The planar dimensions and the depth of each target were measured (Table 3), with centimetre precision.

The surveyed *pila* resulted with a well-preserved squared shape with the following dimension: 14.3 m on the N side; 14.4 m on the E side; 14.5 m on the W side; 14.8 m on the S side; 9.3 m of max height; and 7.1 m in height of the concrete structure.

A direct survey (Figures 22 and 23) of a scuba diver validated the main indirect measurements and integrated the dataset with information on the vertical sides of the *pila*. On the W vertical side, an alignment of seven well-preserved *catenae* traces was detected at a depth of 3.1 m b.s.l. These shapes represent the highest line of *catenae* composing the cofferdam, so they are at a depth greater than the upper limit of the hydraulic concrete (see also Figures 20 and 21) of about 0.6 m.

The vertical sides did not undergo significant erosive effects, preserving the facing in *opus reticolatum* up to the base of the structure at about 10 m b.s.l. In particular, the NW corner of the *pila*, the *opus reticulatum* is well preserved and clearly visible at the base due to the absence of algal vegetation. The *cubilia* composing the *opus reticolatum* are small cubes with sides of about 15 cm (Figure 23).

Table 3. Table of the planar dimensions of the wooden board traces on the top face of the *pila*.

ID	DIMENSION (m)	DEPTH (m)	Type
D1	0.3 × 0.4	−2.4	*Destina*
D2	0.4 × 0.3	−2.3	*Destina*
D3	0.4 × 0.6	−2.4	*Destina*
D4	0.5 × 0.3	−2.5	*Destina*
C1	13.9 × 0.5	−2.9	*Catena*

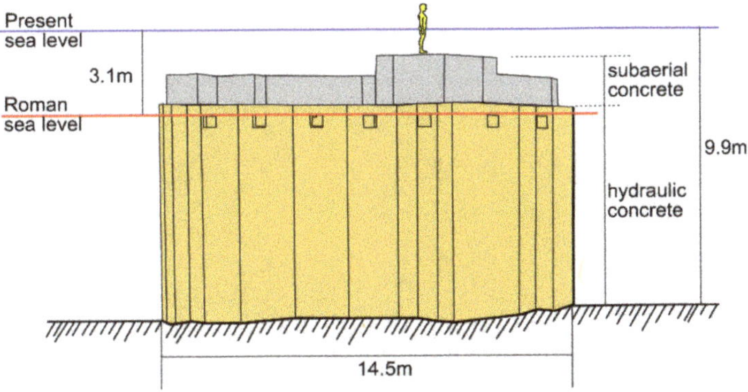

Figure 20. Sketch of the surveyed *pila* with the main construction features detected during the surveys and with the Roman sea level position here deduced.

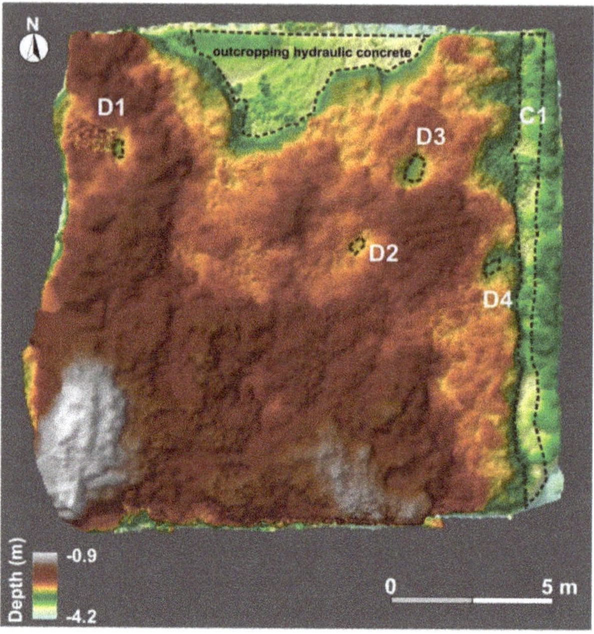

Figure 21. Three-dimensional reconstruction of the upper surface of the more preserved *pila* (pilaDTM) with the traces of the oaken stakes (*catenae* and *destinae*) composing the cofferdam (light-dark dashed line).

Figure 22. Underwater photo taken during the direct survey.

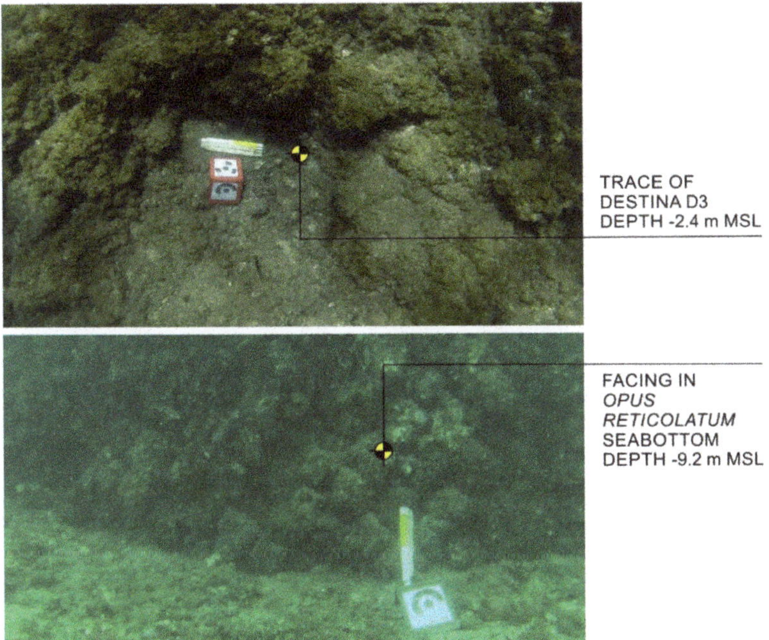

Figure 23. Underwater photo taken during the direct survey.

6. Discussion

The multiscale approach used in this research turned out to be very efficient to reconstruct the underwater archaeological and natural landscape as well as its evolution. The integration between acoustic and photogrammetric data allowed a morphometric analysis of the landscape characteristics as well as a precise mapping of the submerged archaeological structures that can be efficiently used as sea level markers.

The detection of past sea levels represents a challenge for the modern scientific community that studies the coastal changes due to relative sea level variations and the ongoing climate changes. The precise measurement of the height or depth of specific markers of former sea levels is the goal of this kind of studies. The concept "relative" includes both the eustatic sea level variations [56] and the vertical ground movements that have affected a specific area since the construction or formation of sea level markers [57–59].

These studies have a significant impact on both the understanding of coastal evolutive dynamics and the assessment of the ongoing climate changes effects. The main operative problem during the surveys finalised to detect and measure a relative sea level marker is to establish its indicative meaning [60,61]. The indicative meaning is the elevation on which the relative sea level (RSL) indicator was formed or was built with respect to the palaeo-sea level and includes an uncertainty.

The amount of the indicative meaning and its uncertainty are directly connected to the identification of a precise measuring point. In the case of the *pilae*, Auriemma and Solinas [47] proposed an indicative meaning range (or functional height in the specific case of an archaeological marker) between 0.6 and 1 m, depending on the measuring points identifiable on field (i.e., walkways, missing carpentry, or bollards and the mooring rings or stringcourses between building techniques and different coverings).

Thanks to the high-resolution surveys carried out on the remnant *pilae* of Nisida Roman harbour and the subsequent three-dimensional processing, a new type of measuring point—valid in the case of port-like structures built in hydraulic concrete cast and set underwater—is proposed.

In addition, its indicative meaning correction was precisely established, according to Vitruvius, reducing the uncertainty.

The small-scale analysis applied to the geophysical data allowed a geomorphological characterisation of the seabed, useful to understand the landscape morphology when the Nisida Roman harbour was built during the first century BC. In the first instance, by interpolating the bathymetric data, the seabed morphology was reconstructed by detecting a less deep sub-horizontal sector on which the *pilae* were built at a depth about of 10 m b.s.l., and a steep slope sector at a depth greater than 10 m b.s.l.

This seabed was acoustically characterised by means of the backscattering signal analysis as a tufa substrate covered by a thin sediments layer. This landform was interpreted as an ancient abrasion platform cut in Nisida yellow tuff [21] used as a base to build the *pilae* composing the pier. It can be supposed that the hard substrate was not modified in the last 2000 years and—as the direct survey demonstrated—the external *pila* of the *opus pilarum* that made up the pier is still in its original position. Furthermore, the seabed morphology and the *pila* position with its base at a present depth between 9 and 10 m b.s.l., as well as the steep slope immediately off the pier, endorse the hypothesis that the Nisida harbour was used for the landing of cargo ships (according to Gianfrotta et al. [52]).

The large-scale analysis applied to the photogrammetric data allowed a precise evaluation of the *pila* conservation state as well as on its construction properties. Above all, it allowed to accurately measuring the palaeo- sea level during the Roman period. In fact, the morphometric analysis of the photogrammetric data led to the detection of the upper limit of the hydraulic concrete at a depth of 2.6 m b.s.l. By using this value as measuring point, the Roman sea level at -3.1 m at the time of the port construction can be deduced (Figure 20), endorsing the hypothesis that the Neapolitan coast was controlled by a subsiding trend in the last 2.0 ky [5,20,21,25,62].

The effects of this subsiding trend affecting the Neapolitan area were both the submersion of the maritime structures of Roman age, as the case of Nisida harbour (Figure 24), and a coastline retreat of several meters. However, the ancient abrasion platform here detected—cut in Nisida yellow tuff [21]—on which the archaeological structures lay, is more clear evidence of this retreating trend.

This coastal sector was also modified since 1800 AD by the intense anthropic activity, as testified by the construction of several infrastructures such as the Nisida wharf and the breakwater. These infrastructures have partially destroyed the Roman port facilities of which only the two *pilae* remain.

As demonstrated by the direct survey, the *pilae* composing the pier—coated with *opus reticulatum*—is still in place and in a good conservation state as shown by the various archaeological evidence here identified, such as the traces of *catenae* and *destinae*.

The large-scale elaboration also provided relevant information about the erosive effects due to the waves action—after the total submersion of the *pila* in the modern times—evaluating in at least 240 m^3 of eroded material. On the other hand, we can affirm that the material mainly suffering the erosive effects is the concrete laid in the subaerial.

If the erosion effect was quantitatively evaluated by elaborating the z values of the point cloud, several qualitative evaluations were obtained by visualising the 3D point cloud of RGB colours. In particular, the sectors with the outcropping hydraulic concrete were bordered as shown in Figures 18 and 21. Furthermore, the areas most covered by vegetation (light brown in Figure 18a) were identified. This information was very useful in the evaluation of eroded volume by using the slope analysis described in Section 5.2. In fact, the direct survey provided the precise measuring of the vegetation strata in the less eroded point (P) and, consequently, the measurement of the real submersion of the surveyed pila in P. This measurement was used to improve the evaluation of the eroded volume, as described in Section 5.2.

The RGB point cloud is also a three-dimensional documentation of this archaeological structure of considerable dimension (14 m $\times$ 15 m), hardly visible because submerged at least 2 m, but easily viewable with the software that manages the point clouds.

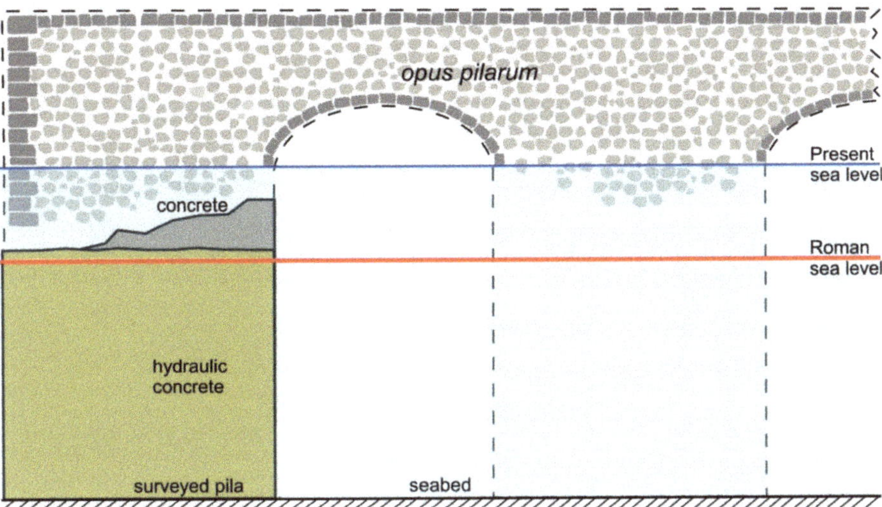

Figure 24. Roman sea level position and pictorial reconstruction of the Nisida *opus pilarum*. The surveyed *pila* is drawn in continuous black line. The structures drawn in dashed line were reconstructed from the historical maps in Figure 3.

On the other hand, the high resolution of the 3D point cloud created a digital archive of this submerged archeo-site that represents a significant Roman harbour in the Gulf of Naples, which is little-known from historical sources. This kind of elaboration made possible the visualisation and presentation of all information in a user-friendly way. Furthermore, the processing of the point cloud made it possible to carry out a morphometric analysis of the archaeological structure that provided important information on both the construction features and the conservation state.

7. Concluding Remarks

This study—thanks to its multidisciplinary vocation—allowed carrying out a series of high-resolution analyses both on the characteristics of the natural and anthropic landscape and on its evolution. Furthermore, it provided a detailed reconstruction of the submerged remains of the Nisida Roman harbour, allowing the recording of a vast amount of four-dimensional—3D points and time—multi-source and multi-format information, with high accuracy.

Regarding operating procedures, the use of a USV during the indirect marine surveys resulted very efficiently to carry out an integrated survey inspired to the multi-modal mapping philosophy. The challenge was to integrate professional sensors, low-cost miniaturised sensors (i.e., gyroscopes, GPS, motion sensors, etc.) and innovative open hardware architecture (Arduino, Raspberry, etc.). Thanks to the excellent cost-performance ratio, this technology experimentation guaranteed a robust solution for the reconstruction of the high-definition archaeological landscape.

The integration of several instruments during the survey has reduced the time of the survey, but above all, has allowed the spatial overlay—through the GPS position—of all measurements deriving from both the acoustic and photogrammetric sensors. This feature made it possible to integrate morphometric analysis with geo-environmental qualitative assessments.

In conclusion, the multiscale approach here proposed to elaborate a transdisciplinary dataset turned out to be very efficient to obtain:

- the qualitative and quantitative characterisation of the underwater landscape in an archaeo-site with submerged structures;

- the morphometric analysis of an archaeological structure with a consequent assessment of its conservation state;
- the detailed four-dimensional documentation of a submerged archaeo-site difficult to access; and
- the definition of a new type of measuring point for a more precise evaluation of the relative sea level variation—in the last 2000 years—in the case of port-like structures built with the cofferdam technique in hydraulic concrete.

Author Contributions: Conceptualisation, G.M., S.T., P.P.C.A., G.P., F.P. and M.S.; methodology, G.M., S.T., and F.P.; investigation, G.M., F.P., and M.S.; software G.M., F.P. and S.T.; data curation, G.M., S.T., and F.P.; writing—original draft preparation, G.M. and M.S.; writing—review and editing, S.T., P.P.C.A., and G.P.; and visualisation, G.M., F.P. and S.T.

Funding: This research received no external funding.

Acknowledgments: This paper is in memory of Professor Raffaele Santamaria, the first to believe in the MicroVeGA project. The authors sincerely thank Alberto Greco, Ferdinando Sposito, Luigi De Luca and Alberto Giordano for their precious support during the marine surveys. Sincere thanks are also due to the 2nd Nucleo Operatori Subacquei of Coast Guard of Naples for the highly specialised support during all phases of the direct and indirect marine surveys. The Ministry of the Environment kindly provided the coastal LIDAR data used in this paper. The authors finally thanks the two anonymous reviewers for their critical reviews which greatly improved the manuscript.

Conflicts of Interest: The authors declare no conflict of interest. The funders had no role in the design of the study; in the collection, analyses, or interpretation of data; in the writing of the manuscript, or in the decision to publish the results.

References

1. Goiran, J.-P.; Morhange, C. Geoarcheology of ancient Mediterranean harbours: Issues and case studies. *Eng. Transl. E Willcox Topoï* **2001**, *11*, 647–669.
2. Parry, M.L.; Canziani, O.F.; Palutikof, J.P.; van der Linden, P.J.; Hanson, C.E. Climate change 2007: Impacts, adaptation, and vulnerability. In *Contributions of Working Group II to the Fourth Assessment Report of the Intergovernmental Panel on Climate Change*; Cambridge University Press: Cambridge, UK, 2007.
3. Solomon, S.; Qin, D.; Manning, M.; Chen, Z.; Marquis, M.; Averyt, K.B.; Tignor, M.; Miller, H.L. Climate change 2007: The physical science basis. In *Contributions of Working Group I to the Fourth Assessment Report of the Intergovernmental Panel on Climate Change*; Cambridge University Press: Cambridge, UK, 2007.
4. Amato, V.; Aucelli, P.P.C.; Mattei, G.; Pennetta, M.; Rizzo, A.; Rosskopf, C.M.; Schiattarella, M. A geodatabase of Late Pleistocene-Holocene palaeo sea-level markers in the Gulf of Naples. *Alpine Mediterr. Quat.* **2018**, *31*, 5–9.
5. Aucelli, P.P.C.; Cinque, A.; Mattei, G.; Pappone, G. Late Holocene landscape evolution of the gulf of Naples (Italy) inferred from geoarchaeological data. *J. Maps* **2017**, *13*, 300–310. [CrossRef]
6. Aucelli, P.P.C.; Cinque, A.; Mattei, G.; Pappone, G. Historical sea level changes and effects on the coasts of Sorrento Peninsula (Gulf of Naples): New constrains from recent geoarchaeological investigations. *Palaeo3* **2016**, *463*, 112–125. [CrossRef]
7. Aucelli, P.P.C.; Cinque, A.; Giordano, F.; Mattei, G. A geoarchaeological survey of the marine extension of the Roman archaeological site Villa del Pezzolo, Vico Equense, on the Sorrento Peninsula, Italy. *Geoarchaeology* **2016**, *31*, 244–252. [CrossRef]
8. McNamee, C.; Cyr, H.; Wilson, L. Multi-Scalar Approaches to Geoarchaeological Questions. *Geoarchaeology* **2013**, *28*, 191–194. [CrossRef]
9. Wynn, R.B.; Huvenne, V.A.; Le Bas, T.P.; Murton, B.J.; Connelly, D.P.; Bett, B.J.; Sumner, E.J. Autonomous Underwater Vehicles (AUVs): Their past, present and future contributions to the advancement of marine geoscience. *Mar. Geol.* **2014**, *352*, 451–468. [CrossRef]
10. Yoerger, D.R.; Bradley, A.M.; Jakuba, M.; German, C.R.; Shank, T.; Tivey, M. Autonomous and remotely operated vehicle technology for hydrothermal vent discovery, exploration, and sampling. *Oceanography* **2007**, *20*, 152–161. [CrossRef]
11. German, C.R.; Yoerger, D.R.; Jakuba, M.; Shank, T.M.; Langmuir, C.H.; Nakamura, K. Hydrothermal exploration with the Autonomous Benthic Explorer. *Deep Sea Res. I* **2008**, *55*, 203–219. [CrossRef]

12. Giordano, F.; Mattei, G.; Parente, C.; Peluso, F.; Santamaria, R. MicroVeGA (micro vessel for geodetics application): A marine drone for the acquisition of bathymetric data for GIS applications. The international archives of photogrammetry. *Remote Sens. Spat. Inf. Sci.* **2015**, *40*, 123–130.

13. Giordano, F.; Mattei, G.; Parente, C.; Peluso, F.; Santamaria, R. Integrating sensors into a marine drone for bathymetric 3D surveys in shallow waters. *Sensors* **2016**, *16*, 41. [CrossRef] [PubMed]

14. Eric, M.; Berginc, G.; Pugelj, M.; Stopinšek, Z.; Solina, F. The impact of the latest 3D technologies on the documentation of underwater heritage sites. In Proceedings of the Digital Heritage International Congress (DigitalHeritage), Marseille, France, 28 October–1 November 2013; Volume 2, pp. 281–288.

15. Remondino, F.; Rizzi, A. Reality-based 3D documentation of natural and cultural heritage sites—Techniques, problems, and examples. *Appl. Geomat.* **2010**, *2*, 85–100. [CrossRef]

16. Yastikli, N. Documentation of cultural heritage using digital photogrammetry and laser scanning. *J. Cult. Herit.* **2007**, *8*, 423–427. [CrossRef]

17. Troisi, S.; Baiocchi, V.; Del Pizzo, S.; Giannone, F. A prompt methodology to georeference complex hypogea environments. *Int. Arch. Photogramm. Remote Sens. Spat. Inf. Sci.* **2017**, *42*, 639–644. [CrossRef]

18. Troisi, S.; Del Pizzo, S.; Gaglione, S.; Miccio, A.; Testa, R.L. 3D models comparison of complex shell in underwater and dry environments. *Int. Arch. Photogramm. Remote Sens. Spat. Inf. Sci.* **2015**, *40*, 215–222. [CrossRef]

19. Drap, P. Underwater Photogrammetry for archaeology. In *Special Applications of Photogrammetry*; Da Silva, D.D., Ed.; InTech Open: London, UK, 2012; pp. 112–135.

20. Aucelli, P.; Cinque, A.; Mattei, G.; Pappone, G.; Rizzo, A. Studying relative sea level change and correlative adaptation of coastal structures on submerged Roman time ruins nearby Naples (southern Italy). *Quat. Int.* **2017**. [CrossRef]

21. Aucelli, P.P.C.; Cinque, A.; Mattei, G.; Pappone, G.; Stefanile, M. Coastal landscape evolution of Naples (Southern Italy) since the Roman period from archaeological and geomorphological data at Palazzo degli Spiriti site. *Quat. Int.* **2018**, *483*, 23–38. [CrossRef]

22. Aucelli, P.P.C.; Cinque, A.; Mattei, G.; Pappone, G.; Stefanile, M. First results on the coastal changes related to local sea level variations along the Puteoli sector (Campi Flegrei, Italy) during the historical times. *Alpine Mediterr. Quat.* **2018**, *31*, 13–16.

23. Cinque, A.; Aucelli, P.P.C.; Brancaccio, L.; Mele, R.; Milia, A.; Robustelli, G.; Romano, P.; Russo, F.; Russo, M.; Santangelo, N.; et al. Volcanism, tectonics and recent geomorphological change in the Bay of Napoli. *Suppl. Geogr. Fis. Din. Quat.* **1997**, *3*, 123–141.

24. Fedele, L.; Morra, V.; Perrotta, A.; Scarpati, C.; Putignano, M.L.; Orrù, P.; Schiattarella, M.; Aiello, G.; D'Argenio, B.; Conforti, A. *Note Illustrative Della Carta Geologica D'Italia Alla Scala 1:50.000, Foglio 465 Isola di Procida*; Istituto Superiore per la Protezione e Ricerca Ambientale: Rome, Italy, 2015.

25. Giordano, F.; Mattei, G.; Milia, A.; Torrente, M. Quaternary faulting off Ischia island (Italy): Preliminary results. *Rend. Online Della Soc. Geol. Ital.* **2013**, *29*, 74–77.

26. Milia, A.; Torrente, M.M.; Russo, M.; Zuppetta, A. Tectonics and crustal structure of the Campania continental margin: Relationships with volcanism. *Mineral. Petrol.* **2003**, *79*, 33–47. [CrossRef]

27. Di Vito, M.A.; Acocella, V.; Aiello, G.; Barra, D.; Battaglia, M.; Carandente, A.; Del Gaudio, C.; de Vita, S.; Ricciardi, G.P.; Ricco, C.; et al. Magma transfer at Campi Flegrei caldera (Italy) before the 1538 AD eruption. *Sci. Rep.* **2016**, *6*, 32245. [CrossRef] [PubMed]

28. Santacroce, R.; Cioni, R.; Marianelli, P.; Sbrana, A.; Sulpizio, R.; Zanchetta, G.; Joron, J.L. Age and whole rock–glass compositions of proximal pyroclastics from the major explosive eruptions of Somma-Vesuvius: A review as a tool for distal tephrostratigraphy. *J. Volcanol. Geotherm. Res.* **2008**, *177*, 1–18. [CrossRef]

29. Milia, A.; Torrente, M.M. The influence of paleogeographic setting and crustal subsidence on the architecture of ignimbrites in the Gulf of Naples (Italy). *Earth Planet. Sci. Lett.* **2007**, *263*, 192–206. [CrossRef]

30. Corrado, G.; Amodio, S.; Aucelli, P.P.C.; Etro Incontri, P.; Pappone, G.; Schiattarella, M. Late Quaternary Geology and morphoevolution of the Volturno Coastal Plain, Southern Italy. *Alpine Mediterr. Quat.* **2018**, *31*, 23–26.

31. Cinque, A.; Irollo, G.; Romano, P.; Ruello, M.R.; Amato, L.; Giampaola, D. Ground movements and sea level changes in urban areas: 5000 years of geological and archaeological record from Naples (Southern Italy). *Quat. Int.* **2011**, *232*, 45–55. [CrossRef]

32. Strabone. *Geographica, V Book, 23 AD*; Loeb Classical Library Edition; Harvard University Press: Cambridge, MA, USA, 1923.

33. Felici, E. La ricerca sui porti romani in cementizio: Metodi e obiettivi', in Archeologia subacquea—Come opera l'archeologo sott'acqua. Storie dalle acque. In *VIII Ciclo di Lezioni Sulla Ricerca Applicata in Archeologi*; Volpe, G., Ed.; All'Insegna del Giglio: Firenze, Itlay, 1998; pp. 275–340.

34. Felici, E. Costruire nell'acqua: I porti antichi. In *Lezioni Fabio Faccenna, Conferenze di Archeologia Subacquea*; Giacobelli, M., Ed.; Edipuglia: Bari, Italy, 2001; pp. 161–178.

35. Felici, E. Ricerche sulle tecniche costruttive dei porti romani. *J. Anc. Topogr.* **2006**, *16*, 59–84.

36. Stefanile, M. The Project PILAE, For an Inventory of the Submerged Roman Piers. A Preliminary Overview. *Int. J. Environ. Geoinform.* **2015**, *2*, 34–39. [CrossRef]

37. Oleson, J.P.; Brandon, C.; Cramer, S.M.; Cucitore, R.; Gotti, E.; Hohlfelder, R.L. The ROMACONS Project: A Contribution to the Historican and Engineering Analysis of the Hydrauilc Concrete in Roman Maritime Structures. *Int. J. Naut. Archaeol.* **2004**, *33*, 199–229. [CrossRef]

38. Hohlfelder, R.L.; Oleson, J.P.; Brandon, C. Building a Roman pila in the sea—Experimental Archaeology of Brindisi, Italy September 2004. *Int. J. Naut. Archaeol.* **2005**, *34*, 124–129.

39. Brandon, C.J.; Hohlfelder, R.L.; Oleson, J.P. The concrete construction of the Roman harbours of Baiae and Portus Iulius, Italy: The ROMACONS 2006 field season. *Int. J. Naut. Archaeol.* **2008**, *37*, 374–379. [CrossRef]

40. Gazda, E.K. Cosa's Contribution to the Study of Roman Hydraulic Concrete: An Historiographic Commentary. In *Classical Studies in Honor of Cleo Rickman Fitch*; Goldman, N.W., Ed.; Peter Lang Pub Inc.: New York, NY, USA, 2001; pp. 145–177.

41. Oleson, J.P.; Bottalico, L.; Brandon, C.; Cucitore, R.; Gotti, E.; Hohlfelder, R.L. Reproducing a Roman Maritime Structure with Vitruvian pozzolanic concrete. *J. Roman Archaeol.* **2006**, *19*, 31–52. [CrossRef]

42. Painter, K. Roman Flasks with Scenes from Baiae and Puteoli. *J. Glass Stud.* **1975**, *17*, 54–67.

43. Blackmann, D.J. Ancient Harbours in the Mediterranean, Part 1. *Int. J. Naut. Archaeol.* **1982**, *11*, 79–104. [CrossRef]

44. Bejarano Osorio, M.A. Una ampolla de vidrio decorada con la planta topográfica de la ciudad de Puteoli. *Mérida Excav. Arqueol.* **2002**, *8*, 513–532.

45. Benjamin, J.; Rovere, A.; Fontana, A.; Furlani, S.; Vacchi, M.; Inglis, R.H.; Mourtzas, N. Late Quaternary sea-level changes and early human societies in the central and eastern Mediterranean Basin: An interdisciplinary review. *Quat. Int.* **2017**, *449*, 29–57. [CrossRef]

46. Lambeck, K.; Anzidei, M.; Antonioli, F.; Benini, A.; Esposito, A. Sea level in Roman time in the Central Mediterranean and implications for recent change. *Earth Planet. Sci. Lett.* **2004**, *224*, 563–575. [CrossRef]

47. Auriemma, R.; Solinas, E. Archaeological remains as sea-level change markers: A review. *Quat. Int.* **2009**, *206*, 134–146. [CrossRef]

48. Morhange, C.; Marriner, N. Archaeological and biological relative sea-level indicators. In *Handbook of Sea Level Research*; Shennan, I., Long, A., Horton, B.P., Eds.; Wiley: Oxford, UK, 2017; pp. 146–156.

49. Severino, N. Recenti ricerche archeologiche sull'isola di Nisida. *Orizz. Rass. Archeol.* **2005**, *6*, 119–133.

50. D'Arms, J. *Romans on the Bay of Naples: A Social and Cultural Study of the Villas and Their Owners from 150 B.C. to A.D. 400*; Harvard University Press: Cambridge, MA, USA, 1970.

51. Gunther, R.T. *Pausilypon, the Imperial Villa near Naples*; Hart, H., Ed.; Oxford University Press: Oxford, UK, 1913.

52. Gianfrotta, P.A. I porti dell'area flegrea. In *Porti, Approdi e Rotte nel Mediterraneo Antico*; Laudizi, G., Marangio, C., Eds.; Studi di Filologia e Letteratura 4: Galatina, Italy, 1998; pp. 155–168.

53. Mattei, G.; Giordano, F. Integrated geophysical research of Bourbonic shipwrecks sunk in the Gulf of Naples in 1799. *J. Archaeol. Sci.* **2015**, *1*, 64–72. [CrossRef]

54. Menna, F.; Nocerino, E.; Troisi, S.; Remondino, F. Joint alignment of underwater and above-the-water photogrammetric 3D models by independent models adjustment. *Int. Arch. Photogramm. Remote Sens. Spat. Inf. Sci.* **2015**, *40*, 143–151. [CrossRef]

55. Lechtman, H.N.; Hobbs, L.W. Roman concrete and the Roman architectural revolution. *Ceram. Civiliz.* **1987**, *3*, 81–128.

56. Lambeck, K.; Antonioli, F.; Anzidei, M.; Ferranti, L.; Leoni, G.; Scicchitano, G.; Silenzi, S. Sea level change along the Italian coast during the Holocene and projections for the future. *Quat. Int.* **2011**, *232*, 250–257. [CrossRef]

57. Antonioli, F.; Ferranti, L.; Fontana, A.; Amorosi, A.; Bondesan, A.; Braitenberg, C.; Dutton, A.; Fontolan, G.; Furlani, S.; Lambeck, K.; et al. Holocene relative sea-level changes and vertical movements along the Italian and Istrian coastlines. *Quat. Int.* **2009**, *221*, 37–51. [CrossRef]

58. Rovere, A.; Stocchi, P.; Vacchi, M. Eustatic and relative sea-level changes. *Curr. Clim. Chang. Rep.* **2016**, *2*, 221–231. [CrossRef]

59. Vacchi, M.; Marriner, N.; Morhange, C.; Spada, G.; Fontana, A.; Rovere, A. Multiproxy assessment of Holocene relative sea-level changes in the western Mediterranean: Sea-level variability and improvements in the definition of the isostatic signal. *Earth Sci. Rev.* **2016**, *155*, 172–197. [CrossRef]

60. Shennan, I.; Long, A.J.; Horton, B.P. (Eds.) *Handbook of Sea-level Research*; John Wiley & Sons: Oxford, UK, 2015; pp. 1–581.

61. Van de Plassche, O. (Ed.) *Sea-Level Research: A Manual for the Collection and Evaluation of Data*; GeoBooks, Galliard (Printers) Ltd: Great Yarmouth, UK, 1986; pp. 1–603.

62. Romano, P.; Di Vito, M.A.; Giampaola, D.; Cinque, A.; Bartoli, C.; Boenzi, G.; Detta, F.; Di Marco, M.; Giglio, M.; Iodice, S.; et al. Intersection of exogenous, endogenous and anthropogenic factors in the Holocene landscape: A study of the Naples coastline during the last 6000 years. *Quat. Int.* **2013**, *303*, 107–119. [CrossRef]

Article

Geomorphological Signature of Late Pleistocene Sea Level Oscillations in Torre Guaceto Marine Protected Area (Adriatic Sea, SE Italy)

Francesco De Giosa [1], Giovanni Scardino [2], Matteo Vacchi [3,4], Arcangelo Piscitelli [1], Maurilio Milella [1], Alessandro Ciccolella [5] and Giuseppe Mastronuzzi [2,*]

[1] Environmental Surveys Srl, Via Dario Lupo 65, 74121 Taranto, Italy; francescodegiosa@ensu.it (F.D.G.); arcangelo.piscitelli@libero.it (A.P.); maurisismo@libero.it (M.M.)
[2] Dipartimento di Scienze della Terra e Geoambientali, Università degli Studi di Bari "Aldo Moro", Via Edoardo Orabona 4, 70125 Bari, Italy; giovanni.scardino@uniba.it
[3] Dipartimento di Scienze della Terra, Università di Pisa, Via Santa Maria 53, 56126 Pisa, Italy; matteo.vacchi@unipi.it
[4] CIRSEC, Center for Climate Change Impact, University of Pisa, Via del Borghetto 80, 56124 Pisa, Italy
[5] Consorzio di Gestione di Torre Guaceto, Via Sant'Anna 6, 72012 Carovigno (Brindisi), Italy; segreteria@riservaditorreguaceto.it
* Correspondence: giuseppe.mastronuzzi@uniba.it

Received: 10 September 2019; Accepted: 12 November 2019; Published: 16 November 2019

Abstract: Morphostratigraphy is a useful tool to reconstruct the sequence of processes responsible for shaping the landscape. In marine and coastal areas, where landforms are only seldom directly recognizable given the difficulty to have eyewitness of sea-floor features, it is possible to correlate geomorphological data derived from indirect surveys (marine geophysics and remote sensing) with data obtained from direct ones performed on-land or by scuba divers. In this paper, remote sensing techniques and spectral images allowed high-resolution reconstruction of both morpho-topography and morpho-bathymetry of the Torre Guaceto Marine Protected Area (Italy). These data were used to infer the sequence of climatic phases and processes responsible for coastal and marine landscape shaping. Our data show a number of relict submerged surfaces corresponding to distinct phases of erosional/depositional processes triggered by the late-Quaternary interglacial–glacial cycles. In particular, we observed the presence of submerged marine terraces, likely formed during MIS 5–MIS 3 relative highstand phases. These geomorphic features, found at depths of ~26–30, ~34–38, and ~45–56 m, represent important evidence of past sea-level variations.

Keywords: morphostratigraphy; sea-level changes; marine terraces; river incisions; Adriatic Sea

1. Introduction

Ice-cores and marine sediment records indicate that the climate of the last 500 ky was characterized by ~100 ky warm–cold cyclicity [1,2] which led to repeated transitions between glacial and interglacial periods. These transitions triggered cycles of accretion and melting of the major ice-sheets with consequent major oscillations of sea-level position [3–5] and significant modifications of the on-land and sea-floor landscapes.

Morphostratigraphy applied to landscape evolution has allowed recognition of relict sequences of past morphogenetic processes associated with these glacial and interglacial climatic phases [6,7]. In particular, past landscapes have been often reconstructed through a combination of geomorphological and sedimentological analyses [8–10]. In the Mediterranean Sea, this multidisciplinary approach has proved to be very useful to understand genesis and evolution of particular landforms and sea-floor features as well as to infer relative sea-level (RSL) changes and their influence on coastal landscape evolution (e.g., [11–14]).

However, many relict landforms are often difficult to observe and to analyze because they are located in submarine areas or covered by thick layers of more recent sediments. This is, for instance, the case of colder climatic phases when sea level was many m below the modern position, considering the tectonic contribution as well (e.g., [15,16]). This issue was progressively resolved thanks to recent technological advances that allow collecting high-resolution data to characterize the morpho-bathymetry and morpho-topography of a specific zone, even in underwater environments [17–19], using remote sensing techniques and spectral images.

In this paper, we investigated the coastal and off-shore zones of the Marine Protected Area (MPA) of Torre Guaceto (Adriatic Sea, SE Italy, Figure 1) through remote sensing techniques, morphological, and stratigraphic surveys in order to recognize and correlate chronologically subaerial and submerged landforms.

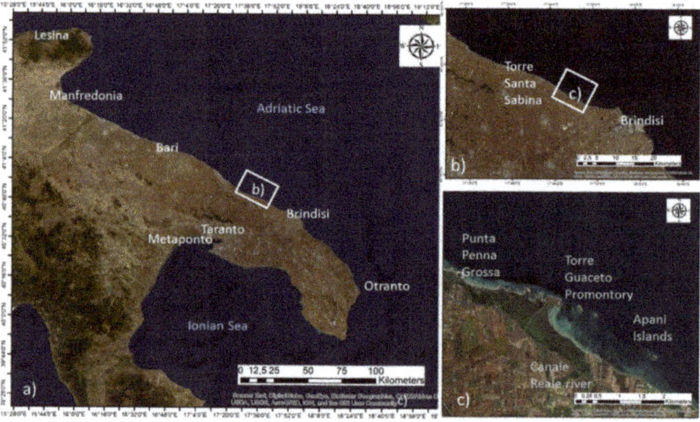

Figure 1. (**a**) Study area located in Apulia region; (**b**) location of Torre Guaceto surveyed area (Adriatic Sea, SE Italy); (**c**) surveys were performed from Punta Penna Grossa to Apani Islands.

Pleistocene marine terraces and Holocene deposits can provide an important insight into late-Quaternary rates of vertical displacements [20,21]. In south-eastern Apulia, evidence of Marine Isotope Stage (MIS) 5 are rare and poorly constrained in terms of chronology and elevation. The sole available indirect age comes from the south-eastern Murge (Torre Santa Sabina locality near Brindisi, Figure 1b). Here, a coastal deposit situated at ~3 m a.s.l. overlies a colluvial deposit bearing Late Paleolithic–Mousterian flints and, thus, it can be correlated to a generic MIS 5 [22,23]. In the northern part of Apulia, near Manfredonia (Figure 1a,b), MIS 5 deposits were found at depth of −22 m [24]; it implies significant subsidence rate (−0.17 mm/y) that are comparable to the areas placed in a very different geodynamic context (e.g., Trieste, Versilia, and Sarno plains [20]).

In this study, we identified the presence of relict marine terraces and other subaerial landforms presently lying along the Apulian continental shelf. We correlated them to the major late-Quaternary climatic phases coupling models with data available on land [20,21]. This analysis allowed reconstructing the coastal evolution of the Torre Guaceto Marine Protected Area (MPA) during the last 150 ky.

2. Geological Settings

Torre Guaceto is a tower of the XV century, placed on a promontory located north of the town of Brindisi, inside of the MPA of Torre Guaceto (Figure 1). The study area overlaps the Canale Reale river, which divides the Murge plateau from the Taranto-Brindisi plain [25,26]. Different lithological units crop out in this area, whose deposition is connected to past sea level stands and tectonic factors during Middle-Late Pleistocene [25,27]. The bedrock is represented by Mesozoic limestone, which is overlain by discontinuous marine deposits of Plio-Pleistocene age (up to 70 m), belonging to the Calcarenite di Gravina Fm, and the etheropic argille subappennine informal unit [28,29]. These units are covered by middle-upper Pleistocene biocalcarenitic beach and dune deposits [11,25,30]. Along the Murgia scarp they crop out as stepped terraces stretching from −400 m to a few m above the present mean sea-level. Although along the coastal area of Ionian Apulia the younger marine terraces are generally characterized by the presence of a senegalensis faunal assemblage that, in combination with U/Th ages, indicates its deposition during the MIS 5.5 (e.g., [23,31–33]), the Adriatic coastal area of Apulia does not permit any chronological attribution due to the lack of geochronological or paleontological markers. Every chronological attribution related to the upper/late Pleistocene derives from the use and applications of the morphostratigraphy principles [6]. All along the Adriatic coast stretching north to Brindisi and in particular between Torre Guaceto and Punta Penne, sea-level drop associated with the last glacial maximum (LGM, ~26 to ~21 ky BP, [34]), caused the incision of the basement, in correspondence on the current hydrographic network, showing features of sapping processes [11,23]. Seaward, sapping valleys cut the upper Pleistocene biocalcarenitic sedimentary cover. Surveys performed on the land allowed forming a hypothesis that the shaping of the paleo-beach-dune system occurred during MIS 5; the sapping valleys were shaped due to the increase of the relief energy caused by the lowering of sea level. Their maximum shaping would correspond to the lowermost sea-level stand (−120 m) at the LGM [11,25,27,30].

The following Holocene marine transgression (last 12 ky BP) and inter-strata dissolution produced a series of sub-circular inlets that host pocket beaches (Figure 2). On their border, a polyphasic dune ridge was recognized, formed by two aeolian sediment generations, the first at ~6.0 ky BP and the second at ~2.5 ky BP [11,25,30]. The south-eastern part of the foredune extends for ~500 m reaching a maximum elevation of −12 m msl. This is composed of brownish sand layers intercalated with brown soil rich in *Helix* spp. [30,35]. Geo-archaeological and bio-stratigraphic analysis indicate that RSL was 2.25 ± 0.2 m below the present one at about 3.5 ky BP and that the total RSL variation in the last ~2.0 ky BP was about of 0.9 m below the present mean sea-level (msl) [36,37].

Figure 2. The promontory of Torre Guaceto is shaped on a sequence of Calcarenite di Gravina Fm. (late Pliocene–Early Pleistocene) and of late Pleistocene–Tyrrhenian biocalcarenite [26].

The comparison of these data with the available GIA (glacial isostatic adjustment) models [38,39] indicates a low rate of tectonic subsidence of this coastal area at least during the last 125 ky. This is further corroborated by the absence of significant historical seismicity and by the GPS data that indicate zero to weakly negative on-going vertical movements [40]. Subsidence rate increases on the northern sector of Apulia, reaching values of −0.3 mm/y as indicated by tectonic structures observed in seismic profiles and boreholes off-shore the Gargano Promontory and Tremiti Islands [41,42]. In this tectonic framework the significant sea-level changes are added, in particular during the last 100 ky, when sea-level falls on MIS 5.4 and MIS 5.2 determined downward and seaward shifts of the shoreline and the decrease of sediment supply, possibly in response to the reduction in basin width that hampered lateral advection [43].

3. Materials and Methods

In this work, we merged high-resolution bathymetry derived by LIDAR remote sensing (with a penetration into water-column of ~50 m) techniques with detailed MIVIS spectral images (with a penetration into water-column of −7 m). We further corroborate these data by scuba diving surveys (e.g., [44,45]).

Remote sensing data consisted LIDAR data (Airborne Laser Terrain Mapper—ALTM Optech's Gemini 167 kHz, near infrared, Teledyne Optech, Toronto, ON, Canada) and Daedalus AA5000 MIVIS (Multispectral Infrared and Visible Imaging Spectrometer, Italian National Council Research CNR, Italy) hyperspectral images both for the on-shore and nearshore coastal zone (e.g., [17]). LIDAR Optech's Gemini 167 data are part of a wider acquisition of remote sensing data in the areas ascribed to the protected marine areas of Calabria, Campania, Apulia, and Sicily. This LIDAR system allows obtaining elevation data to an accuracy of 5 to 10 cm. This LIDAR system flies up to 4000 m to cover large coastal area where a high degree of accuracy and speed is necessary and where accessibility is difficult, as in back-dune zones or in steep coastal slope.

The surveys collected a point distribution, with each point consisting of x-y-z coordinates and associated reflectance value. Acquired points were processed in GIS environment to build digital surface models (DSMs) and digital terrain models (DTMs). In order to characterize the submerged coastal zone particularly for the near shore bathymetry up to −7 m, hyperspectral images have been used and acquired with MIVIS scanner. This instrument, property of the National Research Council of Italy, is a 102 channel scanner covering visible and near infrared (0.43–0.83 μm), middle infrared (1.15–1.55 and 1.98–2.50 μm) and thermal infrared (8.21–12.70 μm) regions of the electromagnetic spectrum, providing a wealth qualitative information of the surveyed area. MIVIS scanner has a geometrically correct scan line that, due to the movement of the aircraft, is displaced with the roll, pitch, yaw, and with changes in velocity and direction. The operational flight heights of the scanner can range from 1500 up to 5000 m above ground; at this height, the nadiral pixel dimension ranges from 3 up to 10 m, integrated with a GPS system and a gyro.

In GIS environment, LIDAR data and hyperspectral images have been combined to build digital terrain model (DTM) and digital surface model (DSM) with a grid cell width of 4 × 4 m, in order to define the major landforms occurring both along the coast and the shallow continental shelf up to a depth of −56 m (Figure 3).

Two scuba diving surveys across the incision immediately to the ESE of the promontory of Torre Guaceto (Figure 4) have been performed. The geomorphological surveys were traced up to a depth of about 18 m along both sides of the incision (Figure 5).

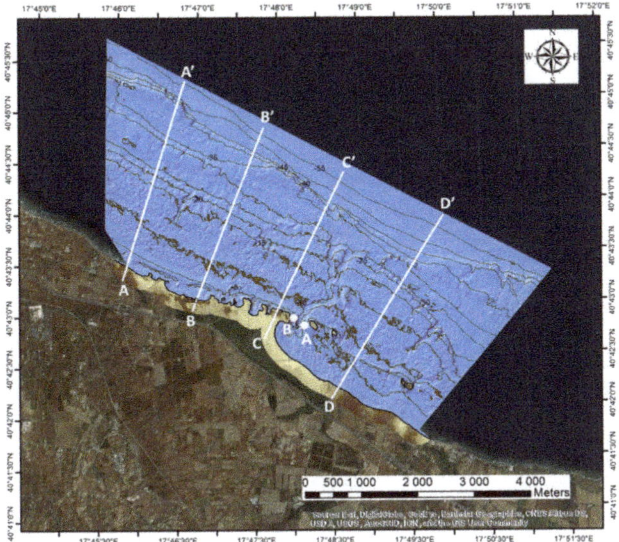

Figure 3. Morpho-topographic and morpho-bathymetric DTM (in blue) and DSM (in brown) of Torre Guaceto area with bathymetric profiles traces (in white) and direct scuba surveyed areas (in white dots).

Figure 4. The channel between Torre Guaceto promontory (reported in aerial view of Figure 3) and homonymous islands (in background); the sea-floor is cut by a sapping valley characterized by classic box profile, shaped in the Calcarenite di Gravina Fm. up to a depth of about 18 m.

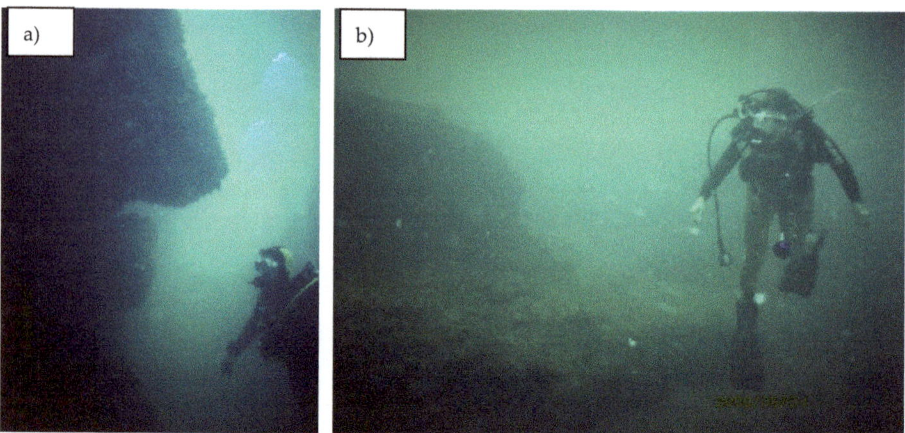

Figure 5. Sapping notch and unstable block (**a**) and two large sub-horizontal steps (**b**) shaped on the southern slope of the submerged sapping valley between the islands and Torre Guaceto Promontory.

4. Results

The coastal tract comprised between Torre Guaceto and Punta Penne (Figure 1b), is characterized by a sequence of small calcarenitic islets (Apani Islands). They are made of cemented beach and dune sediments which likely represent the MIS 5.5 deposits according to the correlation with similar deposits outcropping all along the coast of Apulia. In addition, a further continuous dune belt can be found inland [23,25,31,32].

The composite survey campaign, performed in this study, allowed recognizing a number of significant submarine landforms; in particular, we observed the presence of near-flat surfaces and other morphological features likely related to submerged incisions. Bathymetric reconstruction revealed a staircase of near-flat surfaces which can be observed all over the different surveyed sectors (Figure 6).

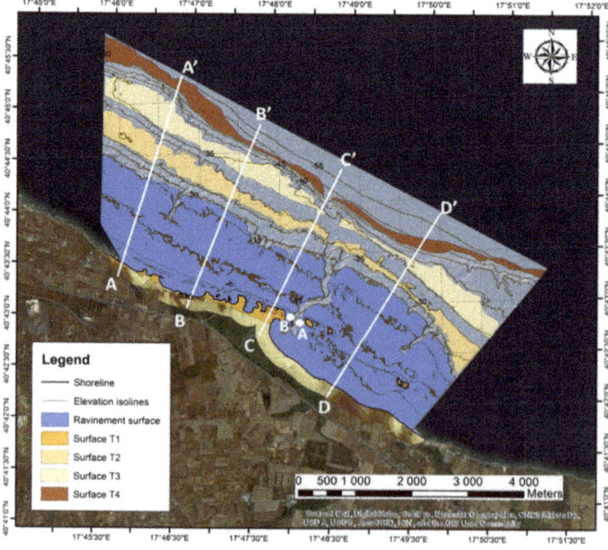

Figure 6. Staircase geometry of near-flat surfaces and ravinement surface observed along the Torre Guaceto shelf.

These surfaces are characterized by gentle slopes which do not exceed 4–5 degrees (about 6%) occurring at four different depth ranges (represented in T1-T2-T3-T4 levels in Figure 7) in all the analyzed transects (A, B, C, and D).

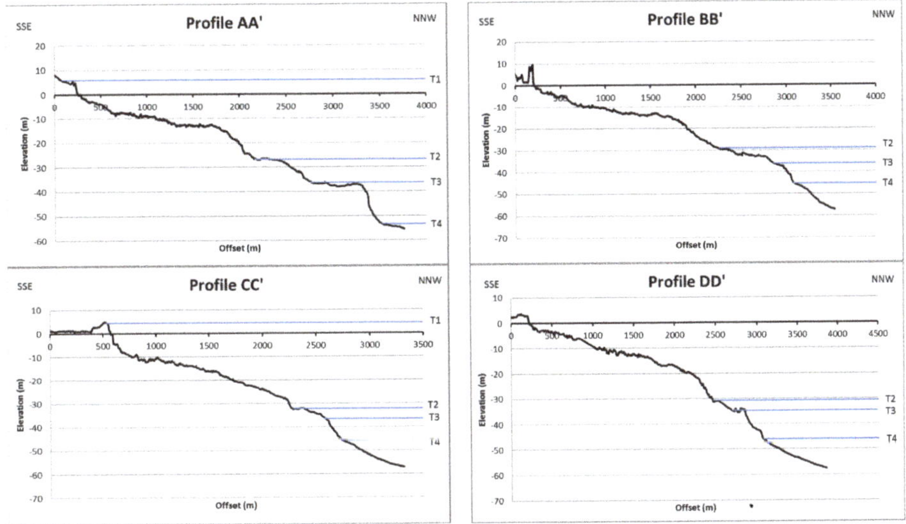

Figure 7. Elevation profiles of Torre Guaceto MPA continental shelf. In blue, the upper limit of the surfaces (T1-T2-T3-T4) connected to the past sea-level stands are represented.

The T1 level occurs on shore; its inner (landward) and outer edge (seaward) are placed at 10 and 5 m, respectively.

The second surface (T2 level) occurs underwater. It develops between −26 (inner edge) and −32 m (outer edge). A third erosive surface (T3 level) occurs between −34 (inner edge) and −38 m (outer edge) while a fourth erosional surface (T4 level) ranges between −45 (inner edge) and −56 m (outer edge), with a great extent in correspondence of the profile AA′ (Figure 7).

Our scuba surveys further documented a ravinement surface. This surface is gently sloping (2–3°) seaward, developing between the present shoreline and −18 m (Figure 8). It is discontinuously covered by coarse and medium sands which, in some sheltered areas, define the present beaches [46].

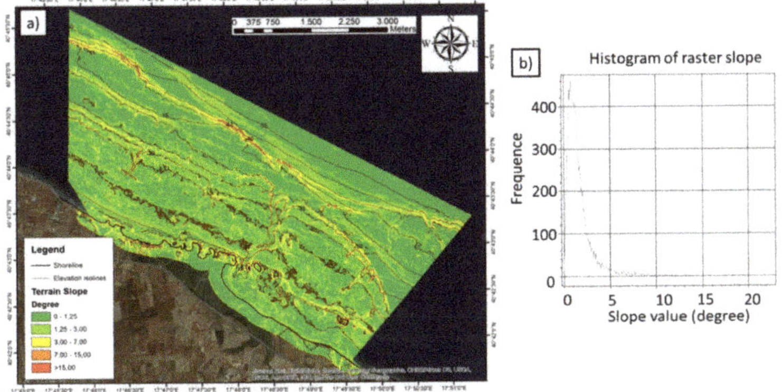

Figure 8. Slope analysis of Torre Guaceto continental shelf; (**a**) slope changes in correspondence of the near-flat surfaces limits; (**b**) the mean morpho-bathymetry slope.

Both bathymetric data and scuba diving surveys showed the presence of submerged incisions orthogonal to surfaces boundaries and connected to the current land hydrographic network. These sea-floor features, recognizable along the whole investigated portion of the sea-bottom (e.g., from 0 to −56 m), cut all the submerged near-flat surfaces (T2 to T4, Figure 7).

5. Discussion

The new sets of topographic and bathymetric data reveal well preserved evidence of past sea-level stands that shaped the coastal and marine landscape near Torre Guaceto. Unfortunately, the geomorphological markers described in this study lack dateable material; therefore, the chronological constraint of their origin can be only speculated on the basis of bathymetric cross-correlations. In the first approximation, our geomorphological markers can be compared with modelled eustatic values [45,47]. The remains of the highest terrace (T1), most likely formed during the MIS 5.5 (~125 ky) when the RSL was −7 m above the present msl [20,23,37,48].

According to Rovere et al. [49] marine terraces can be shaped by marine erosion or can consist of shallow water to slightly emerged accumulations of materials redistributed by shore erosional and depositional processes (e.g., marine-built terraces) [50]. The width of marine terraces ranges from few hundreds of meters to up to 1–2 km and can stretch along many kilometers of coastline. The mapped submerged near-flat surfaces fit well with this morphological description. For this reason, we interpreted the surfaces as relicts of marine terraces. In the absence of any evidence of discontinuity in the sedimentary bodies, it is impossible to find any evidence of their depositional or erosive genesis. However, the hypothesis that these submerged surfaces could represent erosional marine terraces seems to be supported by seismic profiles performed in central Adriatic Sea, where a significant erosion of MIS 5.5–5.1 shelf progradational units during last 100 ky was observed [26,27].

The chronological frame of the submerged surfaces is complex. This is mainly because the presence of submerged features is very seldom reported in the Mediterranean [44]. The sea-level stands that shaped the reconstructed terraces must be located below both the present and the last interglacial ones. According to the available eustatic curves, different Marine Isotope Stages (3, 5.1, 5.3, 6.5, 7.1, 7.3, and 7.5) peaked below both MIS 5.5 and MIS 1 sea-levels. However, the evidence of river incisions shaped through sapping processes [11,23] is observed in all the submerged marine terraces. This incision reached maximum rate during LGM [6], when sea level was −120 m lower. For this reason, all surfaces must necessarily be older than MIS 1 and younger than MIS 5.5.

The submerged terrace found at depths −26 and −32 m (T2 level) can be tentatively attributed to MIS 5.3 (~101 ky BP) that peaked at −30 m on the sea-level curve by Grant et al., 2014 [5]. At MIS 5.1 (~81.5 ky), sea-level stand allowed the genesis of a new erosional marine terrace encountered at a depth variable between −34 and −38 m, corresponding to T3 level at −38 m. In the following phase, a sea-level drop was observed up to a new sea-level stand on MIS 3 at ~54.5 ky, with formation of marine terrace in correspondence of T4 level, at a depth variable between −45 and −56 m, peaked at −52 m on the sea level curve by Grant et al., 2014 [5].

The general tectonic framework of Torre Guaceto area reveals a low subsidence rate of 0.02 mm/y [15,22,23,51]. We corrected the current depth of terraces according to the subsidence rate in order to attribute the actual elevation for each boundary at the moment of their genesis (Table 1).

In the case of the MIS 5.5 surface, a displacement of 2.5 m in 122 ky has been calculated considering a tectonic rate of 0.02 mm/y [22,37,48]. This implies a corrected surface altitude ranging between 12 (inner edge) and 3 m (outer edge) at the moment of surface genesis.

For the MIS 5.3 surface, a displacement of 2.02 m has been calculated in 101 ky. It implies a depth range between −23.98 (inner edge) and −28.98 m (outer edge), while for the MIS 5.1 surface a displacement of 1.63 m has been calculated in 81.5 ky attributing a corrected depth range between −32.37 (inner edge) and −36.37 m (outer edge). Finally, for the MIS 3 surface, a displacement of 1.09 m has been calculated in 54.5 ky which implies a depth range between −43.91 (inner edge) and −54.91 m (outer edge) at the moment of surface genesis.

Table 1. Depth of surfaces detected in each profile corrected for the tectonic displacements.

Surface	Depth Range Profile AA′ (m)	Depth Range Profile BB′ (m)	Depth Range Profile CC′ (m)	Depth Range Profile DD′ (m)	Sea Level Highstand SPECMAP Imbrie & McIntyre 2006	Sea Level Highstand Walbroeck et al., 2002	Sea Level Highstand Grant et al., 2014
MIS 5.5_T1	6.26 and 4.17	-	5.76 and 3.73	-	0.34 m at 122 ky	6.32 m at 123.87 ky	10.62 m at 122 ky
MIS 5.3_T2	−24.57 and −25.11	−27.05 and −31.23	−29.92 and −31.41	−29.02 and −29.6	−35.58 m at 99 ky	−20.86 m at 101 ky	−30 m at 101 ky
MIS 5.1_T3	−34.42 and −36.27	−34.32 and −36.79	−34.74 and −36.06	−33.38 and −33.75	−35.21 m at 80 ky	−18.67 m at 81.5 ky	−38 m at 81.5 ky
MIS 3_T4	−52.43 and −54.11	−44.6 and −47.11	−44.71 and −46.53	−44.28 and −46.82	−76.79 m at 54 ky	−52.08 m at 54.5 ky	−52 m at 54.5 ky
Ravinement surface	−2.8 and −12.92	−3 and −13	−7.7 and -16	−3.3 and −13.7	−26.27 at 7 ky BP	−9.74 m at 8 ky BP	−2.36 m at 7 ky BP

These terraces can be referred to different relative past sea-level stands occurring during MIS 5-3, corresponding to peaks which can be observed in different model curves (e.g., [5,52–55] Figure 9).

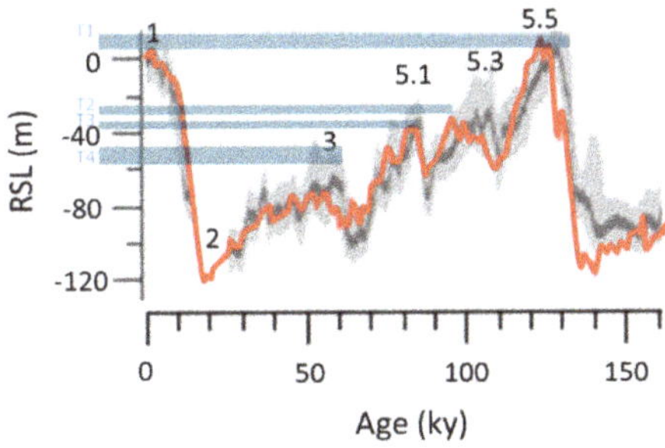

Figure 9. Sea-level changes from 150 ky to the present derived from Red Sea records (modified after Grant et al., 2014). Light blue bands show erosional marine terraces depth ranges surveyed in the Torre Guaceto area.

The position of each marine terrace fits well with evidence of MIS 5 deposits already described by Mastronuzzi et al., 2011; 2018 [25,26]. This correlation was made under the assumption of minimal tectonic movements of this area [52] that show significantly different neotectonics pattern with respect to the northern part of the Apulia [21,24,34–36].

Morphobathymetry suggests that a valley network developed coeval with the sea-level stand through sapping processes [11,23]. In fact, since the evolution of the aquifer and the seawater/fresh water interface is strictly linked to the sea-level, the development of each sapping valley was largely influenced by the Late-Quaternary sea-level changes and the consequent shifts of the coastline [11,30]. Each sea-level (high) stand induced the development of a marine terrace, a shoreline and a number of short valleys. These valleys developed orthogonally to the coastline since the sapping processes was conditioned by structural alignment and/or by general geometry of the local basement [11]. Maximum incision occurred during MIS 2, when sea level was 120 m lower than present.

During the Holocene sea-level rise, incisions were flooded and filled by sediments, until the slowing down of sea-level rising rates (7 ky BP, [38,56], Figure 10), causing the widening of the ravinement surface between −18 m and the present sea level.

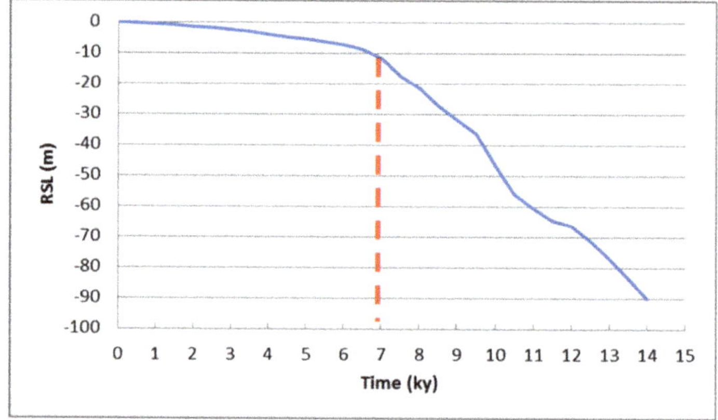

Figure 10. Relative sea-level prediction for Egnatia site [41]; the red dashed line indicates the slowing down of the sea-level rising rate at 7 ky.

Applying the morphostratigraphy principles on the near-flat surfaces depth—both subaerial and submerged—and to the sea-level trend recognizable during the last warming time, it is possible to reconstruct the following morpho-evolutionary steps:

- MIS 5.5—the sea-level highstand at 7 m msl allowed the deposition of beach and dune deposits shaping the marine terrace corresponding to the T1 level;
- MIS 5.3—during the general regression, the relative sea-level stand shaped the marine terrace currently located at a depth range between −26 and −32 m msl (T2 level);
- MIS 5.1—the relative sea-level stand allowed the shaping of the marine terrace currently located between −34 and −38 m msl and corresponding to the T3 level;
- MIS 3—the relative sea-level stand induced the marine terrace shaping currently located between −45 and −56 m msl (T4 level);
- MIS 2—at the LGM, the sea-level placed at −120 m msl facilitate the full incision on valley network;
- MIS 2—the post-LGM sea-level rose at fast rates until 7500 years BP;
- 7000 years BP—the sudden slow in sea-level rising rates produced the widening of the ravinement surface surveyed between −18 m msl and the present msl;
- 3500 years BP—sea-level stands during the Bronze Age (3.5 ky BP) at about −2.25 ± 0.2 m msl;
- 2200 years BP—sea-level stands at about −1.1 ± 0.1 m msl below the present mean sea level; inlets were marked by the presence of beaches with low embryonic dunes and typical back-beach environments;
- 1900 years BP—sea-level stands at about −0.65 ± 0.1 m msl;
- 1700 years BP—sea-level probably stands around −0.3 ± 0.1 m msl

6. Conclusions

Morphostratigraphic approach carried out in Torre Guaceto area allowed recognizing and chronologically correlating a variety of sea-floor features shaped by different sea-level stands in the Late-Quaternary. Last interglacial phases were already described in the literature [25,26,48] while, for the submerged features, new technologies have been useful to individuate the morphodynamic phases following the MIS 5.5 sea-level highstand.

Remote sensing and spectral images allowed detecting the underwater landforms which were used to perform a high-resolution mapping of submarine environments at large spatial scale. In particular, the use of LIDAR and MIVIS instruments allows surveying up to a depth of 55 m, highlighting the main underwater landforms as near-flat surfaces and submerged fluvial incisions.

Morphostratigraphic analysis of the sea-floor features allowed reconstructing the different shaping phases that were correlated to the chronological constrains deriving from the study of the shallow water and subaerial landforms. This analysis clearly indicated that the underwater marine terraces of Torre Guaceto were older than MIS 2; we tentatively attributed their formation in correspondence of the highstand peaks of MIS 5.5–MIS 5.3–MIS 5.1–MIS 3 (Figure 11).

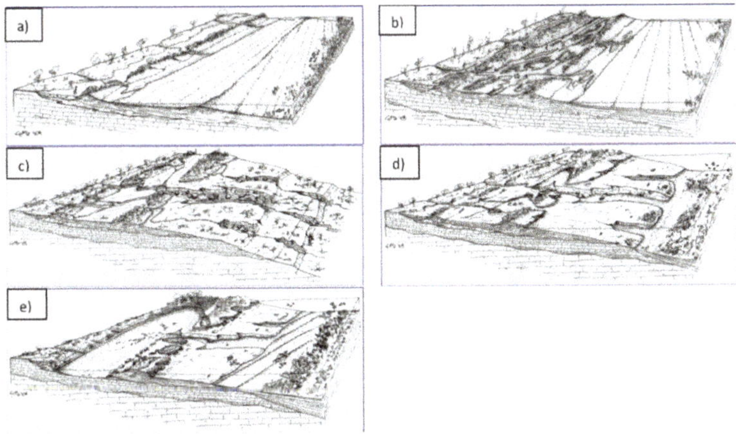

Figure 11. Phases on Torre Guaceto area: (**a**) First Marine Isotope Stage 5.5 phase–(**b**) second Marine Isotope Stage 5.5 phase–(**c**) last glacial maximum (LGM)–(**d**) bronze Age–(**e**) present.

The morphostratigraphic approach presented in this paper represents a fundamental first step to investigate the submerged landforms; further investigation, possibly corroborated by additional coring campaign may provide more precise insights into the genesis of the submerged landforms and into the climatic phases that shaped them.

Author Contributions: Conceptualization: F.D.G., G.S. and M.V.; Data curation: F.D.G. and M.V.; Formal analysis: M.V.; Funding acquisition: G.M.; Investigation: A.P.; Methodology: M.M.; Project administration: G.M.; Resources: A.C.; Visualization: A.C.; Writing—original draft: F.D.G. and G.S.; Writing—review & editing: A.P., M.M. and G.M.

Funding: This research has been realized by a research agreement between the Consortium of Torre Guaceto and the Department of Earth and Geo-environmental Sciences of the University of Bari, approved in the Department Council on 1 March 2019.

Acknowledgments: This paper is the result of studies performed in the framework of the agreement between the Torre Guaceto Marine Protected Area and the Department of Earth and Geo-environmental Sciences of the University of Bari. We thank all collaborators for their logistic and technic support in every phase of this work. We are thankful to the reviewers for their revisions that allow us to improve the quality of the paper. MV is funded by the Rita Levi Montalcini programme of the Italian Ministry of University and Research (MIUR).This work has been carried out under the umbrella of the IGCP Project n. 639 "Sea-level change from minutes to millennia" (Project Leaders: S. Engelhart, G. Hoffmann, F. Yu and A. Rosentau). We extend our gratitude to the MOPP-Medflood (INQUA CMP 1603P) project for fruitful discussions during the workshops.

Conflicts of Interest: The authors declare no conflict of interest. The funders had no role in the design of the study; in the collection, analyses, or interpretation of data; in the writing of the manuscript, or in the decision to publish the results.

References

1. Petit, J.R.; Jouzel, J.; Raynaud, D.; Barkov, N.I.; Barnola, J.-M.; Basile, I.; Bender, M.; Chappellaz, J.; Davis, M.; Delaygue, G.; et al. Climate and atmospheric history of the past 420,000 years from the Vostok ice core, Antarctica. *Nature* **1999**, *399*, 429–436. [CrossRef]
2. Augustin, L.; Barbante, C.; Barnes, P.R.; Barnola, J.M.; Bigler, M.; Castellano, E.; Cattani, O.; Chappellaz, J.; Dahl-Jensen, D.; Delmonte, B.; et al. EPICA community members. Eight glacial cycles from an Antarctic ice core. *Nature* **2004**, *429*, 623–628. [PubMed]

3. Siddall, M.; Chappell, J.; Potter, E.K. Eustatic sea level during past interglacials. *Dev. Quat. Sci.* **2007**, *7*, 75–92. [CrossRef]

4. Rohling, E.J.; Grant, K.; Bolshaw, M.; Roberts, A.P.; Siddall, M.; Hemleben, C.; Kucera, M. Antarctic temperature and global sea level closely coupled over the past five glacial cycles. *Nat. Geosci.* **2009**, *2*, 500–504. [CrossRef]

5. Grant, K.; Rohling, E.; Bronk Ramsey, C.; Cheng, H.; Edwards, L.; Florindo, F.; Heslop, D.; Marra, F.; Roberts, A.P.; Tamisiea, M.E.; et al. Sea-level variability over five glacial cycles. *Nat. Commun.* **2014**, *5*, 5076. [CrossRef] [PubMed]

6. Dramis, F.; Bisci, C. Cartografia geomorfologica. *Pitagora Bologna* **1998**, *215*.

7. Iurilli, V.; Cacciapaglia, G.; Selleri, G.; Palmentola, G.; Mastronuzzi, G. Kyrstmorphogenesis and tectonics in south-eastern Murge (Apulia, Italy). *Geogr. Fis. Din. Quat.* **2009**, *32*, 145–155.

8. Rovere, A.; Vacchi, M.; Firpo, M.; Carobene, L. Underwater geomorphology of the rocky coastal tracts between Finale Ligure and VadoLigure (western Liguria, NW Mediterranean Sea). *Quat. Int.* **2011**, *232*, 187–200. [CrossRef]

9. Melis, R.T.; Depalmas, A.; Di Rita, F.; Montis, F.; Vacchi, M. Mid to late Holocene environmental changes along the coast of western Sardinia (Mediterranean Sea). *Glob. Planet. Chang.* **2017**, *155*, 29–41. [CrossRef]

10. Melis, R.T.; Di Rita, F.; French, C.; Marriner, N.; Montis, F.; Serreli, G.; Sulas, F.; Vacchi, M. 8000 years of coastal changes on a western Mediterranean island: A multiproxy approach from the Posada plain of Sardinia. *Mar. Geol.* **2018**, *403*, 93–108. [CrossRef]

11. Mastronuzzi, G.; Sansò, P. Pleistocene sea-level changes, sapping processes and development of valley networks in the Apulia region (southern Italy). *Geomorphology* **2002**, *46*, 19–34. [CrossRef]

12. Mouslopoulou, V.; Begg, J.; Fülling, F.; Moraetis, D.; Partsinevelos, P.; Oncken, O. Distinct phases of eustatic and tectonic forcing for late Quaternary landscape evolution in southwest Crete, Greece. *Earth Surf. Dyn.* **2017**, *5*, 511–527. [CrossRef]

13. Spampinato, C.R.; Scicchitano, G.; Ferranti, L.; Monaco, C. Raised Holocene paleo-shorelines along the Capo Schisò coast, Taormina: New evidence of recent co-seismic deformation in northeastern Sicily (Italy). *J. Geodyn.* **2012**, *55*, 18–31. [CrossRef]

14. Spampinato, C.R.; Costa, B.; Di Stefano, A.; Monaco, C.; Scicchitano, G. The contribution of tectonics to relative sea-level change during the Holocene in coastal south-eastern Sicily: New data from boreholes. *Quat. Int.* **2011**, *232*, 214–227. [CrossRef]

15. Tropeano, M.; Cilumbriello, A.; Sabato, L.; Gallicchio, S.; Grippa, A.; Longhitano, S.; Bianca, M.; Gallipoli, M.R.; Mucciarelli, M.; Spilotro, G. Surface and subsurface of the Metaponto Coastal Plain (Gulf of Taranto—southern Italy): Present-day- vs. LGM-landscape. *Geomorphology* **2013**, *203*, 115–131. [CrossRef]

16. Valenzano, E.; Scardino, G.; Cipriano, G.; Fago, P.; Capolongo, D.; De Giosa, F.; Lisco, S.; Mele, D.; Moretti, M.; Mastronuzzi, G. Holocene morpho-sedimentaryevolution of the Mar Piccolo basin (Taranto, southernItaly). *Geogr. Fis. Din. Quat.* **2018**, *41*, 119–135. [CrossRef]

17. El Bastawesy, M.; Cherif, O.; Sultan, M. The geomorphological evidences of subsidence in the Nile Delta: Analysis of high resolution topographic DEM and multi-temporal satellite images. *J. Afr. Earthsci.* **2017**, *136*, 252–261. [CrossRef]

18. Rabineau, M.; Berné, S.; Ledrezen, É.; Lericolais, G.; Marsset, T.; Rotunno, M. 3D architecture of lowstand and transgressive Quaternary sand bodies on the outer shelf of the Gulf of Lion, France. *Mar. Pet. Geol.* **1998**, *15*, 439–452. [CrossRef]

19. Scicchitano, G.; Pignatelli, C.; Spampinato, C.; Piscitelli, A.; Milella, M.; Monaco, C.; Mastronuzzi, G. Terrestrial Laser Scanner techniques in the assessment of tsunami impact on the Maddalena Peninsula (south-easternSicily, Italy). *Earth Planets Space* **2012**, *64*, 889–903. [CrossRef]

20. Ferranti, L.; Antonioli, F.; Amorosi, A.; Dai Pra, G.; Mastronuzzi, G.; Mauz, B.; Monaco, C.; Orrù, P.; Pappalardo, M.; Radtke, U.; et al. Elevation of the last interglacialhighstand in Italy: A benchmark of coastal tectonics. *Quat. Int.* **2006**, *145*, 30–54. [CrossRef]

21. Antonioli, F.; Ferranti, L.; Fontana, A.; Amorosi, A.; Bondesan, A.; Braitenberg, C.; Dutton, A.; Fontolan, G.; Furlani, S.; Lambeck, K.; et al. Holocene relative sea-level changes and tectonic movements along the Italian coastline. *Quat. Int.* **2009**, *206*, 102–133. [CrossRef]

22. Marsico, A.; Selleri, G.; Mastronuzzi, G.; Sansò, P.; Walsh, N. Cryptokyrst: A case study of the Quaternary landforms of southern Apulia (Southern Italy). *Acta Carsol.* **2003**, *32*, 147–159. [CrossRef]

23. Mastronuzzi, G.; Sansò, P. Quaternary coastal morphology and sea level changes. In Proceedings of the Puglia 2003, Final Conference—Project IGCP 437 UNESCO—IUGS, Otranto/TarPuglia (Italy), GIS Coast Coast—GruppoInformale di StudiCostieri, Research Publication, Puglia, Italy, 2003; Volume 5, pp. 22–28.

24. De Santis, V.; Caldara, M.; de Torres, T.; Ortiz, J. Stratigraphic units of the Apulian Tavoliere plain (Southern Italy): Chronology, correlation with marine isotope stages and implications regarding vertical movements. *Sediment. Geol.* **2010**, *228*, 255–270. [CrossRef]

25. Mastronuzzi, G.; Caputo, R.; Di Bucci, D.; Fracassi, U.; Milella, M.; Pignatelli, C.; Sansò, P. Middle-Late Pleistocene Evolution of the Adriatic Coastline of Southern Apulia (Italy). In Response to Relative Sea-Level Changes. *Geogr. Fis. Din. Quat.* **2011**, *34*, 207–222.

26. Mastronuzzi, G.; Milella, M.; Piscitelli, P.; Simone, O.; Quarta, G.; Scarano, T.; Calcagnile, L.; Spada, I. Landscape analysis in Torre Guaceto area (Brindisi) aimed at the reconstruction of the late Holocene sea level curve. *Geogr. Fis. Din. Quat.* **2018**, *41*, 65–79. [CrossRef]

27. Mastronuzzi, G.; Quinif, Y.; Sansò, P.; Selleri, G. Middle-Late Pleistocene polycyclic evolution of a stable coastal area (southern Apulia, Italy). *Geomorphology* **2007**, *86*, 393–408. [CrossRef]

28. Ciaranfi, N.; Pieri, P.; Ricchetti, G. Note alla carta geologica delle Murge e del Salento (Puglia centromeridionale). *Mem. Soc. Geol. Ital.* **1988**, *41*, 449–460.

29. Ricchetti, G.; Ciaranfi, N.; Luperto Sinni, E.; Mongelli, F.; Pieri, P. Geodinamica ed evoluzione sedimentaria e tettonica dell'Avampaese Apulo. *Mem. Soc. Geol. Ital.* **1988**, *41*, 57–82.

30. Mastronuzzi, G.; Sansò, P. Holocene coastal dune development and environmental changes in Apulia (southern Italy). *Sediment. Geol.* **2002**, *150*, 139–152. [CrossRef]

31. Amorosi, A.; Antonioli, F.; Bertini, A.; Marabini, S.; Mastronuzzi, G.; Montagna, P.; Negri, A.; Piva, A.; Rossi, V.; Scarponi, D.; et al. The Middle-Late Quaternary Fronte Section (Taranto, Italy): An exceptionally preserved marine record of the Last Interglacial. *Glob. Planet. Chang.* **2014**, *119*, 23–38. [CrossRef]

32. Negri, A.; Amorosi, A.; Antonioli, F.; Bertini, A.; Florindo, F.; Lurcock, P.; Marabini, S.; Mastronuzzi, G.; Regattieri, E.; Rossi, V.; et al. A potential Global Stratotype Section and Point (GSSP) for the Tarentian Stage, Upper Pleistocene: Work in progress. *Quat. Int.* **2015**, *383*, 145–157. [CrossRef]

33. De Santis, V.; Caldara, M.; Torres, T.; Ortiz, J.; Sánchez-Palencia, Y. A review of MIS 7 and MIS 5 terrace deposits along the Gulf of Taranto based on new stratigraphic and chronological data. *Ital. J. Geosci.* **2018**, *137*, 349–368. [CrossRef]

34. Peltier, W.R.; Fairbanks, R.G. Global glacial ice volume and Last Glacial Maximum duration from an extended Barbados sea level record. *Quat. Sci. Rev.* **2006**, *25*, 3322–3337. [CrossRef]

35. Mastronuzzi, G.; Romaniello, L. Holocene aeolian morphogenetic phases in Southern Italy: Problems in 14C age determinations using terrestrial gastropods. *Quat. Int.* **2008**, *183*, 123–134. [CrossRef]

36. Antonioli, F.; Mourtzas, N.; Anzidei, M.; Auriemma, R.; Galili, E.; Kolaiti, E.; Lo Presti, V.; Mastronuzzi, G.; Scicchitano, G.; Spampinato, C.; et al. Millstone quarries along the Mediterranean coast: Chronology, morphological variability and relationships with past sea levels. *Quat. Int.* **2017**, *439*, 102–116. [CrossRef]

37. Mastronuzzi, G.; Antonioli, F.; Anzidei, M.; Auriemma, R.; Alfonso, C.; Scarano, T. Evidence of relative sea level rise along the coasts of central Apulia (Italy) during the late Holocene via maritime archaeological indicators. *Quat. Int.* **2017**, *439*, 65–78. [CrossRef]

38. Lambeck, K.; Anzidei, M.; Antonioli, F.; Benini, A.; Esposito, E. Sea level in Roman time in the Central Mediterranean and implications for recent change. *Earth Planet. Sci. Lett.* **2004**, *224*, 563–575. [CrossRef]

39. Lambeck, K.; Antonioli, F.; Anzidei, M.; Ferranti, L.; Leoni, G.; Scicchitano, G.S.; Silenzi, S. Sea level change along the Italian coast during the Holocene and projections for the future. *Quat. Int.* **2011**, *232*, 250–257. [CrossRef]

40. Serpelloni, E.; Faccenna, C.; Spada, G.; Dong, D.; Williams, S. Vertical GPS ground motion rates in the Euro-Mediterranean region: New evidence of velocity gradients at different spatial scales along the Nubia-Eurasia plate boundary. *J. Geophys. Res. Solid Earth* **2013**, *118*, 6003–6024. [CrossRef]

41. Ridente, D.; Trincardi, F. Active foreland deformation evidenced by shallow folds and faults affecting late Quaternary shelf-slope deposits (Adriatic Sea, Italy). *Basin Res.* **2006**, *18*, 171–188. [CrossRef]

42. Maselli, V.; Trincardi, F.; Cattaneo, A.; Ridente, D.; Asioli, A. Subsidence pattern in the central Adriatic and its influence on sediment architecture during the last 400 kyr. *J. Geophys. Res.* **2010**, *115*, B12. [CrossRef]

43. Ridente, D.; Trincardi, F.; Piva, A.; Asioli, A.; Cattaneo, A. Sedimentary response to climate and sea level changes during the past 400 ka from borehole PRAD1–2 (Adriatic margin). *Geochem. Geophys. Geosyst.* **2008**, *9*. [CrossRef]

44. Vacchi, M.; Rovere, A.; Schiaffino, C.F.; Ferrari, M. Monitoring the effectiveness of re-establishing beaches artificially: Methodological and practical insights into the use of video transects and SCUBA-operated coring devices. *Underw. Technol.* **2012**, *30*, 201–206. [CrossRef]

45. Rovere, A.; Parravicini, V.; Vacchi, M.; Montefalcone, M.; Morri, C.; Bianchi, C.N.; Firpo, M. Geo-environmental cartography of the marine protected area "Isola di Bergeggi"(Liguria, NW Mediterranean Sea). *J. Maps* **2010**, *6*, 505–519. [CrossRef]

46. Moretti, M.; Tropeano, M.; Van Loon, A.J.; Acquafredda, P.; Baldacconi, R.; Festa, V.; Lisco, S.; Mastronuzzi, G.; Moretti, V.; Scotti, R. Texture and composition of the Rosa Marina beach sands (Adriatic coast, southern Italy): A sedimentological/ecological approach. *Geologos* **2016**, *22*, 87–103. [CrossRef]

47. Stocchi, P.; Vacchi, M.; Lorscheid, T.; de Boere, B.; Simms, A.; van de Wal, R.; Vermeersen, B.; Pappalardo, M.; Rovere, A. MIS 5e relative sea-level changes in the Mediterranean Sea: Contribution of isostatic disequilibrium. *Quat. Sci. Rev.* **2018**, *185*, 122–134. [CrossRef]

48. Scarano, T.; Auriemma, R.; Mastronuzzi, G.; Sansò, P. L'archeologia del paesaggio costiero e la ricostruzione delle trasformazioni ambientali: Gli insediamenti di Torre Santa Sabina e Torre Guaceto (Carovigno, Br). In Proceedings of the Secondo Simposio Internazionale "Il Monitoraggio Costiero Mediterraneo: Problematiche e Tecniche di Misura", Napoli, Firenze, Italy, 1–6 June 2008; pp. 391–402.

49. Rovere, A.; Raymo, M.E.; Vacchi, M.; Lorscheid, T.; Stocchi, P.; Gómez-Pujol, L.; Harris, D.L.; Casella, E.; O'Leary, M.J.; Hearty, P.J. The analysis of Last Interglacial (MIS 5e) relative sea-level indicators: Reconstructing sea-level in a warmer world. *Earth Sci. Rev.* **2016**, *159*, 404–427. [CrossRef]

50. Pirazzoli, P.A. Marine erosion features and bioconstructions as indicators of tectonic movements, with special attention to the eastern Mediterranean area. *Z. Geomorphol. Suppl.* **2005**, *137*, 71–77.

51. Di Bucci, D.; Coccia, S.; Iurilli, V.; Mastronuzzi, G.; Palmentola, G.; Sansò, P.; Selleri, G.; Fracassi, U.; Valensise, G. Late Quaternarydeformation of the southernAdriaticforeland (southern Apulia) from mesostructural data: Preliminary results. *Boll. Soc. Geol. Ital.* **2009**, *128*, 33–46.

52. Waelbroeck, C.; Labeyrie, L.; Michel, E.; Duplessy, J.; McManu, J.; Lambeck, K.; Balbon, E.; Labracherie, M. Sea-level and deep water temperature changes derived from benthic foraminifera isotopic records. *Quat. Sci. Rev.* **2002**, *21*, 295–305. [CrossRef]

53. Caputo, R. Sea-level curves: Perplexities of an end-user in morphotectonic applications. *Glob. Planet. Chang.* **2007**, *57*, 417–423. [CrossRef]

54. Shackleton, N.; Chapman, M.; Sanchez Goñi, M.; Pailler, D.; Lancelot, Y. The Classic Marine Isotope Substage 5e. *Quat. Res.* **2002**, *58*, 14–16. [CrossRef]

55. Imbrie, J.D.; McIntyre, A. SPECMAP time scale developed by Imbrie et al., 1984 based on normalized planktonic records (normalized O-18 vs. time, specmap.017). *Pangaea* **2006**. [CrossRef]

56. Vacchi, M.; Ghilardi, M.; Melis, R.T.; Spada, G.; Giaime, M.; Marriner, N.; Lorscheid, T.; Morhange, M.; Burjachs, F.; Rovere, A. New relative sea-level insights into the isostatic history of the Western Mediterranean. *Quat. Sci. Rev.* **2018**, *201*, 396–408. [CrossRef]

Article

Sound Velocity in a Thin Shallowly Submerged Terrestrial-Marine Quaternary Succession (Northern Adriatic Sea)

Ana Novak [1,2,*], Andrej Šmuc [2], Sašo Poglajen [3], Bogomir Celarc [1] and Marko Vrabec [2]

[1] Geological Survey of Slovenia, Dimičeva ulica 14, 1000 Ljubljana, Slovenia; bogomir.celarc@geo-zs.si
[2] Department of Geology, University of Ljubljana, Faculty of Natural Sciences and Engineering, Aškerčeva 12, 1000 Ljubljana, Slovenia; andrej.smuc@geo.ntf.uni-lj.si (A.Š.); marko.vrabec@geo.ntf.uni-lj.si (M.V.)
[3] Sirio d.o.o., Kvedrova cesta 16, 6000 Koper, Slovenia; poglaj@gmail.com
* Correspondence: ana.novak@geo-zs.si; Tel.: +386-128-09-779

Received: 30 December 2019; Accepted: 15 February 2020; Published: 18 February 2020

Abstract: Estimating sound velocity in seabed sediment of shallow near-shore areas submerged after the Last Glacial Maximum is often difficult due to the heterogeneous sedimentary composition resulting from sea-level changes affecting the sedimentary environments. The complex sedimentary architecture and heterogeneity greatly impact lateral and horizontal velocity variations. Existing sound velocity studies are mainly focused on the surficial parts of the seabed sediments, whereas the deeper and often more heterogeneous sections are usually neglected. We present an example of a submerged alluvial plain in the northern Adriatic where we were able to investigate the entire Quaternary sedimentary succession from the seafloor down to the sediment base on the bedrock. We used an extensive dataset of vintage borehole litho-sedimentological descriptions covering the entire thickness of the Quaternary sedimentary succession. We correlated the dataset with sub-bottom sonar profiles in order to determine the average sound velocities through various sediment types. The sound velocities of clay-dominated successions average around 1530 m/s, while the values of silt-dominated successions extend between 1550 and 1590 m/s. The maximum sound velocity of approximately 1730 m/s was determined at a location containing sandy sediment, while the minimum sound velocity of approximately 1250 m/s was calculated for gas-charged sediments. We show that, in shallow areas with thin Quaternary successions, the main factor influencing average sound velocity is the predominant sediment type (i.e. grain size), whereas the overburden influence is negligible. Where present in the sedimentary column, gas substantially reduces sound velocity. Our work provides a reference for sound velocities in submerged, thin (less than 20 m thick), terrestrial-marine Quaternary successions located in shallow (a few tens of meters deep) near-shore settings, which represent a large part of the present-day coastal environments.

Keywords: sound velocity; Quaternary sediment; submerged alluvial plain

1. Introduction

In geophysical (acoustic/seismic) investigations of the subsurface, velocity modeling is essential for converting two-way travel time of the observed reflections into the depth domain. Velocity data are routinely extracted from multi-channel seismic data [1]; however, in many circumstances, particularly in shallow near-shore settings, obtaining offshore multi-channel data is not feasible. Restricted navigation, legal constraints, busy marine traffic, relatively low resolution of the acquired data and the surveying cost itself often make acquisition and maneuvering with streamers and seismic sources impractical or even impossible. In such settings, high-resolution single-channel seismic and acoustic surveys provide a common alternative, but the velocity data must then be obtained by other means, such as in situ measurements, laboratory core logging, and geo-acoustic modeling [2–20]. Due to costly offshore core drilling, these approaches are mostly focused on the upper few meters of the seafloor sediment. Consequently, the acoustic properties of surficial seafloor sediments have been well known for some time [4,5,11,15,19], but sound velocity in thicker (more than 10 m) sedimentary sequences has rarely been investigated (e.g., [17]). Therefore, when velocity data for depth conversion of single-channel seismic or acoustic data are not acquired during surveying, a velocity value corresponding to the surficial sediment grain size (e.g., [19]) or a previously published value from a nearby location is usually used. Whereas this approach is sufficient for geophysical surveys of uniform sedimentary layers, it produces significant uncertainties when dealing with pronounced lateral and vertical variability in sediment composition and architecture.

Typical example of such complex settings are shallow continental shelf areas drowned during the global sea-level rise that followed the Last Glacial Maximum (LGM) lowstand when the global mean sea level was approximately at −130 m [21,22]. During transgression, earlier terrestrial depositional environments (e.g., alluvial plains) were overlain by terrestrial and marine-derived sediments deposited in fluvial, estuarine, and open marine settings, resulting in complex sedimentary architecture and highly variable sedimentary types [22–24], which markedly affect the sub-bottom propagation of acoustic waves [1]. Precise mapping of the 2D and 3D geometries of the sedimentary bodies in these near-shore shallow environments and their appropriate time-to-depth conversion are essential for reliable interpretation of high-resolution acoustic and seismic surveys, not only for unraveling their depth and depositional history, but also for geotechnical site assessments in various engineering projects.

During the LGM lowstand, vast areas of the presently submerged continental shelves were exposed (e.g., Figure 1a) and amounted to approximately 40% of additional landmass in Europe and 5% globally [25,26]. Therefore, the shallow-most near-shore parts of presently submerged continental shelves extend over a considerable area globally and represent an important and often poorly studied geological environment. Due to the steadily increasing interest of the geological and archaeological research communities in shallow, presently submerged, and often buried landscapes [27–41], accurate depth conversion is crucial for future geological studies and paleoenvironmental reconstructions in such settings. We investigate an example of a transgressed and submerged alluvial plain in the Bay of Koper (Gulf of Trieste, northern Adriatic Sea; Figure 1) to provide sound velocity values for thin (up to 20 m thick) Quaternary sediments deposited in terrestrial-marine sedimentary environments located in shallow near-shore environments a few tens of meters deep. Our work ranks among the few studies that are not limited only to the surficial seafloor sediments, but also include the entire sedimentary succession from the seafloor to the base of the sediment on the bedrock.

Setting

The post-LGM sea-level rise induced significant changes in the sedimentary environments of the northern Adriatic Sea with terrestrial environments transitioning in paralic and later shallow marine environments [22,42–56]. In the Gulf of Trieste, where our study area is located (Figure 1), the Late Pleistocene alluvial plain transitioned into a paralic environment until open marine conditions finally prevailed approximately 10,000 years ago [22,42,44,45,55–64].

The Bay of Koper is located in the southeastern part of the Gulf of Trieste, which represents the northeasternmost extension of the Adriatic Sea (Figure 1). The seabed morphology of the Bay is smooth with depths ranging up to 20 m in the open part (Figure 1b). The main fluvial source draining into the Bay is the Rižana river with its mouth located in the reclaimed eastern part of the Bay within the Port of Koper complex. The smaller Badaševica stream is located west of the city of Koper.

The hinterland of the Bay of Koper is composed of Eocene turbidites (flysch) comprising interbedded sandstones and marlstones, which are overlain by Quaternary alluvial and paralic sediments in the valleys of Rižana and Badaševica (Figure 1b; [65,66]). Offshore the Eocene succession is unconformably overlain by Quaternary terrestrial and paralic sediments topped by Holocene marine sediments [55,56,59,62,67–71]. The Quaternary succession in the Bay of Koper, which was recognized as a submerged fluvial valley of the Rižana river (Figure 1b, [68–71]), is up to a few tens of meters thick and is composed of a lower alluvial part and an upper paralic part; however, alternations of terrestrial and paralic sedimentary environments have also been observed [68,72]. The alluvial sediments are generally composed of fine-grained clastic sediments with occasional gravelly and sandy horizons, whereas the paralic sediments are mostly composed of silty clay [68–70]. The Holocene marine cover comprises fine-grained bioclastic sediment with the surficial sediments showing a clear zonation: sandy silt near the coastline, clayey silt in the central part, and silt in the outer part of the Bay [62,68–71]. Holocene sedimentation rates in the Gulf of Trieste are relatively low and amount to a few millimeters per year [44,57,69,73]. Repeated multibeam bathymetric surveys in the Bay of Koper do not show significant changes in the seafloor morphology [74] and therefore imply a low-energy sedimentary environment.

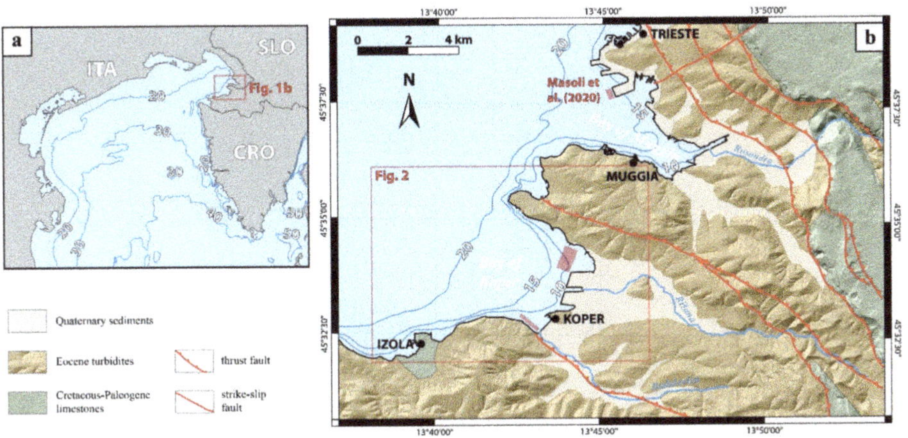

Figure 1. Geographical location of the study area. (**a**) Regional map of the northern Adriatic, which was entirely subaerially exposed during the Last Glacial Maximum (LGM). Bathymetry data from [75]. (**b**) Geological map of the hinterland of the Bay of Koper. Two red rectangles in the Bay of Koper mark the areas investigated in this study. The red rectangle in the Bay of Muggia marks the study area of [76]. Bathymetric data are simplified after [77] and [78]. Geological data are simplified after [65,66,79–82].

2. Materials and Methods

We used archive geotechnical reports from the borehole database of the Geological Survey of Slovenia. Boreholes were mainly located in the NE part of the Bay (Figure 2) and were drilled in the late 1980s and early 1990s for geotechnical investigations supporting various infrastructure projects of the Port of Koper. Borehole metadata and descriptions are provided in Table 1 and Figure 5. The boreholes were drilled with rotary drilling and were cored. Sediment core samples were used for geomechanical testing and were not preserved. We therefore reconstructed the borehole sedimentary logs (Figure 5) from borehole descriptions contained in the geotechnical reports.

Sub-bottom sonar profiles were acquired in June 2016 on board vessel Lyra with the Innomar SES-2000 Compact sub-bottom sonar. Profile transects were designed to directly cross the borehole locations. Navigation north of the second pier was obstructed by a containment boom and very shallow water depths near the coastline. For this reason, some of the sub-bottom profiles are located at some distance from the borehole locations (Table 1 and Figure 2b). We used a transmitter frequency of 8 kHz. A total of 7 sub-bottom sonar profiles were acquired (Figure 2). At shallow depths, the seafloor was also observed visually to distinguish between sedimentary, rocky, and seagrass-covered seabed. The sub-bottom profiles were visualized and interpreted in the IHS Markit Kingdom software (Version

2018, IHS Markit, London, UK). Two-way travel times (TWT^s; in milliseconds) of the seafloor (sf^{TWT}) and the top of the weathered/compact bedrock (d^{TWT}) were determined at borehole locations (Table 2).

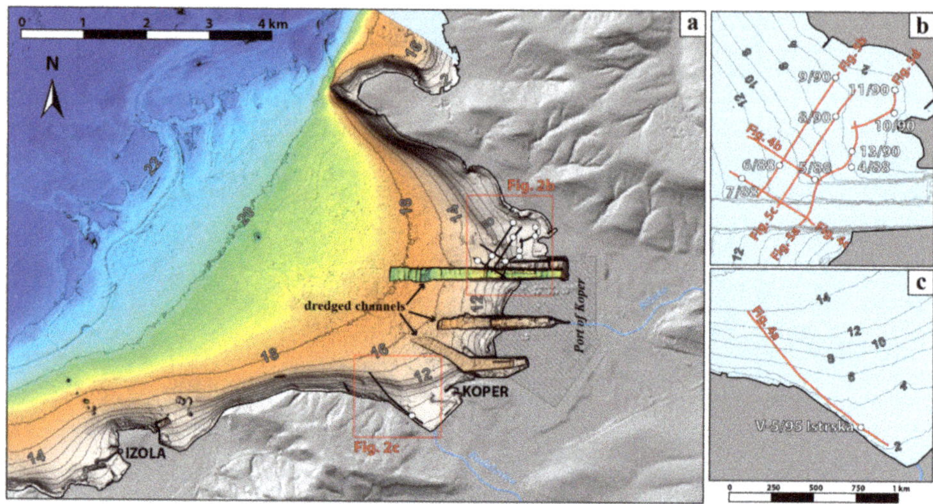

Figure 2. Study area. (**a**) Bay of Koper with borehole (white circles) and sub-bottom sonar profile locations (thick black lines). (**b**) and (**c**) close-ups of the studied areas with bathymetry indicated by thin grey lines. Red lines show the locations of sub-bottom sonar profiles presented in Figures 3 and 4. White circles indicate borehole locations. For clarity the "A-III-" prefix is not shown in (**b**).

Table 1. Boreholes used in this study (GeoZS: Geological Survey of Slovenia; IGGG: Institute for Geology, Geotechnics and Geophysics Ljubljana).

Borehole Name	Geographical Coordinates		Seafloor Depth [m b.s.l.]	Borehole Length [m]	Drill Period	Orthogonal Distance to the Nearest Sonar Profile [m]	Contractor
	Latitude	Longitude					
A-III-4/88	45°34′4.2119″	13°44′11.0875″	4.8	22.0	December 1988	22.0	GeoZS
A-III-5/88	45°34′1.6538″	13°44′1.1537″	7.0	24.0	November 1988	0.5	GeoZS
A-III-6/88	45°34′4.2390″	13°43′51.4528″	8.0	23.0	December 1988	1.5	GeoZS
A-III-7/88	45°34′1.5963″	13°43′41.2889″	15.0	20.0	December 1988	1.0	GeoZS
A-III-8/90	45°34′14.0408″	13°44′6.4254″	5.1	21.0	November 1990	1.0	GeoZS
A-III-9/90	45°34′21.5862″	13°44′6.2099″	4.4	18.0	November 1990	1.5	GeoZS
A-III-10/90	45°34′14.9625″	13°44′22.4093″	2.4	20.0	November 1990	23.5	GeoZS
A-III-11/90	45°34′19.4662″	13°44′22.4931″	2.1	16.0	November 1990	8.5	GeoZS
A-III-13/90	45°34′7.2908″	13°44′11.1891″	4.9	27.0	November 1990	3.5	GeoZS
V-5/95 Istrska	45°32′40.5603″	13°42′56.0674″	4.5	10.0	November 1995	2.5	IGGG

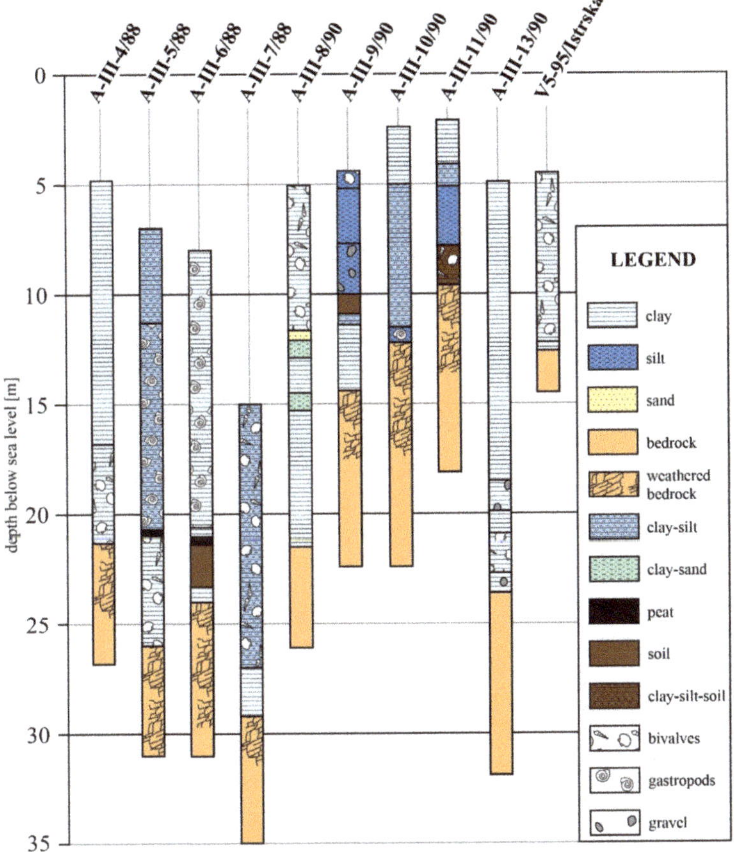

Figure 3. Sub-bottom sonar profiles (with superimposed boreholes) where the top of the Eocene bedrock is expressed as a single undulating medium-to-high amplitude reflection ((**a**), (**b**) and (**c**)). For profile locations, see Figure 2. Blue overlay marks the Quaternary sediment and orange overlay marks the bedrock. Peat layers are indicated with brown arrows. Red overlay within the Quaternary section marks isolated diffraction hyperbolas. Red overlay in the water column marks reflection events above the seafloor. Green overlay marks the extent of seagrass meadows on the seafloor. Dredged areas are marked with grey arrows. Multiples are indicated by thin black arrows.

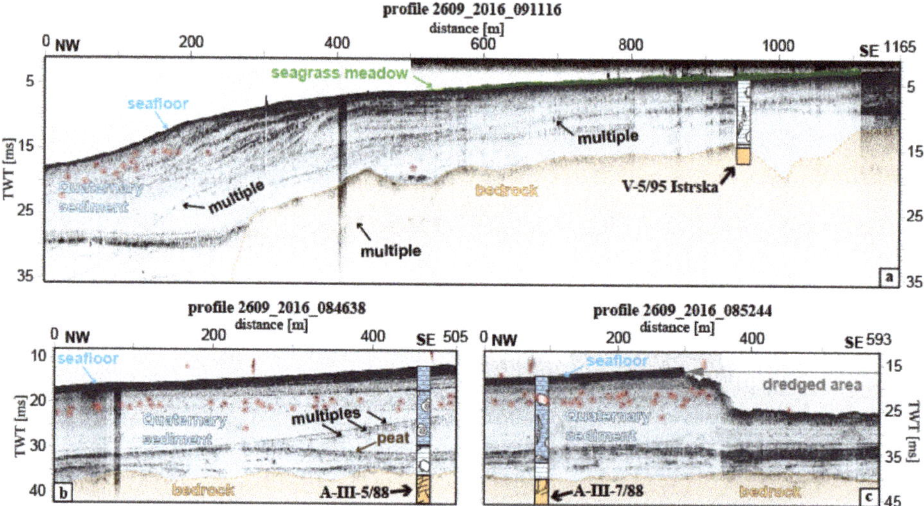

Figure 4. Sub-bottom sonar profiles (with superimposed boreholes) where the top of the Eocene succession is expressed as a medium-to-high amplitude reflection unit with a well-definable top from which downward shallow-dipping reflections emerge ((**a–c**)). For profile locations, see Figure 2. For an explanation of the color overlays, the reader is referred to Figure 3.

Table 2. Thickness of Quaternary sediments at borehole locations from borehole logs (th^b) and sonar profiles (th^{TWT}), the depth of the seafloor (sf^{TWT}), and the top of the bedrock (d^{TWT}) from the sonar profiles and the calculated average sound velocity in Quaternary sediments at the borehole location.

Borehole Name	From Borehole Logs	From Sonar Profiles			Average Sound Velocity in Quaternary Sediments at the Borehole Location [m/s]	Orthogonal Distance to the Nearest Sonar Profile [m]
	th^b [m]	sf^{TWT} [ms]	d^{TWT} [ms]	th^{TWT} [ms]		
A-III-4/88	16.5	6.4	29.9	23.5	1404.3	22.0
A-III-5/88	19.0	11.5	36.0	24.5	1551.0	0.5
A-III-6/88	16.0	15.0	36.0	21.0	1523.8	1.5
A-III-7/88	14.2	17.5	40.2	22.7	1251.1	1.0
A-III-8/90	16.4	6.4	25.4	19.0	1726.3	1.0
A-III-9/90	10.0	5.1	17.7	12.6	1587.3	1.5
A-III-10/90	9.8	2.8	17.0	14.2	1380.3	23.5
A-III-11/90	7.5	3.0	11.9	8.9	1685.4	8.5
A-III-13/90	18.7	5.0	28.4	23.4	1598.3	3.5
V-5/95 Istrska	8.1	4.0	14.6	10.6	1528.3	2.5

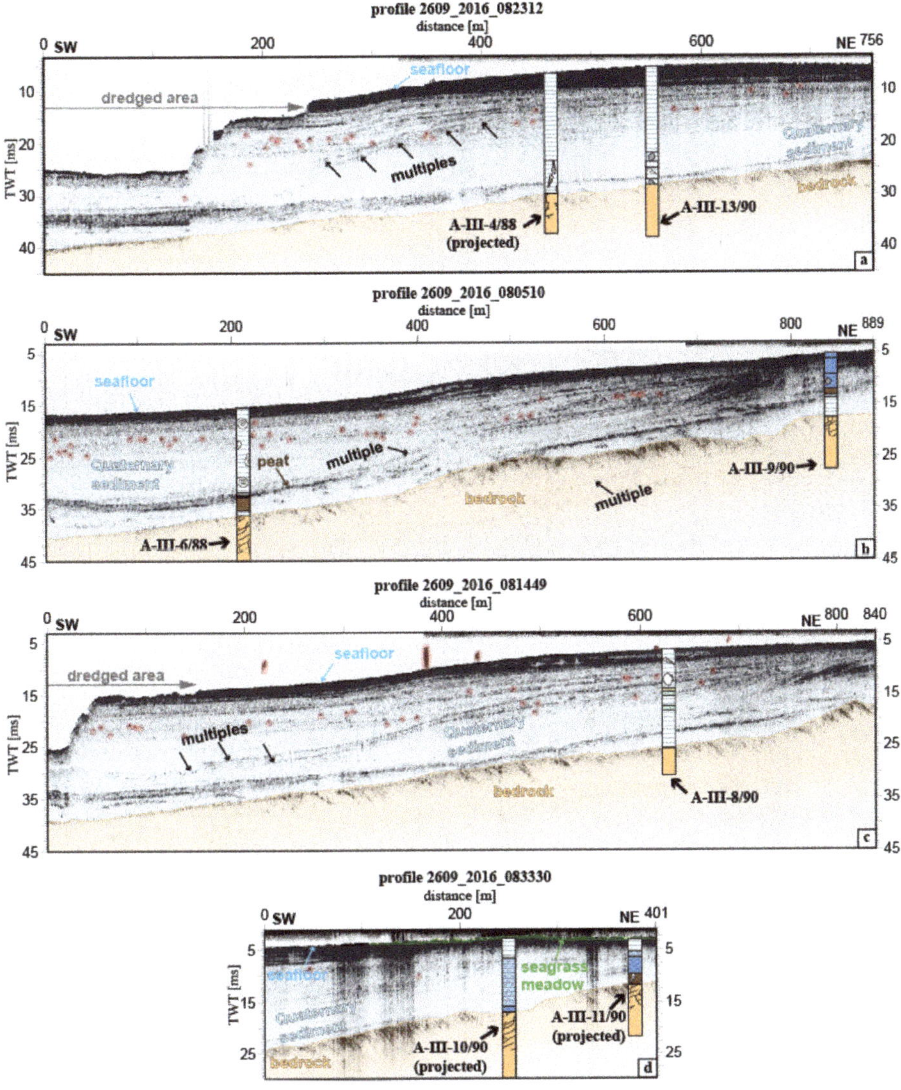

Figure 5. Borehole logs determined from geotechnical reports (see Section 2).

The thickness of Quaternary sediments in the borehole (th^b, in meters) was derived from the geotechnical reports and was calculated as the depth from the top of the core to the top of the bedrock represented by weathered or compact Eocene turbidites (see Figure 5). The thickness of Quaternary sediments from the sub-bottom profiles (th^{TWT}, in miliseconds) was obtained by subtracting the TWTs from the top of the bedrock (d^{TWT}, in miliseconds) and the seafloor (sf^{TWT}, in miliseconds) at the borehole location (see Figures 3 and 4; Table 2). When the profiles did not directly overlie the borehole (Figure 2b and Table 1), the part of the profile closest to the borehole was used to determine th^{TWT}

(Figure 4a,d). The average sound velocity in Quaternary sediments at the borehole location (v, in meters per second; Table 2) was calculated using the following formula:

$$v = \frac{2 \times 10^3 \times th^b}{th^{TWT}}.$$

(1)

3. Results

3.1. Boreholes

Borehole logs are provided in Figure 5. The Quaternary succession (including the Holocene marine sediment) is composed of fine-grained clastics with occasional gravelly horizons. Only core A-III-8/90 contains sandy horizons. Soil-rich and peat horizons are present in some of the boreholes. Horizons with bivalves and/or gastropods occur in all boreholes; however, no remarks on the species or their environment are provided in the borehole geotechnical reports. Due to the lack of detailed descriptions of the boreholes, we did not attempt to interpret the sedimentary environments. However, it is clear that the Quaternary sediments comprise terrestrial-marine deposits. The bottom parts of all the boreholes consist of weathered and/or compact bedrock built of Eocene interbedded sandstones and marlstones.

3.2. Sub-Bottom Sonar Profiles

The sub-bottom profiles with superimposed boreholes are shown in Figures 3 and 4. Boreholes more than 5 m away from the nearest sub-bottom profile were projected orthogonally to the profile (Table 1, Figure 4a,d). The seafloor is marked by the first strong sub-horizontal reflection and is indicated by a blue arrow. The Quaternary sequence is indicated by a light blue overlay. Quaternary sediments are seen as (1) acoustically transparent units (Figures 3 and 4), (2) units containing onlapping or concordant reflection geometries (Figures 3 and 4), and (3) units with sigmoidal (prograding) reflection configurations (Figures 3a and 4b,c). Eocene bedrock is indicated by a light orange overlay. The often undulating unconformity at the top of the bedrock is expressed as (1) a medium-to-high amplitude reflection under which deeper reflections are not observed (Figure 3) or (2) an up to 5 ms TWT thick medium-to-high amplitude reflection unit with a well-definable top from which downward short shallow-dipping reflections emerge (Figure 4). The acoustic record does not discriminate between weathered or compact bedrock.

In addition to the two described units, the sub-bottom profiles contain several other features. Columnar-shaped reflections (gas flares) within the seawater column are located above the seafloor and are indicated by a light red overlay in Figure 3b,c and Figure 4c. Rough seafloor morphologies with plentiful diffraction hyperbolas are observed above the dredged areas (Figure 2a,b, Figures 3c and 4a,c), which were excavated to accommodate ships with larger drafts in the Port of Koper. Slightly rougher seafloor morphologies are also produced by seagrass (most commonly *Posidonia* sp.) meadows (marked by a light green overlay in Figures 3a and 4d), which were visually recognized during sub-bottom sonar acquisition. Significant, yet variable reflection degradation is seen directly beneath the areas covered by seagrass. Within the Quaternary sedimentary column, peat layers produce medium-to-high amplitude, sub-horizontal, 1–2 ms TWT thick reflections (Figures 3b and 4b). All sonar profiles show many small diffraction hyperbolas scattered between 25 and 15 ms TWTs within the sedimentary column (Figures 3 and 4). Seafloor multiples appear on the majority of the profiles (Figure 3a,b and Figure 4a–c) due to the shallow acquisition depths. Figure 3a also contains a multiple of the top of the bedrock.

3.3. Average Sound Velocity in Quaternary Sediments

Thicknesses of Quaternary sediments from drilling reports and geophysical data along with the calculated average sound velocities at borehole locations are provided in Table 2. The mean, median, and standard deviation for the whole dataset are 1523.6 m/s, 1539.7 m/s, and 144.0, respectively.

4. Discussion

4.1. Sound Velocity Variation

The sound velocity variation in the presented dataset is significant (Section 3.3). Some of the calculated velocity values can be considered less reliable due to significant distance between the respective borehole and its closest sub-bottom sonar profile (Table 2). If we omit the two boreholes that are separated by more than 20 m from their nearest sub-bottom profile (A-III-4/88 and A-III-10/90), the mean, median, and standard deviation for the dataset become 1556.4 m/s, 1569.2 m/s, and 143.1, respectively. Since the standard deviation remains relatively high, we discuss the principal influences on the sound velocity scattering below.

4.1.1. Influence of Overburden

Sound velocity in sediments strongly depends on porosity, which is in turn influenced by overburden and compaction [1,4]. Here, the overburden comprises the combined weight of the water and sedimentary columns at the borehole locations. In Figure 6, we plot the calculated velocity against the thickness of Quaternary sediments taken from borehole logs and sonar profiles (a proxy for the weight of the sedimentary column), the depth of the seafloor taken from sonar profiles (a proxy for the weight of the water column), and the depth of the top of the bedrock taken from sonar profiles (a proxy for the weight of the water and sedimentary column). A strong scattering of plotted data points demonstrates that there is no relation between the determined sound velocity and these parameters. Additionally, in the correlation plots, the *x*-axis values of the minimum (red) and maximum sound velocities (green) are often quite similar (Figure 6a,c,d), again suggesting that velocities are uncorrelated to the overburden thickness. This shows that, in thin sedimentary successions located in shallow water depths, overburden does not significantly influence the sound velocity.

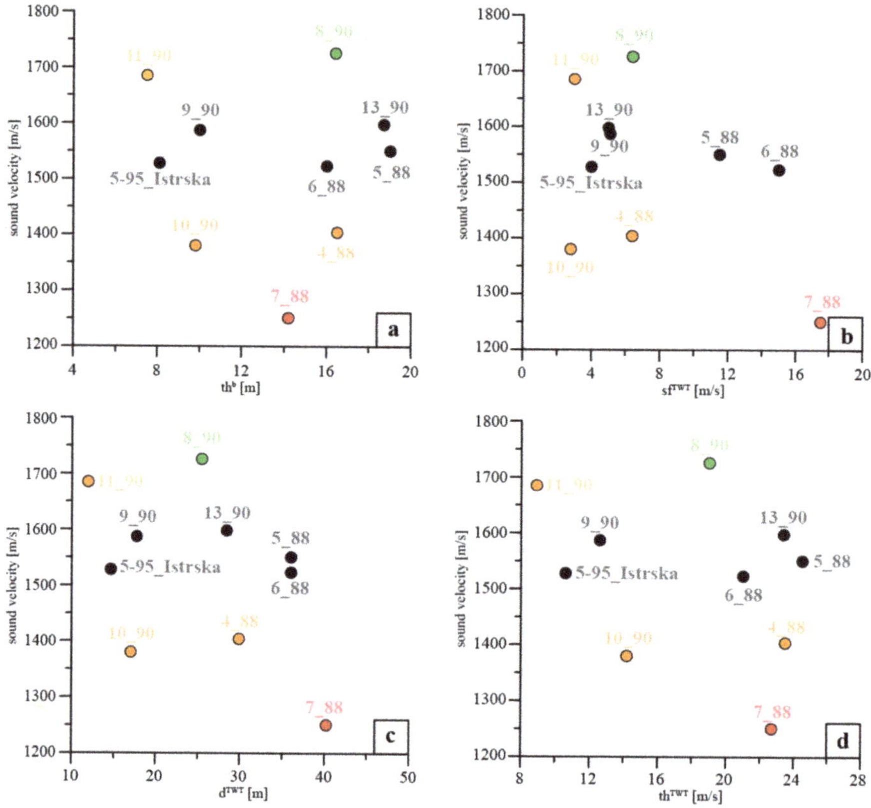

Figure 6. Plots of sound velocity versus (**a**) the thickness of Quaternary sediments (taken from borehole logs), (**b**) the depth of the seafloor (taken from sonar profiles), (**c**) the depth of the top of the bedrock (taken from sonar profiles), and (**d**) the thickness of Quaternary sediments (taken from sonar profiles). For clarity, A-III- and V- prefixes are removed from borehole labels. Data points in orange mark less reliable velocities (boreholes more than 5 m away from the nearest sub-bottom sonar profile; Table 2 and Figure 2b). Data points in green and red mark the maximum and minimum calculated velocities (Table 2).

4.1.2. Influence of Grain Size

The influence of mean grain size on sound velocity in surficial marine sediments is well known [4,5,19,83,84]. Although granulometric analyses of the cored sediments used in our study were unavailable, general grain size classes could still be determined from borehole geotechnical logs (see Section 2). Therefore, an estimation of the influence of grain size on the sound velocity in our dataset is possible. In Figure 7, the calculated sound velocities at our study site were added to the plot, correlating sound velocity with the mean grain size of surficial sediments of continental shelves from [19]. Clearly, our calculated velocities (at boreholes located close to the acquired profiles) correspond well with the expected sound velocity range of the predominant grain size class determined from the borehole logs (Figures 5 and 7), even though the dataset of [19] is based on surficial sediment samples. Boreholes penetrating exclusively clay (A-III-6/88 and V5/95 Istrska) are in the lower sound velocity spectra, whereas boreholes mainly encountering silt (A-III-5/88, A-III-9/90 and A-III-11/90) are in the middle velocity spectra, corresponding to silt mean grain size velocities. Maximum sound velocity was calculated for Borehole A-III-8/90, which is the only core containing sandy sediment. Although clay dominates in this borehole, the amount of sandy sediment seems to be sufficient enough to significantly increase the average velocity in the Quaternary succession. A special case is presented by Borehole A-III-13/90, which penetrated clay; however, its calculated velocity corresponds to the mean grain sizes of silt. This discrepancy can be attributed to (1) an inadequate geotechnical description of the core, (2) an abrupt increase in sound velocity within the gravelly horizons (Figure 5), or (3) the presence of an overconsolidated layer within the sequence. In conclusion, the generally good agreement between the calculated velocities and expected velocities for the predominant grain size classes indicates that the composition of the stratal succession is a major factor influencing velocity variations in our study area.

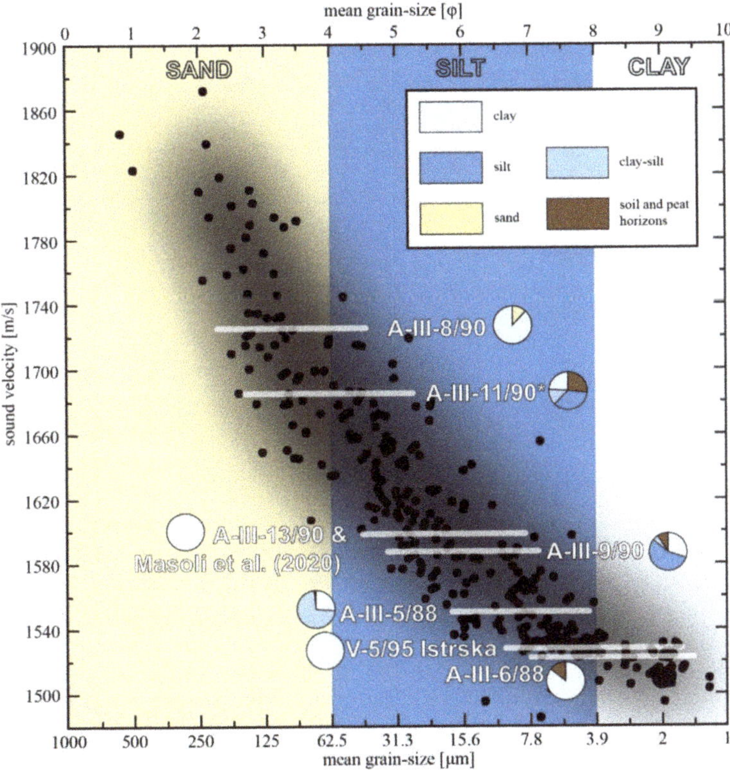

Figure 7. The relation between sound velocity and mean grain size for continental shelf sediments (after [19]) with added velocities from our study. Our estimated sound velocities and the velocity value from [76] for a similar setting in the Gulf of Trieste (see Section 4.3) are shown by white horizontal overlays. Asterisk at Borehole A-III-11/90 indicates that this borehole is separated by more than 5 m from the nearest sub-bottom profile. Pie charts display the proportions of sediment types in each borehole core (for details, see Figure 5).

4.1.3. Influence of Gas Presence

Abundant diffraction hyperbolas are present within the Quaternary succession (Figures 3 and 4; Section 3.2). They could be produced by reflections from gravel horizons; however, their occurrence does not correlate with gravel layers determined in the boreholes (Figures 3 and 4). The diffraction hyperbolas more likely indicate low concentrations of gas in the sedimentary column, commonly encountered in high-resolution geophysical profiles [85–87]. Reflective features in the water column (Figure 3b,c and Figure 4c; Section 3.2), which commonly result from gas-bubble plumes emitting from the seafloor [88–90], further indicate gas occurrence. The lowest sound velocity in our dataset was calculated for Borehole A-III-7/88, which penetrates a hyperbola-dense zone clearly visible on the corresponding sub-bottom sonar profile (Table 2 and Figure 3c). Since even minor gas concentrations (1–2%) dramatically reduce sound velocity in sediments [83,84,91–93], we attribute this significantly lower velocity value to gas in the Quaternary sediment. In the northern Adriatic seabed, gas seeps are commonly observed and are attributed to both deep and shallow sources [94–98]. Since the diffraction hyperbolas in our sonar profiles are constrained only to a narrow zone in the uppermost part of the Quaternary sequence (Figures 3 and 4), we propose that the gas (probably methane) originates from a degradation of organic matter contained in the Holocene paralic and marine sediments and/or Late Pleistocene terrestrial sequences [62,97,99]. Gas production related to biological processes in seagrass meadows can also greatly hinder the propagation of acoustic signals [100–102]; however, the contribution of this effect is difficult to determine from our dataset since only a single borehole (V-5/95) is located within a meadow (Figure 3a). Nevertheless, significant signal attenuation below the meadows can be observed in the geophysical data (Section 3.2, Figures 3a and 4d).

4.2. Are Average Sound Velocities an Oversimplification?

Using average sound velocities for depth conversion of a highly heterogeneous Quaternary succession can be considered a gross oversimplification as the velocity strongly varies with grain size [4,5,19,83,84]. However, when comparing the sub-bottom sonar profiles and the borehole logs used in our study (Figure 5,Figure 3, and Figure 4), a good alignment between the reflections and the main sedimentological boundaries from the borehole logs is apparent (Figure 3b,c and Figure 4a,b). Especially peat layers prove to be very effective reflectors, which has already been noted in the northern Adriatic Sea by other authors [43,47,50–52,76,97,103,104]. This demonstrates that average velocity can be quite effectively used for robust depth conversion of sonar profiles in thin and shallow Quaternary successions.

4.3. Comparison with the Sound Velocity from the Bay of Muggia

An earlier study in a similar geological setting [76] reported sound velocity in Late Quaternary sediments from the neighboring Bay of Muggia (Figure 1b), which comprise Rosandra river deposits submerged in the Holocene transgression. There, the Quaternary succession is between 20 and 30 m thick and is composed of clay and silt with occasional sand and peat horizons. The water depth extends between 18 and 21 m. The Bay of Muggia site is therefore quite similar to our study site both in sediment thickness and composition. Using P-wave seismic refraction [76] led to a sound velocity value of 1595 m/s, which fits within the range of velocities estimated in our study (Figure 7). This implies that, also in the Bay of Muggia, the sound velocity in the Quaternary sediment is largely controlled by the sediment type and further corroborates sedimentary type as the major factor influencing sound velocity in shallow, thin, terrestrial-marine Quaternary sedimentary environment successions.

4.4. Choosing the Appropriate Velocity for the Depth Conversion of Geophysical Data

The results of our study show that sound velocity in thin (up to 20 m thick) submerged terrestrial-marine Quaternary successions located in near-shore areas few tens of meters deep is mostly controlled by the predominant grain size class of the succession (Section 4.1.2). The sound velocity for depth conversion in these settings can be chosen based on the predominant grain size class. We show that the sound velocity vs. grain size relationships previously documented in surficial sediments [19] are also valid for buried and submerged Quaternary successions (Section 4.1.2). Therefore, published values for sound velocity of surficial sediments can be utilized for depth conversion of shallow offshore high-resolution geophysical data, as long as the selected grain size corresponds to the predominant grain size class of the Quaternary succession.

Terrestrial-marine Quaternary successions often contain significant amounts of degrading organic matter; consequently, locally present gas further influences sound velocity in these settings (Section 4.1.3). Different gas indicators can easily be recognized from high-resolution geophysical data [85–93], facilitating the mapping of low-velocity areas. As the velocity decrease associated with the presence of gas is often quite variable e.g., [93], we suggest avoiding detailed velocity analysis in gas-rich areas.

5. Conclusions

We used geophysical and borehole data to determine sound velocities through the Quaternary fill of a submerged alluvial plain containing terrestrial, paralic, and marine sediments. Our study shows that an average sound velocity through the Quaternary sedimentary column is sufficient for depth conversion of high-resolution geophysical profiles acquired in thin (up to 20 m thick) Quaternary successions in shallow (up to 20 m) water depths. We find that, in these settings, the main factor influencing sound velocity is the sediment type (i.e. mean grain size) contained within the studied sedimentary column, whereas overburden effects do not show any influence. However, where gas is present in the sedimentary column, it reduces sound velocity by a few hundred meters per second and becomes the dominant factor influencing sound velocity.

We found that, for a good approximation of the average sound velocity at a borehole, the velocity typical for the most represented sediment type in the borehole column can be employed. Nevertheless, in highly heterogeneous sedimentary settings, such as the Bay of Koper investigated in this study, significant lateral variations in average velocity will occur within a small area, necessitating a careful selection of multiple, most representative values, if relying on velocities published in the literature.

Using our study area in the northern Adriatic, we provided reference values for sound velocity in thin, mud-dominated Quaternary sedimentary successions in shallow coastal areas. Velocity values determined in our study correlate well with the sound velocity vs. grain size relationships previously documented in surficial sediments [19], showing that these published values can also be used for shallow sub-bottom sedimentary sequences.

Author Contributions: All authors have read and agree to the published version of the manuscript. Conceptualization: A.N. and M.V.; methodology: A.N.; validation: A.N.; formal analysis: A.N.; investigation: A.N., S.P., and B.C.; resources: S.P. and B.C.; writing—original draft preparation: A.N.; writing—review and editing: A.N., A.Š., S.P., B.C., and M.V.; visualization: A.N.; supervision: M.V.; project administration: A.N. and M.V.; funding acquisition: M.V.

Funding: This research was funded by: the Slovenian Research Agency, Young Researcher grant number 38136; the Slovenian Research Agency and Harpha Sea d.o.o., grant number L1-5452; the Slovenian Research Agency, grant number J1-1712; the Slovenian Research Agency, research programme P1-0195. The APC was funded by the Slovenian Research Agency, research programme P1-0195.

Acknowledgments: This work was supported by the Slovenian Research Agency (Young Researcher grant Nr. 38136), by joint funding of the Slovenian Research Agency and Harpha Sea d.o.o. within the project L1-5452 (Application of sonar in research of active tectonics and paleoseismology in low-strain environments) and by funding of the Slovenian Research Agency within the project J1-1712 (Record of environmental change and human impact in Holocene sediments, Gulf of Trieste) and within the research programme P1-0195 (Geoenvironment and Geomaterials). We would like to acknowledge IHS Markit Kingdom and their University Education Grant, which provided us with IHS Markit Kingdom software licences. The crew of the vessel Lyra (Iztok Rant and Rok Soczka Mandac) is acknowledged for their assistance and hospitality during sub-bottom acquisition. We would also like to thank Karoly Nemeth and an anonymous reviewer for their comments, which allowed us to improve the manuscript.

Conflicts of Interest: The authors declare that there is no conflict of interest. The funders had no role in the design of the study; in the collection, analyses, or interpretation of data; in the writing of the manuscript; or in the decision to publish the results.

References

1. Sheriff, R.E.; Geldart, L.P. *Exploration Seismology*, 2nd ed.; Cambridge University Press: Cambridge, UK, 1995.
2. Endler, M.; Endler, R.; Bobertz, B.; Leipe, T.; Arz, H.W. Linkage between acoustic parameters and seabed sediment properties in the south-western Baltic Sea. *GeoMar. Lett.* **2015**, *35*, 145–160. [CrossRef]
3. Endler, M.; Endler, R.; Wunderlich, J.; Bobertz, B.; Leipe, T.; Moros, M.; Jensen, J.B.; Arz, H.W. Geo-acoustic modelling of late and postglacial sedimentary units in the Baltic Sea and their acoustic visibility. *Mar. Geol.* **2016**, *376*, 86–101. [CrossRef]

4. Jackson, D.R.; Richardson, M.D. *High-Frequency Seafloor Acoustics*; Springer: Berlin/Heidelberg, Germany, 2007; ISBN 0387369457.

5. Hamilton, E.L. Geoacoustic modeling of the sea floor. *J. Acoust. Soc. Am.* **1980**, *68*, 1313–1340. [CrossRef]

6. Brandes, H.G.; Silva, A.J.; Sadd, M.H. Physical and acoustic measurements on cohesionless sediments from the northwest Florida Sand Sheet. *Geophys. Res. Lett.* **2001**, *28*, 823–826. [CrossRef]

7. Kan, G.; Liu, B.; Wang, J.; Meng, X.; Li, G.; Hua, Q.; Sun, L. Sound speed dispersion characteristics of three types of shallow sediments in the southern yellow sea. *Mar. Georesources Geotechnol.* **2018**, *36*, 853–860. [CrossRef]

8. Kim, D.C.; Sung, J.Y.; Park, S.C.; Lee, G.H.; Choi, J.H.; Kim, G.Y.; Seo, Y.K.; Kim, J.C. Physical and acoustic properties of shelf sediments, the South Sea of Korea. *Mar. Geol.* **2001**, *179*, 39–50. [CrossRef]

9. Kim, G.Y.; Park, K.J.; Lee, G.S.; Yoo, D.G.; Kong, G.S. Physical property characterization of quaternary sediments in the vicinity of the paleo-Seomjin River of the continental shelf of the South Sea, Korea. *Quat. Int.* **2018**, *503*, 153–162. [CrossRef]

10. Kim, S.R.; Lee, G.-S.; Kim, D.C.; Bae, S.H.; Kim, S.-P. Physical properties and geoacoustic provinces of surficial sediments in the southwestern part of the Ulleung Basin in the East Sea. *Quat. Int.* **2017**, *459*, 35–44. [CrossRef]

11. Richardson, M.D.; Lavoie, D.L.; Briggs, K.B. Geoacoustic and physical properties of carbonate sediments of the Lower Florida Keys. *Geo Mar. Lett.* **1997**, *17*, 316–324. [CrossRef]

12. Zhang, Y.; Guo, C.; Wang, J.; Hou, Z.; Chen, W. Relationship between in situ sound velocity and granular characteristics of seafloor sediments in the Qingdao offshore region. *Chin. J. Oceanol. Limnol.* **2017**, *35*, 704–711. [CrossRef]

13. Zheng, J.; Liu, B.; Kan, G.; Li, G.; Pei, Y.; Liu, X. The sound velocity and bulk properties of sediments in the Bohai Sea and the Yellow Sea of China. *Acta Oceanol. Sin.* **2016**, *35*, 76–86. [CrossRef]

14. Wang, J.; Guo, C.; Liu, B.; Hou, Z.; Han, G. Distribution of geoacoustic properties and related influencing factors of surface sediments in the southern South China Sea. *Mar. Geophys. Res.* **2016**, *37*, 337–348. [CrossRef]

15. Richardson, M.D.; Briggs, K.B. In situ and laboratory geoacoustic measurements in soft mud and hard-packed sand sediments: Implications for high-frequency acoustic propagation and scattering. *Geo Mar. Lett.* **1996**, *16*, 196–203. [CrossRef]

16. Gorgas, T.J.; Wilkens, R.H.; Fu, S.S.; Frazer, L.N.; Richardson, M.D.; Briggs, K.B.; Lee, H. In situ acoustic and laboratory ultrasonic sound speed and attenuation measured in heterogeneous soft seabed sediments: Eel River shelf, California. *Mar. Geol.* **2002**, *182*, 103–119. [CrossRef]

17. Kim, G.Y.; Narantsetseg, B.; Lee, J.Y.; Chang, T.S.; Lee, G.S.; Yoo, D.G.; Kim, S.P. Physical and geotechnical properties of drill core sediments in the Heuksan Mud Belt off SW Korea. *Quat. Int.* **2018**, *468*, 33–48. [CrossRef]

18. Orsi, T.H.; Dunn, D.A. Correlations between sound velocity and related properties of glacio-marine sediments: Barents sea. *GeoMar. Lett.* **1991**, *11*, 79–83. [CrossRef]

19. Hamilton, E.L.; Bachman, R.T. Sound velocity and related properties of marine sediments. *J. Acoust. Soc. Am.* **1982**, *72*, 1891–1904. [CrossRef]

20. Bae, S.H.; Kim, D.C.; Lee, G.S.; Kim, G.Y.; Kim, S.P.; Seo, Y.K.; Kim, J.C. Physical and acoustic properties of inner shelf sediments in the South Sea, Korea. *Quat. Int.* **2014**, *344*, 125–142. [CrossRef]

21. Waelbroeck, C.; Labeyrie, L.; Michel, E.; Duplessy, J.C.; Lambeck, K.; McManus, J.F.; Balbon, E.; Labracherie, M. Sea-level and deep water temperature changes derived from benthic foraminifera isotopic records. *Quat. Sci. Rev.* **2002**, *21*, 295–305. [CrossRef]

22. Lambeck, K.; Rouby, H.; Purcell, A.; Sun, Y.; Sambridge, M. Sea level and global ice volumes from the Last Glacial Maximum to the Holocene. *Proc. Natl. Acad. Sci. USA* **2014**, *111*, 15296–15303. [CrossRef]

23. Allen, G.P.; Posamentier, H.W. Sequence Stratigraphy and Facies Model of an Incised Valley Fill: The Gironde Estuary, France. *Sepm J. Sediment. Res.* **1993**, *63*, 378–391. [CrossRef]

24. Allen, G.P. Sedimentary processes and facies in the Gironde estuary: A recent model for macrotidal estuarine systems. *Clastic Tidal Sedimentol.* **1991**, *16*, 29–39. [CrossRef]

25. Cattaneo, A.; Steel, R.J. Transgressive deposits: A review of their variability. *Earth Sci. Rev.* **2003**, *62*, 187–228. [CrossRef]

26. Sturt, F.; Flemming, N.C.; Carabias, D.; Jöns, H.; Adams, J. The next frontiers in research on submerged prehistoric sites and landscapes on the continental shelf. *Proc. Geol. Assoc.* **2018**, *129*, 654–683. [CrossRef]

27. Harff, J.; Bailey, G.N.; Lüth, F. Geology and archaeology: Submerged landscapes of the continental shelf: An introduction. *Geol. Soc. Lond. Spec. Publ.* **2016**, *411*, 1–8. [CrossRef]

28. Brunović, D.; Miko, S.; Hasan, O.; Papatheodorou, G.; Ilijanić, N.; Miserocchi, S.; Correggiari, A.; Geraga, M. Late Pleistocene and Holocene paleoenvironmental reconstruction of a drowned karst isolation basin (Lošinj Channel, NE Adriatic Sea). *Palaeogeogr. Palaeoclimatol. Palaeoecol.* **2020**, *544*, 109587. [CrossRef]

29. Benjamin, J.; Rovere, A.; Fontana, A.; Furlani, S.; Vacchi, M.; Inglis, R.H.; Galili, E.; Antonioli, F.; Sivan, D.; Miko, S.; et al. Late Quaternary sea-level changes and early human societies in the central and eastern Mediterranean Basin: An interdisciplinary review. *Quat. Int.* **2017**, *449*, 29–57. [CrossRef]

30. Bailey, G.N.; Harff, J.; Sakellariou, D. *Under the Sea: Archaeology and Palaeolandscapes of the Continental Shelf*; Springer: Berlin/Heidelberg, Germany, 2017. [CrossRef]

31. Prins, L.T.; Andresen, K.J. Buried late Quaternary channel systems in the Danish North Sea – Genesis and geological evolution. *Quat. Sci. Rev.* **2019**, *223*, 105943. [CrossRef]

32. Wiberg-Larsen, P.; Bennike, O.; Jensen, J.B. Submarine Lateglacial lake deposits from the Kattegat, southern Scandinavia. *J. Quat. Sci.* **2019**, *34*, 165–171. [CrossRef]

33. De Clercq, M.; Missiaen, T.; Wallinga, J.; Zurita Hurtado, O.; Versendaal, A.; Mathys, M.; De Batist, M. A well-preserved Eemian incised-valley fill in the southern North Sea Basin, Belgian Continental Shelf-Coastal Plain: Implications for northwest European landscape evolution. *Earth Surf. Process. Landf.* **2018**, *43*, 1913–1942. [CrossRef]

34. Cooper, J.A.G.; Green, A.N.; Compton, J.S. Sea-level change in southern Africa since the Last Glacial Maximum. *Quat. Sci. Rev.* **2018**, *201*, 303–318. [CrossRef]

35. García-Moreno, D.; Gupta, S.; Collier, J.S.; Oggioni, F.; Vanneste, K.; Trentesaux, A.; Verbeeck, K.; Versteeg, W.; Jomard, H.; Camelbeeck, T.; et al. Middle–Late Pleistocene landscape evolution of the Dover Strait inferred from buried and submerged erosional landforms. *Quat. Sci. Rev.* **2019**, *203*, 209–232. [CrossRef]

36. Kirkpatrick, L.H.; Green, A.N.; Pether, J. The seismic stratigraphy of the inner shelf of southern Namibia: The development of an unusual nearshore shelf stratigraphy. *Mar. Geol.* **2019**, *408*, 18–35. [CrossRef]

37. Mattei, G.; Troisi, S.; Aucelli, P.; Pappone, G.; Peluso, F.; Stefanile, M. Sensing the Submerged Landscape of Nisida Roman Harbour in the Gulf of Naples from Integrated Measurements on a USV. *Water* **2018**, *10*, 1686. [CrossRef]

38. Perkins, E.J.; Gorman, A.R.; Tidey, E.J.; Wilson, G.S.; Ohneiser, C.; Moy, C.M.; Riesselman, C.R.; Gilmer, G.; Ross, B.S. High-resolution seismic imaging reveals infill history of a submerged Quaternary fjord system in the subantarctic Auckland Islands, New Zealand. *Quat. Res.* **2019**, *93*, 255–266. [CrossRef]

39. Laws, A.W.; Maloney, J.M.; Klotsko, S.; Gusick, A.E.; Braje, T.J.; Ball, D. Submerged paleoshoreline mapping using high-resolution Chirp sub-bottom data, Northern Channel Islands platform, California, USA. *Quat. Res.* **2019**, *93*, 1–22. [CrossRef]

40. García-Moreiras, I.; Cartelle, V.; García-Gil, S.; Muñoz Sobrino, C. First high-resolution multi-proxy palaeoenvironmental record of the Late Glacial to Early Holocene transition in the Ría de Arousa (Atlantic margin of NW Iberia). *Quat. Sci. Rev.* **2019**, *215*, 308–321. [CrossRef]

41. Pretorius, L.; Green, A.N.; Cooper, J.A.G.; Hahn, A.; Zabel, M. Outer- to inner-shelf response to stepped sea-level rise: Insights from incised valleys and submerged shorelines. *Mar. Geol.* **2019**, *416*, 105979. [CrossRef]

42. Lambeck, K.; Antonioli, F.; Purcell, A.; Silenzi, S. Sea-level change along the Italian coast for the past 10,000yr. *Quat. Sci. Rev.* **2004**, *23*, 1567–1598. [CrossRef]

43. Correggiari, A.; Roveri, M.; Trincardi, F. Late Pleistocene and Holocene evolution of the North Adriatic Sea. *Quaternario* **1996**, *9*, 697–704.

44. Ogorelec, B.; Mišič, M.; Šercelj, A.; Cimerman, F.; Faganeli, J.; Stegnar, P. Sediment sečoveljske soline. *Geologija* **1981**, *24*, 180–216.

45. Vacchi, M.; Marriner, N.; Morhange, C.; Spada, G.; Fontana, A.; Rovere, A. Multiproxy assessment of Holocene relative sea-level changes in the western Mediterranean: Sea-level variability and improvements in the definition of the isostatic signal. *Earth Sci. Rev.* **2016**, *155*, 172–197. [CrossRef]

46. Ronchi, L.; Fontana, A.; Correggiari, A. Characteristics and potential application of Holocene Tidal Inlets in the Northern Adriatic Shelf (Italy). *Alp. Mediterr. Quat.* **2018**, *31*, 31–34.

47. Ronchi, L.; Fontana, A.; Correggiari, A.; Asioli, A. Late Quaternary incised and infilled landforms in the shelf of the northern Adriatic Sea (Italy). *Mar. Geol.* **2018**, *405*, 47–67. [CrossRef]

48. Zecchin, M.; Ceramicola, S.; Lodolo, E.; Casalbore, D.; Chiocci, F.L. Episodic, rapid sea-level rises on the central Mediterranean shelves after the Last Glacial Maximum: A review. *Mar. Geol.* **2015**, *369*, 212–223. [CrossRef]

49. Zecchin, M.; Gordini, E.; Ramella, R. Recognition of a drowned delta in the northern Adriatic Sea, Italy: Stratigraphic characteristics and its significance in the frame of the early Holocene sea-level rise. *Holocene* **2015**, *25*, 1027–1038. [CrossRef]

50. Moscon, G.; Correggiari, A.; Stefani, C.; Fontana, A.; Remia, A. Very-high resolution analysis of a transgressive deposit in the Northern Adriatic Sea (Italy). *Alp. Mediterr. Quat.* **2015**, *28*, 121–129.

51. Correggiari, A.; Field, M.E.; Trincardi, F. Late Quaternary transgressive large dunes on the sediment-starved Adriatic shelf. *Geol. Soc. Lond. Spec. Publ.* **1996**, *117*, 155–169. [CrossRef]

52. Trincardi, F.; Correggiari, A.; Roveri, M. Late Quaternary transgressive erosion and deposition in a modern epicontinental shelf: The Adriatic semienclosed basin. *GeoMar. Lett.* **1994**, *14*, 41–51. [CrossRef]

53. Storms, J.E.A.; Weltje, G.J.; Terra, G.J.; Cattaneo, A.; Trincardi, F. Coastal dynamics under conditions of rapid sea-level rise: Late Pleistocene to Early Holocene evolution of barrier–lagoon systems on the northern Adriatic shelf (Italy). *Quat. Sci. Rev.* **2008**, *27*, 1107–1123. [CrossRef]

54. Ronchi, L.; Fontana, A.; Correggiari, A.; Remia, A. Anatomy of a transgressive tidal inlet reconstructed through high-resolution seismic profiling. *Geomorphology* **2019**, *343*, 65–80. [CrossRef]

55. Novak, A.; Šmuc, A.; Poglajen, S.; Vrabec, M. Linking the high-resolution acoustic and sedimentary facies of a transgressed Late Quaternary alluvial plain (Gulf of Trieste, northern Adriatic). *Mar. Geol.* **2020**, *419*, 106061. [CrossRef]

56. Trobec, A.; Šmuc, A.; Poglajen, S.; Vrabec, M. Submerged and buried Pleistocene river channels in the Gulf of Trieste (Northern Adriatic Sea): Geomorphic, stratigraphic and tectonic inferences. *Geomorphology* **2017**, *286*, 110–120. [CrossRef]

57. Covelli, S.; Fontolan, G.; Faganeli, J.; Ogrinc, N. Anthropogenic markers in the Holocene stratigraphic sequence of the Gulf of Trieste (northern Adriatic Sea). *Mar. Geol.* **2006**, *230*, 29–51. [CrossRef]

58. Trincardi, F.; Argnani, A.; Correggiari, A.; Foglini, F.; Rovere, M.; Angeletti, L.; Asioli, A.; Campiani, E.; Cattaneo, A.; Gallerani, A.; et al. *Note illustrative della Carta Geologica dei mari italiani alla scala 1:250.000 foglio NL 33-7 Venezia*; Istituto di Scienze Marine Consiglio Nazionale delle Ricerche: Ancona, Italy, 2011.

59. Trobec, A.; Busetti, M.; Zgur, F.; Baradello, L.; Babich, A.; Cova, A.; Gordini, E.; Romeo, R.; Tomini, I.; Poglajen, S.; et al. Thickness of marine Holocene sediment in the Gulf of Trieste (northern Adriatic Sea). *Earth Syst. Sci. Data* **2018**, *10*, 1077–1092. [CrossRef]

60. Chiocci, F.L.; Casalbore, D.; Marra, F.; Antonioli, F.; Romagnoli, C. *Relative Sea Level Rise, Palaeotopography and Transgression Velocity on the Continental Shelf*; Springer Cham: Berlin/Heidelberg, Germany, 2017; pp. 39–51. [CrossRef]

61. Mautner, A.K.; Gallmetzer, I.; Haselmair, A.; Schnedl, S.M.; Tomašových, A.; Zuschin, M. Holocene ecosystem shifts and human-induced loss of Arca and Ostrea shell beds in the north-eastern Adriatic Sea. *Mar. Pollut. Bull.* **2018**, *126*, 19–30. [CrossRef] [PubMed]

62. Ogrinc, N.; Faganeli, J.; Ogorelec, B.; Čermelj, B. The origin of organic matter in Holocene sediments in the Bay of Koper (Gulf of Trieste, northern Adriatic Sea). *Geologija* **2007**, *50*, 179–187. [CrossRef]

63. Tomašových, A.; Gallmetzer, I.; Haselmair, A.; Kaufman, D.S.; Mavrič, B.; Zuschin, M. A decline in molluscan carbonate production driven by the loss of vegetated habitats encoded in the Holocene sedimentary record of the Gulf of Trieste. *Sedimentology* **2019**, *66*, 781–807. [CrossRef] [PubMed]

64. Gallmetzer, I.; Haselmair, A.; Tomašových, A.; Mautner, A.-K.; Schnedl, S.-M.; Cassin, D.; Zonta, R.; Zuschin, M. Tracing origin and collapse of Holocene benthic baseline communities in the northern Adriatic Sea. *Palaios* **2019**, *34*, 121–145. [CrossRef]

65. Pleničar, M. *Osnovna geološka karta SFRJ. L 33-88, Trst*; Zvezni Geološki Zavod: Beograd, Yugoslavia, 1969.

66. Pleničar, M.; Polšak, A.; Šikić, D. *Osnovna Geološka Karta SFRJ. 1:100.000. Tolmač Za List Trst: L 33-88*; Zvezni Geološki Zavod: Beograd, Yugoslavia, 1973.

67. Vrabec, M.; Busetti, M.; Zgur, F.; Facchin, L.; Pelos, C.; Romeo, R.; Sormani, L.; Slavec, P.; Tomini, I.; Visnovic, G.; et al. Refleksijske Seizmične Raziskave v Slovenskem Morju Slomartec 2013. In Proceedings of the Raziskave s Področja Geodezije in Geofizike 2013, Ljubljana, Slovenia, 30 January 2014; Kuhar, M., Čop, R., Gosar, A., Kobold, M., Kralj, P., Malačič, V., Rakovec, J., Skok, G., Stopar, B., Vreča, P., Eds.; University of Ljubljana: Ljubljana, Slovenia, 2014; pp. 97–101.

68. Ogorelec, B.; Faganeli, J.; Mišič, M.; Čermelj, B. Reconstruction of paleoenvironment in the Bay of Koper (Gulf of Trieste, Northern Adriatic). *Annales* **1997**, *11*, 187–200.

69. Ogorelec, B.; Mišič, M.; Faganeli, J. Marine geology of the Gulf of Trieste (northern Adriatic): Sedimentological aspects. *Mar. Geol.* **1991**, *99*, 79–92. [CrossRef]

70. Ogorelec, B.; Mišič, M.; Faganeli, J.; Šercelj, A.; Cimerman, F.; Dolenec, T.; Pezdič, J. Kvartarni sediment vrtine V-3 v Koprskem zalivu. *Slov. Morje Zaled.* **1984**, *7*, 165–186.

71. Ogorelec, B.; Mišič, M.; Faganeli, J.; Stegnar, P.; Vrišer, B.; Vukovič, A. Recentni sediment Koprskega zaliva. *Geologija* **1987**, *30*, 87–121.

72. Faganeli, J.; Ogorelec, B.; Mišič, M.; Dolenec, T.; Pezdič, J. Organic geochemistry of two 40-m sediment cores from the Gulf of Trieste (Northern Adriatic). *Estuar. Coast. Shelf Sci.* **1987**, *25*, 157–167. [CrossRef]

73. Covelli, S.; Faganeli, J.; Horvat, M.; Brambati, A. Mercury contamination of coastal sediments as the result of long-term cinnabar mining activity (Gulf of Trieste, northern Adriatic Sea). *Appl. Geochem.* **2001**, *16*, 541–558. [CrossRef]

74. Žerjal, A.; Kolega, N.; Poglajen, S.; Rant, I.; Jeklar, M.; Lovrič, E.; Vranac, D.; Mozetič, D.; Slavec, P.; Berden Zrimec, M.; et al. *Zajem Naravnih Geomorfoloških Značilnosti Morskega dna, Analiza Antropogenih Fizičnih Poškodb Morskega dna in Klasifikacija Tipov Morskega dna z Določitvijo Obsežnejšega Morskega Rastja na Morskem Dnu*; Harpha Sea: Koper, Slovenia, 2014.

75. EMODnet Bathymetry Consortium. EMODnet Digital Bathymetry (DTM 2018). **2018**. [CrossRef]

76. Masoli, C.A.; Petronio, L.; Gordini, E.; Deponte, M.; Boehm, G.; Cotterle, D.; Romeo, R.; Barbagallo, A.; Belletti, R.; Maffione, S.; et al. Near-shore geophysical and geotechnical investigations in support of the Trieste Marine Terminal extension. *Near Surf. Geophys.* **2020**, *18*, 73–89. [CrossRef]

77. *Carta Nautica Da Punta Tagliamento a Pula 1:100 000*; Istituto Idrografico della Marina: Genova, Italy, 2004.

78. Ministry of Transport of the Republic of Slovenia. *Male karte Tržaški zaliv Merilo 1:100 000*; Ministry of Transport of the Republic of Slovenia: Ljubljana, Slovenia, 2005.

79. Placer, L.; Vrabec, M.; Celarc, B. The bases for understanding of the NW Dinarides and Istria Peninsula tectonics. *Geologija* **2010**, *53*, 55–86. [CrossRef]

80. Jurkovšek, B.; Biolchi, S.; Furlani, S.; Kolar-Jurkovšek, T.; Zini, L.; Jež, J.; Tunis, G.; Bavec, M.; Cucchi, F. Geology of the Classical Karst Region (SW Slovenia–NE Italy). *J. Maps* **2016**, *12*, 352–362. [CrossRef]

81. Furlani, S.; Finocchiaro, F.; Boschian, G.; Lenaz, D.; Biolchi, S.; Boccali, C.; Monegato, G. Quaternary evolution of the fluviokarst Rosandra valley (Trieste, NE Italy). *Alp. Mediterr. Quat.* **2016**, *29*, 169–179.

82. Biolchi, S.; Furlani, S.; Covelli, S.; Busetti, M.; Cucchi, F. Morphoneotectonics and lithology of the eastern sector of the Gulf of Trieste (NE Italy). *J. Maps* **2016**, *12*, 936–946. [CrossRef]

83. Anderson, A.L.; Hampton, L.D. Acoustics of gas-bearing sediments. II. Measurements and models. *J. Acoust. Soc. Am.* **1980**, *67*, 1890–1903. [CrossRef]

84. Anderson, A.L.; Hampton, L.D. Acoustics of gas-bearing sediments I. Background. *J. Acoust. Soc. Am.* **1980**, *67*, 1865–1889. [CrossRef]

85. Kim, D.C.; Lee, G.S.; Lee, G.H.; Park, S.C. Sediment echo types and acoustic characteristics of gas-related acoustic anomalies in Jinhae Bay, southern Korea. *Geosci. J.* **2008**, *12*, 47–61. [CrossRef]

86. Visnovitz, F.; Bodnár, T.; Tóth, Z.; Spiess, V.; Kudó, I.; Timár, G.; Horváth, F. Seismic expressions of shallow gas in the lacustrine deposits of Lake Balaton, Hungary. *Near Surf. Geophys.* **2015**, *13*, 433–447. [CrossRef]

87. Missiaen, T.; Murphy, S.; Loncke, L.; Henriet, J.P. Very high-resolution seismic mapping of shallow gas in the Belgian coastal zone. *Cont. Shelf Res.* **2002**, *22*, 2291–2301. [CrossRef]

88. Anderson, A.L.; Bryant, W.R. Gassy sediment occurrence and properties: Northern Gulf of Mexico. *GeoMar. Lett.* **1990**, *10*, 209–220. [CrossRef]

89. Fonseca, L.; Mayer, L.; Orange, D.; Driscoll, N. The high-frequency backscattering angular response of gassy sediments: Model/data comparison from the Eel River Margin, California. *J. Acoust. Soc. Am.* **2002**, *111*, 2621–2631. [CrossRef]

90. Judd, A.G.; Hovland, M. *Seabed Fluid Flow: The Impact on Geology, Biology and the Marine Environment*; Cambridge University Press: Cambridge, UK, 2007; ISBN 0521819504.

91. Fu, S.S.; Wilkens, R.H.; Frazer, L.N. In situ velocity profiles in gassy sediments: Kiel Bay. *GeoMar. Lett.* **1996**, *16*, 249–253. [CrossRef]

92. Lee, G.-S.; Kim, D.-C.; Lee, G.-H.; Park, S.-C.; Kim, G.-Y.; Yoo, D.-G.; Kim, J.-C.; Cifci, G. Physical and Acoustic Properties of Gas-bearing Sediments in Jinhae Bay, the South Sea of Korea. *Mar. Georesources Geotechnol.* **2009**, *27*, 96–114. [CrossRef]

93. Tóth, Z.; Spiess, V.; Mogollón, J.M.; Jensen, J.B. Estimating the free gas content in Baltic Sea sediments using compressional wave velocity from marine seismic data. *J. Geophys. Res. Solid Earth* **2014**, *119*, 8577–8593. [CrossRef]

94. Donda, F.; Forlin, E.; Gordini, E.; Panieri, G.; Buenz, S.; Volpi, V.; Civile, D.; De Santis, L. Deep-sourced gas seepage and methane-derived carbonates in the Northern Adriatic Sea. *Basin Res.* **2015**, *27*, 531–545. [CrossRef]

95. Conti, A.; Stefanon, A.; Zuppi, G.M. Gas seeps and rock formation in the northern Adriatic Sea. *Cont. Shelf Res.* **2002**, *22*, 2333–2344. [CrossRef]

96. Gordini, E.; Falace, A.; Kaleb, S.; Donda, F.; Marocco, R.; Tunis, G. Methane-Related Carbonate Cementation of Marine Sediments and Related Macroalgal Coralligenous Assemblages in the Northern Adriatic Sea. In *Seafloor Geomorphology as Benthic Habitat*; Elsevier: Amsterdam, Netherlands, 2012; pp. 185–200. [CrossRef]

97. Zecchin, M.; Caffau, M.; Tosi, L. Relationship between peat bed formation and climate changes during the last glacial in the Venice area. *Sediment. Geol.* **2011**, *238*, 172–180. [CrossRef]

98. Gordini, E.; Marocco, R.; Tunis, G.; Ramella, R. I depositi cementati del Golfo di Trieste (Adriatico Settentrionale): Distribuzione areale, caratteri geomorfologici e indagini acustiche ad alta risoluzione. *Quaternario* **2004**, *17*, 555–563.

99. Ogrinc, N.; Fontolan, G.; Faganeli, J.; Covelli, S. Carbon and nitrogen isotope compositions of organic matter in coastal marine sediments (the Gulf of Trieste, N Adriatic Sea): Indicators of sources and preservation. *Mar. Chem.* **2005**, *95*, 163–181. [CrossRef]

100. McCarthy, E.M.; Sabol, B. Acoustic characterization of submerged aquatic vegetation: Military and environmental monitoring applications. In Proceedings of the OCEANS 2000 MTS/IEEE Conference and Exhibition, Providence, RI, USA, 11–14 September 2000; 3, pp. 1957–1961. [CrossRef]

101. Komatsu, T.; Igarashi, C.; Tatsukawa, K.; Sultana, S.; Matsuoka, Y.; Harada, S. Use of multi-beam sonar to map seagrass beds in Otsuchi Bay on the Sanriku Coast of Japan. *Aquat. Living Resour.* **2003**, *16*, 223–230. [CrossRef]

102. Lee, K.M.; Ballard, M.S.; McNeese, A.R.; Wilson, P.S. Sound speed and attenuation measurements within a seagrass meadow from the water column into the seabed. *J. Acoust. Soc. Am.* **2017**, *141*, EL402–EL406. [CrossRef]

103. Zecchin, M.; Baradello, L.; Brancolini, G.; Donda, F.; Rizzetto, F.; Tosi, L. Sequence stratigraphy based on high-resolution seismic profiles in the late Pleistocene and Holocene deposits of the Venice area. *Mar. Geol.* **2008**, *253*, 185–198. [CrossRef]

104. Zecchin, M.; Brancolini, G.; Tosi, L.; Rizzetto, F.; Caffau, M.; Baradello, L. Anatomy of the Holocene succession of the southern Venice lagoon revealed by very high-resolution seismic data. *Cont. Shelf Res.* **2009**, *29*, 1343–1359. [CrossRef]

 water

 MDPI

Article

Submarine Geomorphology of the Southwestern Sardinian Continental Shelf (Mediterranean Sea): Insights into the Last Glacial Maximum Sea-Level Changes and Related Environments

Giacomo Deiana [1,3], Luciano Lecca [1], Rita Teresa Melis [1], Mauro Soldati [2], Valentino Demurtas [1,*] and Paolo Emanuele Orrù [1,3]

1   Department of Chemical and Geological Sciences, University of Cagliari, 09042 Monserrato, Italy; giacomo.deiana@unica.it (G.D.); leccal@unica.it (L.L.); rtmelis@unica.it (R.T.M.); orrup@unica.it (P.E.O.)
2   Department of Chemical and Geological Sciences, University of Modena and Reggio Emilia, 41125 Modena, Italy; mauro.soldati@unimore.it
3   CoNISMa Interuniversity Consortium on Marine Sciences, 00126 Roma, Italy
*   Correspondence: valentino.demurtas@unica.it

Abstract: During the lowstand sea-level phase of the Last Glacial Maximum (LGM), a large part of the current Mediterranean continental shelf emerged. Erosional and depositional processes shaped the coastal strips, while inland areas were affected by aeolian and fluvial processes. Evidence of both the lowstand phase and the subsequent phases of eustatic sea level rise can be observed on the continental shelf of Sardinia (Italy), including submerged palaeo-shorelines and landforms, and indicators of relict coastal palaeo-environments. This paper shows the results of a high-resolution survey on the continental shelf off San Pietro Island (southwestern Sardinia). Multisensor and multiscale data—obtained by means of seismic sparker, sub-bottom profiler chirp, multibeam, side scan sonar, diving, and uncrewed aerial vehicles—made it possible to reconstruct the morphological features shaped during the LGM at depths between 125 and 135 m. In particular, tectonic controlled palaeo-cliffs affected by landslides, the mouth of a deep palaeo-valley fossilized by marine sediments and a palaeo-lagoon containing a peri-littoral thanatocenosis (18,983 ± 268 cal BP) were detected. The Younger Dryas palaeo-shorelines were reconstructed, highlighted by a very well preserved beachrock. The coastal paleo-landscape with lagoon-barrier systems and retro-littoral dunes frequented by the Mesolithic populations was reconstructed.

Keywords: submarine geomorphology; morphostratigraphy; sea-level changes; Last Glacial Maximum; Sardinia; Italy

Citation: Deiana, G.; Lecca, L.; Melis, R.T.; Soldati, M.; Demurtas, V.; Orrù, P.E. Submarine Geomorphology of the Southwestern Sardinian Continental Shelf (Mediterranean Sea): Insights into the Last Glacial Maximum Sea-Level Changes and Related Environments. *Water* **2021**, *13*, 155. https://doi.org/10.3390/w13020155

Received: 27 November 2020
Accepted: 4 January 2021
Published: 11 January 2021

1. Introduction

Sea-level variations connected to climatic oscillations [1] cause changes in the landscape of coastal areas and continental shelves [2]. The comparative geomorphological analysis of emerged and submerged areas is particularly effective for revealing the land- and seascape changes [3,4]. Landscape evolutionary phases can be reconstructed considering morphostructural and morphostratigraphic settings and using geomorphological, seismic, sedimentological, palaeontological, and isotopic data. The detailed reconstruction of the submerged coastal palaeo-landscape is useful to understanding the dynamics of the human population during the Last Glacial Maximum (LGM) [5,6]. As such, marine and continental geomorphological analyses are crucial for better representing and understanding the Pleistocene landscape evolution [3,7–11].

This study aims to obtain new insights into the palaeo-geographic evolution of the San Pietro continental shelf of southwestern Sardinia (Figure 1) during the last cold stage

(MIS 2) by analysing erosional and depositional landforms formed during the LGM sea-level lowstand, as well as the palaeo-geographic coastal evolution connected to the LGM sea-levels. Several studies used different methodological approaches and analysed various palaeo-sea-level indicators (e.g., palaeo-cliffs, lowstand depositional terraces, beachrocks, fossiliferous deposits) to evaluate the post-glacial sea levels in the Mediterranean Sea in the past 20 ka. Previous studies also successfully applied the glacial-hydro-isostatic adjustment (GIA) models [12–17].

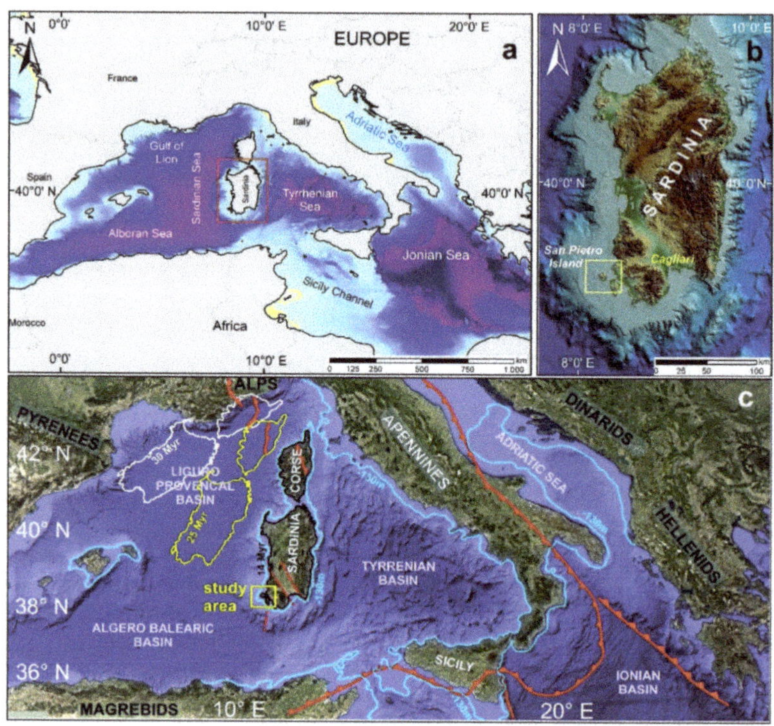

Figure 1. Geographic location and structural setting of the study area: (**a**) Sardinia Island within the Mediterranean Sea; (**b**) San Pietro Island on the SW side of Sardinia; (**c**) structural sketch map of the Mediterranean area. Red lines mark thrust fronts; white line the Sardinian-Corse block translation 30 Myr BP; yellow line Sardinian-Corse block translation 25 Myr BP (mod. after Carminati and Doglioni, 2008 [18]); Black line the Sardinian-Corse block translation 14 Myr (Gattacceca et al., 2007 [19]); Light blue color line, isobath of −130 m represents the coastline during the Last Glacial Maximum (LGM).

LGM shorelines are known from other areas of the Mediterranean Sea, including the Adriatic continental shelf [17,20–22], southern Tyrrhenian margin offshore Sicily [22–26], Calabria [27], and Malta [2,11,28,29].

For example, morphobathymetric data acquisition (i.e., high-resolution multi-beam and seismic data) integrated with direct survey methods (i.e., remotely operated vehicle (ROV) and diving) allowed scientists to obtain a particularly rich database of the southwestern Sardinian continental shelf [30–35].

Herein, we analysed the structural and volcanic geological settings linked to the Oligo-Miocene rifting of the western Mediterranean and Sardinian-Corsican blocks to highlight the geomorphological features of the continental shelf surrounding San Pietro Island. These data contribute to the knowledge of the coastal palaeo-landscape and its evolution from LGM to the Holocene (Figure 1a,b). In particular, submerged high rocky coast morphotypes, a large palaeo-valley, a palaeo-lagoon and the successive phases of post-glacial sea level stationing were analyzed.

2. Geological and Structural Settings

The southwestern continental margin of Sardinia is characterised by normal faults that define intrashelf and intraslope basins [36]. This part of the Sardinian continental margin has been explored using geophysical surveys and deep drills, defining the order and geometry of the depositional sequences [36–41] (Figure 1c). High-angle normal fault systems characterised the western Sardinia continental margin setting between the Middle-Upper Oligocene and Miocene when, owing to the Apennine-Maghrebian chain orogeny, the intra-back arc basins opening caused the formation of an extensive system of rifts [42–45]. The genesis of the margin was clarified based on the ECORS-CROP Programme seismic data by examining the extensional tectonic inversion of a compressive structure of the Pyrenean western branch (Figure 1c) [39].

The margin formed as the transition between the western Mediterranean rift and the western branch of the Sardinian rift system and later assumed the structural and evolutionary characteristics of a divergent margin [36] (Figure 2). The kinematic analysis of the central Mediterranean shows that the Sardinia-Corsica block rotated until about 15 Ma later it became almost stable [18,19]. However, in the western part of the base of the margin, a significant earthquake (38.21°–08.21°; 5.4 Mw) was recorded on in August 1977 [46]. Furthermore, the INGV (Istituto Nazionale di Geofisica e Vulcanologia) earthquake catalogue, which contains the seismic records for the past 25 years, shows three other major earthquakes in southern Sardinia: one earthquake with a magnitude of 5.5 in August 1988 along the Sant'Antioco active fault, from Toro Island to Quirino Seamount, and two earthquakes with a magnitude of 4.5 in March 2006 at the sea prolongation of a major fault NW–SE Campidano graben that marks the western edge of this Plio-Quaternary graben. Therefore, slight fault movements that produce an occasional seismicity are still present and affect the margin.

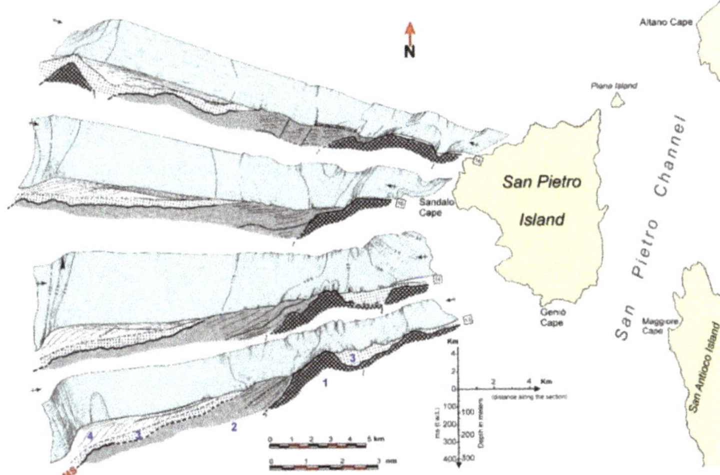

Figure 2. Sectioned block diagram of the Sardinian southwestern continental shelf off the San Pietro Island. (**1**) Acoustic basement (volcanic complex—Lower-Middle Miocene); (**2**) lower sedimentary sequence (Middle-Upper Miocene); MS) Messinian erosional surface; (**3**) sedimentary sequence poorly or not stratified in the lower part, with undulating stratification in the upper part (Lower-Middle Pliocene); (**4**) upper sedimentary sequence, prograding complex of the external platform, superficial deposits in the proximal platform (Upper Pliocene—Quaternary) (after Lecca, 2000 modified [36]).

The sedimentary units preceding the Oligo-Miocene Sardinian rifting stage are represented by the Palaeozoic basement, marine clastic Eocene series, and fluvial sandstones and claystones of the Cixerri Formation (Upper Eocene to Lower Oligocene). The initiation of the Oligo-Miocene rifting was accompanied by the andesitic volcanism (Upper Oligocene to Aquitanian) of the Sulcis block. The subsiding basin was filled by the fluvial sediments of the Ussana Formation and the marine marly-arenaceous and carbonate sediments of the Lower Miocene [42].

The continental margin off San Pietro Island is characterised by a steep slope, which extends to the Sardinian-Balearic abyssal plain to a depth of approximately 2800 m [41]. The inner and intermediate continental shelf is characterised by the extensive outcrops of volcanic rocks, consisting of ignimbrites (comendites) and pyroclasts [35]. From a geochemical point of view, the rhyolites predominate, while the dacites characterise the basal volcanic formations [43,47]. Explosive volcanic eruptions occurred on San Pietro Island during the Burdigalian, Miocene (15–17 Ma). From a morphostructural point of view, the ignimbrite outcrops are characterised by wide mega-cuestas, calderas, necks, and dikes (Figure 2) [35].

The presence of the Oligo-Miocene volcanites at the tectonic block boundaries is marked by magnetic anomalies on the inner continental shelf [43] and was documented by analysing the rock samples from the lower margin of the Seamount Quirino [36] (Figure 3). On the distal shelf, the Miocene volcano-sedimentary and sedimentary strata rest on the volcanic substrate. The Miocene sedimentary sequence, up to the pre-evaporitic Tortonian marls, tends to be characterised by an erosional surface tied up to the Messinian eustatic fall [48]. In the lower part of the Miocene sedimentary sequence, the clinoform reflections are spaced wider, and the ages close to the Burdigalian are suggested [36] (Figure 4).

3. Geomorphological Setting

The major factor controlling the evolution of submarine canyons in the Mediterranean basin is the Messinian salinity crisis, which induced a significant forced sea-level fall of approximately 2000 m from the present-day sea level [42]. The consequent emergence of the continental margin led to intense erosion [37,48,49]. The following Pliocene flooding event deposited a thick mud drape over the entire continental shelf [7].

The shelf break is located at depths of 190–220 m and hosts the Plio-Quaternary prograding sedimentary wedge [35]. Both the shelf break and the Upper continental slope are eroded by the canyon heads formed via the retrogressive erosion processes. Intrachannel landslides are observed in the canyon sidewalls, while the Upper continental slope is distinguished by creeping areas and complex landslides often associated with pockmark fields due to fluid emissions [34,50].

The Messinian eustatic sea-level fall has been recognised on the Sparker seismic tracks acquired during the MAGIC (Marine Geohazard Along Italian Coasts) Sardinia Channel 2009 survey off Cala Fico. That study identified a palaeo-valley with polycyclic evolution that engraved both the volcanic substrate and the lower sedimentary sequence. Convoluted and plane parallel reflectors seem to characterise the Quaternary sequence.

Arenaceous beachrocks are represented by two extensive outcrops located to the north of La Punta and Piana Island at depths of 45–50 m. The outcrops display prominent erosion features both on the top surface and at their edges [51] (Figures 3 and 4).

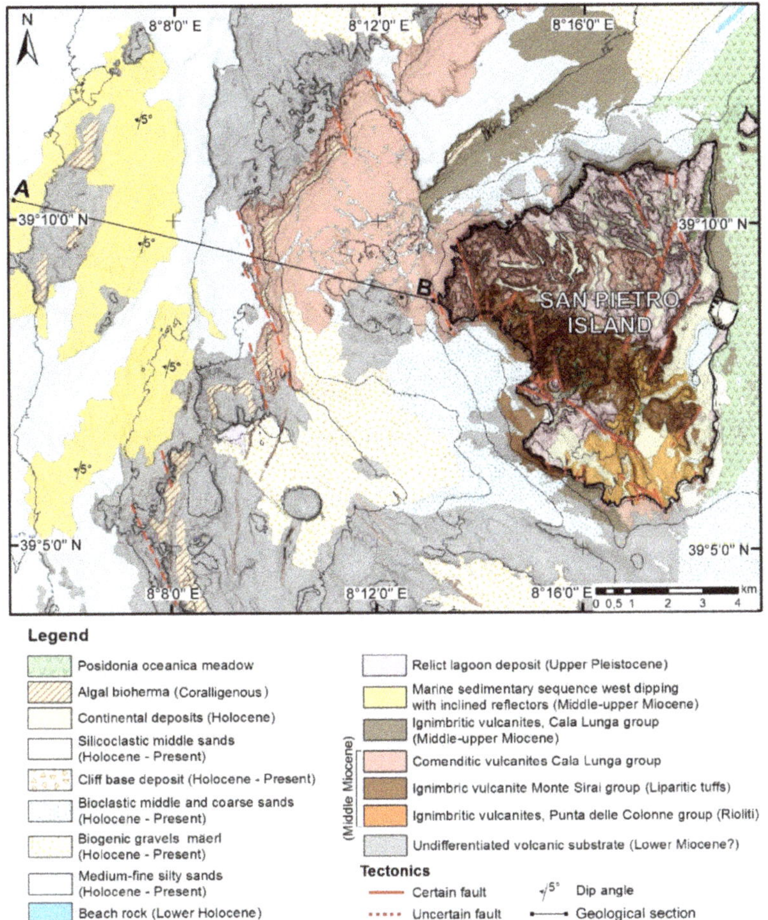

Figure 3. Geolithological sketch map of study area from the Geological Map of Italy. Scale 1:50,000—Sheet 563 "Isola di San Pietro" (Rizzo et al., 2015 [35]).

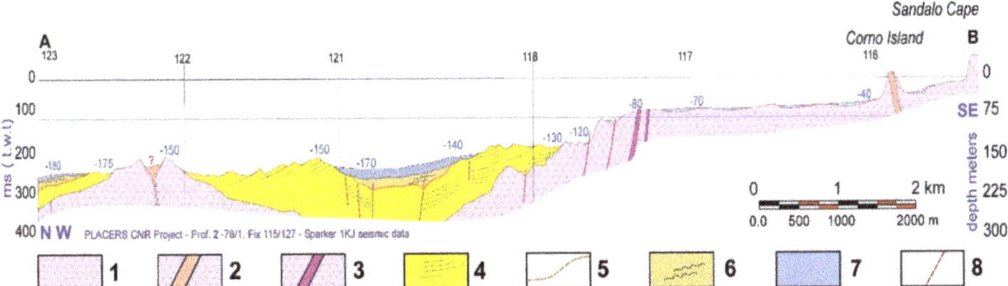

Figure 4. Geological section from seismic data of: Oceanography and Seabed—Mineral Resources—PLACERS Project CNR—Profilo 2-78/1, Fix 115/127—Sparker 1KJ: (1) Volcanites (Lower-Middle Miocene; (2) emission chimney; (3) dikes; (4) marine sedimentary sequence with inclined reflectors (Middle-Upper Miocene; (5) Messinian erosional surface; (6) marine sedimentary sequence with undulated reflectors. (Pliocene-Pleistocene); (7) Holocene-current drape; (8) fault.

The beachrocks are slightly tilted seaward, presenting a typical character of beach sand bodies, with the sedimentary structures (e.g., parallel lamination and wedge-shaped, sigmoidal, and inclined stratification) common for coastal environments [52].

Considering the outcrop depth, these beachrocks are attributed to the end of the Younger Dryas event and are interpreted to be formed when the eustatic sea level dropped during the Pleistocene–Holocene marine transgression.

The actual and subactual sediments on the distal continental shelf off San Pietro Island are represented by pelitic sands and sandy pelites. These deposits contain variable bioclastic fractions, composed of foraminifera and the degradation products of algal bioherms. These algal bioherms colonise the rocky substrate and can be found both outcropping and sub-outcropping [35] (Figure 2).

The Middle continental shelf has medium-grained, slightly pelitic sands, which bioclastic component increases towards the lower limit of *Posidonia oceanica* meadows. The areas farther offshore are dominated by biogenic gravels consisting of red algae (mäerl). These gravels form patches and hydraulic dunes. The near-continental shelf and the peri-littoral area are dominated by deposits linked to the retreat of high rocky coasts. In particular, base cliff deposits consist of sub-rounded heterometric blocks of volcanic lithology and landslide deposits with isolated sub-angular mega-blocks. The submerged beaches are characterised by medium- to coarse-grained sands with a predominantly quartz composition, whereas medium- and fine-grained sands are present in the bays of the southeastern sector. The sandy deposits with a predominantly quartz composition are located near the shoreline and Upper limit of the *Posidonia oceanica* prairie and have an important carbonate bioclastic fraction.

The first studies published on the LGM palaeo-shoreline of the western Sardinian shelf were conducted northward of our study area and indicated the existence of both erosional landforms and sedimentary sequences, in distal continental shelf, at depths between −120 and −140 m [53–56].

4. Materials and Methods

4.1. Seismic Data

The dataset used herein includes the seismic analogic data (Sparker 0.8 KJ) and by a high-resolution 3.5 kHz seismic sub-bottom profiler. These data were purchased from R/V Bannock (CNR) and collected during the oceanographic cruise "Placers 78/1" as part of the "Oceanografia e Fondi Marini" project. These data allowed the reconstruction of the Upper continental margin geological structure [36].

In order to reconstruct the palaeo-geomorphological setting, in particular, the intermediate continental shelf palaeo-hydrography, digital seismic surveys were carried out by R/V "Universitatis" during the oceanographic cruise "Canale di Sardegna 2009" in the frame of the MAGIC Project. The seismic surveys used a seismic energy source (Sparker 100/1000 J, Applied Acoustic CSP 20200, Great Yarmouth, United Kingdom), while the sub-bottom surveys aimed to reveal the structure of the surface deposits and were carried out using a geoacoustic source (Geochirp II-CP931, GeoAcoustics–Kongsberg, Great Yarmouth, United Kingdom) (Figure 5).

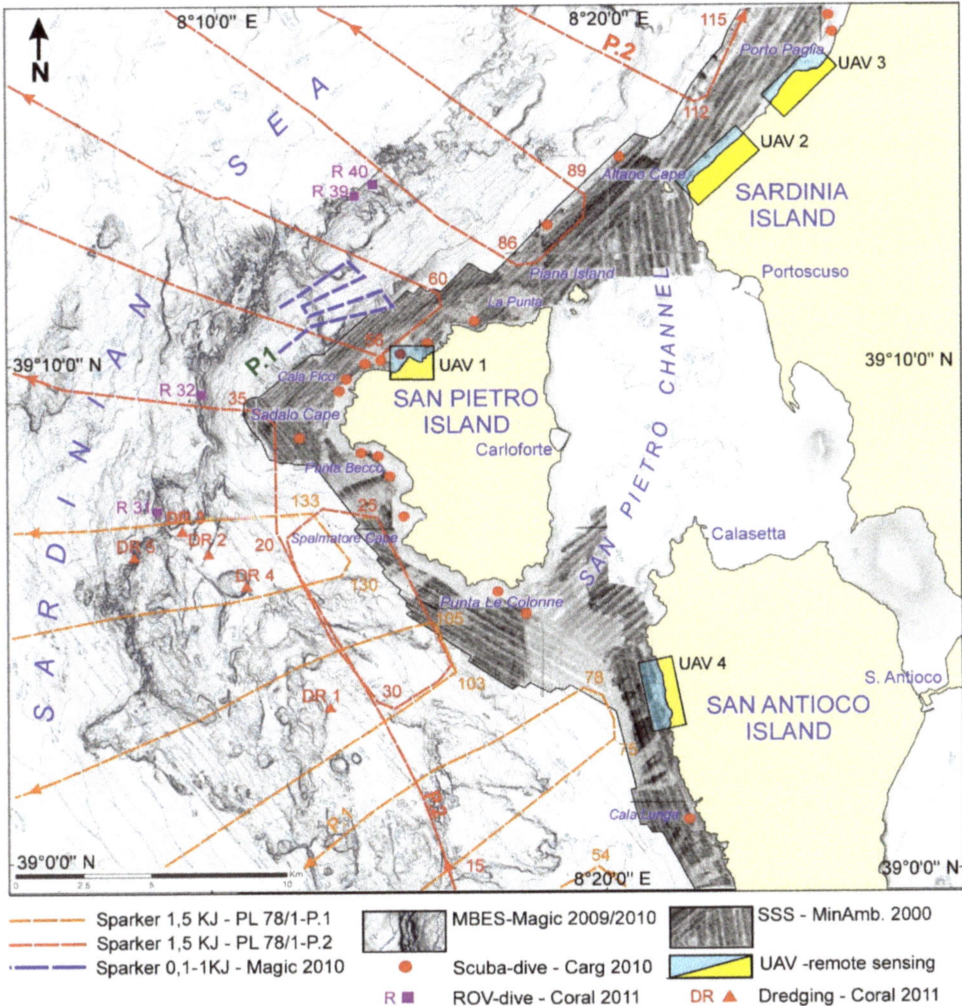

Figure 5. Data locations: spatial coverage multi-beam echosounder (MBES), side scan sonar (SSS) and uncrewed aerial vehicle (UAV); seismic profiles, scuba-dive and ROV stations; dredging sampling points.

4.2. Multibeam, Singlebeam, and Side-Scan Sonar Data

Morphobathymetric data were acquired during the oceanographic cruises "Canale di Sardegna 2009" and "Sardegna 2010" using R/V "Universitatis CoNISMa" as part of the Marine geohazard along Italian coasts (MAGIC) Project. The 50 kHz multi-beam echosounder (MBES, RESON SEABAT 8160) was calibrated with continuous sound velocity detection lines and vertical profiles. Onboard R/V Universitatis, the integrated system contained a motion sensor and gyro (IXSEA OCTANS) and a satellite differential GPS (Global Positioning System). The geocentric datum WGS84 and the UTM projection were chosen for navigation and display. The data collected during the survey were integrated with the Official National Italian Geological Cartography (CARG) project data.

Side-scan sonar data acquisition was performed on the proximal continental shelf with depths ranging from 10 to 50 m as part of the "Mapping of *Posidonia oceanica* meadows along the coasts of Sardinia" project funded by the Italian Ministry for the Environment on

R/V "Copernaut Franca". A 100–500 kHz, dual-frequency sensor was used with a towfish (Model 272/T, EG&G Marine Instruments, Massachusetts, USA) connected to the Triton Elics system (Triton Elics International, Portland, OR, USA) with ISIS software (Triton) for geo-referenced acquisition and Delf Map for the construction and correction of the mosaic. The correct positioning of the acquired data was ensured by a GPS receiver with differential correction (Trimble 5007, Sunnyvale, CA, USA).

In the coastal areas with depths of 5–20 m, single-beam echosounder and lateral sonar data were acquired using a towed sensor (1 MHz, Starfish 990, Tritech, Aberdeenshire, Scotland) and sonar (200/800 kHz, Lowrance Elite 12, Tulsa, Oklahoma) with a hull transducer (Lowrance Simrad Active, Tulsa, Oklahoma) (Figure 5).

4.3. Direct Seabed Observations

On the internal shelf, diving surveys and sampling were carried out during the surveys "San Pietro Sub 2006" and "San Pietro 2010" as part of the "Official National Italian Geological Cartography" project. Fifteen underwater survey stations were set down to a depth of 50 m. Direct observations aimed to elaborate the interpretative keys for the geophysical data. Two teams of four geologists were engaged in the underwater surveys. The first geomorphological survey data were reported on tablets equipped with a depth gauge.During the underwater survey, sediment sampling was carried out with a vacuum core, whereas rocks were sampled with a chisel and heavy hammer. Six cores of unconsolidated sediments, five sedimentary rock samples (beachrocks and eolianites), and 20 samples of acid volcanites were collected. The data were synthesised using special survey cards (Figure 6). Direct seabed observations in the distal shelf areas and, in particular, the exploration of the palaeo-cliff walls at depths of 85–140 m were conducted using ROV *Polluce* III R/V Astrea (ISPRA, Istituto Superiore Protezione e Ricerca Ambientale). These surveys were carried out as part of the "CORALLIUM RUBRUM" and "MARINESTRATEGY" projects, being sponsored by Italian Environmental Ministry—Autonomous Region of Sardinia. High-definition ROV images supported the habitat mapping of deep rocky bottoms dominated by red algal coralligenous assemblages and coral settlements (*Corallium rubrum* and *Leiopathes glaberrima*) [31,32]). These images allowed scientists to calibrate the geomorphological interpretation of palaeo-cliffs, especially regarding gravity-induced processes (Figure 5).

4.4. Dredging and Shell Sampling for Radiocarbon Analysis

The "SULCIS dredging survey" (2011) was conducted on the distal continental shelf, onboard R/V Gisella, using a classic submerged cylindrical dredger and two-cylinder experimental dredger. The dredging route was planned upon the analysis of morpho-bathymetric data and seismic profiles. The coordinates for the core sampling sites were determined using a differential GPS onboard the ship. The seabed depth at each core sampling point was acquired from the digital terrain models (DTM) processed using sonar multi-beam data. The volcanic rocks were not sampled because massive coralligenous bioconstructions with thicknesses greater than 50 cm covered them (Figure 5).

Figure 6. (**a**) Diver engaged in underwater geomorphological survey, using tablet with compass, clinometer, depth gauge and collimator at −15; (**b**) Fault mirror exhumed by erosion in Cala Fico at 10 m; (**c**) Foot cliff deposit with subspheroidal blocks at −18 m; (**d**) Lamination of pyroclastic lavas (Comenditi) in Cala Vinagra at 13 m.

4.5. Aerial and Uncrewed Aerial Vehicle Inland Remote Sensing

To obtain high-resolution aerial photos and topography suitable for the mapping of the onshore Sardinian coastal sector, we analysed the available topographic data produced by LiDAR (light detection and ranging) surveys. These aerial photogrammetric surveys were carried out by the Autonomous Region of Sardinia in 2008. The high-resolution aerial photos allowed us to analyse the coastal sector with high precision, down to a depth of 15 m. A cell size of 1 m and a mean vertical resolution DTM of approximately 30 cm were extracted.

In the most important sectors, such as the Capo Altano landslide, the surveys were performed with uncrewed aerial vehicles (UAVs, DJI Matrice 200, Shenzhen, Guangdong) equipped with a megapixel camera (ZENMUSE X5S 20.8). The survey was conducted by the UAVs flying at altitudes of 40–80 m above the ground level and maintaining a stable speed of 2.5 m/s. The acquired images were analysed and processed using the photogrammetric PhotoScan software (Agisoft, St. Petersburg, Russia). Being constrained by 12 ground control points, the resulting orthorectified mosaic and digital elevation model (WGS 84 datum and UTM 32N projection) had a cell size of 5 cm/pixel and were deemed precise enough to be used for geomorphological analysis (Figure 5).

4.6. Data Processing and Cartography

The MBES data covered 500 km², with track lines parallel to the coast. The multi-beam data cleaning and filtering were performed using the PDS2000 software package, while the Global Mapper software was used to construct the bathymetric map of the UTM (WGS84) Zone 32 N projection. The bathymetry was plotted on a grid at 5 m node spacing as a contour plot to display detailed bathymetric information. It was also plotted as a slope value and an illuminated 3-D perspective view to visualise prominent features within the investigated area.

Sub-bottom profiler (SBP) data processing was performed using the Triton Elics Information suite software package. The navigation data were plotted in a geographic information system (GIS) application (DelphMap). We exported the processed seismic data in the GeoTIFF format. The side scan sonar (SSS) data processing provided the georeferenced grey-tone acoustic images of the seafloor at a resolution of 1 m.

Bathymetry was investigated by analysing the acquired multibeam data, while for the Sardinia emerged coastal sector, DTM with a 5×5 m cell size of the Autonomous Region of Sardinia was used.

High-resolution multibeam bathymetry was combined with the echo-types of chirp sonar data, documenting the high morphological complexity of the study area. The geomorphological map of the study area was created using ArcGIS by analysing and interpreting data at a scale of 1:5000 to obtain a highly detailed and accurate final map.

5. Results

The San Pietro Island continental shelf morphology presents a strong structural control, in accordance with the tectonic style of the passive continental margin of southwestern Sardinia. A system of normal faults, including low-angle faults, predominates and likely led to the evolution of both intraplatform and intraslope basins (Figures 2 and 4). However, on the proximal shelf, structural morphologies predominate and are often linked to volcanic processes. The distal shelf transition is abrupt and is represented by a normal fault system trending 40° N in the northern sector and N-S in the central-southern sector at depths of 80–140 m. The fault walls show the morphological evidence of a polycyclic evolution in the marine, coastal, and continental environments. The continuity of the rocky outcrops is interrupted by extensive areas with very low slopes, where the surface deposits are represented by medium-grained sands with bioclastic components and mäerl biogenic gravels. These deposits are affected by hydraulic dunes, and their granulometric features are highlighted in the backscatter side-scan sonar images. *Posidonia oceanica* is nearly absent in the western and northern coastal strips and is limited to small discontinuous areas, where seagrasses are visible on the rocks at 10–25 m depth. Starting from Punta delle Colonne and towards the east and northeast, a large *Posidonia oceanica* prairie almost completely colonizes the San Pietro Channel [57].

High rocky coasts dominate on San Pietro Island, and the highest cliffs characterize the coast exposed to the NW waves. During extreme marine events, the waves in this area reach a height of 10 m and a length of over 200 m. In the western sector (Sandalo Cape), plunging cliffs consisting compact lava rocks prevail, while pseudo-stratified pyroclastic volcanites and cliffs with abrasion platforms often masked by large subangular rockfall deposits or cones with subspheroidal boulders characterize the northern and southern coasts.

Off the fault walls, at depths of 150–170 m, a small intraplatform basin is filled with the onlapping Miocene sedimentary strata and features an isolated outcrop of volcanites. The distal platform has a very low slope located at depths of 170–200 m and composed of fine-grained sands. The sand pelitic component increases towards the open sea up to the net topographic convexity of the shelf edge located at an average depth of 220 m.

In a context dominated by volcanic and tectonic-controlled morphologies, we detected several sea-level and climate-change indicators dating back to the Upper Pleistocene and Holocene. As such, morphometric data refers to the palaeo-stages when a basal platform was located at depths of 125–135 m; morphometric and palaeontological data indicate the existence of a palaeo-lagoon at depths of 120–127 m; seismic data allows the identification of a buried palaeo-valley with a base level at a depth of 130 m; and side-scan sonar and petrographic data reveal the presence of beachrocks at depths of 45–48 m (Figure 7).

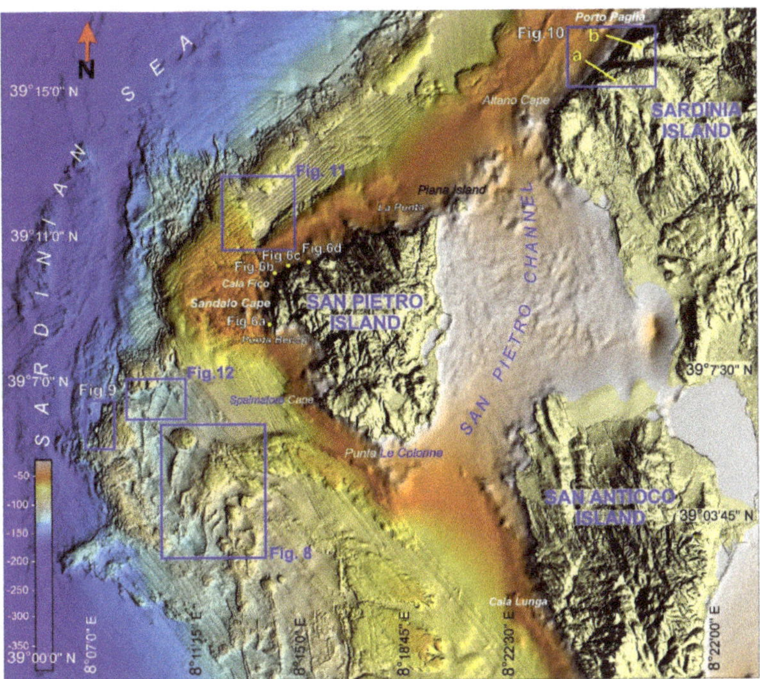

Figure 7. Location of sites depicted in the following figures.

5.1. Structural Landforms

Tectonic control on morphology is evident both in the coastline area and on the continental shelf. A fault-controlled slope affecting the Cala Vinagra comendites was recognised at the base of Punta di Cala Fico promontory, with an edge located at a depth 5 m and a base at depths of 15–25 m (Figure 6b). Two fault-controlled slopes with the same lithology were found 800 m off Sandalo Cape and Punta Becco, with an edge at 10 m depth and a base at 40 m depth. In the western sector, fault wall alignment is controlled by the tectonic lineaments trending N and 345° N, long between 1 and 3 km, revealing an organised subparallel pattern. In the northern sector, the fault walls follow a 60° N line for approximately 5 km in the same direction as the tectonic lines that control the present-day high coastline from Capo Altano to Porto Paglia (Figure 8a).

Structural surfaces, linked to the submerged ignimbrite bedrock, characterize the entire intermediate continental shelf. They are irregular and are covered by superficial sediments up to 12 km off the coast of Punta Spalmatore. The open-sea limit is represented by the edge of the palaeo-cliffs controlled by the N–S trending faults at 90 m depth. These surfaces are interrupted using their relief due to differential erosion, necks, and dikes. The structural surfaces that are slightly inclined towards SW characterise monoclinal "cuesta" reliefs found off the coast of Cala Lunga (Island of Sant'Antioco). Some sectors (e.g., off the coast of Cala Fico) present fault control. The distal platform at depths of 150–190 m presents the Miocene sedimentary sequence outcrop (Figure 3).

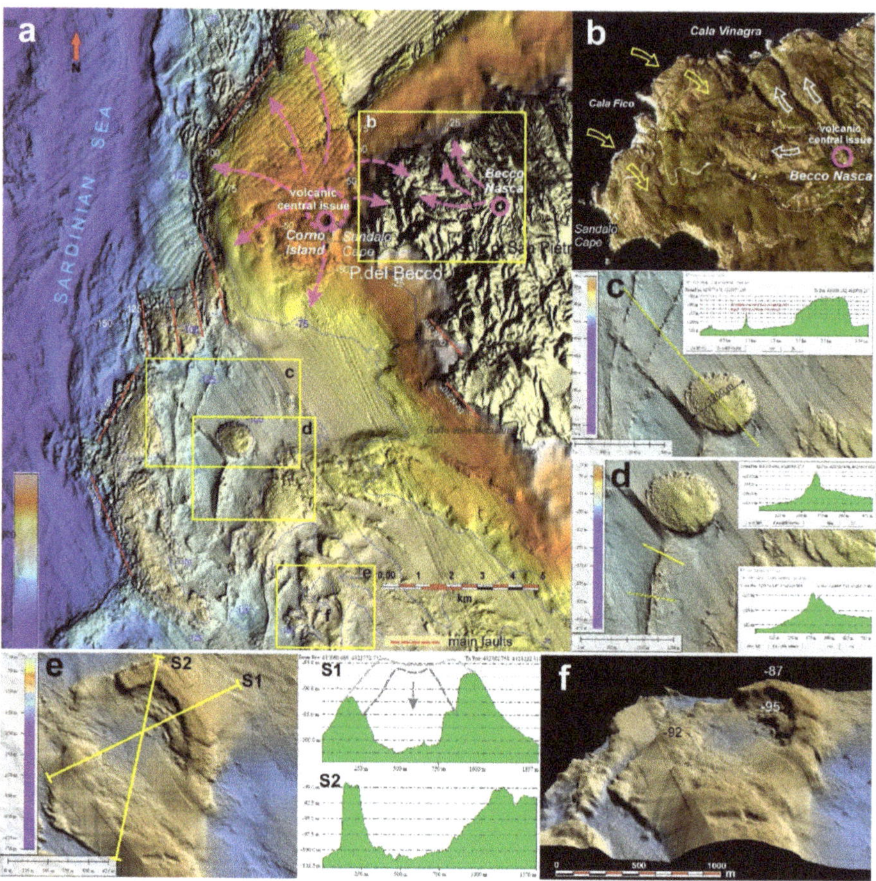

Figure 8. (**a**) Digital terrain model (DTM) from MBES data showing the drowned volcanic landforms in the continental shelf of San Pietro Island. The great volcanic edifice off Corno Island and the main fault sistems are highlighted. (**b**) Aerial photo of NW sector of San Pietro Island; the lava flow structures are evident from Becco Nasca source area (white arrows) and the lava structures from the volcanic edifice off Corno Island, currently submerged (yellow arrows). (**c**) DTM shadow relief and morphometric sections of a tabular volcanic structure (neck?). (**d**) DTM shadow relief and morphometric sections of mega-dikes, in relief due to differential erosion. These morphologies rise up to 12 m above the basal erosion surface, with longitudinal development up to 10 km. (**e**) DTM shadow relief and morphometric sections of a volcanic crater, showing a double collapse-rim structure. (**f**) Crater 3D model, the lowered internal flank and the central depression are evident.

5.1.1. Volcanic Landforms

The proximal portion of the continental shelf is dominated by the medium-Upper Miocene outcrops of acid volcanites. From a morphological point of view, numerous volcanic landforms were noted, including primary (e.g., craters or calderic depressions) and secondary landforms highlighted by differential erosion processes. The most important volcanic edifice on San Pietro Island was found on the continental shelf and occupies 37 km^2, expanding up to 6.5 km off Capo Sandalo (Figure 8a). The emission centre corresponds to the Islet of Corno, where the ignimbrite lavas (Cala Lunga Group, Middle Miocene) of Cala Vinagra were sampled. Marine erosion processes partially eroded summit members, including the Becco Nasca's comendites (Figure 8b). The residual deposit is represented by the hills of Capo Sandalo and Monte della Borrona, where the undulated morphologies of rope lavas attest that the lavas flowed towards SE (from the sea towards the interior of the island).

A larger emission centre was found 8 km from the Gulf of Mezzaluna. It presents a sub-circular and tabular mesa morphology with a diameter of approximately 1 km (Figure 8c). The relief basis is located at a depth of 100 m, and the top is found at 82 m depth. We interpret this landform as a volcanic neck; however, it should have been a huge volcanic chimney, the largest in Sardinia. This landform might also be interpreted as a volcanic plateau, similar to those recently recognised in the western Sardinian continental shelf [58], approximately 60 km north of the study area. These landforms are typical of basic volcanites and, therefore, appear unsuitable for the volcanic context of San Pietro Island. Regardless, this hypothesis requires further investigation. We attempted sampling the rocks of the volcanic neck by dredging. However, red algae bioconstructions completely covered the rocks and prevented us from sampling. A system of eight emission centres, showing neck morphologies, is distributed along a strip extending for 5 km to the SSE of the main volcanic edifice. A crater was found about 10 km off the coast of Punta Geniò, where only the eastern half-rim is preserved. It rises from the seabed at depths of 84–100 m. This volcanic edifice indicates two phases of activity (Figure 8e), and its morphology is similar to that of the volcanic features found on the seabed in front of the Phlegraean Fields, with their unlithified light grey pumiceous cinerite [59,60] (Figure 8f). A depression with a sub-circular perimeter and a diameter of 1200 m was found 1 km to the east the crater off the coast of Punta Geniò. The depression starts at a depth of 105 m and reaches 135 m. A depression with a similar morphology was classified as a caldera [61]. The entire group of emission centres following a tangential trend is crossed by a system of mega-dikes affected by differential erosion for more than 10 m [62]. The mega-dikes have a slightly sinuous form and can extend for up to 5 km without interruption, trending from 5° N to 350° N. The only exception is represented by a dike of considerable thickness, which follows a tangential trend near the main emission centre and is oriented NW-SE. This dike was likely emplaced subsequently to the N-S-trending dike system (Figure 8d). Such extensive mega-dikes are either contemporary or were formed immediately after the rifting phase [63]. The basement rocks are often draped by thin layers of mobile sediments, which partially cover the erosional landforms engraved in the volcanic substrate.

5.1.2. Palaeo-Cliffs and Related Landforms

The morphobathymetric DTM of the studied continental shelf shows a clear discontinuity between the proximal and distal shelves. The discontinuity follows the offshore limit of the volcanites. This limit is represented by the alignments of rocky walls up to 50 m high with evident tectonic control and a prevalent orientation of 340° N.

The fault walls were subjected to polycyclic processes due to variations in the eustatic sea level in the cliff environment. Judging by the depth of the basal platforms (125–140 m), the last phase of subaerial erosion can be attributed to the LGM sea-level fall. These palaeo-cliffs are set in the volcanites, with their base locally reaching a depth of 145 m, and form plunging cliffs, similar to the modern cliffs along the Sandalo Cape coast [64].

The palaeo-cliff base is predominantly located at the shore platforms and is often affected by iso-oriented shallow erosive channels in line with the main tectonic lineaments. The basal abrasion platform has an irregular shape and is frequently covered by large sub-angular blocks of multi-decametric dimensions or by rockfall deposits. Some large blocks were found hundreds of metres from the detachment areas and recalled the evolutionary model of the block slides diagnosed in other submerged areas, such as the continental sector of the Gulf of Cagliari [34], the southern Apulian margin off the coast of Santa Maria di Leuca [65], and the Malta continental shelf [3,29,66,67] (Figure 9).The palaeo-cliff surfaces are often sub-vertical and are affected by sub-vertical fracture systems, which run parallel to the main tectonic lineaments. In some areas, sub-orthogonal joints are present and probably represent columnar cooling fracturing, similar to that in the southern coastal sector of Sardinia. The cliff summit edges are developed at depths of 80–90 m, exhibiting a palaeo-cliff system 30–50 m high, on average. In some palaeo-cliff sectors, double ridges can be observed, pinpointing to the extensional trenches with counter-slope flanks. They were interpreted as distensional landforms and correlated with mass movement involving rotational kinematics (Figures 9b and 10).

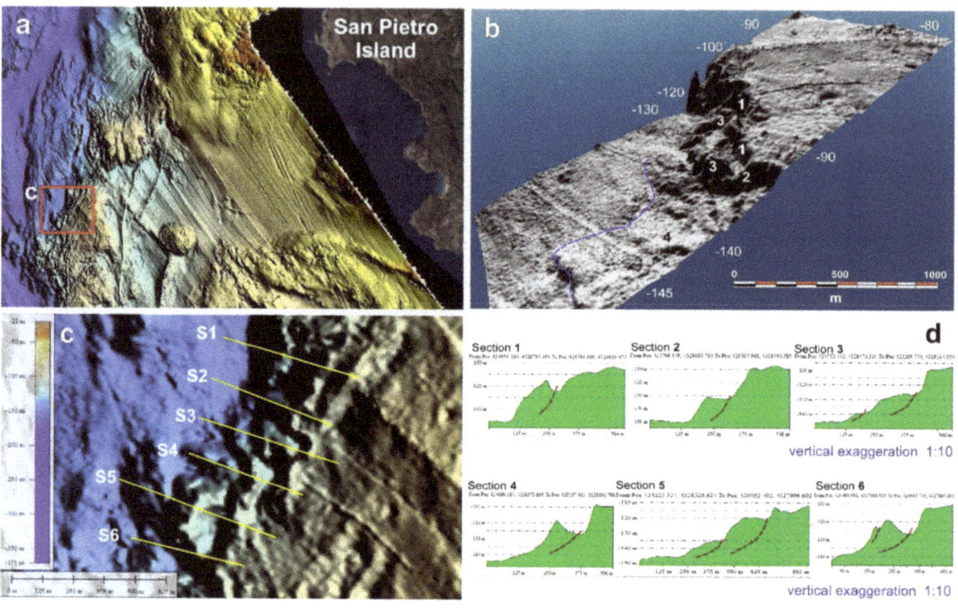

Figure 9. (**a**) DTM from MBES data showing the Last Glacial Maximum (LGM) palaeo-cliffs. (**b**) DTM 3D from MBES data showing some landslides affecting the submerged palaeo-cliffs: 1—main scarp; 2—distensional trench; 3—landslide bodies; LGM palaeo-sea-level (blue line). (**c**) detail of tectonic controlled palaeo-cliffs with localization of morphometric profiles. (**d**) Morphometric sections and hypothesized sliding surfaces (red lines).

The interpretation of the kinematics of these drowned-landslides was based on the geomorphological surveys of similar palaeo-landslides located along the coast (between Capo Altano and Porto Paglia. In this sector, large landslides with rotational kinematics were systematically observed [68]. The first landslide is located 500 m north of Altano Cape, while the second landslide has been recently found to the south of Porto Paglia (Figure 10A). Both palaeo-landslides have their foot fossilised by regression eolianites (MIS 4, MIS 3). Therefore, their movement likely occurred at a high sea-level stand during the last interglacial period (MIS 5) [9] (Figure 10D). From a morphological point of view, the first palaeo-landslide is distinguished by a complex detachment niche and is organized in two scarps. A wide trench and a counter-slope terrace are considerably lowered and are partially covered by colluvial deposits. This landslide shows the evidence of recent reactivation. The second palaeo-landslide has a detachment niche with a single scarp, a counter-slope terrace at the base of the niche, and a trench partially buried by collapsed blocks (Figure 10C,D).

In order to correlate the shapes of the submerged palaeo-cliffs with the modern (subaerial) cliffs and relate them to the landforms associated with rock falling and toppling, we carried out proximity remote sensing surveys by UAVs on some modern cliffs engraved in the same volcanic lithologies on Sant'Antioco Island and Altano Cape.

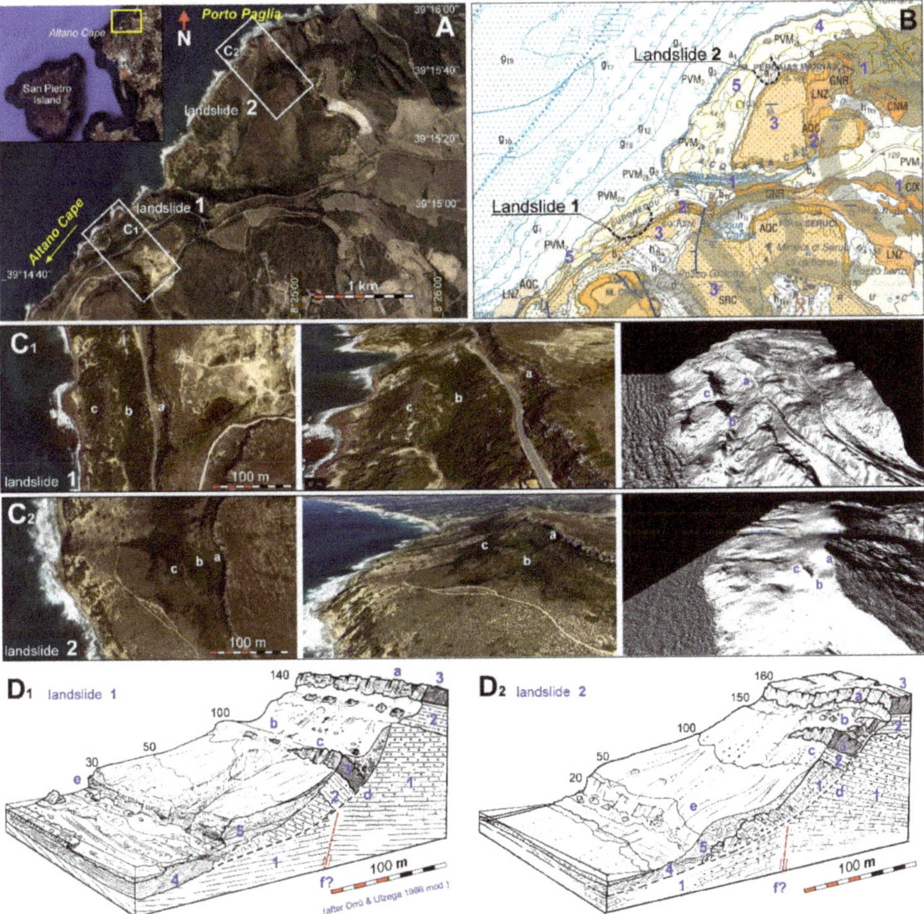

Figure 10. (**A**) Location of the studied coastal palaeo-landslides; (**B**) Excerpt of Geological Map of Italy—scale 1:50,000 Sheet 555 "Iglesias" (Pasci et al., 2015 [33]), with landslides location; (**C**) Aerial photo, photo 3D view and Lidar DTM 3D of landslide 1 (**C$_1$**) and landslide 2 (**C$_2$**); (**D**) Palaeo-landslide 1 3D interpretive models (**D$_1$**) and Palaeo-landslide 2 (**D$_2$**), showing rotational kinematics, probably due to the basal erosion during the high-stand MIS 5.5 and the subsequent foot fossilization by continental deposits of the Upper Pleistocene (MIS 4,3,2). Geolithological legend: (1) sandstones and conglomerates, Cixerri Formation—CIX (Eocene-Oligocene); (2) ignimbrites, tuffs—AQC (Middle Miocene); (3) ignimbrites, lavas—SRC (Middle Miocene); (4) eolianites (Upper Pleistocene—MIS 4-3?); (5) eolianites and colluvia—PVM (Upper Pleistocene—MIS2). Morphological legend: (a) detachment niche; (b) trench; (c) counter-slope terrace; (d) rotational sliding surface; (f) cliff engraved in the Pleistocene aeolian deposits that fossilize the landslide foot.

The ROV surveys allowed us to explore the palaeo-cliff morphology, particularly, the extensional trenches. We observed that erosional channels interrupted the continuity of the cliff, whereas the niches and hollows, formed by differential erosion, created the environments protected from the coelenterate colonies of *Corallium rubrum* (SDC—the biocoenosis of semi-dark caves). By contrast, the top surfaces of the cliffs were almost completely colonised by incrusting algae *Pseudolitophillum expansum* (coralligenous biocoenosis).

5.1.3. Fossil Palaeo-Valleys

The LGM palaeo-hydrographic network has only been partially recognised because the valley incisions are buried by very coarse-grained and gravelly bioclastic sands, especially the mäerl facies, which inhibit the penetration of the chirp elastic signal. The only buried palaeo-riverbed that was completely identified starts from the Ria di Cala Fico and is demarcated by a fault wall that continues for approximately one km offshore trending 280° N. The palaeo-riverbed top is located at depths of 5–10 m, and the base is demarcated at depths of 20–35 m (Figure 6b). Offshore, the palaeo-drainage system is deflected by an orthogonal fault system (Figure 11).

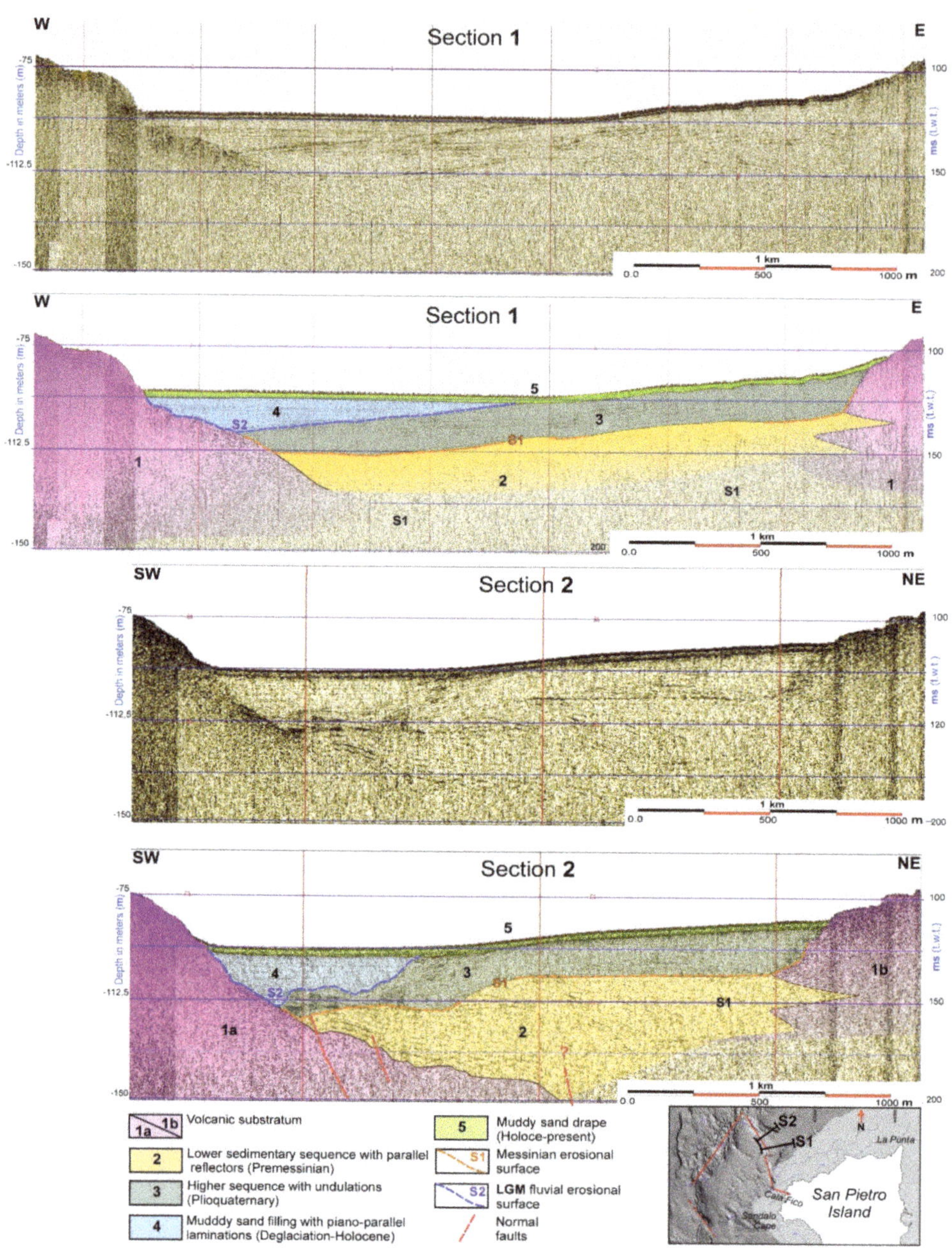

Figure 11. Seismic data of LGM palaeo-valley.

Beyond the intermediate platform with depths of 80–90 m, the morphobathymetric DTM shows a significant incision with steep rocky slopes, which is partially filled with sediments. This surface morphology possibly masks a buried palaeo-valley. To investigate this incision further, a Sparker 0.1–1 kJ seismic survey was planned and carried out. The survey proved the existence of a palaeo-valley, whose incision started in the Middle Miocene, immediately after the emplacement of the lava volcanites (Comenditi Cala Lunga Group). Subsequently, the incised valley was filled by marine sediments with inclined stratification and partially interbedded with the predominantly pyroclastic volcanites of the Middle-Upper Miocene (ignimbrites of the Cala Lunga Group) (Figure 11). The next erosive event was probably related to the Messinian low eustatic sea level [69]. In Plio-Quaternary, the palaeo-valley was completely filled by shallow-marine deposits with wavy laminations. The last identified incision down to a depth of 115 m occurred during the LGM sea-level low stand. The valley was filled by sediments originating from an environment with low wave energy, where low lighting caused a decrease in bioclastic productivity. These deep palaeo-valleys were discovered in a lower to Upper offshore environment due to the fast post-glacial to Holocene sea-level rise. In seismic images, the valley infill is represented by semi-transparent sandy mud alternating with more reflective sands of episodic storm nature (Figure 11—Section 2).

The terminal section of the palaeo-valley is enclosed within a narrow incision with walls approximately 10 m high, where we interpreted a submerged palaeo-delta at depths of 130–140 m.

5.1.4. Palaeo-Lagoon

Approximately 7 km off the coast of Punta Spalmatore, the palaeo-cliff is interrupted by a deep incision that is connected shorewards to a large depressed area at depths of 120–130 m. During the LGM sea-level low stand, this depressed area could have formed a Ria with a head bay lagoon. The palaeo-lagoon is asymmetrical, being characterised by a southern arched bank with a low slope and a deeper northern rectilinear bank (Figure 12a).

To verify the existence of this lagoon, we sampled the relevant deposits by dredging. We used a two-cylinder dredger and started from a depth of 125 m (lat 4329430.040 N, long 426914.014 E) to a depth of 129.8 m. The base was located close to the rocky wall (lat 4329755.209 N, long 426723.042 E), following a 200 m long cross-shaped church (Figure 12a,b).

The dredger sampled compact greenish-grey sandy silt, which, when washed, revealed a significant fossil content with both intact and fragmented lamellibranchs, gastropods, and serpulids (Figure 12b), marking the transition from the meso-littoral to the infra-littoral planes (Figure 12c) [30]. The sampled fauna comprised species common for lagoon and meso-littoral environments, such as the Bivalvia (e.g., *Mytilus* cfr. *Edulis*, *Mytilus* cfr. *Galloprovincialis*, *Glycymeris* sp., *Parvicardium exiguum*, *Pitar* cfr. *Rudis*, *Venus* cfr. *Casina*), Gastropoda (e.g., *Tectura virginea*, *Calliostoma laugieri*) (Figure 12d) [70], and Annelida (e.g., *Serpula vermicularis*).

Several samples were subjected to AMS (accelerator mass spectometry) ^{14}C radiocarbon analysis at the Beta Analytic laboratories (Florida, USA). The radiocarbon analysis results confirmed that the sampled rocks were deposited during MIS 2, at the beginning of the deglaciation period (Table 1) [71]. The dating of *Tectura Virginia* was the closest to that of LGM, and this species is still present in the Mediterranean Sea and some areas of the Aegean and North Adriatic Seas. The most consistent populations of *Tectura virginea* are currently present in the eastern Atlantic (e.g., Scotland, Iceland) and the North Sea (e.g., Norway, Svalbard).

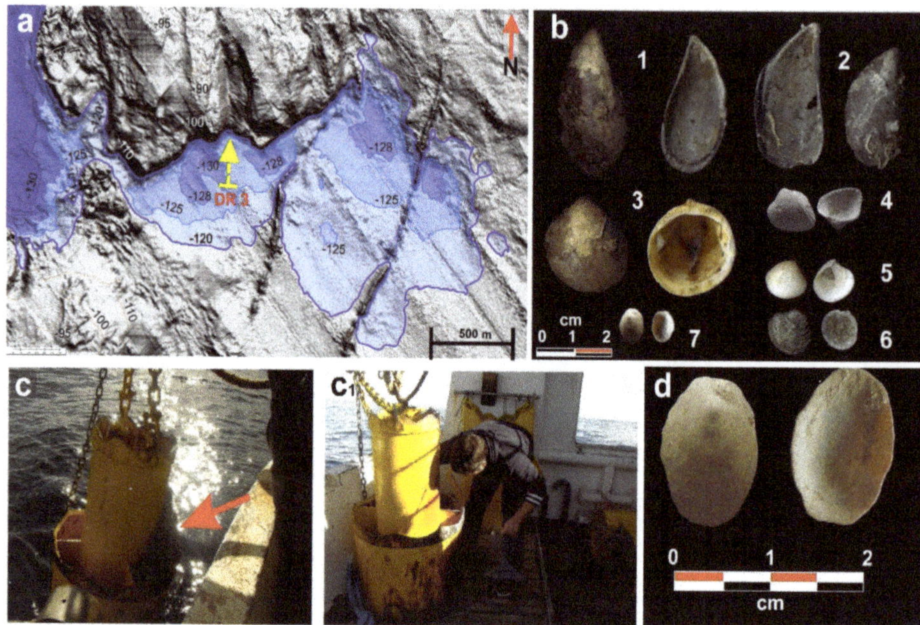

Figure 12. (**a**) Morphobatimetric DTM shadow relief showing the depression that hosted the palaeo-lagoon in the LGM; the arrow shows dredging (DR3); (**b**) sampled thanatocenosis. Bivalvia: (1) *Mytilus* cfr. *Edulis*, (2) *Mytilus* cfr. *Galloprovincialis*, (3) *Glycymeris* sp., (4) *Parvicardium exiguum*, (5) *Pitar* cfr. *Rudis*, (6) *Venus* cfr. *Casina*; Gasteropoda: (7) *Tectura virginea* (López Correa et al., 2010 [70] for the distribution of the *Tectura virginea* species). (**c**, **c₁**) Ascending two-cylinder experimental dredger; the arrow indicates part of the dark grey sandy silt deposit; (**d**) enlarged image of *Tectura virginea*.

Table 1. Radiocarbon dating results. The ^{14}C data were calibrated using an online version of Calib 8.1.0 (http://calib.org) and a standard marine reservoir age Bastia, Corse [71].

Lab. Code	Material	Species	^{13}C/^{12}C Ratio	Calibration Dataset	^{14}C Age (BP)	2-σ Interval of Calibrated Age (cal yr BP)
Beta-310989	shell	*Mytilus* cfr. *Galloprovincialis*	+1.7 o/oo	Calib 8.1.0	13,380 ± 60	15,014–15,688
Beta-310992	shell	*Tectura virginea*	+0.6 o/oo	Calib 8.1.0	16,350 ± 70	18,714–19,251

5.1.5. Beachrocks

Limited cemented conglomerate and sandstone outcrops were noted in the northern sector of the shelf at depths varying between 45 and 49 m (Figure 13a). These outcrops are represented by polygenic and heterometric conglomerates alternating with arenaceous microconglomerates with a feldspar-quartz matrix. The fossiliferous content is high, predominantly including lamellibranchs and gastropods, with the evidence of radiolarians and echinoids. Carbonate cementation was determined to be polyphasic, with the initial formation of magnesian calcite cement precipitated from seawater in the form of acicular crystals and followed by the cryptocrystalline globules deposited via bio-precipitation. After partial dissolution, the cementation was completed by the idiomorphic crystals of calcite deposited from freshwater. The dynamics of cementation are linked to the sea-level oscillations during an overall sea-level rise (Figure 13d). These deposits show depositional characteristics and cementation typical of a beachrock and, consequently, can be referred to as palaeo-submerged shorelines. They define the exact position of an intertidal zone. The main beachrock deposits were observed and mapped at the same depth, approximately

3 km NW of the La Punta promontory. The side-scan sonar images showed weakly inclined banks dipping seaward. Owing to the gradual basal undermining of these beachrock outcrops by strong traction currents, extensive plains are covered by unconsolidated sediments (sand) distributed in patches (Figure 13a). The top outcrop surface is denoted by a typical sub-orthogonal fracture system linked to diagenesis, favouring the occurrence of landslides at the edges (Figure 13b,c).

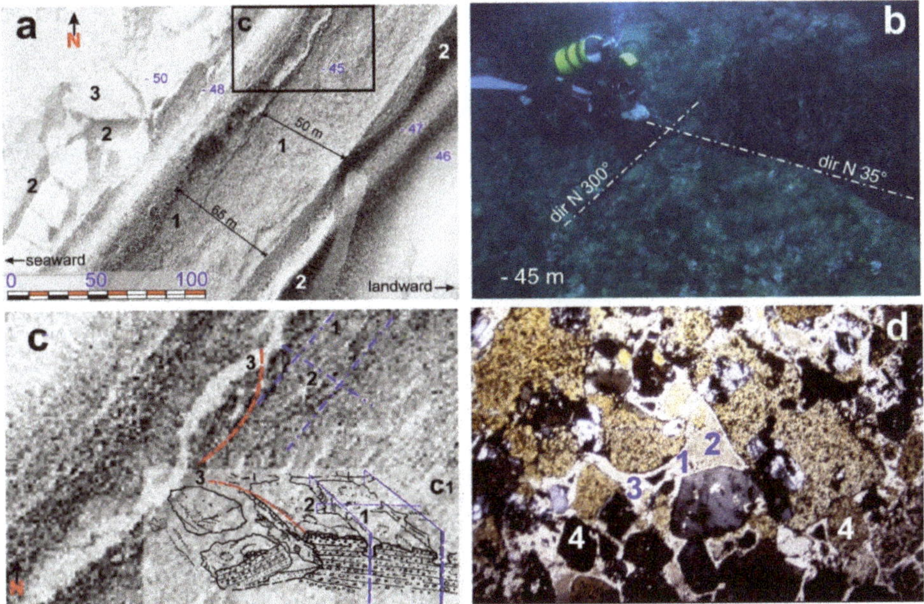

Figure 13. (**a**) The 100 kHz side scan sonar image of the beachrock a depth of 45 m: (1) Outcrop of sandstones and micro-conglomerates with carbonate cement; (2) sandpatches sedimentary structures (mäerl); (3) medium and fine grained bioclastic sands. (**b**) Seabed picture a depth of 45 m, underwater survey of beachrock fracture system. (**c**) Detail of the beachrock outcrops affected by a sub-orthogonal fracture system and by block toppling due to basal erosion: 1—35° N fracture system; 2—300° N fracture system; 3—detachment niche; (**c₁**) block diagram by diving survey. (**d**) Petrographic thin section of the beachrock—45 m: (1) acicular magnesian calcite coating: (2) micritic globular filling; (3) dissolution cavity with secondary clastic-micritic filling; (4) accretion of idiomorphic calcite in dissolution cavity. N.I. × 20.

6. Discussion

6.1. Last Glacial Maximum (LGM) Coastal Palaeo-Landscape

The collected data allowed us to identify the geomorphological evidence of a drowned palaeo-landscape attributable to a base level at approximately 125–145 m depth below the present-day sea level.

Continental landforms (e.g., river valleys and coastal plains), transitional environments (e.g., palaeo-lagoons and coastal landforms represented by cliffs), and associated depositional features were recognized. Different marine indicators observed at various depths helped identify a palaeo-sea-level, suggesting that the drowned palaeo-landscape formed between the LGM and early deglaciation stages [22,24]. These observations testify to the particular mobility of this continental shelf to hydro-isostatic rebounds [12], similar to those reported from the northern sector of San Pietro Island, from Fontanammare Bay to Capo Frasca (Figure 14) [54]. In our study area, the MIS 5 palaeo-sea-level indicators show relative tectonic stability [72,73], while those modelled by GIA on the southwestern Sardinian continental shelf display lower vertical displacement rates, with glacio-hydro-isostasy constituting 0.62 mm/yr off the coast of Oristano and 0.60 mm/yr in the Gulf of Palmas [12]. Therefore, the theoretical sea-level drop attained during LGM (−120 m) could be extended by approximately 10 m (−130 m). Some research carried out on the continental shelf between Capo Pecora and Oristano indicates that the LGM palaeo-shoreline was at depths of 125–140 m [54] (Figure 14).

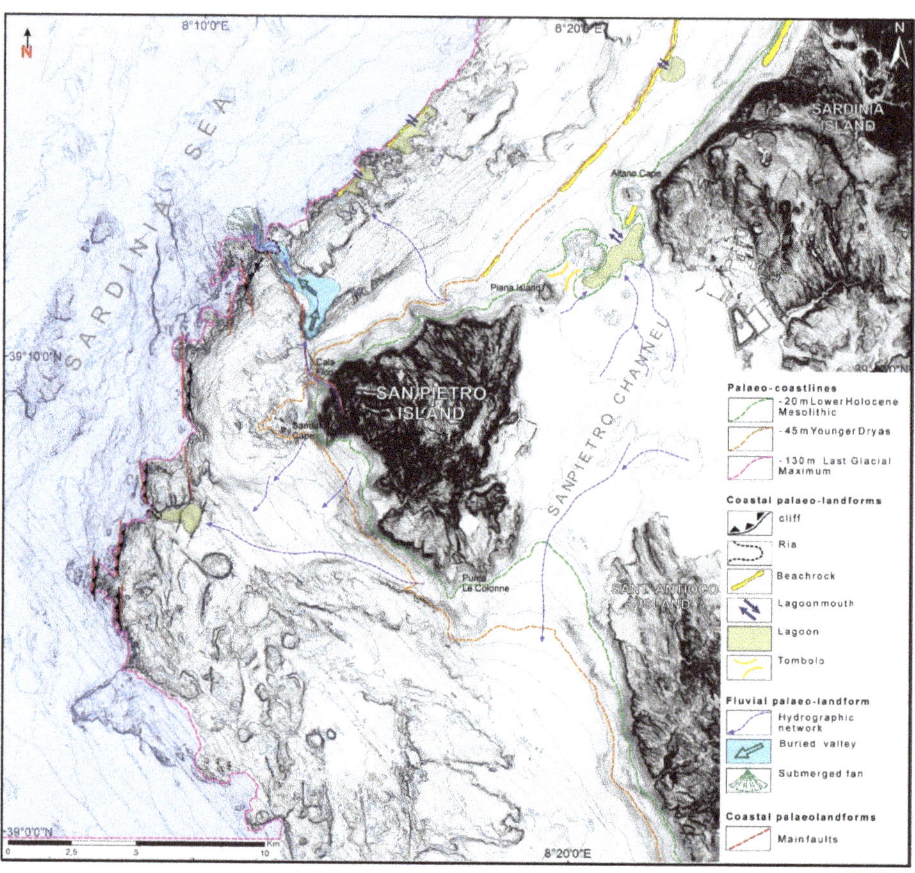

Figure 14. Geomorphological sketch of the San Pietro continental shelf. Submerged palaeo-landscape since LGM (20 ka) to 9 ka.

The LGM coastal palaeo-landscape of the study area can be subdivided into three sectors. The western sector is dominated by cliffs affected by landslides, where several large isolated blocks represent the islets and a deep Ria interrupts the cliffs' continuity and incises the head bay lagoon. The northwestern sector represents a high rocky coast with a gentle slope, probably coinciding with the lava flow fronts originating from the emission centre of Corno Island. An extensive basal wave-cut platform is situated at the base of these rocky coasts and affected by a dense system of iso-oriented erosional channels, with a prevalent direction of 330° N. The northern sector of Altano Cape is characterized by a low rocky coast with a series of beachrocks at depths of 115–120 m. It is located adjacent to two back-littoral areas with palaeo-lagoons. These beachrocks lie in a sub-parallel way and have a morphological response identical to MBES, in contrast to the beachrocks sampled at a depth of 45 m in the same sector (Figure 14).

6.2. Post-Glacial Palaeo-Landscape Evolution

Rapid sea-level rise from 130 m below the present-day sea level during the deglaciation period [22,65] likely contributed to the destabilisation of the palaeo-cliffs [74], with their geomechanical characteristics being worsened by periglacial processes. Such destabilisation could explain many massive rockfall deposits that currently cover the basal abrasion platforms.

The comparison between the geomorphological features of the subaerial coastal palaeo-landslides (Figure 10) and the identified submerged palaeo-landslides (Figure 9) has shown that both types of landslides were affected by rotational kinematics. However, the submerged palaeo-landslides have low-angle sliding surfaces, and their landslide bodies are stacked (Figure 9c), dissimilar to the present-day landslides. These differences can be linked to instability after drowning, in wave energy conditions probably very different from the present ones [75,76]. To better understand the coastal landslides kinematics in this area, it is important underline that the waves currently interacting on the coasts of western Sardinia are the most energetic in the entire Mediterranean basin. In fact, the waves measured in Alghero (northwestern Sardinia) during extreme meteorological events have a maximum height of over 10 m [77,78].

6.3. Younger Dryas Coastal Landscape

At depths of 45–48 m, we diagnosed a littoral spit in the beachrock facies. It is well preserved and extends for approximately 15 km in the northern part of the study area (from La Punta to offshore of Capo Altano). The outcrops are represented by polygenic and heterometric conglomerates with a sandy matrix and carbonate cement. The fossil content varies, and fully bioclastic levels were observed. Rock cementation indicates that a palaeo-depositional environment changed from intertidal to supratidal [52]. The outcrops are characterised by apparent erosional landforms both on the top surface and the edges (Figure 13a).

The strata are tilted slightly seaward, representing an arrangement typical of the beach sedimentary body [51]. The sedimentary structures (e.g., parallel lamination and wedge-shaped, sigmoidal, and inclined stratification) are also common in coastal environments (Figure 13b).

In Sardinia, beachrocks are found at different bathymetric levels. The deepest beachrocks are located at 95–110 m depth off La Maddalena Island, whereas the shallowest beachrocks are found at a depth of 1 m in the Gulf of Palmas and incorporate Roman pottery [30]. The latter type of outcrop constitutes thin (1–2 m) and discontinuous strata.

The beachrocks found at depths of 45–50 m are particularly thick (4–5 m) and continuous. They start in southern Sardinia, pass along eastern Sardinia, and reach the Aléria Platform in central-eastern Corsica. These littoral spits are often associated with retro-littoral areas (i.e., palaeo-lagoons and palaeo-dunes). The same beachrocks were sampled at a depth of 45 m in the Gulf of Palmas, Gulf of Cagliari, Island of Serpentara, and Gulf of Orosei. The ^{14}C analysis indicated that the beachrock ages range between 11 and 9.5 ky cal BP. In particular, the Cagliari beachrock revealed a date of 10,835 $\pm$ 170 ky cal BP [51]. As such, these palaeo-shorelines are attributed to the Younger Dryas cycle of eustatic oscillations. Off the Porto Paglia coast, the beachrock is interrupted, and the area behind it is distinguished by a sub-elliptical depression, interpreted as a palaeo-lagoon approximately 1 km in diameter (Figure 14).

6.4. Lower Holocene Coastal Palaeo-Landscape

We reconstructed the LGM coastal palaeo-landscape corresponding to the Upper Palaeolithic period. During this period, the evidence of human presence is still rare in Sardinia [79,80] (Figure 14), despite the short distance from the continent and the continuity with Corsica due to the lowered sea level during the LGM. [56]. The first evidence of human settlement in Sardinia is attributed to the Holocene [81]. The discovery of the Mesolithic site of S'Omu e S'Orku (SOMK) along the southwestern Sardinian coast, about

40 km from San Pietro Island, is of particular interest, being one of the few Mesolithic coastal sites in the western Mediterranean (Figure 15). By examining the sea-level rise curve [15,16] (Figure 16), we placed the ancient Holocene shoreline at 20 m depth. In the Holocene, the palaeo-landscape represented a vast coastal plain that extended between the present-day islands of San Pietro and Sant'Antioco and the mainland. High rocky coast was interspersed with extensive beaches with coastal dunes, as evidenced by a strip of coastal desert near Capo Altano [9]. A narrow bay was bordered by two rocky promontories between the islands of San Pietro and Capo Altano, while instead of the San Pietro canal, there existed a large bay and river mouth (Figure 14). The coastal morphological context of this part of Sardinia probably influenced the movements of the last Mesolithic groups. The human remains discovered in SOMK were dated to approximately 9 ky cal BP. The site is completely covered by red ocher and jasper artefacts (Figure 15b), the outcrops of which are found on San Pietro Island [81]. These artefacts testify to the movement of Mesolithic groups along wide beach areas close to the reliefs. Those Mesolithic groups benefited from the emerged land area between the mainland and the two small islands, where coastal lagoons favoured not only the mining of jasper and ocher but also offered food resources from sea and lagoon, especially in a period, when relatively scarce terrestrial fauna existed. The terrestrial fauna was mainly represented by the now-extinct genus *Prolagus* [82].

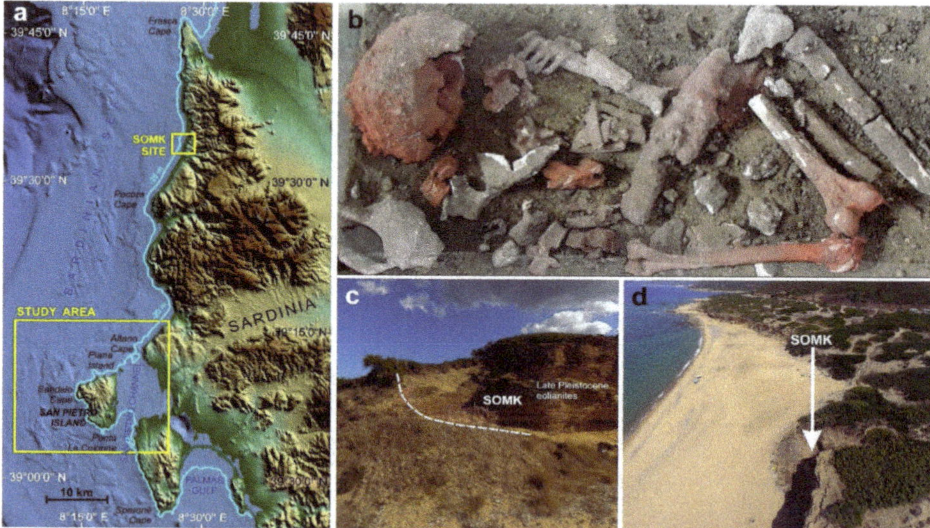

Figure 15. Mesolithic man in Sardinia west coast: (**a**) Location of S'Omu e S'Orku (SOMK) site and study area, palaeo-coastline of Mesolithic period (9 kyr BP) at 20 m. (**b**) The heavily ochre-stained skeletal remains (after Melis and Mussi, 2016 [81]). (**c**) View of SOMK Mesolithic site. (**d**) General view of the present-day position of Mesolithic SOMK site along the Sardinian west coast.

7. Conclusions

The integrated study of new geomorphological, seismic, MBS ultrasound, direct and remote-sensing UAV data allowed the evolution of the coastal palaeo-landscape of the continental shelf off San Pietro Island (southwestern Sardinia) since the LGM to be reconstructed. We found robust evidence that during LGM, the sea level was approximately 130 m depth below the present-day level, in agreement with GIA model indications (Figure 17) [12].

The morphostratigraphic investigation carried out around San Pietro Island allowed recognizing sea-floor features and landforms related to different sea-level stands during the last 22 kyr. In particular, the research highlighted the following (Figure 14):

- During the LGM the central-southern sector of the investigated area was characterized by a large promontory showing tectonically controlled high rocky coasts affected by intense fracturing and rotational landslides.
- Deep rias, set along faults, interrupt the cliffs' continuity during the same period. A large lagoon, with a peri-littoral thanatocenosis that hosted species of cold waters, formed at the bottom of one of these Ria bays (Figure 12).
- In the northern sector was there a wide river valley, while continuing towards north -east the coast became low and sandy with lagoon-barrier systems (LGM).
- The rapid eustatic sea-level rise during deglaciation, probably associated with extreme weather and sea conditions, has favoured the development of rotational landslides and debris avalanches on the cliffs (Figure 9).
- During the Younger Dryas, the eustatic oscillations between the depths of 50 and 45 m led to the construction of very thick littoral spits that extended continuously for tens of kilometers to the north (beachrocks).
- During the Holocene, the landscape comprised a vast coastal plain with lagoons that extended between the present-day islands of San Pietro and Sant'Antioco and the mainland. This landscape played an important role in the movements of Mesolithic groups along the coast. In fact, the presence of Mesolithic burials (about 40 km away), including ocher and jasper artifacts from the Island of San Pietro, shows that the Mesolithic inhabitants could reach the jasper outcrops by walking along a coast characterized by long beaches, lagoons and back-littoral dunes (Figure 15).
- Volcanic morphologies, currently not present onshore, have also been described for the first time. In particular, alignments of mega-dikes similar to those of active rift areas (Figure 8) [62].

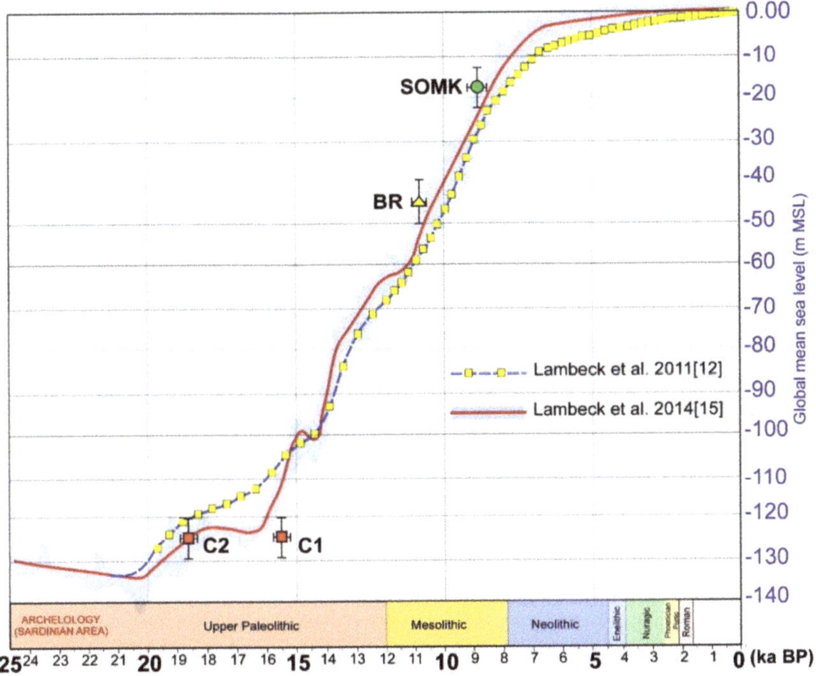

Figure 16. Relative sea-level prediction curve after Lambeck et al., 2011 [12] and Lambeck et al., 2014 [15], associated with palaeo-sea-levels indicators of San Pietro continental shelf: lagunar shells, (C1) *Acmea virginea* (gasteropoda)—18,982 ± 338; (C2) *Mytilus galloprovincialis* (Bivalvia)—15,350 ± 338 yr cal BP; BR) beachrock—10,835 ± 170 ky cal BP (De Muro and Orrù, 1998 [51]); SOMK) Mesolithic site (Melis and Mussi, 2016 [81]).

In literature, LGM palaeo-shorelines are normally investigated by focusing on single research aspects, such as seismic stratigraphy [20,21,27], lowstand depositional terraces [23] and high rocky coast evolutionary models [22]. Within our research, an effort was made to integrate a series of datasets including geological data and geomorphological evidence from both emerged and submerged areas of southwestern Sardinia. This made it possible to reconstruct the drowned palaeo-landscape in its complexity providing the means to infer evolutionary phases from the deglaciation to the present. With reference to the continental margin, the research allowed for the first time the description of geomorphological features and palaeo-landscapes associated with the LGM shoreline. Furthermore, chronological constraints for the development of peculiar landforms were achieved, thanks to the dating of correlative fossiliferous deposits.

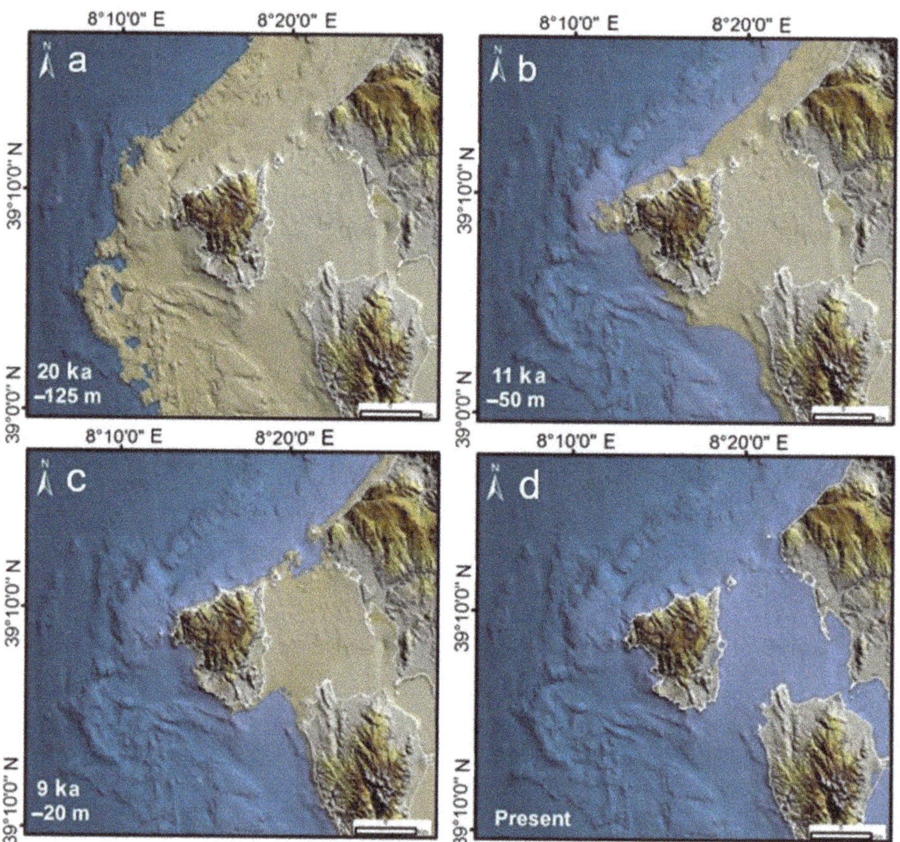

Figure 17. Reconstruction of the palaeo-landscape evolution of the San Pietro continental shelf since the LGM to present: (**a**) LGM; (**b**) Younger Dryas; (**c**) Holocene, Mesolithic; (**d**) Present.

Author Contributions: Conceptualization, G.D., L.L., M.S. and P.E.O.; methodology, G.D. and V.D.; validation, R.T.M.; formal analysis, V.D.; investigation, G.D., P.E.O. and V.D.; resources: G.D. and P.E.O.; writing—original draft preparation, G.D., L.L., R.T.M., M.S. and P.E.O.; writing—review and editing, L.L., M.S., V.D. and P.E.O.; visualization, G.D., P.E.O. and V.D.; supervision, P.E.O. and M.S.; project administration, P.E.O. and M.S.; funding acquisition, P.E.O., M.S. and R.T.M. All authors have read and agreed to the published version of the manuscript.

Funding: The study was carried out in the frame of the Project "MAGIC Marine Geohazard along Italian Coasts" funded by Italian National Civil Protection (Resp.: F.L. Chiocci—Resp. CoNISMa Unit: P.E. Orrù), of the CARG Project Geological Map of Italy, Scale 1:50,000 (Marine area)—Sheet 563 "Isola di San Pietro" and Sheet 555 "Iglesias" (Resp.: P.E. Orrù), of the SOMK Parco Geominerario 150-29.12.2017 and FdS grant number F74I19000960007 "Geogenic and anthropogenic sources of minerals and elements: fate and persistency over space and time in sediments" (Resp.: Rita Teresa Melis). The research is also part of the Project "Coastal risk assessment and mapping" funded by the EUR-OPA Major Hazards Agreement of the Council of Europe (2020–2021). Grant Number: GA/2020/06 no 654503 (Unimore Unit Resp.: Mauro Soldati).

Institutional Review Board Statement: Not applicable.

Informed Consent Statement: Not applicable.

Data Availability Statement: Not applicable.

Acknowledgments: We are thankful to Margherita Mussi for her precious suggestions about human heritage and georcheological implications. We are also grateful to Soprintendenza Archeologica di Cagliari for their valued support. The precious contribution of the three anonymous reviewers and of the journal editors is acknowledged.

Conflicts of Interest: The authors declare no conflict of interest.

References

1. Shackleton, N.J. The 100,000-year ice-age cycle identified and found to lag temperature, carbon dioxide and orbital eccentricity. *Science* **2000**, *289*, 1897–1902. [CrossRef] [PubMed]
2. Micallef, A.; Foglini, F.; Le Bas, T.; Angeletti, L.; Maselli, V.; Pasuto, A.; Taviani, M. The submerged palaeolandscape of the Maltese Islands: Morphology, evolution and relation to Quaternary environmental change. *Mar. Geol.* **2013**, *335*, 129–147. [CrossRef]
3. Prampolini, M.; Foglini, F.; Biolchi, S.; Devoto, S.; Angelini, S.; Soldati, M. Geomorphological mapping of terrestrial and marine areas, northern Malta and Comino (central Mediterranean Sea). *J. Maps* **2017**, *13*, 457–469. [CrossRef]
4. Prampolini, M.; Savini, A.; Foglini, F.; Soldati, M. Seven good reasons for integrating terrestrial and marine spatial datasets in changing environments. *Water* **2020**, *12*, 2221. [CrossRef]
5. Tallavaaraa, M.; Miska, L.; Korhonenc, N.; Järvinend, H.; Seppäb, H. Human population dynamics in Europe over the Last Glacial Maximum. *Proc. Natl. Acad. Sci. USA* **2015**, *112*, 8232–8237. [CrossRef]
6. Benjamin, J. Submerged prehistoric landscapes and underwater site discovery: Reevaluating the 'danish model' for international practice. *J. Island Coast. Archaeol.* **2010**, *5*, 253–270. [CrossRef]
7. Lecca, L.; Carboni, S.; Scarteddu, R.; Sechi, F.; Tilocca, G.; Pisano, S. Schema stratigrafico della piattaforma continentale occidentale e meridionale della Sardegna. *Mem. Soc. Geol. Ital.* **1986**, *36*, 31–40.
8. Orrù, P.; Ulzega, A. Carta geomorfologica della piattaforma continentale e delle coste del Sulcis—Sardegna sud occidentale. Scala 1:100,000. *STEF* **1990**, (in press).
9. Orrù, P.; Ulzega, A. Geomorfologia costiera e sottomarina della baia di Funtanamare (Sardegna sud-occidentale). *Geogr. Fis. Din. Quat.* **1986**, *9*, 59–67.
10. Miccadei, E.; Orru, P.E.; Piacentini, T.; Mascioli, F.; Puliga, G. Geomorphological map of the Tremiti Islands (Puglia, Southern Adriatic Sea, Italy), scale 1:15,000. *J. Maps* **2012**, *8*, 74–87. [CrossRef]
11. Foglini, F.; Prampolini, M.; Micallef, A.; Angeletti, L.; Vandelli, V.; Deidun, A.; Soldati, M.; Taviani, M. Late Quaternary coastal landscape morphology and evolution of the Maltese Islands (Mediterranean Sea) reconstructed from high-resolution seafloor data. In *Geology and Archaeology: Submerged Landscapes of the Continental Shelf*; Harff, J., Bailey, G., Lüth, L., Eds.; Special Publication; Geological Society: London, UK, 2016; Volume 411, pp. 77–95.
12. Lambeck, K.; Antonioli, F.; Anzidei, M.; Ferranti, L.; Leoni, G.; Scicchitano, G.; Silenzi, S. Sea level change along italian coast during Holocene and a projection for the future. *Quat. Int.* **2011**, *232*, 250–257. [CrossRef]
13. Stocchi, P.; Spada, G. Glacio and hydro-isostasy in the Mediterranean Sea: Clark's zones and role of remote ice sheets. *Ann. Geophys.* **2007**, *50*, 741–761.
14. Antonioli, F.; Ferranti, L.; Fontana, A.; Amorosi, A.; Bondesan, A.; Braitenberg, C.; Dutton, A.; Fontolan, G.; Furlani, S.; Lambeck, K.; et al. Holocene relative sea-level changes and vertical movements along the Italian and Istrian coastlines. *Quat. Int.* **2009**, *206*, 102–133. [CrossRef]
15. Lambeck, K.; Rouby, H.; Purcell, A.; Sun, Y.; Sambridge, M. Sea level and global ice volumes from the last glacial maximum to the Holocene. *Proc. Natl. Acad. Sci. USA* **2014**, *111*, 15296–15303. [CrossRef] [PubMed]
16. Rovere, A.; Stocchi, P.; Vacchi, M. Eustatic and relative sea level changes. *Curr. Clim. Chang. Rep.* **2016**, *2*, 221–231. [CrossRef]
17. Mann, T.; Bender, M.; Lorscheid, T.; Stocchi, P.; Vacchi, M.; Switzer, A.; Rovere, A. Relative sea-level data from the SEAMIS database compared to ICE-5G model predictions of glacial isostatic adjustment. *Data Brief* **2019**, *27*, 112–125. [CrossRef] [PubMed]

18. Carminati, E.; Doglioni, C. Mediterranean tectonics. In *Encyclopedia of Geology*; Elsevier: Amsterdam, The Netherlands, 2005; pp. 135–146.

19. Gattacceca, J.; Deino, A.; Rizzo, R.; Jones, D.S.; Henry, B.; Beaudoin, B.; Vadeboin, F. Miocene rotation of Sardinia: New paleomagnetic and geochronological constraints and geodynamic implications. *Earth Planet. Sci. Lett.* **2007**, *258*, 359–377. [CrossRef]

20. Maselli, V.; Trincardi, F. Large-scale single incised valley from a small catchment basin on the western Adriatic margin (central Mediterranean Sea). *Glob. Planet. Chang.* **2013**, *100*, 245–262. [CrossRef]

21. Sikora, M.; Mihanović, H.; Vilibić, I. Palaeo-coastline of the Central Eastern Adriatic Sea, and Palaeo-Channels of the Cetina and Neretva rivers during the last glacial maximum. *Acta Adriat.* **2014**, *55*, 3–18.

22. Zecchin, M.M.; Ceramicola, S.; Lodolo, E.; Casalbore, D.; Chiocci, F.L. Episodic rapid sea-level rises on the central Mediterranean shelves after the last glacial maximum: A review. *Mar. Geol.* **2015**, *369*, 212–223. [CrossRef]

23. Chiocci, F.L.; Romagnoli, C. Terrazzi deposizionali sommersi nelle Isole Eolie (Sicilia). In *Atlante dei Terrazzi Deposizionali Sommersi Lungo le Coste Italiane. Memorie Descrittive Della Carta Geologica d'Italia*; Chiocci, F.L., D'Angelo, S., Romagnoli, C., Eds.; APAT: Rome, Italy, 2004; Volume 58, pp. 81–114.

24. Caruso, A.; Cosentino, C.; Pierre, C.; Sulli, A. Sea level change during the last 41 ka in the outher shelf of southern Tyrrhenian Sea. *Quat. Int.* **2011**, *232*, 122–131. [CrossRef]

25. Casalbore, D.; Falese, F.; Martorelli, E.; Romagnoli, C.; Chiocci, F.L. Submarine depositional terraces in the Tyrrhenian Sea as a proxy for paleo-sea level reconstruction: Problems and perspective. *Quat. Int.* **2017**, *439*, 169–180. [CrossRef]

26. Lo Presti, V.; Antonioli, F.; Palombo, M.R.; Agnesi, V.; Biolchi, S.; Calcagnile, L.; Di Patti, C.; Donati, S.; Furlani, S.; Merizzi, J.; et al. Palaeogeographical evolution of the Egadi Islands (western Sicily, Italy). Implications for late Pleistocene and early Holocene sea crossings by humans and other mammals in the western Mediterranean. *Earth. Sci. Rev.* **2019**, *194*, 160–181. [CrossRef]

27. Pepe, F.; Bertotti, G.; Ferranti, L.; Sacchi, M.; Collura, A.M.; Passaro, S.; Sulli, A. Pattern and rate of post-20 ka vertical tectonic motion around the Capo Vaticano Promontory (W Calabria, Italy) based on offshore geomorphological indicators. *Quat. Int.* **2014**, *332*, 85–98. [CrossRef]

28. Furlani, S.; Antonioli, F.; Biolchi, S.; Gambin, T.; Gauci, R.; Lo Presti, V.; Anzidei, M.; Devoto, S.; Palombo, M.; Sulli, A. Holocene sea level change in Malta. *Quat. Int.* **2013**, *288*, 146–157. [CrossRef]

29. Prampolini, M.; Foglini, F.; Micallef, A.; Soldati, M.; Taviani, M. Malta's submerged landscapes and landforms. In *Landscapes and Landforms of the Maltese Islands. World Geomorphological Landscapes*; Gauci, R., Schembri, J.A., Eds.; Springer: Cham, Switzerland, 2019; pp. 117–128.

30. Orrù, P.E.; Deiana, G.; Taviani, M.; Todde, T. Palaeoenvironmental reconstruction of the Last Glacial Maximum coastline on the San Pietro continental shelf (Sardinia SW). *Rend. Online Soc. Geol. Ital.* **2012**, *21*, 1182–1184.

31. Cau, A.; Follesa, M.C.; Cannas, R.; Sacco, F.; Orrù, P.E.; Deiana, G.; Todde, S.; Cau, A.; Enrico, P. Preliminary data on habitat characterization relevance for red coral conservation and management. *Ital. J. Geosci.* **2013**, *134*, 60–68. [CrossRef]

32. Cau, A.; Follesa, M.C.; Moccia, D.; Alvito, A.; Bo, M.; Angiolillo, M.; Canese, S.; Paliaga, M.E.; Orrù, P.E.; Sacco, F.; et al. Deepwater corals biodiversity along roche du large ecosystems with different habitat complexity along the south Sardinia continental margin (CW Mediterranean Sea). *Mar. Biol.* **2015**, *162*, 1865–1878. [CrossRef]

33. Pasci, S.; Pertusati, P.C.; Salvadori, I.; Medda, F.; Murtas, A.; Rizzo, R.; Uras, V.; Orrù, P.E.; Deiana, G.; Puliga, G. *Geological Map of Italty. Scale 1:50,000. Sheeet 555 "Iglesias"*; ISPRA-Servizio Geologico Nazionale: Roma, Italy, 2015.

34. Deiana, G.; Meleddu, A.; Paliaga, E.; Todde, S.; Orrù, P. Continental slope geomorphology: Landslides and pockforms of Southern Sardinian margin (italy). *Geogr. Fis. Dinam. Quat.* **2016**, *39*, 129–136.

35. Rizzo, R.; Garbarino, C.; Salvadori, I.; Patta, D.; Orrù, P.E.; Deiana, G.; Pulga, G. *Geological Map of Italty. Scale 1:50,000. Sheeet 563 "Isola di San Pietro"*; ISPRA-Servizio Geologico Nazionale: Roma, Italy, 2016.

36. Lecca, L. La piattaforma continentale miocenico-quaternaria del margine occidentale sardo: Blocco diagramma sezionato. *Rend. Sem. Fac. Sci. Univ. Cagliari.* **2000**, *1*, 49–70.

37. Finetti, I.; Morelli, C. Geophysical Exploration of the Mediterranean Sea. *Boll. Geof. Teor. Appl.* **1973**, *60*, 263–342.

38. Ryan, W.B.F.; Hsü, K.J.; Cita, M.B.; Dumitrica, P.; Lort, J.; Maync, W.; Nesteroff, W.D.; Pautot, G.; Stradner, H.; Wezel Forese, C. Boundary of Sardinia Slope with Balearic Abyssal Plain—Sites 133 and 134. *Deep Sea Drill. Proj. Rep.* **1973**, *XIII*, 465–514.

39. Fanucci, F.; Fierro, G.; Ulzega, A.; Gennesseaux, M.; Rehault, J.P.; Viaris De Lesegno, L. The continental shelf of Sardinia: Structure and sedimentary characteristics. *Boll. Soc. Geol. Ital.* **1976**, *95*, 1201–1217.

40. Carta, M.; Lecca, L.; Ferrara, C. La piattaforma continentale della Sardegna. Studi geociacimentologici e di valorizzazione dei minerali contenuti. CNR, P.F. "Oceanografia e fonfi marini". *Final Tech. Rep.* **1986**, 119–218.

41. Finetti, I.R.; Del Ben, A.; Fais, S.; Forlin, E.; Klingelè, E.; Lecca, L. Crustal tectono-stratigraphic setting and geodynamics of the Corso-Sardinian Block from new CROP seismic data. *Crop Proj.* **2005**, *1*, 413–446.

42. Cherchi, A.; Montadert, L. Oligo-Miocene rift of Sardinia and the early history of the Western Mediterranean Basin. *Nature* **1982**, *298*, 736–739. [CrossRef]

43. Fais, S.; Klingele, E.E.; Lecca, L. Structural features of the south-western Sardinian shelf (Western Mediterranean) deduced from aeromagnetic and high-resolution reflection seismic data. *Eclogae Geol. Helv.* **2002**, *95*, 169–182.

44. Casula, G.; Cherchi, A.; Montadert, L.; Murru, M.; Sarria, E. The Cenozoic graben system of Sardinia (Italy): Geodynamic evolution from new seismic and field data. *Mar. Petrol. Geol.* **2001**, *18*, 863–888. [CrossRef]

45. Faccenna, C.; Speranza, F.; D'Ajello Caracciolo, F.; Mattei, M.; Oggiano, G. Extensional tectonics on Sardinia (Italy): Insights into the arc-back-arc transitional regime. *Tectonophysics* **2002**, *356*, 213–232. [CrossRef]

46. Pondrelli, S.; Salimbeni, S.; Ekström, G.; Morelli, A.; Gasperini, P.; Vannucci, G. The Italian CMT dataset from 1977 to the present. *Phys. Earth Planet. Inter.* **2006**, *159*, 286–303. [CrossRef]

47. Cioni, R.; Salaro, L.; Pioli, L. The Cenozoic volcanism of San Pietro Island (Sardinia, Italy). *Rend. Sem. Fac. Sci. Univ. Cagliari.* **2001**, *71*, 149–163.

48. Cita, M.B.; Ryan, W.B.F. Messinian erosional surfaces in the Mediterranean. *Mar. Geol.* **1978**, *27*, 193–365.

49. Haq, B.U.; Hardenbol, J.; Vail, P. Chronology of fluctuating sea levels since the Triassic (250 million years ago to present). *Science* **1987**, *235*, 1156–1167. [CrossRef] [PubMed]

50. Ceramicola, S.; Praeg, D.; Cova, A.; Accettella, D.; Zecchin, M. Seafloor distribution and last glacial to postglacial activity of mud volcanoes on the Calabrian accretionary prism, Ionian Sea. *Geo Mar. Lett.* **2014**, *34*, 111–129. [CrossRef]

51. De Muro, S.; Orrù, P. Il contributo delle beachrock nello studio della risalita del mare olocenico. Le beachrock post-glaciali della Sardegna nord orientale. *J. Quat. Sci.* **1998**, *11*, 1–21.

52. Kelletat, D. Beachrock as sea-level indicator? Remarks from a geomorphological point of view. *J. Coast. Res.* **2006**, *22*, 1555–1564. [CrossRef]

53. Carboni, S.; Lecca, L.; Ferrara, C. La discordanza Versiliana sulla piattaforma continentale occidentale della Sardegna. *Boll. Soc. Geol. Ital.* **1989**, *108*, 503–519.

54. Carboni, S.; Lecca, L. Upper Pleistocene sea-level lowstands in the continental shelf of western Sardinia (Italy). *Int. Union Quat. Res. Comm. Quat. Shorel. Subcomm. Mediterr. Black Sea Shorel.* **1992**, *14*, 57–65.

55. Lecca, L.; De Muro, S.; Pascucci, V.; Carboni, S.; Tilocca, G.; Andreucci, S.; Pusceddu, G. Note illustrative della Carta Geologica d'Italia alla scala 1:50.000, Foglio 528 Oristano. *ISPRA* **2016**, 152–156, in press.

56. Palombo, M.R.; Antonioli, F.; Lo Presti, V.; Mannino, M.A.; Melis, R.T.; Orrù, P.; Stocchi, P.; Talamo, S.; Quarta, G.; Calcagnile, L.; et al. The late Pleistocene to Holocene palaeogeographic evolution of the Porto Conte area: Clues for a better understanding of human colonization of Sardinia and faunal dynamics during the last 30 ka. *Quat. Int.* **2017**, *439*, 117–140. [CrossRef]

57. Di Gregorio, F.; Orrù, P.E.; Piras, G.; Puliga, G. Carta geomorfologica costiera e marina. Isola di San Pietro (Sardegna sud-occidentale)—Scala 1:25.000. *Boll. Assoc. Ital. Cartogr.* **2010**, *138*, 311–326.

58. Conforti, A.; Budillon, F.; Tonielli, R.; De Falco, G. A newly discovered Pliocene volcanic field on the western Sardinia continental margin (western Mediterranean). *Geo Mar. Lett.* **2016**, *36*, 1–14. [CrossRef]

59. Putignano, L.; Orrù, P.E. Note Illustrative della Carta Geologica d'Italia 1:50.000. Foglio 465—Isola di Procida. Area Marina. *ISPRA* **2010**, in press.

60. Putignano, M.L.; Orrù, P.E.; Schiattarella, M. Palaeoenvironmental reconstruction of Holocene coastline of Procida Island, Bay of Naples. *Quat. Int.* **2012**, *332*, 115–125. [CrossRef]

61. Di Vito, M.A.; Isaia, R.; Orsi, G.; Southon, J.; De Vita, S.; D'Antonio, M.; Pappalardo, L.; Piochi, M. Volcanism and deformation since 12,000 years at the Campi Flegrei caldera (Italy). *J. Volcanol. Geotherm. Res.* **1999**, *91*, 221–246. [CrossRef]

62. Keir, D.; Pagli, C.; Bastow, I.D.; Ayele, A. The magma-assisted removal of Arabia in Afar: Evidence from dike injection in the Ethiopian rift captured using InSAR and seismicity. *Tectonics* **2011**, *30*, 3–13. [CrossRef]

63. Ayele, A.; Keir, D.; Ebinger, C.; Tim, J.; Stuart, W.G.; Roger, B.W.; Jacques, E.; Ogubazghi, G.; Sholan, J. Mega-dike emplacement in the Manda-Harraro nascent oceanic rift (Afar depression). *Geophys. Res. Lett.* **2009**, *36*, 1–5. [CrossRef]

64. Sunamura, T. *The Geomorphology of Rocky Coasts*; Wiley: Chichester, UK, 1992.

65. Savini, A.; Corselli, C. High-resolution bathymetry and acoustic geophysical data from Santa Maria di Leuca Cold Water Coral province (Northern Ionian Sea—Apulian continental slope). *Deep Sea Res. Part II Top. Stud. Oceanogr.* **2010**, *57*, 326–344. [CrossRef]

66. Prampolini, M.; Gauci, C.; Micallef, A.S.; Selmi, L.; Vandelli, V.; Soldati, M. Geomorphology of the north-eastern coast of Gozo (Malta, Mediterranean Sea). *J. Maps* **2018**, *14*, 402–410. [CrossRef]

67. Soldati, M.; Barrows, T.T.; Prampolini, M.; Fifield, K.L. Cosmogenic exposure dating constraints for coastal landslide evolution on the Island of Malta (Mediterranean Sea). *J. Coast. Conserv.* **2018**, *22*, 831. [CrossRef]

68. Castedo, R.; Paredes, C.; De la Vega-Panizo, R.; Santos, A.P. *The Modelling of Coastal Cliffs and Future Trends*; Hydro-Geomorphology-Models and Trends; InTech: Houston, TX, USA, 2017.

69. Ohneiser, C.; Florindo, F.; Stocchi, P.; Roberts, A.P.; De Conto, R.M.; Pollard, D. Antarctic glacio-eustatic contributions to late Miocene Mediterranean desiccation and reflooding. *Nat. Commun.* **2015**, *6*, 1–10. [CrossRef] [PubMed]

70. López Correa, M.; Montagna, P.; Vendrell-Simón, B.; McCulloch, M.; Taviani, M. Stable isotopes ($\delta^{18}O$ and $\delta^{13}C$), trace and minor element compositions of Recent scleractinians and Last Glacial bivalves at the Santa Maria di Leuca deep-water coral province, Ionian Sea. *Deep Sea Res. Part II Top. Stud. Oceanogr.* **2010**, *57*, 471–486. [CrossRef]

71. Siani, G.; Paterne, M.; Arnold, M.; Bard, E.; Métivier, B.; Tisnerat, N.; Bassinot, F. Radiocarbon Reservoir Ages in the Mediterranean Sea and Black Sea. *Radiocarbon* **2000**, *42*, 271–280. [CrossRef]

72. Ferranti, L.; Antonioli, F.; Amorosi, A.; Dai Prà, G.; Mastronuzzi, G.; Mauz, B.; Monaco, C.; Orrù, P.E.; Pappalardo, M.; Radtke, U.; et al. Markers of the last interglacial sea-level high stand along the coast of Italy: Tectonic implications. *Quat. Int.* **2006**, *145*, 30–54. [CrossRef]

73. Antonioli, A.; Lo Presti, V.; Rovere, A.; Ferranti, L.; Anzidei, M.; Furlani, S.; Mastronuzzi, G.; Orru, P.E.; Scicchitano, G.; Sannino, G.; et al. Tidal notches in Mediterranean Sea: A comprehensive analysis. *Quat. Sci. Rev.* **2015**, *119*, 1–19. [CrossRef]

74. Castedo, R.; William, M.; Lawrence, J.; Paredes, C. A new process–response coastal recession model of soft rock cliffs. *Geomorphology* **2012**, *177*, 128–143. [CrossRef]
75. Sunamura, T. A relationship between wave-induced cliff erosion and erosive force of waves. *J. Geol.* **1977**, *85*, 613–618. [CrossRef]
76. Kageyama, M.; Valdes, P.J.; Ramstein, G.; Hewitt, C.; Wyputta, U. Northern Hemisphere Storm Tracks in Present Day and Last Glacial Maximum Climate Simulations: A Comparison of the European PMIP models. *J. Clim.* **1999**, *12*, 742–760. [CrossRef]
77. Atzeni, A. Effetti idrodinamici sulle spiagge della costa occidentale della Sardegna. *Studi Costieri* **2003**, *7*, 61–80.
78. Sulis, A.; Annis, A. Morphological response of a sandy shoreline to a natural obstacle at Sa Mesa Longa Beach, Italy. *Coast. Eng.* **2014**, *84*, 10–22. [CrossRef]
79. Bini, C.; Martini, F.; Pitzalis, G.; Ulzega, A. *Sa Coa de Sa Multa e Sa Pedrosa-Pantallinu: Due "paleosuperfici" clactoniane in Sardegna*; Atti della XXX Riunione Scientifica "Paleosuperfici del Pleistocene e dell'Olocene in Italia": Firenze, Italy, 1993; pp. 179–196.
80. Mussi, M.; Melis, R.T. Santa Maria in Acquas e le problematiche del Paleo/itico superiore in Sardegna. *Origini* **2004**, *XXIV*, 67–94.
81. Melis, R.T.; Mussi, M. Mesolithic burials at S'Omu e S'Orku (SOMK) on the south-western coast of Sardinia. In *Mesolithic Burials— Rites, Symbols and Social Organisation of Early Postglacial Communities*; Grünberg, J.M., Gramsch, B., Larsson, L., Orschiedt, J., Meller, H., Eds.; International Conference Haale (Saale): Halle, Germany, 2016; pp. 733–740.
82. Melis, R.; Galheb, B.; Boldrini, R.; Palombo, M.R. The Grotta dei Fiori (Sardinia, Italy) stratigraphical successions: A key for inferring palaeoenvironment evolution and updating the biochronology of the Pleistocene mammalian fauna from Sardinia. *Quat. Int.* **2013**, *288*, 81–96. [CrossRef]

Article

Terraced Landforms Onshore and Offshore the Cilento Promontory (South-Eastern Tyrrhenian Margin) and Their Significance as Quaternary Records of Sea Level Changes

Alessandra Savini [1,2], Valentina Alice Bracchi [1], Antonella Cammarosano [3], Micla Pennetta [3,*] and Filippo Russo [4]

1   Department of Earth and Environmental Sciences, University of Milano Bicocca (DISAT), 20126 Milan, Italy; alessandra.savini@unimib.it (A.S.); valentina.bracchi@unimib.it (V.A.B.)
2   CoNISMa, Local Research Unit, University of Milano Bicocca, 20126 Milan, Italy
3   Independent Researcher, 80138 Naples, Italy; anto.camm@hotmail.it
4   Dipartimento di Scienze e Tecnologie, Università del Sannio, 82100 Benevento, Italy; filrusso@unisannio.it
*   Correspondence: micla.pennetta@gmail.com

check for **updates**

Citation: Savini, A.; Bracchi, V.A.; Cammarosano, A.; Pennetta, M.; Russo, F. Terraced Landforms Onshore and Offshore the Cilento Promontory (South-Eastern Tyrrhenian Margin) and Their Significance as Quaternary Records of Sea Level Changes. *Water* **2021**, *13*, 566. https://doi.org/10.3390/w13040566

Academic Editor: Giorgio Anfuso

Received: 31 December 2020
Accepted: 17 February 2021
Published: 23 February 2021

Abstract: Climate change and tectonic uplift are the dominant forcing mechanisms responsible for the formation of long and narrow terraced landforms in a variety of geomorphic settings; and marine terraces are largely used to reconstruct the Quaternary glacial and interglacial climates. Along the Mediterranean coast, a considerable number of popular scientific articles have acknowledged a range of marine terraces in the form of low-relief surfaces resulting from the combined effects of tectonic uplift and eustatic sea-level fluctuations, as relevant geomorphological indicators of past sea-level high-stands. With the exception of a few recent studies on the significance of submarine depositional terraces (SDT), submerged terraced landforms have been less investigated. By integrating different marine and terrestrial datasets, our work brings together and re-examines numerous terraced landforms that typify the Cilento Promontory and its offshore region. In this area, studies since the 1960s have allowed the recognition of well-defined Middle to Upper Pleistocene marine terraces on land, while only a few studies have investigated the occurrences of late Pleistocene SDT. Furthermore, to date, no studies have consistently integrated findings. For our work, we correlated major evidence of emerged and submarine terraced landforms in order to support an improved understanding of the tectono-geomorphological evolution of the Cilento Promontory and to further clarify the geomorphological significance of submerged terraces.

Keywords: marine terraces; submarine geomorphology; coastal geomorphology; sea level oscillation; Tyrrhenian margin

1. Introduction

A terraced landform is any relatively flat horizontal or gently inclined surface bounded by a steeper ascending slope on one side and a steeper descending slope on the opposite side [1,2]. Terraces can be formed in many ways and in different geologic and environmental settings. In geomorphology, tectonic uplift and climate change are the dominant forcing mechanisms responsible for the formation of long and narrow terraced landforms. Resulting terraces can, therefore, be used for studying variations in tectonic, climate, and erosion, and for investigating how processes have interacted in the past and how they currently interact. The recognition of late Pleistocene uplifted coral platforms as indicators of past sea levels (i.e., reef terraces) was, for example, a significant finding in sea-level research [3]. Terrestrial, fluvial-glacial counterpart [4], coral reef terraces [3], and marine terraces [5,6] have been (and still are) largely used for reconstructing Quaternary glacial and interglacial climates. Where tectonic uplift considerably impacts coastal regions, sub-aerial marine terraces clearly document high-stands of sea level during interglacial stages [6], alternating with low levels during glacial stages. In temperate regions, Pirazzoli [7] noted

313

"marine-cut" terraces (or shore platforms) resulting from marine erosion and "marine-built" terraces formed by shallow-water and slightly emerged accumulations of materials removed by shore erosion. Along the Mediterranean coast, a range of sub-aerial marine terraces have been acknowledged to be relevant geomorphological indicators of past sea-level highstands (see [8] among other references). The study of Pirazzoli [7] has even allowed the definition of still popular marine stratotypes, outlining the first Quaternary chronostratigraphy (i.e., Calabrian, Emilian, Sicilian, Milazzian and Tyrrhenian for the Pleistocene, and Versilian for the Holocene). Although their work has been revised and refined, the chronostratigraphy is still used in gray literature and open discussions. The geomorphological significance of submerged terraced landforms [9–14], as evidence of Quaternary sea-level variations, has been less investigated and has only recently gained attention [15], thanks to advances in seafloor mapping techniques [16] and the resulting recognition of Submarine Depositional Terraces (SDT) (defined as sedimentary bodies with a clinostratified internal structure and a prograding growth towards the sea [17–21]). Minor studies have investigated the possible erosive nature of submerged terraced landforms formed on bedrock outcropping on the shelf and their relationship to sea-level oscillations, among them Bilbao-Lasa et al. [22]. Additionally, by taking into account both emerged and submerged terraces, a few investigations have integrated their findings to support an improved understanding of the tectono-geomorphological evolution of coastal areas and the physiography of the margin [15,23], as we have undertaken in our work for the Cilento Promontory.

In this region, systematic studies since the 1960s [24–34] have determined well-defined Middle to Upper Pleistocene marine terraces on land. The submarine sector was the subject of minor research during the 1990s. Among such studies, Trincardi and Field [35] investigated the origin and forming mechanisms of remnants of late Pleistocene prograded coastal deposits, locally preserved on the middle and outer portions of the shelf.

2. The Cilento Promontory and Its Offshore: Geological Setting and Stratigraphic Framework

The coastal area of the Cilento region (Southern Italy) (Figure 1), included between Agropoli and Pioppi, represents the western end of one of the most important peri-Tyrrhenian, morpho-structures belonging to the Campano-Lucano arch of the southern Apennine orogenic thrust system. Compressive tectonogenesis and structuring, initiated in the lower Miocene, appear to have ended in the Lower Pleistocene [29,36–40], through displacement of the Mesozoic-Tertiary bedrock of the Cilento coast during major Quaternary (Lower to Middle Pleistocene) tectonic activity [41–43]. The complex lithogenic history of the Cilento region has thus been shaped by numerous tectono-sedimentary events and orogenic shifts [44] that today allow us to distinguish different lithostratigraphic units outcropping along the coastline (Figure 1). Both siliciclastic and calcareous units outcrop on the Cilento Promontory. Siliciclastic units are primarily represented by the Cilento Flysch Unit [45,46] or Cilento Flysch Group "*Auct*" [41,42] that dominates in the north-western sector, and secondarily by the Ligurian and Northern Calabrese *Auct* tectonic units. Such units are often indicated as the "Internidi" [42,43] (Figure 1), the highest structural tectonic unit (thickness 1300 m) that emerges for a few hundred meters in the central-southern portion of the promontory. Calcareous bedrock outcrops on the south-eastern sector of the Cilento region within the Monte Bulgheria Unit [47]. The general structural setting is dominated by low-angle overthrust surfaces that are clipped and folded by subvertical transcurrent and extensional surfaces with a variable orientation from NW–SE to E–W. The Internidi have been described as tectonic overlap on calcareous units. However, such overlap is sometimes reworked and masked by recent tectonics that are responsible for major displacements that caused carbonate uplift and relief inversion. Indeed, all along the Cilento Promontory the highest peaks are formed by carbonate units, while the most erodible siliciclastic units are found in places only preserved along valleys and on morphological and structural lows [48].

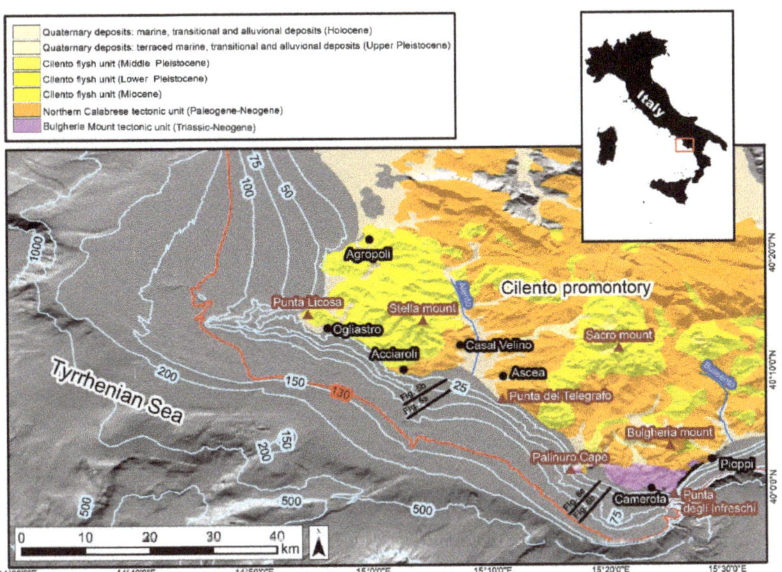

Figure 1. The study area. The on-land portion of the Cilento Promontory is represented by a schematic geological map of the Cilento region (The map was adapted from a geological map containing thematic elements and underwater landscapes at a 1:110,000 scale) overlaid on a Digital Elevation Model (DEM) adapted from Campania Region Technical Cartography at a 1:5000 scale. Isobaths and offshore shaded relief were obtained from the EMODNET portal (https://www.emodnet-bathymetry.eu/, accessed on15 February 2021).

Quaternary deposits formed by marine, transitional, or alluvial sediments, where preserved, are found in angular unconformity on the described bedrock. Outcrops of coastal marine sediments (biocalcarenites, fossiliferous beaches and aeolian sands), often associated with typical forms of marine abrasion and bioerosion (shore platforms, fossil eroded notches, "Lithophaga" holes, etc.), are found in very small and discontinuous terraced strips (at various heights between 2 m and 300 m above sea level (a.s.l.)) along the entire Cilento coastal zone. Such deposits, sometimes alternating with "colluvial", pyroclastics, and paleosoils, have been the subject of careful and systematic studies since the 1960s [24–34], allowing chronological attribution to the Middle and Upper Pleistocene.

Offshore, the shelf of the eastern Tyrrhenian Margin lies between the uplifting Apennine chain on land and the Tyrrhenian offshore basin that has been subsiding at a rate of 1 mm/yr since the end of the Lower Pleistocene [49]. Due to the multifaceted tectonic history determined by the opening of the Tyrrhenian Sea, that is also associated with limited Plio-Quaternary sedimentation at places interrupted on the shelf by bedrock outcrops [48], the seabed topography is extremely complex. Seismic data has revealed a deformed acoustic basement, displaced by quaternary faults with very sharp and steep scarps similar to the ones detected on land. The continental shelf is wider in the northern portion, extending for almost 30 km to the north of Punta Licosa [50], and delimited seaward by an uncertain shelf break, from 180 to more than 200 m in depth. To the south, continental shelf width is reduced to less than 10 km, and in offshore Acciaroli the shelf break is sharper and is located at a water depth (w.d.) of roughly 130 m. In the southern sector (Figure 1), the shelf further narrows to 6 km and a transition between the shelf and the slope is evident, but located at variable depth, gradually decreasing from 140 m off Punta del Telegrafo to 130 m in .w.d. offshore of the south-eastern corner of the promontory, where the upper continental slope is much steeper and likely coincident with a tectonic escarpment [51–53]. Different sub-horizontal surfaces bounded by fairly continuous and slightly sinuous escarpments have also been determined on the shelf. They have been interpreted as submarine terraces [48,54,55] that formed by local and prolonged low sea level stationing, occurring between the regression of the Last Glacial Maximum (LGM) and subsequent rapid post-glacial sea-level rise (i.e., the Flandrian Transgression). Only a few of the submarine terraced landforms located near sea level (8 m to 12 m below sea level (b.s.l.)) are ascribed to Marine oxygen Isotopic Stage (MIS) 5 [32,56,57], corresponding to the Last Interglacial period.

The continental slope is marked by depressions and topographic highs of variable dimension, down to a depth of 1600 m. Numerous escarpments document the existence of simple to complex landslide scars, testifying to the dominant role of mass-wasting phenomena in shaping the continental margin. A complex tectonic framework of bedrock is also still visible along the slope, where it has created local, intra-slope reliefs and marked tectonic lineaments [53–55].

3. Data and Methods

Our study was driven by the collection of major evidences of terraced landforms, both on land and in offshore areas of the Cilento Promontory coastal zone, recovered from scientific literature and detected on available Digital Elevation Models (DEM—i.e., Emodnet database—https://portal.emodnet-bathymetry.eu/ (accessed on 20 February 2021)—grid cell size 50 × 50 m and Magic project: http://dati.protezionecivile.it/geoportalDPC/rest/document#MagicFoglio10/ (accessed on 20 February 2021)—grid cell size 50 × 50 m) for the offshore sector. Submarine terraced landforms were also manually and automatically detected by applying a geomorphometric analysis performed using Spatial Analysis Tool available in ArcGis®. All terraced landforms were then collected in a proper database by including information regarding dating, altitude or depth, and references (Tables 1 and 2). Landform spatial and temporal distributions were analysed in order to detect the role of associated bedrock and the structural framework in controlling distributions and geomorphological differentiation.

Terraces were grouped according to lithostratigraphic units of the corresponding bedrock (i.e., siliciclastic or calcareous), namely, from North to South: (1) Cilento Flysch and "Internidi" Units; and (2) the Carbonate Unit of Mount Bulgheria, with outcroppings regions located in the area surrounding Palinuro (Palinuro Cape, Mingardo river mouth and Camerota) on the southern coast of the Cilento Promontory (Figure 1).

Study of the offshore region was also supported by the availability of high resolution seismic data collected using a GeoAcoustic GEOCHIRPII (GeoAcoustic Limited, Shuttleworth Close, Doncaster, UK) Subbottom Profiler System (SBP) in 2003, between 10 and 130 m in w.d., as well as by results obtained from a sedimentological analysis performed on 16 gravity cores and 32 grab samples, as described in [58]. An interpretation of depositional and erosional processes, as detected from a seismo-stratigraphic analysis, was performed using the concepts of sequence stratigraphy [59,60].

3.1. On Land Terraced Landforms

To understand traces of described ancient marine deposits and sea-level markers, background knowledge of the study area was obtained from an extensive literature review [24,26,28,30–34,61–69], and a field survey. Table 1 provides on-land terraced landforms according to their altitude and dating (as ascribed in the scientific literature).

3.2. Offshore Terraced Landforms

Background knowledge for the submarine sector of the study area was obtained by collecting public bathymetry (EMODnet portal) and high-resolution seismic data, as described in Savini et al. [58], along with analogous remote data and evidence of direct observations as reported in the scientific literature [32,48,50,54–57]. To detect flat surfaces, basic geomorphometric analysis were performed in ArcGIS®. All areas with a slope value ≤ 1 and confined by a marked break of slope (according to [1]) were segmented and converted in polygons (Figure 2B).

Table 1. List, reported dating in the scientific literature, altitude, and referenced literature for on-land terraced landforms.

	Geological Unit	Dating	Altitude (m)	Notes and References
Siliciclastic bedrock	Cilento Flysch Unit	Lower Pleistocene	?	No clear evidences [64]
		Middle Pleistocene (MIS 9)	60	[30]
		Middle Pleistocene (MIS 7)	25	[30,67]
		Upper Pleistocene (MIS 5e)	6.5–10	[28,30,67]
		Upper Pleistocene (MIS 5c)	4–5	[30,66,67]
		Upper Pleistocene (MIS 5a)	1.5?	[30,66,67]
	Cilento Group and Internidi Units	Upper Pleistocene (MIS 5e)	6	Beach-ridge deposits [24,28,65]
Mt. Bulgheria Carbonatic Unit	Palinuro Cape	Lower Pleistocene	350	[29,32,64]
		Middle Pleistocene	170–180	[32]
		Middle Pleistocene	130–140	[32]
		Middle Pleistocene	75–65	[32]
		Middle Pleistocene	50	[32]
		Upper Pleistocene	8–7	[32]
		Upper Pleistocene	3–2	[32]
	Mingardo river	Middle Pleistocene	Many orders 75–15	[34]
		Upper Pleistocene	12–10	[34]
		Upper Pleistocene	4–3	[34]
	Camerota	Lower Pleistocene	0–350	[29,30,32,64]
		Middle Pleistocene	50–200	[68]
		Upper Pleistocene	15	[32,34]
		Upper Pleistocene	12–10	[32,34,69]
		Upper Pleistocene (MIS 5)	8.5–8 7.5–5 4.5–4 3.5–3	[31,32,34]

?: Uncertain value or not confirmed by consistent data.

Table 2. List, reported dating in the scientific literature, depth, and referenced literature for offshore terraced landforms. The table takes into account terraces cited in the scientific literature. The correspondence with terraces detected by geomorphometric analysis is reported in the last column on the right.

		Geological Unit	Dating	Depth (m)	Notes and References	Correspondence on Slope Value
Siliciclastic bedrock	**Offshore Cilento Flysch Unit**		Upper Pleistocene	−8	[55]	Not evident
			Upper Pleistocene	−10/14	[55]	Not evident
			Upper Pleistocene	−17/27	[55]	Yes (−21/26)
			Upper Pleistocene	−43/50	[55]	Yes (−47/52)
			Upper Pleistocene	−86	[48]	Yes (−86)
			Upper Pleistocene	−107	[48]	Yes (−108)
			Upper Pleistocene (MIS 2)	−120	[50]	Not evident
			Upper Pleistocene (MIS 2)	−160	[50]	Not verified

Table 2. *Cont.*

Geological Unit		Dating	Depth (m)	Notes and References	Correspondence on Slope Value
Siliciclastic bedrock	Internidi Unit	Flandrian transgression	−46/51	[58]	Yes (−47/52)
Mt. Bulgheria Carbonatic Unit	Palinuro Cape	Upper Pleistocene	−7/8	Notches [32]	Not evident
		Upper Pleistocene	−12/14	Wave-cut platform [32]	Yes (−12/21)
		Upper Pleistocene (MIS 3)	−18/24	Wave-cut platform [32]	Not evident
		Flandrian transgression	−44/46	Notches [32]	Not evident

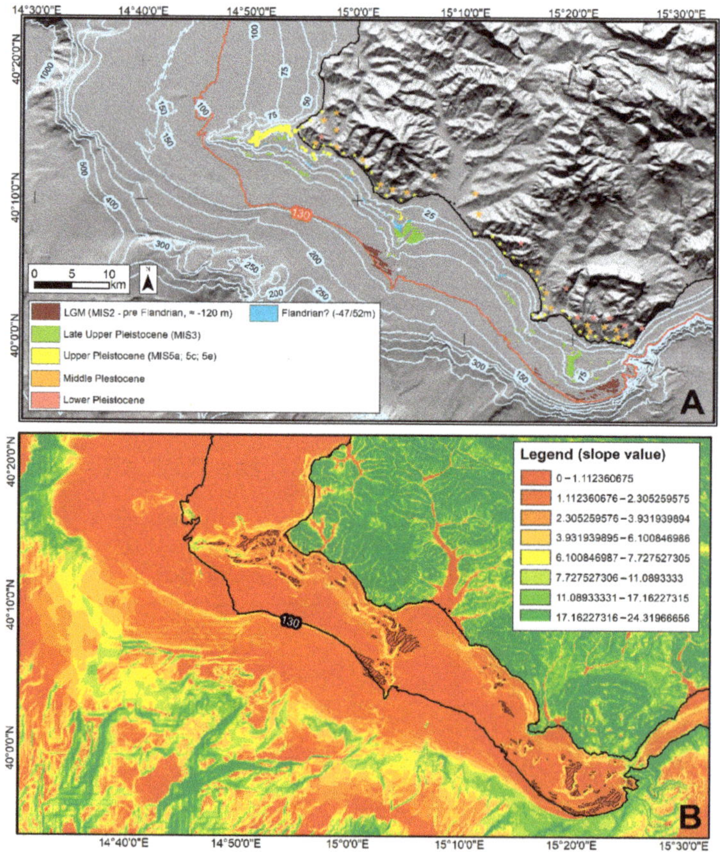

Figure 2. (**A**): A map of major terraced landforms detected on land and offshore in the Cilento Promontory, as reported in Tables 1 and 2, with the exception of the submarine depositional terraces (SDT) reported in [48,50]. (**B**): A slope map with red polygons indicating flat areas (slope ≤ 1) delimited by marked ascending and descending breaks in slope.

4. Results

4.1. Terraced Landforms: Temporal and Spatial Distribution

The presence of terraced landforms in the Cilento coastal area marked the emerged and submerged sectors (Tables 1 and 2). Scientific literature has documented at least seven orders of Pleistocene terraced surfaces on land (Table 1 and Figure 2), spanning from the Lower to Upper Pleistocene [32].

Offshore terraced landforms (Figure 2) were, instead, first detected using geomorphometric techniques (Figure 2B), then correlated with evidence in the scientific literature [32,48,50,54–58], and then grouped according to depth range of occurrence (Figure 2A). For terraced surfaces deeper than 50 m, we also refer to [50] and [48]. As discussed below, the distribution of both on-shore and off-shore terraced landforms was then resumed according to geological unit.

4.1.1. Northern Cilento Group

On-land five terraced surfaces have been identified (Table 1 and Figure 2). Two were from the Middle Pleistocene (MIS 7 and 9) with clear evidence of uplift, apparently sealed by Upper Pleistocene deposits. The remaining three were from the Upper Pleistocene (MIS 5a, MIS 5c and MIS 5e), with no significant contribution from tectonics.

The submerged sector is typified by prolongation towards the sea of the "Punta Licosa" Promontory (Figures 1 and 2) that (1) provides an EW aligned spur formed by an outcrop of the acoustic basement that rests over more than 16 km^2, between 25 and 80 m of w.d.; (2) likely originated from the Cilento Group synorogenic unit (or "Flysch del Cilento") and; (3) was bounded by direct faults. The spur rises from the surrounding seafloor through several sharp escarpments bounded by flat terraced surfaces (Figures 2 and 3).

According to the scientific literature, terraces are positioned at −8 m, −10/14 m, −17/27 m, and −43/50 m and reportedly range from MIS 5a or 5c up to MIS 3 [50,55,57,66]. A performed geomorphometric analysis distinctly outlined the marked stepped profile of the Licosa spur and several submarine terraces (slope ≤ 1), with a prevalence at −21/26 m, −47/52 m, and −76/86 m, having consistent lateral continuity (especially toward the south and for depth intervals of −21/26 m and −47/52 m) (Figures 3A and 4), were located. Small scale landforms resembling tension fractures at the crowning areas of modest landslides are frequent on the southern slope of the spur [55], an isolated group of terraced surfaces downward of small landslide scars was identified at −28/30 m (Figure 4). Ferraro et al. [48] detected additional terraced landforms at 86 m and even deeper at 107 m of w.d., both of an erosional origin, and indicated that the terraces formed at the outcrop of the acoustic basement. Further offshore, biogenic coarse sandy depositional bodies, bounded at their top by a ravinement surface, were described at 120 m and 160 m in w.d. Such bodies are developed over more than 20 km along-slope [48]. A small fragment of "*Arctica islandica*" (Linneo, 1767) was also recovered from a core sample ([48]; Pennetta pers. com.) allowing attribution of their formation to the last low-stand period (i.e., MIS 2). Trincardi and Field [50] also reported the occurrence of depositional bodies in the form of shelf-margin deposits, truncated at their tops by an outer-shelf ravinement surface at −150/160 m. The shelf-margin is reported to occur at −200 m in w.d. (Figures 1 and 2) (i.e., deeper than the sea level low-stand reported for the Last Glacial Maximum (namely 120 m in w.d.). Marani et al. [70] indicated that these sandy bodies appear to have formed in a shallow (<30 m deep) marine setting. This evidence, together with the detection of an outer limit for the ravinement surface generated by the post-Würmian Transgression at a deeper depth than the one reported for the eustatic minimum (i.e., 120 m), suggests that the outer continental shelf has been subject to important (tectonic) subsidence phenomena over the Holocene. Considering Mediterranean wave-base level in the order of 10/15 m [71], we speculate that the subsidence rate reported for the Tyrrhenian sea by Kastens et al. [49] (i.e., 1 mm/yr.) allowed a merger of the geodynamics of the outer portion of the continental shelf with the structural system that currently controls evolution of the Tyrrhenian basin.

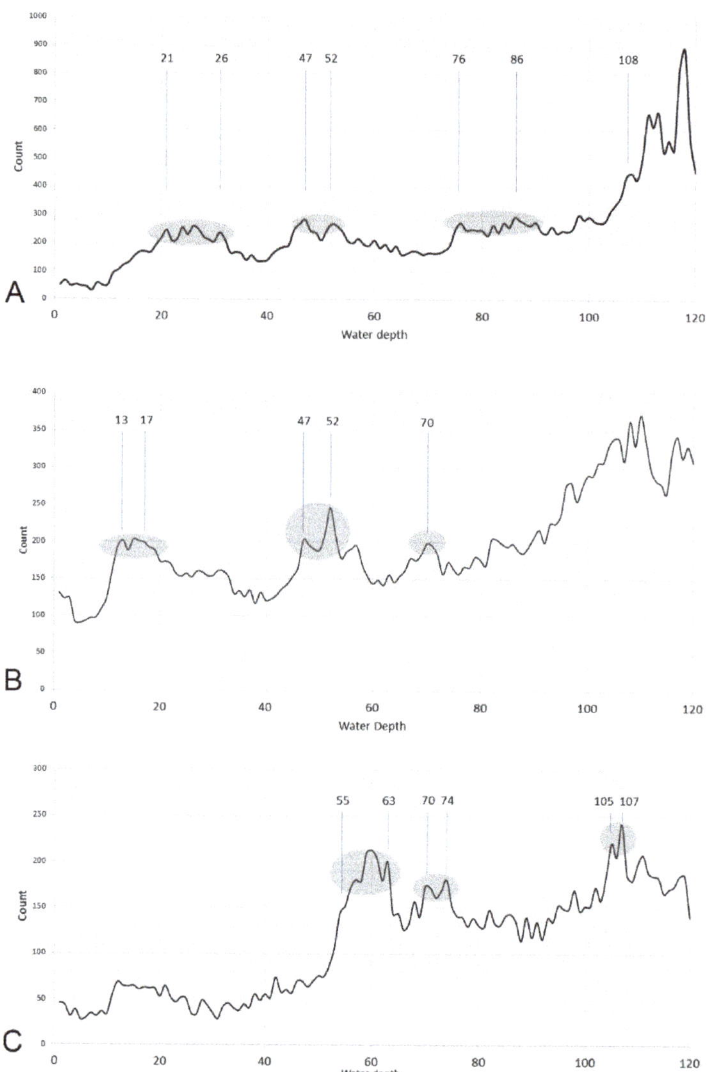

Figure 3. Histograms of depth values (depth range 0–120 m in w.d.) for the offshore areas of Northern Cilento group (**A**); Cilento Group and Internidi Units (**B**); and Southern Bulgheria Mount (**C**). The gray ellipses show the depth ranges that, on the map, are clearly delimited by sharp breaks in slope.

4.1.2. Cilento Group and Internidi Units

The central continental coastal area of Cilento, between Acciaroli and Palinuro, is essentially composed by deformed units of Mesozoic-Tertiary Bedrock (members of the Cilento Group and Internidi Units) covered in angular discordance by Quaternary alluvial and coastal deposits. In places, coastal deposits represent the filling of localized morpho-tectonic depressions of actual alluvial (i.e., the plain of the Alento River) and coastal (i.e., plain of Casalvelino-Ascea—Figure 1) plains consisting of fluvial sediments, dune and beach-ridge deposits sometimes covered by continental colluviums, and slope debris. Marine beach-ridge deposits, emerging up to six m a.s.l. along the coast at Ogliastro [28] and Acciaroli [24], have been attributed to the Upper Pleistocene (Table 1 and Figure 2). In this sector, no deposits or forms have been found that can be attributed, with certainty, to the Lower or Middle Pleistocene (Figure 2).

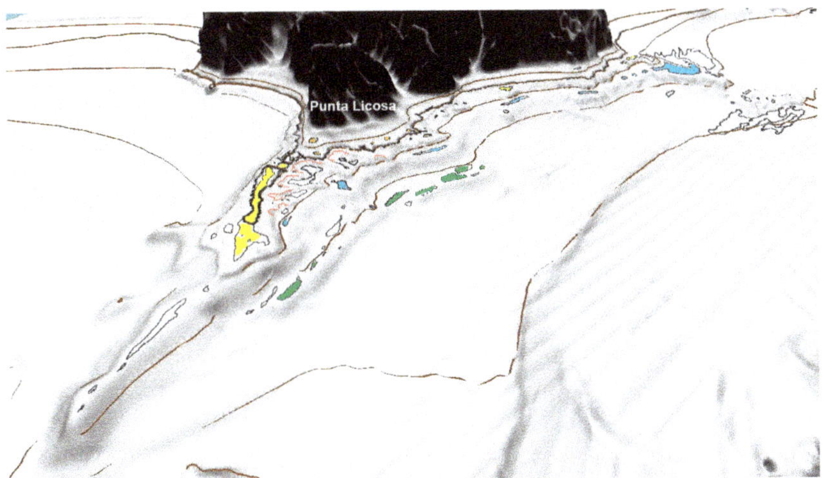

Figure 4. A 3D view of the seaward prolongation of the Punta Licosa Promontory, with polygons detected by geomorphometric analysis, representing terraced landforms according to [1] and distinguished in different colors according to depth range (yellow: 21/26 m; light blue: 47/52 m; green: 76/86 m). Small-scale landslide scars are also mapped (red lines) on the southern side of the submerged Punta Licosa spur. The Digital Terrain Model data products used to provide the 3D view have been derived from http://dati.protezionecivile.it/geoportalDPC/rest/document#MagicFoglio10 (accessed on 20 February 2021).

Offshore, the morpho-structural depression, marked on land by the coastal and alluvial plain of Casal Velino-Ascea (Figure 1), is filled by sandy-silty deposits [58] over almost all of the central and southern continental shelf, in continuity with terrestrial physiography (Figure 2). The depression is confined to the north by the outcrop of the acoustic basement, that forms a southward-elongated ridge offshore Acciaroli, interrupted at the shelf break (Figures 1 and 2). The coupling of bathymetric and high-resolution seismic data clearly indicates marked terraced landforms along the ridge. Terraced landforms are particularly evident at a depth interval between 47 and 52 m (Figure 3B), in the form of an erosional surface (wave-cut platforms [58]) sculpted within the acoustic basement (i.e., bedrock, Figure 5A). As shown by the marine DEM and the associated slope value (Figure 2B), the terraced landforms are still in continuity with those detected northward at Punta Licosa, especially for the depth range 47/52 m. A north-south elongated depositional body (with a lenticular section formed by poorly defined sloping depositional units that resemble shoreface clinoforms [58]) rest at 55 m in w.d. in overlap above the southeastern edge of the acoustic basement that outcrops to the south of Acciaroli (Figure 5). According to sediment composition reported in [58,72], the depositional body likely formed in a shallow (<10 m deep) marine setting. Since an older origin would have resulted in aerial exposure due to sea level drop during the LGM, the absence of an obvious erosional surface at the top of the deposit and partial burial towards the sea due to the high-stand drape (Figure 5B), warrants ascription to the transgression that followed MIS 2.

4.1.3. Southern Bulgheria Mount

On land, the best-preserved Quaternary landforms and deposits of the Cilento Promontory are found in Palinuro Cape (the Monte Bulgheria Carbonatic Unit), an area intensively studied and well described within the scientific literature [32]. The oldest evidence of flat eroded surfaces within the region are dated to the Upper and Lower Pliocene. Landforms are found at altitudes between 1200 and 400 m a.s.l. and have been associated with sub-aerial surfaces of fluvial-karst erosion, although some authors do not exclude marine abrasion as a potential origin [69].

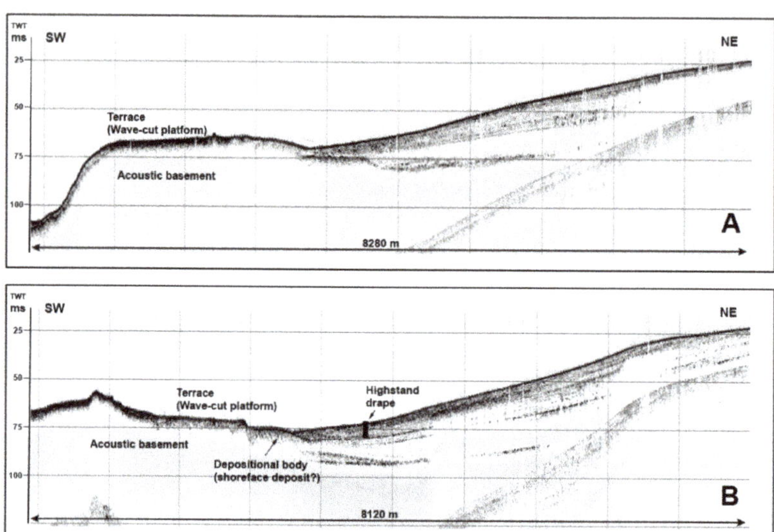

Figure 5. High-resolution seismic profiles acquired offshore Cilento Group and Internidi Units. See Figure 1 for location. (**A**) A submerged terraced landform (wave-cut platform) is visible at 47–51 m in w.d. (corresponding to 63/68 ms). (**B**) A submerged terraced landform (wave-cut platform) is visible at roughly 51 m in w.d. (corresponding to 68 ms).

The oldest Pleistocene terrace outcrops around 300 m a.s.l. and corresponds to the Lower Pleistocene [64].

Overall, the area records several order of terraces positioned at variable altitudes that are not easily correlated to the Middle Pleistocene. Four order of terraces, at an altitude range between 20 m and 200 m a.s.l., have been ascribed to the Middle Pleistocene. Additionally, five orders of Middle Pleistocene marine abrasion terraces were carved along the coastal slopes at altitudes of 180/170 m, 140/130 m, 100 m, 75/65 m, and 50 m, for which important tectonization cannot be excluded.

Upper Pleistocene terraced surfaces outcrop continuously along the cost and they are located between 1.5 m and 10 m a.s.l. (Table 1). Two marine terraces located along a sea-cliff that marks the coastal area at 8/7 m and 3/2 m a.s.l., and beach-ridge deposits containing fragments of *Thetystrombus latus* Gmelin 1791 (=*Strombus bubonius*) are found at 3/2 m a.s.l. The location of these Upper Pleistocene landforms suggested a small (few meters) tectonic lowering of the area. In general, since good lateral continuity is preserved and since the terraces are quite well correlated, terraces created by Upper Pleistocene sea-level oscillations seem to document a relatively stable tectonic period [73].

As for the submarine portion between Palinuro Cape and Bulgheria Mount, four orders of submarine terraces, located at 7/8 m, 12/14 m, 18/24 m, and 44/46 m in w.d., have been extensively described within the scientific literature [32]. Since they show evidence of subaerial erosion associated with a regressive period, the first two terraces were ascribed to the Last Interglacial (MIS5), or to an earlier period. Evidence of former sea-level positions at 7/8 m in w.d. have been attributed to MIS5a with good confidence.

Terraces detected in the form of wave-cut platforms at 18/24 m in w.d. were, instead, ascribed to the last phase of MIS3, which seems to be characterized by long stationing that occurred during the post-Last Interglacial regression [56]. According to Antonioli et al. [32], good conservation of the deposit and the absence of subaerial erosion for terraces located at 44/46 m in w.d. leads attribution to a lower standing period that occurred during the last, post glacial transgression (i.e., the Flandrian Transgression). Sparce terraces found at deeper depth were, in contrast, formed by depositional bodies and are ascribed to the last glacial low-stand period (MIS2). The geomorphometric analysis performed on the DEM detected quite large terraces at three main depth range: 55/63 m, 70/77 m, and 105/107 m. Seismic data well confirmed the erosional origin of mapped terraces located at 50/55 m in w.d. (Figure 6). Due to DEM resolution, which cannot resolve submarine terraces of small dimensions, many shallower terraces were likely difficult to detect.

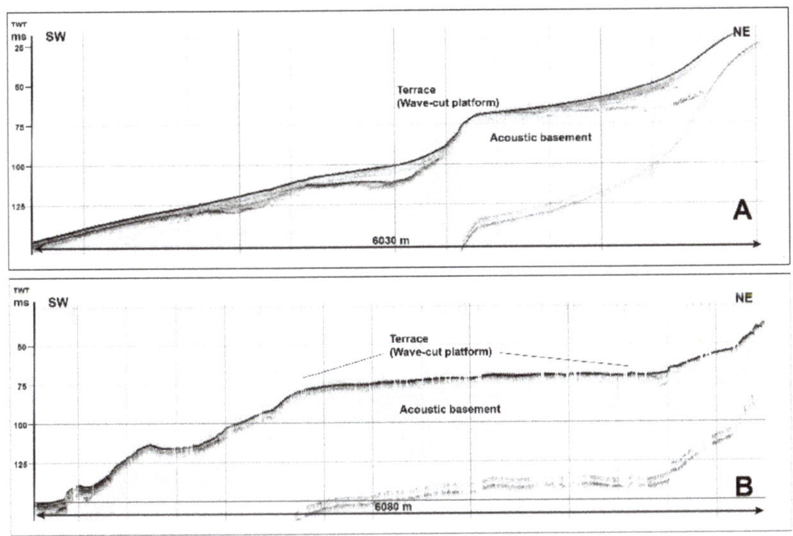

Figure 6. High-resolution seismic profiles acquired offshore Southern Bulgheria Mount.. See Figure 1 for location. (**A**) A submerged terraced landform (wave-cut platform) is visible at roughly 50 m in w.d. (corresponding to 67 ms); (**B**) A submerged terraced landform (wave-cut platform) is visible between 51 and 55 m in w.d. (corresponding to 68/75 ms).

5. Discussion

Abundant evidence of former Quaternary sea-level stationing, in the form of terraced landforms, occurs on the Cilento Promontory from north to south and on its offshore counterpart. One of the first observations obtained by grouping various terraces according to their lithological unit was consequent variation in the degree of conservation of erosive forms created by former sea level positions (i.e., marine terraces of dominant erosional origin). Such forms were, indeed, better preserved when associated with carbonate rocks of the Mount Bulgheria Unit. Bedrock of terrigenous origin (Cilento Flysch and Internidi Units) hinders the conservation of Quaternary erosive landforms [26,74], that were smaller and poorly represented. On the Mount Bulgheria Unit, a more widespread and continuous conservation of the Quaternary deposits was evident, allowing plentiful geological and geomorphological information to be obtained for reconstructing the age and alternation of former sea-level positions [25,28,29,31,32,34,61,62].

From analyzed data, it appears that older terraces can be detected up to 300 m a.s.l. Several Lower and Middle Pleistocene terraces have been cataloged. However, the complex tectonic history of the region makes it difficult to perform accurate correlations, although focused dating could improve our understanding of the post-orogenic tectonic differentiation that typified the variuos uplifting rates of the Cilento Units. Additionally, some sector of the promontory were tectonically displaced upwards by approximately 400 m during the late Lower and the Middle Pleistocene. The Upper Pleistocene experienced reactivations, but on a smaller scale and were differentiated from north to south.

A different Upper Pleistocene geodynamic behavior seems to characterize the offshore region where tectonic movements or relative stability, documented by the position of dated marine terraces on-land, are well correlated with the submarine sector for areas shallower than 120/130 m in w.d. In the offshore sector, the shallower orders of terraced landforms, likely generated as wave-cut shore-platforms, did not record relevant tectonic/vertical movements during the Upper Pleistocene. Since their position seems to be well correlated with former sea-level positions, as reported in the global mean sea-level curves (Figure 5), the result is in good agreement with documented research on land [75,76]. Seismic data additionally indicates that during the last period of sea level rise, a transgressive erosional surface (i.e., ravinement surface) formed in the area [50] and that its relationship to depositional bodies detected on the shelf critically improved constraints for ascribing the relative position of sea-level to detected submerged terraces. We gave considerable importance to the curve of isotopic stratigraphy in [77], and to other evidences reported for Holocene relative sea-level curves [78,79], where a short stasis is reported for the rapid Flandrian Transgression between 45 and 40 m in w.d.. This depth range is close to some of the mapped offshore terraces in the form of a wave-cut shore platform (Table 2, Figures 2 and 3) and displays strong lateral continuity all along the offshore sector of the Cilento Promontory (Figures 2B and 4). In contrast, as speculated on the basis of relative sea-level fluctuations documented by low-stand depositional bodies that formed SDT at 160 m in w.d. [50], the outer shelf appears to have experienced an important tectonic subsidence.

Using all collected evidence, we observed that submarine terraced landforms offshore of the Cilento Promontory can be distinguished as erosional and depositional, respectively representing paleo wave-cut shore platforms (see [80] for a comprehensive definition and differentiation from marine terraces) and SDTs (as described in [9] and references therein). In our study area, terraced landform distinction is marked by the depths at which they occur (Figure 5). On the outer zone of the continental shelf, and especially in areas deeper than 120 m in w.d., SDTs have been described by Ferraro et al. [48] and have been interpreted by Trincardi and Field [50] as shelf-margin deposits, with a different configuration according to physiographic shelf-break depth during the last sea-level low-stand (i.e., MIS2). Shelf margin deposits particularly occur offshore of the Cilento Promontory where the physiographic shelf-break is deeper than the position of the low-stand shoreline of the Last Glacial Maximum (MIS2). Trincardi and Field [50] highlighted the absence of such deposits, where the shelf-break was close to the shoreline during MIS2. The different configuration of SDTs, located on the outer shelf (as described in [50]) and the concurrent deepening and widening of the physiographic shelf break toward the north, warrants a distinction between the two main morpho-structural elements forming the shelf, as follows:

- A shelf sensu stricto, extendeding from the coastline down to 130 m in w.d., where there is an almost continuous break in slope, that, south of the Cilento Promontory, is sharp and coincides with the physiographic shelf break (i.e., offshore Mount Bulgheria) that progressively leads to a slightly deeper and flat outer shelf toward the north (Figure 1).
- An outer shelf, particularly evident offshore of Punta Licosa from 130 m down to more than 200 m in w.d. (Figure 1).

The two morpho-structural elements seem to represent the components of a regional fault system. The system is defined by NW-SE and NNE-SSW lineaments, marking the core area that separates the uplifting morpho-structural high forming the Cilento Promontory on the margin (interposed between the coastal depressions of the Sele Plain-Salerno Gulf to the north and of the Policastro Gulf to the south-east), and the Tyrrhenian basin offshore. The offshore tyrrhenian basin has been subsiding at a rate of 1 mm/yr since the end of the Lower Pleistocene [48] and from the Last Glacial Maximum until present could have been responsible for the lowering of the shelf break. The shelf s.s., represents a sector that experienced the same tectonic of the on-land system. Such a result is confirmed by a good correlation between the depth of marine terraces of erosive origin and eustatic sea level variations recorded for the last 200 ka (Figure 7) that are attributed, for the most part, to shallower submarine terraces of the stationing of the Flandrian Transgression (40/46 m in w.d. —as reported for the offshore of the southern Bulgheria Mount [32]) and the stationing of MIS 3 (50/55 m and 70/76 m in w.d.), MIS 5c, and MIS5a (10/15 m and 18/24 m in w.d.). In the offshore the Cilento Flysch and Internidi Units, the strong lateral continuity that characterises the terraces located at 47/52 m in w.d. (Figures 2B and 3) suggests that they could also have an origin associated to the Flandrian Transgression.

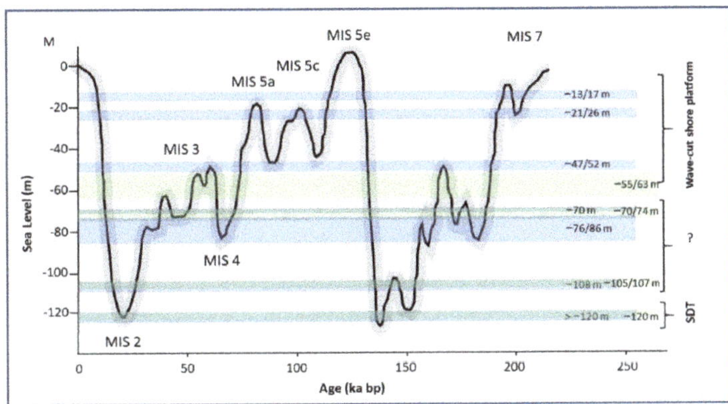

Figure 7. A graph with eustatic sea level variations, expressed in meters, recorded during the last 200 ka cal (adapted from [76]). Thick horizontal lines indicate major depth intervals where submarine terraced landforms were located in the offshore region of the Cilento Promontory. Different colors refer to different regions as distinguished in Table 2 (light transparent blue for the offshore Cilento Group and Internidi Units, light transparent green for the offshore of the Bulgheria Mount); SDT: Submarine Depositional Terraces.

Sub-aerial marine terraces [7] have, indeed, been traditionally acknowledged to be relevant geomorphological indicators of past sea-level high-stands in regions subjected to tectonic uplift [8]. With time, uplift determines the formation of terraced coastlines, often with a step-like profile, where older terraces are higher and farther from today's coastline [7]. In this work, we focused on understanding the occurrence of paleo, wave-cut shore platforms (forming marine terraces) within the submarine domain [22]. Here, it is important to note that phases of relatively high sea-level or stationary phases during transgressive periods, in the end, determine the most favorable condition for wave-cut shore platform formations on rocky cliffs (because they cause marine processes to prevail over sub-aerial processes). A relative decrease in sea level would, instead, lead to a decrease in the efficiency of marine processes, with the formation of beaches or debris at the base of a cliff, preventing wave-cut shore platform formation. For this reason, wave-cut shore platforms are unlikely to form during low stands or regressive conditions. Therefore, for our study area, we conclude that the occurrence of marine terraces of erosive origin within the submarine domain resulted because shelf s.s. was subjected to the same geodynamics that impacted the Cilento Margin on-land; and because the area was relatively stable during the Upper Pleistocene and, therefore, during earlier high-stand (MIS5c and MIS5a) and stationary periods of the Flandrian Transgression. In contrast, the outer shelf has been lowering, at least since the end of the Lower Pleistocene, and has been involved in the same geodynamics that are altering the Tyrrhenian Basin, promoting the formation of SDTs.

Deeper submarine terraces of erosive origin, as described by [48], at 80/86 m and 100/107 m in w.d., are more difficult to interpret. Therefore, further investigation is required to confirm their actual association to bedrock outcrops.

A higher resolution of the DEM would also lead to a more effective morphometric analysis. The more accurate list of terraced surfaces that would result from the analysis, combined with an adequate reconstruction of late Quaternary environmental conditions that controlled formation of wave-cut shore platforms (e.g., wave climatology, as performed in [22]), would provide more precise information to confirm the ascription of the terraces to a defined high-stand period.

6. Conclusions

An interesting finding of our study, obtained by coupling terrestrial and submarine terraced landforms [81], is the detection of two main types of submarine terraced landforms in the surveyed sector of the south-eastern Tyrrhenian Margin: (1) erosional terraces (wave-cut or abrasion platforms) formed on outcropping bedrock and (2) depositional terraces (i.e., SDTs) generated by late Quaternary depositional sequences. A distinction between the two types of landforms actually depends on a set of parameters dominated by regional geologic settings (the type of bedrock, and geodynamic and sediment inputs) subject to sea-level oscillation. Submarine marine terraces that result from the generation of wave-cut shore platforms were predominantly generated during interglacial periods or during relevant stasis occurring in transgressive events. On the surveyed sector of the south-eastern Tyrrhenian Margin, they formed when bedrock outcrops were exposed on the shelf, making them vulnerable to substantial erosion due to marine processes. Bedrock outcrops also contributed to a disruption of late Quaternary sedimentation on the inner shelf. Based on this evidence, submarine terraces of a strictly erosional nature could not have been formed on a traditional passive margin subject to subsidence. In contrast, subsidence, provides accommodation for the formation of depositional bodies, that according to sediment availability, shelf morphology, and local sea level history provide a suitable condition for a variety of SDTs.

Author Contributions: Conceptualization, M.P., F.R. and A.S.; methodology, V.A.B., A.C. and A.S.; software, V.A.B. and A.S.; investigation, M.P., F.R. and A.C.; data curation, M.P., F.R., V.A.B. and A.S.; writing—original draft preparation, M.P. and F.R. writing—review and editing, A.S., F.R., M.P. and V.A.B.; visualization, V.A.B.; supervision, M.P. and A.S.; project administration, M.P. and A.S.; funding acquisition, A.S. All authors have read and agreed to the published version of the manuscript.

Funding: This article is an outcome of the Project MIUR—Dipartimenti di Eccellenza 2018–2022, Department of Earth and Environmental Sciences, University of Milano-Bicocca.

Institutional Review Board Statement: Not applicable.

Informed Consent Statement: Not applicable.

Data Availability Statement: The bathymetric metadata and Digital Terrain Model data products have been derived from the EMODnet Bathymetry portal—http://www.emodnet-bathymetry.eu (accessed on 20 February 2021). and from http://dati.protezionecivile.it/geoportalDPC/rest/document#MagicFoglio10 (accessed on 20 February 2021).

Conflicts of Interest: The authors declare no conflict of interest.

References

1. International Hydrographic Organization (IHO). *Standardization of Undersea Feature Names: Guidelines Proposal form Terminology,* 4th ed.; International Hydrographic Organisation(IHO): Monaco City, Monaco; Intergovernmental Oceanographic Commission (IOC): Paris, France, 2008.
2. Jackson, J.A. *Glossary of Geology;* American Geological Institute: Alexandria, VA, USA, 1997.
3. Chappell, J.M. Geology of coral terraces, Huon Peninsula, New Guinea: A study of Quaternary tectonic movements and sea-level changes. *Geol. Soc. Am. Bull.* **1974**, *85*, 553–570. [CrossRef]
4. Penck, A.; Brückner, E. *Die Alpen im Eiszeitalter;* Tauchnitz: Leipzig, German, 1909; Volume 3, p. 1200.
5. Pirazzoli, P.A. Global sea level changes and their measurement. *Glob. Plan. Chang. Lett.* **1993**, *8*, 135–148. [CrossRef]
6. Zazo, G. Interglacial sea levels. *Quat. Int.* **1999**, *55*, 101–113. [CrossRef]
7. Pirazzoli, P.A. Marine Terraces. In *Encyclopedia of Coastal Science;* Schwartz, M.L., Ed.; Encyclopedia of Earth Science Series; Springer: Berlin/Heidelberg, Germany, 2005. Available online: https://doi.org/10.1007/1-4020-3880-1_209 (accessed on 20 February 2021).
8. Selli, R. Il Paleogene nel quadro della Geologia dell'Italia Meridionale. *Mem. Soc. Geol. Ital.* **1962**, *3*, 737–790.

9. Rabineau, M.; Bern e, S.; Olivet, J.L.; Aslanian, D.; Guillocheau, F.; Joseph, P. Paleo sea levels reconsidered from direct observation of paleoshoreline position during Glacial Maxima (for the last 500,000 yr). *Earth Planet. Sci. Lett.* **2006**, *252*, 119–137. [CrossRef]

10. Casalbore, D.; Romagnoli, C.; Chiocci, F.; Frezza, V. Morpho-sedimentary characteristics of the volcanic apron around Stromboli volcano. *Mar. Geol.* **2010**, *269*, 132–148. [CrossRef]

11. Fraccascia, S.; Chiocci, F.L.; Scrocca, D.; Falese, F. Very high-resolution seismic stratigraphy of Pleistocene eustatic minima markers as a tool to reconstruct the tectonic evolution of the northern Latium shelf (Tyrrhenian Sea, Italy). *Geology* **2013**, *41*, 375–378. [CrossRef]

12. Romagnoli, C. Characteristics and morphological evolution of the Aeolian volcanoes from the study of submarine portions. In *The Aeolian Islands Volcanoes*; Lucchi, F., Peccerillo, A., Keller, J., Tranne, C.A., Rossi, P.L., Eds.; Geological Society: London, UK, 2013; Volume 37, pp. 13–26.

13. Romagnoli, C.; Casalbore, D.; Bosman, A.; Braga, R.; Chiocci, F.L. Submarine structure of Vulcano volcano (Aeolian islands) revealed by high-resolution bathymetry and seismo-acoustic data. *Mar. Geol.* **2013**, *338*, 30–45. [CrossRef]

14. Quartau, R.; Hipolito, A.; Romagnoli, C.; Casalbore, D.; Madeira, J.; Tempera, F.; Roque, C.; Chiocci, F.L. The morphology of insular shelves as a key for understanding the geological evolution of volcanic islands: Insights from Ter-ceira Island (Azores). *Geochem. Geophys. Geosystems* **2014**, *15*, 1801–1826. [CrossRef]

15. Casalbore, D.; Falese, F.; Martorelli, E.; Romagnoli, C.; Chiocci, F.L. Submarine depositional terraces in the Tyrrhenian Sea as a proxy for paleo-sea level reconstruction: Problems and perspective. *Quat. Int.* **2020**, *544*, 1–11. [CrossRef]

16. Micallef, A.; Krastel, S.; Savini, A. Introduction. In *Submarine Geomorphology*; Micallef, A., Krastel, S., Savini, A., Eds.; Springer: Berlin/Heidelberg, Germany, 2017.

17. Chiocci, F.L.; Orlando, L. Lowstand terraces on Tyrrhenian Sea steep continental slopes. *Mar. Geol.* **1996**, *134*, 127–143. [CrossRef]

18. Chiocci, F.L.; D'Angelo, S.; Romagnoli, C. Atlas of submerged depositional terraces along the Italian coasts. *Mem. Descr. Della Carta Geol. D'italia* **2004**, *58*, 197.

19. De Pippo, T.; Pennetta, M. Submerged depositional terraces in the Gulf of Policastro (Southern Tyrrhenian sea, Italy). *Mem. Descr. Carta Geol. D'italia* **2004**, *58*, 35–38.

20. Sulli, A.; Lo Presti, V. Gasparo Morticelli, M., Antonioli, F.,. Vertical movements in NE Sicily and its offshore: Outcome of tectonic uplift during the last 125 ky. *Quat. Int.* **2013**, *288*, 168–182. [CrossRef]

21. Pepe, F.; Bertotti, G.; Ferranti, L.; Sacchi, M.; Collura, A.M.; Passaro, S.; Sulli, A. Pattern and rate of post-20 ka vertical tectonic motion around the Capo Vaticano Promontory (W Calabria, Italy) based on offshore geomorphological indicators. *Quat. Int.* **2014**, *322*, 85–98. [CrossRef]

22. Bilbao-Lasa, P.; Jara-Muñoz, J.; Pedoja, K.; Álvarez, I.; Aranburu, A.; Iriarte, E.; Galparsoro, I. Submerged Marine Terraces Identification and an Approach for Numerical Modeling the Sequence Formation in the Bay of Biscay (Northeastern Iberian Peninsula). *Front. Earth Sci.* **2020**, *8*, 47. [CrossRef]

23. Ricchi, A.; Quartau, R.; Ramalho, R.S.; Romagnoli, C.; Casalbore, D.; Ventura da Cruz, J.; Fradique, C.; Vinhas, A. Marine terrace development on reefless volcanic islands: New insights from high-resolution marine geophysical data offshore Santa Maria Island (Azores Archipelago). *Mar. Geol.* **2018**, *406*, 42–56. [CrossRef]

24. Blanc, A.C.; Segre, A.G. Les formations Quaternaires et les gîsements paléolithiques de la côte de Salerno. In *Excursion dans les Abruzzes, les Pouilles et sur la cote de Salerne, Procedings of the Actes du IV Congres Internationale du Quaternaire, Roma and Pisa, Italy, August–September 1953*; Stanford Library: Stanford, CA, USA, 1956.

25. Lirer, L.; Pescatore, T.; Scandone, P. Livello di piroclastici nei depositi continentali post-Tirreniani del litorale sud-tirrenico. *Atti Accad. Gioenia Sci. Nat. Catania* **1967**, *18*, 85–115. (In Italian with English abstract)

26. Baggioni, M. Les côtes du Cilento (Italie du Sud): Morphogénèse littorale actuelle et héritée. *Méditerranée* **1975**, *3*, 35–52. (In French with English abstract) [CrossRef]

27. Palma Di Cesnola, A. Il Paleolitico inferiore in Campania. In Proceedings of the 23rd Scientific Meeting, Firenze, Italy, 7–9 May 1980; Istituto Italiano di Preistoria e Protostoria: Firenze, Italy, 1982; pp. 207–224.

28. Brancaccio, L.; Cinque, A.; Russo, F.; Belluomini, G.; Branca, M.; Delitala, L. Segnalazione e datazione di depositi marini tirreniani sulla costa campana. *Boll. Soc. Geol. Ital.* **1990**, *109*, 259–265. (In Italian with English abstract)

29. Romano, P. La distribuzione dei depositi marini pleistocenici lungo le coste della Campania. Stato delle conoscenze e prospettive di ricerca. *Studi Geol. Camerti N. Sp.* **1992**, *1*, 265–269. (In Italian with English abstract)

30. Cinque, A.; Romano, P.; Rosskopf, C.; Santangelo, N.; Santo, A. Morfologie costiere e depositi quaternari tra Agropoli e Ogliastro Marina (Cilento, Italia meridionale). *Il Quat.* **1994**, *7*, 3–16. (In Italian with English abstract)

31. Russo, F. Segnalazione di un livello fossilifero riferibile al Tirreniano a Cala Bianca (Marina di Camerota). *Mem. Descr. Carta Geol. D'italia* **1994**, *52*, 395–398. (In Italian with English abstract)

32. Antonioli, F.; Cinque, A.; Ferranti, L.; Romano, P. Emerged and submerged quaternary marine terraces of Palinuro Cape (southern Italy). *Mem. Descr. Carta Geol. D'italia* **1994**, *52*, 237–260.

33. Ascione, A.; Romano, P. Vertical movements on the eastern margin of the Tyrrhenian extensional basin. New data from Mt Bulgheria (Southern Appenines, Italy). *Tectonophysics* **1999**, *315*, 337–358. [CrossRef]

34. Esposito, C.; Filocamo, F.; Marciano, R.; Romano, P.; Santangelo, N.; Scarciglia, F.; Tuccimei, P. Late Quaternary shorelines in Southern Cilento (Mt. Bulgheria): Morphostratigraphy and chronology. *Il Quat. It. J. Quater. Sci.* **2003**, *16*, 3–14.

35. Guida, D.; Valente, A. Terrestrial and marine landforms along the Cilento coastland (Southern Italy): A framework for landslide hazard assessment and environmental conservation. *Water* **2019**, *11*, 2618. [CrossRef]

36. Bartole, R.; Savelli, C.; Tramontana, M.; Wezel, F.C. Structural and sedimentary features in the Tyrrhenian margin off Campania, Southern Italy. *Mar. Geol.* **1984**, *55*, 163–180. [CrossRef]

37. Cinque, A.; Patacca, E.; Scandone, P.; Tozzi, M. Quaternary kinematic evolution of the southern Apennines. Relationships between surface geological features and deep lithospheric structures. *Ann. Geof.* **1993**, *36*, 249–259.

38. Casciello, E.; Cesarano, M.; Pappone, G. Extensional detachment faulting on the Tyrrhenian margin of the southern Apennines contractional belt (Italy). *J. Geol. Soc.* **2006**, *163*, 617–629. [CrossRef]

39. Cammarosano, A.; Danna, M.; De Rienzo, F.; Martelli, L.; Miele, F.; Nardi, G. Il substrato del Gruppo del Cilento tra il M. Vesole e il M. Sacro (Cilento, Appennino Meridionale). *Boll. Soc. Geol. Ital.* **2000**, *119*, 395–405, (In Italian with English abstract).

40. Cammarosano, A.; Cavuoto, G.; Danna, M.; De Capoa, P.; De Rienzo, F.; Di Staso, A.; Giardino, S.; Martelli, L.; Nardi, G.; Sgrosso, A. Nuovi dati e nuove interpretazioni sui flysch terrigeni del Cilento (Appennino meridionale, Italy). *Boll. Soc. Geol. Ital.* **2004**, *123*, 253–273. (In Italian with English abstract)

41. Amore, F.O.; Bonardi, G.; Ciampo, G.; De Capoa, P.; Perrone, V.; Sgrosso, I. Relazioni tra flysch Interni e domini appenninici: Reinterpretazione delle formazioni di Pollica, S. Mauro e Albidona e l'evoluzione infra-medio miocenica delle zone esterne sudappenniniche. *Mem. Soc. Geol. Ital.* **1988**, *41*, 285–297. (In Italian with English abstract)

42. Bonardi, G.; Amore, F.O.; Ciampo, G.; De Capoa, P.; Miconnet, P.; Perrone, V. Il Complesso Liguride Auct.: Stato delle conoscenze e problemi aperti sulla sua evoluzione pre-appenninica ed i suoi rapporti con I'Arco Calabro. *Mem. Soc. Geol. Ital.* **1988**, *41*, 7–35. (In Italian with English abstract)

43. Zuppetta, A.; Mazzoli, S. Deformation history of a synorogenic sedimentary wedge, northern Cilento area, southern Apennines thrust and fold belt, Italy. *Geol. Soc. Am. Bull.* **1997**, *109*, 698–708. [CrossRef]

44. Ciarcia, S.; Vitale, S.; Di Staso, A.; Iannace, A.; Mazzoli, S.; Torre, M. Stratigraphy and tectonics of an Internal Unit of the southern Apennines: Implications for the geodynamic evolution of the peri-Tyrrhenian mountain belt. *Terra Nova* **2009**, *21*, 88–96. [CrossRef]

45. Cocco, E. *Note Illustrative della Carta Geologica d'Italia alla Scala 1:100.000. Foglio 209 Vallo della Lucania*; Servizio Geologico d'Italia: Rome, Italy, 1971; p. 45.

46. Cieszkowski, M.; Oszczypko, N.; Pescatore, T.S.; Slaczka, A.; Senatore, M.R. Megatorbiditi calcareo-marnose nelle successioni flyscioidi dell'Appennino Meridionale (Cilento, Italia) e dei Carpazi Settentrionali (Polonia). *Boll. Soc. Geol. Ital.* **1995**, *114*, 67–88. (In Italian with English abstract)

47. D'Argenio, B.; Pescatore, T.S.; Scandone, P. Schema geologico dell'Appennino Meridionale (Campania e Lucania). *Atti Accad. Naz. Lincei* **1973**, *183*, 49–72.

48. Ferraro, L.; Pescatore, T.; Russo, B.; Senatore, M.R.; Vecchione, C.; Coppa, M.G.; Di Tuoro, A. Studi di geologia marina del margine tirrenico: La piattaforma continentale tra Punta Licosa e Capo Palinuro (Tirreno meridionale). *Boll. Soc. Geol. Ital.* **1997**, *116*, 473–485. (In Italian with English abstract)

49. Kastens, K.; Mascle, J.; Auroux, C.; Bonatti, E.; Broglia, C.; Channel, J.; Curzi, P.; Emeis, K.; Glacon, G.; Hasegawa, S.; et al. ODP Leg 107 in the Tyrrhenian sea: Insights into passive margin and back-arc basin evolution. *Geol. Soc. Am. Bull.* **1988**, *100*, 1140–1156. [CrossRef]

50. Trincardi, F.; Field, M.E. Geometry, lateral variation and preservation of downlapping regressive shelf deposits: Eastern Tyrrhenian Sea margin, Italy. *J. Sediment. Petrol.* **1991**, *61*, 775–790.

51. Pennetta, M. Margine tirrenico meridionale: Morfologia e sedimentazione tardo pleistocenica–olocenica del sistema di piattaforma-scarpata. *Boll. Soc. Geol. Ital.* **1996**, *115*, 339–354. (In Italian with English abstract)

52. Pennetta, M. Evoluzione morfologica quaternaria del margine tirrenico sud-orientale tra Capo Palinuro e Capo Bonifati. *Il Quat.* **1996**, *9*, 353–358. (In Italian with English abstract)

53. Budillon, F.; Conforti, A.; Tonielli, R.; De Falco, G.; Di Martino, G.; Innangi, S.; Marsella, E. The Bulgheria canyon-fan: A small-scale proximal system in the eastern Tyrrhenian Sea (Italy). *Mar. Geophys. Res.* **2011**, *32*, 83–97. [CrossRef]

54. De Pippo, T.; Pennetta, M. Late Quaternary morphological evolution of a continental margin based on emerged and submerged morphostructural features: The south-eastern Tyrrhenian margin (Italy). *Zeit. Geomorph. N. F.* **2000**, *44*, 435–448.

55. Aiello, G. Elaborazione ed interpretazione geologica di sismica di altissima risoluzione nell'offshore del promontorio del Cilento (Tirreno meridionale, Italia). *Quad. Geof.* **2019**, *155*, 7–19. (In Italian with English abstract)

56. Alessio, M.; Allegri, F.; Antonioli, F.; Belluomini, G.; Ferranti, L.; Improta, S.; Manfra, L.; Proposito, A. Risultati preliminari relativi alla datazione di speleotemi sommersi nelle fasce costiere del Tirreno centrale. *G. Di Geol.* **1992**, *54*, 165–193. (In Italian with English abstract)

57. Senatore, M.R. Terrazzi deposizionali sommersi al largo di Punta Licosa. *Mem. Descr. Carta Geol. D'italia* **2004**, *58*, 153–154. (In Italian with English abstract)

58. Savini, A.; Basso, D.; Bracchi, V.A.; Corselli, C.; Pennetta, M. Maerl-bed mapping and carbonate quantification on submerged terraces offshore the Cilento promontory (Tyrrhenian Sea, Italy). *Geodiversitas* **2012**, *34*, 77–98. [CrossRef]

59. Posamentier, H.W.; Jervey, M.T.; Vail, P.R. Eustatic control on clastic deposition. I. conceptual framework. *Sepm Sp. Publ.* **1988**, *42*, 109–124.

60. Swift, D.J.P.; Stanley, D.J.; Curray, J.R. Relict sediments on continental shelf: A reconsideration. *J. Geol.* **1971**, *79*, 322–346. [CrossRef]

61. Sgrosso, I.; Ciampo, G. Sulla presenza di terreni calabriani nei dintorni di Camerota. *Boll. Soc. Natur. Napoli* **1966**, *75*, 561–587. (In Italian with English abstract)

62. Baggioni, M. Le Mont Bulgheria (Italie méridionale): Morphologie littorale et néotectonique. *Méditerranée* **1978**, *32*, 33–46. (In French with English abstract) [CrossRef]

63. Baggioni, M.; Suc, J.P.; Vernet, J.L. Le Plio-Pleistocene du Camerota (Italie meridionale): Geomorphologie et paleoflores. *Geobios* **1981**, *14*, 229–237. (In French with English abstract) [CrossRef]

64. Borrelli, A.; Ciampo, G.; De Falco, M.; Guida, D.; Guida, M. La morfogenesi del Monte Bulgheria (Campania) durante il Pleistocene inferiore e medio. *Mem. Soc. Geol. Ital.* **1988**, *41*, 667–672. (In Italian with English abstract)

65. Brancaccio, L.; Cinque, A.; Romano, P.; Rosskopf, C.; Russo, F.; Santangelo, N. L'evoluzione delle pianure costiere della Campania: Geomorfologia e neotettonica. *Mem. Soc. Geogr. Ital.* **1995**, *53*, 313–336. (In Italian with English abstract)

66. Iannace, A.; Romano, P.; Santangelo, N.; Santo, A.; Tuccimei, P. The OIS 5c along Licosa Cape promontory (Campania region, Southern Italy): Morphostratigraphy and U/Th dating. *Zeit. Geomorph. N. F.* **2001**, *45*, 307–319.

67. Marciano, R.; Munno, R.; Petrosino, P.; Santangelo, N.; Santo, A.; Villa, I. Late quaternary tephra layers along the Cilento coastline (southern Italy). *J. Volcan Geotherm. Res.* **2008**, *177*, 227–243. [CrossRef]

68. Gambassini, P.; Ronchitelli, A. Linee di sviluppo dei complessi del Paleolitico inferiore—Medio nel Cilento. *Riv. Sc. Preist.* **1998**, *49*, 357–376. (In Italian with English abstract)

69. Baggioni-Lippmann, M. Néotectonique et géomorphologie dans l'Apennin campanien (Italie méridionale). *Rev. Géol. Dynam. Géogr. Phys.* **1981**, *23*, 41–54. (In French with English abstract)

70. Marani, M.; Tavani, M.; Trincardi, F.; Argnani, A.; Borsetti, A.M.; Zitellini, N. Pleistocene progradation and postglacial events of the NE Tyrrhenian continental shelf between the Tiber river delta and Capo Circeo. *Mem. Soc. Geol. Ital.* **1988**, *36*, 67–89.

71. Pranzini, E. *La Forma Delle Coste*; Zanichelli: Modena, Italy, 2004.

72. Savini, A.; Pennetta, M.; Corselli, C. Geophysical Investigation for exploring marine sand deposits (Cilento Peninsula–Southern Italy). In Proceedings of the Seventh International Conference on Mediterranean Coastal Environment, MEDCOAST 05, Kusadasi, Turkey, 24–29 October 2005; Volume 2, pp. 973–984.

73. Ferranti, L.; Antonioli, F.; Mauz, B.; Amorosi, A.; Dai Pra, G.; Mastronuzzi, G.; Monaco, C.; Orrù, P.; Pappalardo, M.; Radtke, U.; et al. Markers of the last interglacial sea level high stand along the coast of Italy: Tectonic implications. *Quat. Int.* **2006**, *145*, 30–54. [CrossRef]

74. Budetta, P.; Santo, A.; Vivenzio, F. Landslide hazard mapping along the coastline of Cilento region (Italy) by means of a GIS-based parameter rating approach. *Geomorphology* **2008**, *94*, 340–352. [CrossRef]

75. Moore, W.S. Late Pleistocene sea level history. In *Uranium Series Disequilibrium: Application to Environmental Problems*; Ivanovich, M., Harmon, R.S., Eds.; Oxford University Press: Oxford, UK, 1982.

76. Waelbroeck, C.; Labeyrie, L.; Michel, E.; Duplessy, J.; McManu, J.; Lambeck, K.; Balbon, E.; Labracherie, M. Sea-level and deep water temperature changes derived from benthic foraminifera isotopic records. *Quat. Sci. Rev.* **2002**, *21*, 295–305. [CrossRef]

77. Martinson, D.G.; Pisias, N.G.; Hays, J.D.; Imbrie, J.; Moore, T.C.; Shackleton, N.J. Age dating and the orbital theory of the ice ages: Development of a high-resolution 0 to 300.000 year chronostratigraphy. *Quat. Res.* **1987**, *27*, 1–29. [CrossRef]

78. Lambeck, K.; Rouby, H.; Purcell, A.; Sun, Y.; Sambridge, M. Sea level and global ice volumes from the Last Glacial Maximum to the Holocene. *Proc. Natl. Acad. Sci. USA* **2014**, *111*, 15296–15303. [CrossRef] [PubMed]

79. Benjamin, J.; Rovere, A.; Fontana, A.; Furlani, S.; Vacchi, M.; Inglis, R.H.; Galili, E.; Antonioli, F.; Sivan, D.; Miko, S.; et al. Late Quaternary sea-level changes and early human societies in the central and eastern Mediterranean Basin: An interdisciplinary review. *Quat. Int.* **2017**, *449*, 29–57. [CrossRef]

80. Rovere, A.; Raymo, M.E.; Vacchi, M.; Lorscheid, T.; Stocchi, P.; Gómez-Pujol, L.; Harris, D.L.; Casella, E.; O'Leary, M.J.; Hearty, P.J. The analysis of Last Interglacial (MIS 5e) relative sea-level indicators: Reconstructing sea-level in a warmer world. *Earth Sci. Rev.* **2016**, *159*, 404–427. [CrossRef]

81. Prampolini, M.; Savini, A.; Foglini, F.; Soldati, M. Seven Good Reasons for Integrating Terrestrial and Marine Spatial Datasets in Changing Environments. *Water* **2020**, *12*, 2221. [CrossRef]

MDPI

St. Alban-Anlage 66

4052 Basel

Switzerland

Tel. +41 61 683 77 34

Fax +41 61 302 89 18

www.mdpi.com

Water Editorial Office

E-mail: water@mdpi.com

www.mdpi.com/journal/water